W9-BMX-025

LINEAR ALGEBRA

WITH APPLICATIONS

LINEAR ALGEBRA

WITH APPLICATIONS

JOHN W. AUER

PRENTICE-HALL CANADA INC., Scarborough, Ontario

Canadian Cataloguing in Publication Data

Auer, John W. (John Willem), 1942–
Linear algebra with applications

ISBN 0-13-538349-8

1. Algebras, Linear. I. Title.
QA184.A93 1990 512′.5 C90-093218-X

Prentice-Hall, Inc., Englewood Cliffs, New Jersey
Prentice-Hall International, Inc., London
Prentice-Hall of Australia, Pty., Ltd., Sydney
Prentice-Hall of India Pvt., Ltd., New Delhi
Prentice-Hall of Japan, Inc., Tokyo
Prentice-Hall of Southeast Asia (Pte.) Ltd., Singapore
Editora Prentice-Hall do Brasil Ltda., Rio de Janeiro
Prentice-Hall Hispanoamericana, S.A., Mexico

ISBN 0-13-538349-8

Production Editor: Cecilia Chan
Editorial Coordinator: Doris Wolf
Production Coordinator: Crystale Chalmers
Text and Cover Design: Bruce Farquhar
Composition and Technical Art: Q Composition Inc.

1 2 3 4 5 JD 95 94 93 92 91

Printed and bound in Canada by John Deyell Company

To the memory of
Professor J.W. Reed
Founding Chairman
of the Brock University
Mathematics Department

CONTENTS

PREFACE XI

1 ANALYTIC AND VECTOR GEOMETRY 1

1-1 Lines and Planes in R^2 and R^3 2
1-2 Vectors in R^2 and R^3 15
1-3 The Dot Product and Vector Geometry 31
1-4 Further Vector Geometry 42

2 SYSTEMS OF LINEAR EQUATIONS AND MATRICES 53

2-1 Gauss-Jordan Elimination 54
2-2 Matrices and Gauss-Jordan Reduction 64
2-3 Matrices: Some Examples and Applications (Optional) 78
2-4 Matrix Algebra 85
2-5 Elementary Matrices and Matrix Inversion 101
2-6 Computational Considerations I (Optional) 118

3 DETERMINANTS 129

3-1 Introduction 130
3-2 Properties of Determinants 144
3-3 Cramer's Rule 160
3-4 The Cross Product (Optional) 166

4 VECTOR SPACES 175

4-1 Two Basic Examples 176
4-2 Vector Spaces and Subspaces 181
4-3 Generating Sets, Linear Independence, and Bases 195
4-4 Computational Techniques — Rank of a Matrix 216
4-5 Electrical Networks (Optional) 234
4-6 Complex Vector Spaces (Optional) 240

5 LINEAR TRANSFORMATIONS 250

5-1 Introduction 251
5-2 Properties of Linear Transformations 262
5-3 Isomorphism (Optional) 281
5-4 Basis Change and Similarity 299
5-5 Linear Functionals and Applications (Optional) 310

6 THE EIGENVALUE PROBLEM 322

6-1 Classification of Linear Operators 323
6-2 The Characteristic Polynomial 329
6-3 Diagonalization of Linear Operators 339
6-4 Computational Considerations II (Optional) 350
6-5 Application to Economics (Optional) 359

7 INNER PRODUCTS 374

7-1 Inner Products 375
7-2 Inner Product Spaces 380
7-3 Linear Operators on Inner Product Spaces 396
7-4 Self-adjoint Operators 408
7-5 Bilinear and Quadratic Forms (Optional) 418
7-6 Complex Eigenvalues and Eigenvectors (Optional) 441

APPENDIX A: COMPLEX NUMBERS 455

APPENDIX B: POLYNOMIALS 474

APPENDIX C: INTRODUCTION TO LINEAR PROGRAMMING 486

ANSWERS TO SELECTED PROBLEMS 507

SYMBOLS AND NOTATIONS INDEX 539

INDEX 540

PREFACE

This book has been written as a first text in linear algebra for university students. There are many problems in teaching students in this group: mathematical backgrounds and interests vary widely. This book assumes only a knowledge of elementary analytic geometry of lines in the plane R^2. This is reviewed in Chapter 1 and extended to the geometry of lines and planes in three-space R^3, to simple systems of equations from a geometrical and qualitative point of view, and then to the various notions of vectors in R^2 and R^3. The treatment here differs from that of most comparable texts in that some analytic geometry is placed before discussions of vector geometry, matrices, and so on. I feel that the notions of lines and planes are easier and less abstract than the idea of a vector. Also, the problem of describing the direction of a line in R^2 and R^3 provides an excellent motivation for the concept of a vector. Once the notion of vector has been introduced, its power is demonstrated in solving further geometrical problems, including the usual vector approach to lines and planes.

The very leisurely pace of the first two chapters is intended to capture those students whose backgrounds may not include this material. *At the other extreme are those students who have already studied the material in Chapter 1. They should of course skip this chapter and start with Chapter 2.*

Computers continue to increase in importance in many undergraduate courses, including linear algebra. While many texts provide linear algebra software for use on one type of microcomputer or another, this text takes a more universal approach by providing instruction in the solution of linear algebra problems using the MAPLE symbolic manipulation package. The use of this software package has become quite widespread in many calculus courses at universities, including Brock. It also contains machinery for doing most of the computation found in linear algebra courses at this level. In addition, the fact that MAPLE can do *algebra* as opposed to strictly numerical computation is a real asset here. MAPLE is available not only for microcomputers but also for mainframes such as the VAX. In order to make the use of this material *optional and independent* of the text itself, it has been included in the *Maple Solutions Manual for Linear Algebra with Applications* accompanying this book, in the form of actual printouts and discussion of MAPLE sessions used to solve selected problems from the text.

Another difficulty that faces instructors in teaching the large group of students who are required to take linear algebra is that of *motivation*. Since the students' primary interest is usually not the mathematics itself, they want to know "why are we doing this?" more often than mathematics majors do. In view of this, one of my main goals in writing this book has been to provide a *continuous flow of motivation* of ideas. Thus, when a new idea is introduced, I do so by means of examples, links with earlier ideas, or concrete applications, where feasible. In this regard, the book has been *written for the student*, not the instructor. This is reflected in the informal and conversational style which would obviously be less appropriate for a seasoned mathematician.

So, for example, Chapter 2 examines the problem of linear equations and their solution by means of an informal discussion and the solution of a linear equilibrium model of supply and demand from economics. This example and the steps used in its solution lead gradually to the general definition of a system of m linear equations in n unknowns, the formalization of Gauss-Jordan elimination as an algorithm, equivalence of systems, and so on.

Systems of equations, in turn, motivate the definition of a matrix. And, as is so often the case in mathematics where economy of notation is sought, the expression $AX = B$ for a system of equations is used to motivate the notion of products of matrices. That is why products are discussed before sums and scalar multiples of matrices. Of course, there are many other important applications of matrices. A few of these are discussed in a broader context in section 2-3.

While Gauss-Jordan elimination is valuable for discussing the theory of systems of linear equations, inverses of matrices, and related ideas in linear algebra, it is, of course, not a practical method for solving "real" systems. The optional section 2-6 is included to alert the reader to some of the problems involved in the computer solution of such systems, including a comparative discussion of Gaussian elimination with and without pivoting, rounding error, and iterative methods.

I regard determinants (Chapter 3) largely as a device for discussing eigenvalues. The cross product of vectors in R^3 and its applications to vector geometry are included here. The material is not really needed until the discussion of eigenvalues, but I have placed it after Chapter 2 because many instructors like to cover this material at an early stage in their courses.

Again, with respect to motivation of ideas, in Chapter 4, the object is to make the central definitions relating to generating sets, linear independence, and bases seem as natural as they actually are. So, one tries to "describe" a vector space V (an infinite set) in an *economical* way. The model is R^2 or (R^3), where one can "describe" these sets of points using two (respectively three) axes. This leads to the notion of generators and then to linear independence: one wants a *most economical* (smallest possible) set of generators S for V (no proper subsets of S generate V).

The rather theoretical flavour of section 4-3 containing these central ideas about vector spaces contrasts with the next section where the whole process of finding bases, testing for linear independence, etc., is made mechanical using matrices and Gauss-Jordan reduction.

While the book deals almost exclusively with real vector spaces, an optional section 4-6 gives an introduction to complex vector spaces so that some discussion of complex eigenvalues and eigenvectors can be included in the optional section 7-6. The basic concepts of complex numbers are covered in Appendix A.

At various points throughout the text I have indicated that some readers may wish to skip the proofs of certain theorems on first reading (or altogether). This decision naturally depends on the priorities of the reader (or the instructor). It has been my preference, based on the trend to higher mathematical sophistication in the sciences, to provide as complete a treatment as is reasonable for this level.

Applications here are a means to an end, not an end in themselves. They are included at various points throughout the text to motivate concepts, stimulate the students' interest, and to provide a background for developing computational skills. Among those presented are (1) a transportation problem with its linear programming formulation (an introduction to the solution of LP problems is given in Appendix C), (2) some graph theory and its associated matrix theory, (3) simple electrical network problems, (4) some applications from economics, and (5) a discussion of quadratic forms as they apply to the geometry of conic sections. Needless to say, *they should be considered as optional reading; none of the main material depends on them.*

Chapter 5 contains more material on linear transformations than most instructors will be able to use in a first short course on linear algebra. The basics are contained in sections 5-1, 5-2, and 5-4. The other sections are not required for the rest of the text, but contain a fairly complete study of isomorphism, linear functionals, and some applications.

Eigenvalues and eigenvectors (Chapter 6) are discussed from the point of view raised by the question: Can we choose the basis for a vector space wisely so that the matrix of a linear operator is as "simple" as possible? Once again, a section is included to alert the reader to some of the numerical problems in the actual computation of eigenvalues. Some techniques for finding zeros of polynomials are presented in Appendix B.

The discussion of inner products (Chapter 7) has as its goal the diagonalization of symmetric matrices and (*optionally*) symmetric bilinear forms, quadratic forms, and applications to geometry.

No calculus is assumed anywhere in the text. However, readers familiar with calculus will notice that some examples are motivated by concepts from that discipline (e.g., Examples 5-8, 5-9).

Instructors may, of course, select from the material in a variety of ways to suit their needs. For example, the *shortest* introduction to linear algebra having as its goal the eigenvalue problem could consist of

Sections	1-1,	1-2,	1-3,	1-4,	(if needed)
	2-1,	2-2,	2-4,	2-5,	
	3-1,	3-2,	3-3,		
	4-1,	4-2,	4-3,	4-4,	
	5-1,	5-2,	5-4,		
	6-1,	6-2,	6-3.		

In a one-semester course, this program should leave time to cover some applications and discussion of numerical methods.

On the other hand, a course having as its goal the orthogonal diagonalization of symmetric matrices would add sections 7-1 to 7-4. The interdependence of sections is indicated at the end of this preface.

The end of a proof (or discussion of a Theorem or Corollary, when no proof is given) is indicated by ■, and the end of an example by ❑. Important points in a discussion are noted in **boldface** or in *italics*. Vectors are also denoted by **boldface**.

Interdependence of Sections

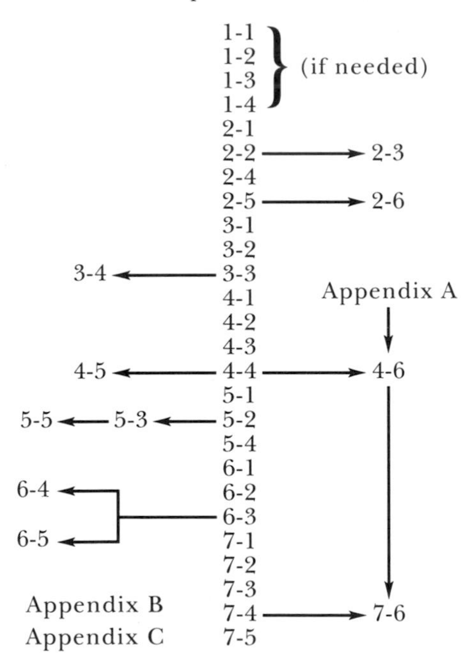

ACKNOWLEDGEMENTS

I wish to express my sincerest appreciation to the following reviewers for their invaluable comments and suggestions: Dietrich Burbulla, University of Toronto; David Gates, Vanier College; Ole A. Nielsen, Queen's University; Donald Pelletier, York University; Robert W. Quackenbush, University of Manitoba; and John A. Suvak, Memorial University of Newfoundland.

I am indebted to the Department of Mathematics, Brock University, for its support of this project, and to the Computer Centre and Department of Computer Science for their assistance in the preparation of parts of this manuscript. I wish to record my particular gratitude to Jack Sentineal and Marg Bernat for their contributions in this regard. Special thanks are also due to Howard E. Bell, David McCarthy, John Radue, Darrel Short, and Kelly Hall for their comments on parts of the manuscript and for other assistance. Finally, I am very grateful to the College Division of Prentice-Hall Canada Inc., particularly Rich Ludlow, Acquisitions Editor, and Ed O'Connor, Project Editor, for their enthusiasm, encouragement, and constructive criticism at various stages of this project. I also wish to thank Cecilia Chan, Doris Wolf, and Marta Tomins for their scrupulous and creative editorial work.

J.W. Auer

Mathematics Department
Brock University
October 1990

1

ANALYTIC AND VECTOR GEOMETRY

1-1 Lines and Planes in R^2 and R^3
1-2 Vectors in R^2 and R^3
1-3 The Dot Product and Vector Geometry
1-4 Further Vector Geometry

1-1 Lines and Planes in R^2 and R^3

It is reasonable to say that linear algebra is a sophisticated theory of **linear equations**, which we shall define below. Historically, it is certainly true that modern linear algebra evolved from the study of collections of such equations. Linear algebra has become increasingly important in the study of mathematics and its applications to the sciences (both physical and social) and to engineering. This is so, in part, because often when studying a process that is described (or **modelled**) by a set of equations, it is convenient, as a first approximation, to assume that they are linear in the variables pertaining to the process. Such a model is then said to be **linear**.

DEFINITION 1-1

An equation in variables $x_1, x_2, ..., x_n$ is called a **linear equation** if it has the form

$$a_1x_1 + a_2x_2 + ... + a_nx_n = b$$

where $a_1, a_2, ..., a_n$, and b are constants (i.e., real numbers). A **solution** of this equation is a set of numbers $x_1', x_2', ..., x_n'$ that satisfy the equation when substituted for $x_1, x_2, ..., x_n$ respectively. We often say "$x_1 = x_1', ..., x_n = x_n'$ is a solution". The **solution set** of the equation is the set of all solutions.

Notice that each term in a linear equation contains at most one variable and that variable has exponent 1 or 0.

EXAMPLE 1-1

(a) The following equations are linear:

(i) $2x_1 + 3x_2 - \sqrt{2}x_3 = 1.5$

In this equation, there are $n = 3$ variables, x_1, x_2, and x_3. The constants are $a_1 = 2, a_2 = 3, a_3 = -\sqrt{2}$, and $b = 1.5$. An example of a solution is $x_1 = \sqrt{2}, x_2 = 0.5, x_3 = 2$. Another example of a solution is $x_1 = 3/4, x_2 = x_3 = 0$. On the other hand, for example, $x_1 = x_2 = x_3 = 0$ is not a solution.

(ii) $5x - 6y = 9$

In this equation, there are $n = 2$ variables, $x_1 = x$ and $x_2 = y$. The constants are $a_1 = 5, a_2 = -6$, and $b = 9$.

(iii) $-5x + 4cy + 7z = 2$, where c is a constant.

Here the variables are understood to be x, y, and z, since c is specified to be a constant.

(b) The following equations are not linear:

(i) $x^2 - 2y + 3z = -4$

The variable x appears with exponent 2.

(ii) $x - 5xy + 2y - 4w + 3z = 0$

The second term contains two variables, x and y. ❑

EXAMPLE 1-2

The area A of a rectangle of dimensions x and y is given by the equation $A = xy$. This is not a linear equation in x and y. We consider the change in the area, ΔA, when the sides are changed by *small* amounts Δx and Δy (i.e., very much less than 1 in absolute value).

$$\Delta A = (x + \Delta x)(y + \Delta y) - xy = y\Delta x + x\Delta y + \Delta x\Delta y$$

The term $\Delta x\Delta y$ is very much smaller than the other terms when Δx and Δy are small. For example, if $\Delta x = 0.02$, $\Delta y = 0.03$, $x = 3$, and $y = 7$, then $y\Delta x = 0.14$, $x\Delta y = 0.09$, while $\Delta x\Delta y = 0.0006$. Thus, we may neglect $\Delta x\Delta y$ in approximating the change in the area ΔA. Therefore, ΔA is approximately given by a linear equation in the variables Δx and Δy when the dimensions x and y are fixed (constant).

$$\Delta A \approx y\Delta x + x\Delta y$$

For example, with the preceding numbers, $\Delta A \approx 0.23$ is the approximate change in area when the sides of a rectangle with sides 3 and 7 units are increased by 0.02 and 0.03 units respectively. ❑

One of the most important uses for linear equations is in the analytic geometry of lines and planes. We assume that the reader is familiar with the basics of plane analytic geometry, and, in particular, with the representation of points of the plane by ordered pairs of numbers (x,y). We denote the set of all such ordered pairs by R^2 and call it the **plane**. The beauty of analytic geometry is that we can study geometrical entities (such as lines) through algebraic properties of their equations (and vice-versa).

Recall that a **line** in the plane is the solution set of a linear equation in x and y of the form

$$ax + by = c \tag{1-1}$$

where a, b, and c are constants. Since the solution set is a set of ordered pairs (x,y) of numbers, we may think of it as a subset of the plane. This subset is the **graph** of the equation. Some examples of straight lines and their equations are illustrated in Figure 1-1.

FIGURE 1-1

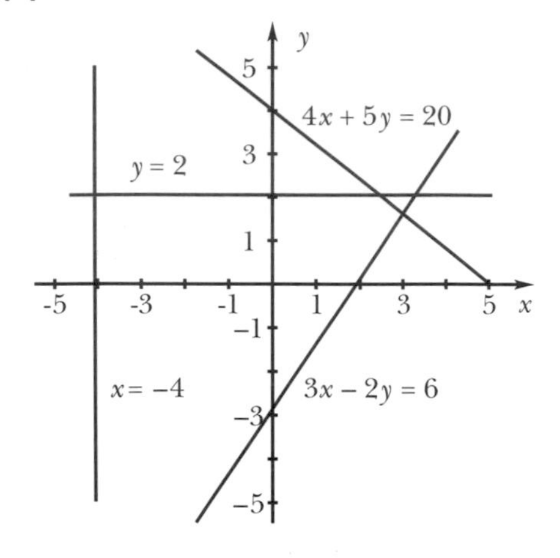

As shown in Figure 1-2, intuitively, we can see that two or more lines in the plane may intersect in

(1) **no** points (e.g., $3x + 2y = -4$ and $3x + 2y = 6$);
(2) exactly **one** point (e.g., $3x + 2y = 6$ and $2x - y = 4$);
(3) **infinitely** many points (e.g., $4x - 2y = 8$ and $2x - y = 4$).

FIGURE 1-2

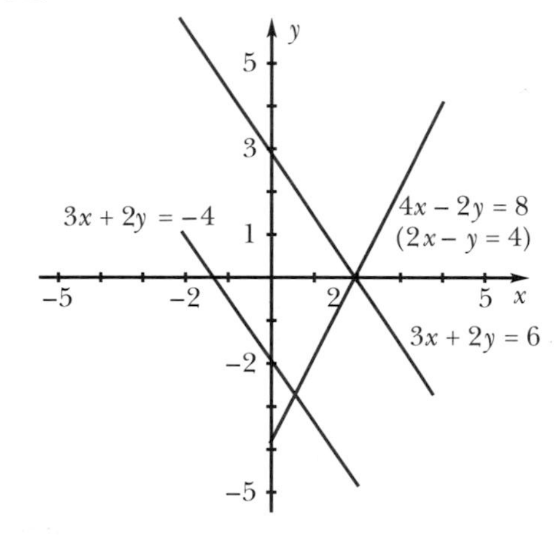

Consider the two lines with equations $3x + 2y = 6$ and $2x - y = 4$. The coordinates (x,y) of the points that lie on *both* lines are those that satisfy both equations. These can be obtained from the graphs of the equations: in Figure

1-2, we read from the graph that $x = 2$ and $y = 0$, as accurately as can be determined by the scale of the graph.

Alternatively, we can solve analytically for this point of intersection. We make use of the fact that if (x,y) is on both lines, then it must be on the second and

$$2x - y = 4$$

is true. By rearranging, we have

$$y = 2x - 4.$$

We then substitute this expression for y into the first equation to get

$$3x + 2(2x - 4) = 6$$

or,

$$\begin{aligned} 7x - 8 &= 6 \\ 7x &= 14 \\ x &= 2. \end{aligned}$$

From this and the expression $y = 2x - 4$, we now get

$$y = 2(2) - 4 = 0,$$

so the only point of intersection of the two lines is $(2,0)$ as found before. We say that the **system** of two linear equations in the two variables (or unknowns) x and y has the unique solution $x = 2$, $y = 0$.

DEFINITION 1-2

A **system** of m linear equations in two variables x and y is a collection of m equations of the form $ax + by = c$. A **solution** of the system is an ordered pair (x',y') that satisfies **all** equations of the system when x' and y' are substituted for x and y respectively. The **solution set** is the set of all solutions.

As seen from the preceding example, the solutions of such a system are exactly the points lying on all m lines which are the graphs of the m equations in the system.

Clearly one can consider systems of linear equations in more than two variables. Indeed, as we shall see in the next chapter, there are many practical applications where systems of equations in many variables arise, and where it is necessary to determine the solutions of those systems. A systematic procedure for solving such systems is one of the main concerns of Chapter 2. Our purpose in this section is to gain some geometrical insight by considering some simple cases.

EXAMPLE 1-3

Consider the system of three equations

$$
\begin{aligned}
x + y &= 4 && \text{(i)} \\
x - 2y &= 0 && \text{(ii)} \\
4x + y &= 4. && \text{(iii)}
\end{aligned}
$$

These lines are graphed in Figure 1-3, from which it is clear that no point lies on all three lines, although any two of the lines have a point in common. Thus, the system of three linear equations given above has no solution. Analytically we may see this as follows. From equation (i), we obtain

$$x = 4 - y,$$

so that substitution into equation (iii) yields

$$4(4 - y) + y = 4$$

or,

$$
\begin{aligned}
-3y &= -12 \\
y &= 4.
\end{aligned}
$$

FIGURE 1-3

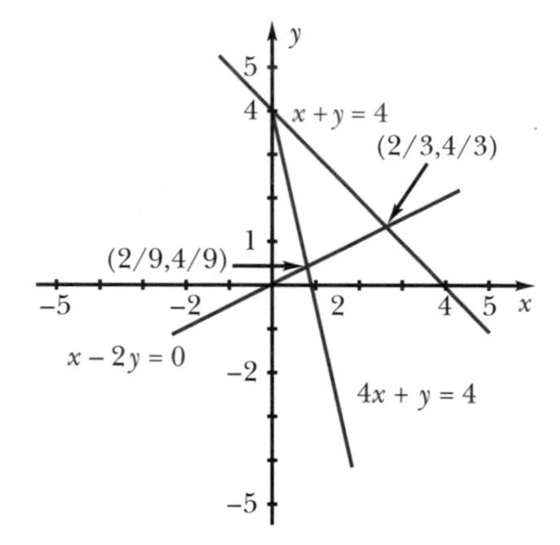

Alternatively, equation (ii) gives

$$x = 2y,$$

and substitution into equation (iii) gives

$$4(2y) + y = 4$$

or,

$$
\begin{aligned}
9y &= 4 \\
y &= 4/9.
\end{aligned}
$$

Since y cannot be 4 and 4/9 at the same time, the system can have no solution. ❑

EXAMPLE 1-4

Consider the three lines shown in Figure 1-4, and having equations

$$\begin{aligned} 4x + 6y &= 6, \\ 6x + 9y &= 18, \\ x &= y. \end{aligned}$$

FIGURE 1-4

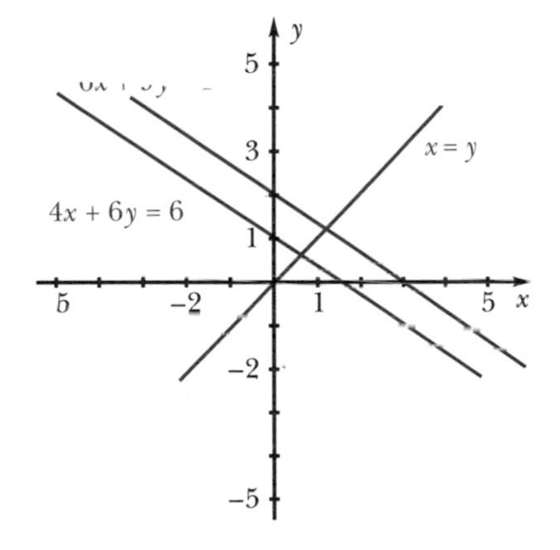

From the graphs, we see that these lines have no points in common as the first two are parallel. It follows that this system of three equations in x and y has no solution. ❑

A system of linear equations in x and y has **infinitely** many solutions exactly when all lines which are the graphs of the equations coincide. This happens when the equations of the system are simply multiples of one another.

EXAMPLE 1-5

Consider the system of two linear equations

$$-0.5x + 2.3y = 7 \qquad \text{(i)}$$

$$3.5x - 16.1y = -49. \qquad \text{(ii)}$$

Equation (ii) is -7 times equation (i). The system has infinitely many solutions, namely, any point (x,y) on their common graph. Since $x = 4.6y - 14$ by equation (i), the solution set consists of the infinitely many ordered pairs $(4.6y - 14, y)$ for $y \in R$. ❑

Let us now consider lines and planes in 3-space, $R^3 = \{(x,y,z) \mid x, y, z \in R\}$. This is the mathematical representation of the immediate space in which we live and where we have three directions in which to move. Every ordered triple, (x,y,z), corresponds to an unique point in this space, and vice-versa. When we draw pictures of this space on paper or on a blackboard, we need three axes to locate points. These three axes are usually taken at right angles to one another. Figure 1-5 is a drawing of some points in R^3. The usual location of the axes is shown in the figure, where the positioning of the x, y, and z axes is said to constitute a right-handed coordinate system (or right-handed orientation). Although we shall have no occasion to explore this notion of orientation, it is worth noting that, for example, if we interchanged the x and y axes, the coordinate system would be called a left-handed one. A similar situation exists in the plane R^2, where the usual positions of the x and y axes could be interchanged to yield a different orientation of the plane.

FIGURE 1-5

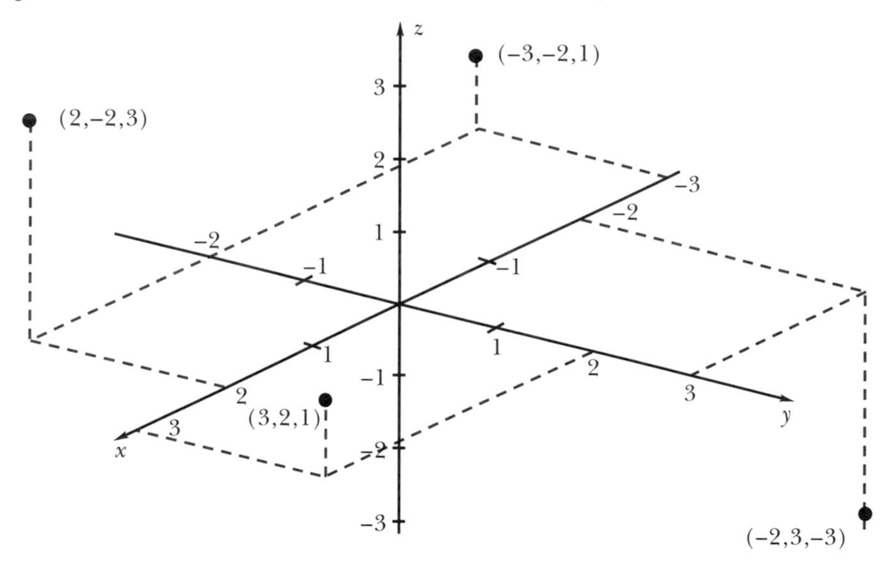

In contrast to the situation in R^2, the graph of a linear equation in three variables,

$$ax + by + cz = d \tag{1-2}$$

where a, b, c, and d are constants, is a **plane** in R^3, rather than a straight line. This can be made plausible by considering the **coordinate planes**, which are the planes in R^3 each containing exactly two of the axes. Thus, the xy plane is the plane containing the x and y axes, and is therefore perpendicular to the (remaining) z axis. It has equation $z = 0$. Similarly, the xy and yz planes have equations $y = 0$ and $x = 0$ respectively. Notice that each equation is a linear equation in x, y, and z.

More generally, for example, it is easy to visualize that the equation $x = 2$ describes all points lying 2 units in front of the yz plane ($x = 0$). Hence it is a plane parallel to the yz plane. Note that $x = 2$ is a linear equation of the form of 1-2 above, with $a = 1$, $b = 0$, $c = 0$, and $d = 2$.

Similarly, if k, l, m are constants, the equations $x = k$, $y = l$, and $z = m$ are the equations of planes parallel to the yz, xz, and xy planes respectively, and at distances of k, l, and m units from these planes.

Returning now to the general equation 1-2, we note that if we set any one of the variables x, y, or z equal to zero, we obtain the equation of a straight line in the remaining variables. The graph of this equation is a straight line in the coordinate plane whose equation is defined by that variable being set to zero. We illustrate in a specific case.

EXAMPLE 1-6

The equation $2x + 3y + 4z = 12$ has the form of 1-2. First, set $z = 0$ to get the equation $2x + 3y = 12$, which is a line L in the xy plane (Figure 1-6). It may be sketched in the usual manner, for example, by noting that the x and y intercepts are $x = 6$ and $y = 4$ respectively. Thus, this line L consists of all points in R^3 satisfying *both* equations

$$\begin{aligned} 2x + 3y + 4z &= 12 \\ z &= 0, \end{aligned}$$

or, after simplification,

$$\begin{aligned} 2x + 3y &= 12 \\ z &= 0, \end{aligned}$$

so that it is the set of points on *both* planes: it is their **intersection**. We call L the xy **trace** of the plane. ❑

FIGURE 1-6

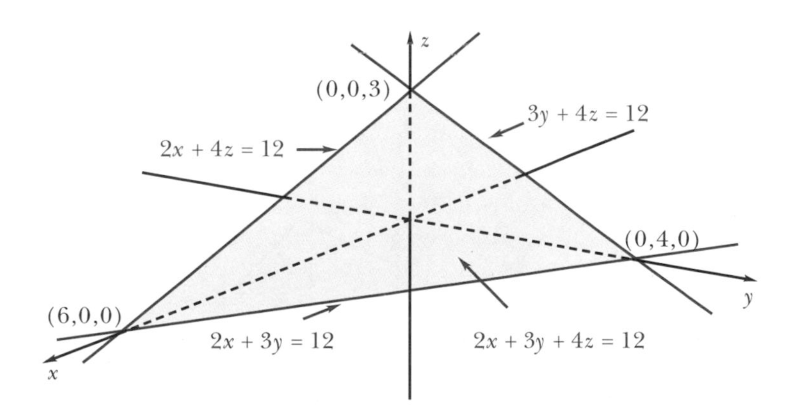

DEFINITION 1-3

The xy **trace** of a plane in R^3 is the line of intersection of the plane with the xy coordinate plane. Similarly, the xz and yz traces are the lines of intersection of the plane with the xz and yz coordinate planes respectively.

It should be emphasized that we know the traces are straight lines because they are the graphs of linear equations in two variables in a coordinate plane. However, as sets of points in R^3, they are the solution sets of *pairs* of linear equations in the three variables x, y, and z. Thus, in Example 1-6, the other two traces are:

$$\begin{aligned} 2x + 4z &= 12 \\ y &= 0 \end{aligned}$$

and

$$\begin{aligned} 3y + 4z &= 12 \\ x &= 0. \end{aligned}$$

Keeping in mind the fact that the traces are subsets of a given plane, it is clear that in order to sketch the plane, we need only draw its traces. This has been done in Figure 1-6. Of course, it may happen that a plane doesn't intersect a certain coordinate plane (in which case the plane must be parallel to it). In this case, the equation of the plane must be one of the three simple types discussed earlier: $x = k$, $y = l$, or $z = m$.

FIGURE 1-7

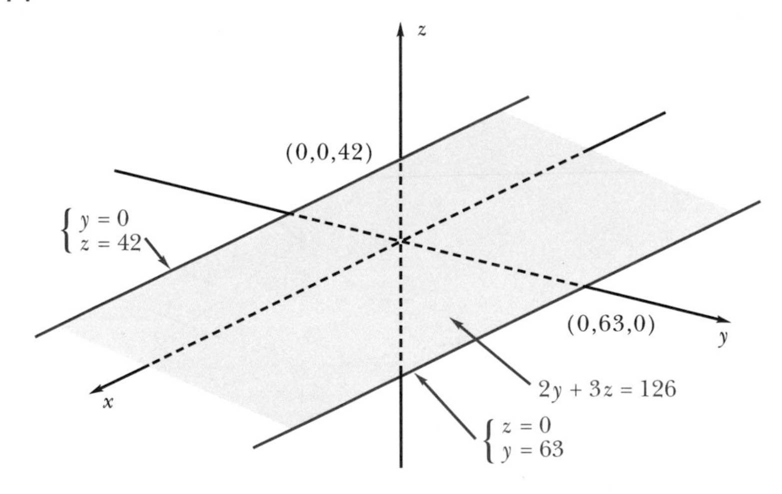

Figure 1-7 illustrates the plane $2y + 3z = 126$. Notice that it is parallel to the x axis. In general, a plane is parallel to a coordinate axis exactly when its equation doesn't contain the corresponding variable (in this case x).

Figure 1-8 illustrates the plane $2x + y - z = 0$, which is a plane through the origin.

FIGURE 1-8

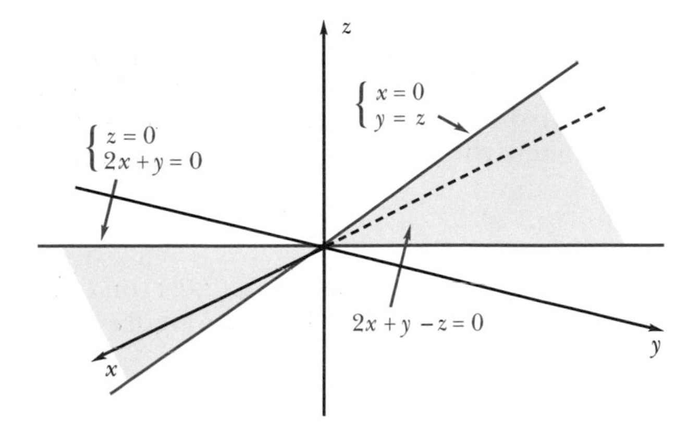

We have seen above that certain lines in R^3 (namely, the traces of planes) are the solution sets of systems of two linear equations in x, y, and z, since they arise as the intersections of two planes. Intuitively it is clear that any straight line can be obtained as the intersection of two suitable planes (in many different ways). It follows that one way of describing a line in R^3 analytically is as the solution set of the corresponding system of two linear equations.

DEFINITION 1-4

If $ax + by + cz = d$ and $a'x + b'y + c'z = d'$ are equations of planes which intersect in a line L, then the system of two equations

$$\begin{aligned} ax + by + cz &= d \\ a'x + b'y + c'z &= d' \end{aligned}$$

is called a **general system of equations** for the line L.

EXAMPLE 1-7

Draw the line in R^3 with the general system of equations

$$\begin{aligned} x + y + z &= 3 \\ 2x - y + z &= 5. \end{aligned}$$

Solution We solve this problem by recalling that *any straight line is completely determined when we know two points on it*. Analogous to the situation in R^2, where we often use the intercepts to sketch lines, in R^3 we can locate lines by obtaining the points where they intersect the coordinate planes. For example, the line L which we want to draw intersects the yz plane when $x = 0$. Putting $x = 0$ in the equations for L we get

$$\begin{aligned} y + z &= 3 \\ -y + z &= 5. \end{aligned}$$

These two equations represent lines in the yz plane which intersect at $y = -1$ and $z = 4$, as we can see by solving the system as in earlier examples. Thus, the point $(0,-1,4)$ is the point on L where L intersects the yz plane. Similarly, by putting $y = 0$, we obtain the equations

$$\begin{aligned} x + z &= 3 \\ 2x + z &= 5 \end{aligned}$$

which, when solved, yield a second point $(2,0,1)$ on L. We may then draw L by joining these two points as shown in Figure 1-9. ❑

FIGURE 1-9

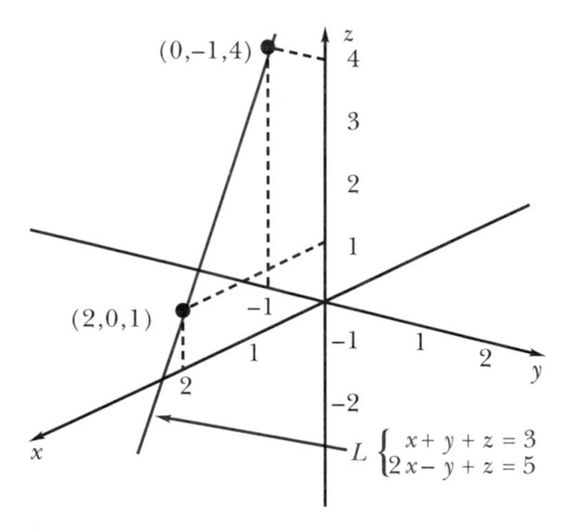

We conclude this section by noting that a line in R^2 or R^3 can also be characterized in other ways. For example,

(1) by specifying two points P' and P'' on it;
(2) by specifying a point P' on L and the "direction" of L.

Intuitively, one way of specifying the "direction" of a line is in terms of the angles it makes with the coordinate axes. However, with the help of a sketch, the reader will realize that a line makes many angles with the axes, and that some additional structure or convention is necessary to make this notion precise. The appropriate notion is that of a **vector**, to be discussed in the next section. Vectors are very useful for formulating the equations of lines as suggested in (1) and (2) above.

EXERCISES 1-1

1. Which of the following equations are linear in the variables employed? For each linear equation, give **(i)** an example of a solution, and **(ii)** an example of a set of values for the variables that is not a solution.

(a) $x + 3y - 2z = 6$

(b) $2x - 3y - 1 = 0$

(c) $a - \sqrt{2}b + 3c = 0$

(d) $-2x - \sqrt{y} + 4z = 3$

(e) $-2x_1 + 1/x_2 - 3x_3 + x_4 = 0$

(f) $x_1^2 + 2x_2x_3 - x_4 = 5$

(g) $\log q + 2\log K - 3\log L = 4$

2. Find the solution(s) for each of the following systems of equations by drawing the graphs of the lines represented by the equations and determining their intersection(s).

(a) $x - 2y = 1$
$x + y = 0$

(b) $3x - 2y = 6$
$6x = 4y + 4$

(c) $4x + 3y = 12$
$y = -(4/3)x + 4$
$x - 2y = 0$

(d) $2x + 4y = 8$
$3x - 2y = 4$
$x = 2$

3. In each of the following, solve the second equation for y (as a function of x) and then substitute this expression for y into the first equation to obtain the solution(s) for x (if possible). Then obtain the solution(s) for the system.

(a) $x - 2y = 1$
$x + y = 0$
(compare with 2(a))

(b) $3x - 2y = 6$
$6x - 4y = 4$
(compare with 2(b))

(c) $x - y = 4$
$2y + 8 = 2x$

4. Determine the value(s) of b for which the system of equations

$$x + y = 2$$
$$y = b - x$$

has **(i)** exactly one solution; **(ii)** no solution; **(iii)** infinitely many solutions.

5. Consider the function $y = f(x) = (1 - x)^{-1/2}$, where $0 < x < 1$.

(a) Plot the graph of this function.

(b) An important linear approximation for $y = f(x)$ can be obtained from the binomial expansion:

$$y = f(x) = 1 + (1/2)x + (1)(1/2)(-1/2)(-1/2 - 1)x^2 + \dots$$
$$\approx 1 + (1/2)x$$

when x is much less than 1. Compare the values of $f(x)$ and this linear approximation for $x = 0.01$, 0.1, and 0.5.

(c) By the theory of special relativity, the time t kept by a clock moving at speed v with respect to a clock keeping time t' is slowed down according to $t/t' = (1 - v^2/c^2)^{-1/2}$ where c is the speed of light. Determine the value of this ratio for a clock moving at one-half the speed of light. Find the linear approximation for this ratio as in (b) and compare its value with the value of the ratio.

6. Use the binomial theorem to find the approximate volume of a cube whose side measures 1.005 m.

7. Frequently the production q in units of a commodity can be modelled by a so-called Cobb-Douglas function of the capital K and the labour L:

$$q = a_0 K^{a_1} L^{a_2} \text{ where } a_0, a_1, a_2 \text{ are positive constants.}$$

This is a nonlinear equation in the variables q, K, and L.

(a) Show that $\log q$, $\log K$, and $\log L$ are related by a linear equation.

(b) If the commodity is sold at a price p dollars per unit, write a formula for the revenue R and show that $\log R$, $\log p$, $\log K$, and $\log L$ are related by a linear equation.

8. Sketch each of the following planes in R^3 by graphing its traces on the three coordinate planes.

(a) $2x + 4y - 6z = 12$

(b) $2x - 6y - 6z = 12$

(c) $2x + 6y + 6z = 12$

9. Find the equation of the plane through the point $(-1,3,-2)$ and that is **(a)** parallel to the xz plane; **(b)** perpendicular to the x axis; **(c)** parallel to both the x and y axes.

10. Find the coordinates of the points of intersection of the plane

$$2x - 3y + 5z - 15 = 0$$

with the coordinate axes.

11. Write down a general system of equations for the line of intersection of the following pairs of planes from question 8 above; then sketch each line by finding the points of intersection with the coordinate planes.

(i) planes (a) and (b)
(ii) planes (b) and (c)
(iii) planes (a) and (c)

12. Sketch the line with general system of equations

$$\begin{aligned} 2x + 5y + 4z - 4 &= 0 \\ 28x - 4y + 19z + 92 &= 0. \end{aligned}$$

13. Sketch the line with general system of equations

$$\begin{aligned} x - 2y + 4z - 14 &= 0 \\ x + 20y - 18z + 30 &= 0. \end{aligned}$$

1-2 Vectors in R^2 and R^3

We saw in the previous section that there are occasions when it is useful to consider the notion of lines having a direction specified on them. Roughly speaking, a **vector** is a line segment with a direction specified on it. There are many examples in the sciences where the values of variables can be regarded as vectors. The traditional example is that of the velocity of a body, such as a ship steaming a certain course on the ocean. If we wish to show the velocity of the ship on a map at any instant in time, we can draw an arrow pointing in the direction of travel of the ship and having its length proportional to the speed of the ship (with respect to some arbitrary scale). Then, the resulting arrow is a vector that represents the velocity of the ship at that instant.

The notion of a vector is attributed to August Ferdinand Moebius (1790–1860) and William Rowan Hamilton (1805–1865). Moebius, a German mathematician, made numerous contributions to geometry and what is now known as topology and combinatorics. He was the inventor of the paradoxical one-sided surface called the "Moebius strip". Hamilton was an Irish mathematician famous for his profound contributions to algebra and mechanics. He was so precocious as a student that he was offered a professorship at the age of 22. His formulation of mechanics has been fundamental to the development of contemporary physics.

DEFINITION 1-5

Given points A and B in R^3, the **vector AB** in R^3 is the directed line segment (arrow) from point A to point B. A is called the **initial point** of **AB** and B is called the **terminal point**. The arrow points from A to B and we also say that the vector **AB** is **bound** at A to indicate that the initial point is A. The **length** of the vector **AB**, denoted $|\mathbf{AB}|$, is the distance from A to B as given by the Pythagorean formula:

$$|\mathbf{AB}| = \sqrt{(b-a)^2 + (b'-a')^2 + (b''-a'')^2}$$

when $A = (a,a',a'')$ and $B = (b,b',b'')$.

Note that the order of writing A and B is important: **BA** is the vector with initial point B.

Some examples are sketched in Figure 1-10, where we see that

$$|\mathbf{EF}| = \sqrt{(-1-2)^2 + (0-(-3))^2 + (-4-1)^2} = \sqrt{43} \approx 6.56.$$

In many books, the notation for the vector **AB** is $\overrightarrow{AB}$, in order to avoid confusion with the common notation AB for the undirected segment from A to B. In this book, we will always use **AB** to denote the vector from A to B.

In this connection, it is probably not premature to warn the reader that the word "vector" is used in mathematics in many different senses. For example, what we have defined here as vectors are sometimes called "bound" vectors. Later in this section we introduce another meaning for the term. This will then be replaced in Chapter 4 by another definition, of which the earlier definition will be a particular case.

FIGURE 1-10

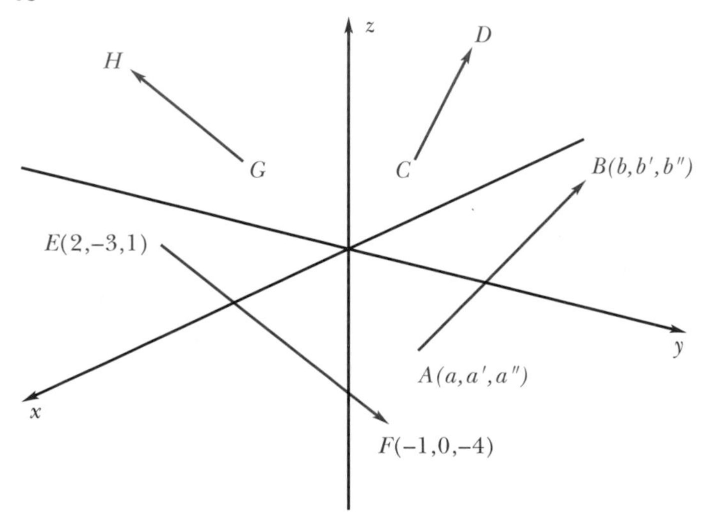

Frequently we shall be interested in vectors all lying in the same plane, which we speak of as vectors in the plane, or vectors in R^2. In fact, it is convenient to start our discussion of the algebra of vectors by considering vectors in the plane (Figure 1-11).

FIGURE 1-11

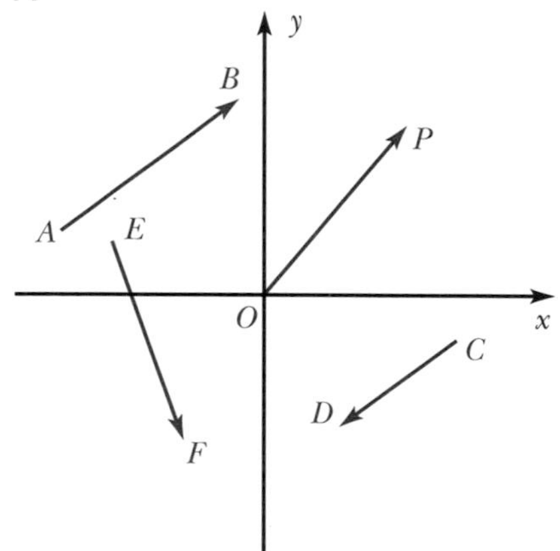

Given two vectors **AB** and **AC** with the same initial point A (Figure 1-12), there is a natural geometrical procedure for producing a new vector

AD from them: **AD** is simply the arrow which is the diagonal of the parallelogram as shown. This vector can be constructed using ruler and compass only, but, as can be seen from the figure, the coordinates of D are $(b + c - a, b' + c' - a')$, when $A = (a,a')$, $B = (b,b')$, and $C = (c,c')$. Note that **AD** also has initial point A.

FIGURE 1-12

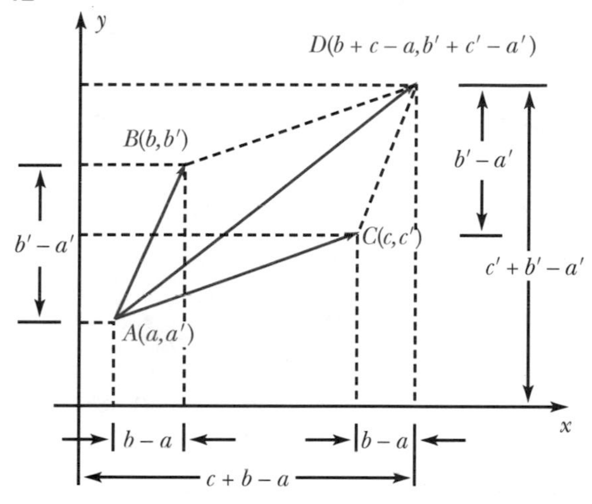

> **DEFINITION 1-6**
>
> The vector **AD** obtained from the diagonal AD of the parallelogram $ABDC$ is called the **sum** of **AB** and **AC**, and we also denote **AD** by **AB** + **AC**.

Other examples of sums of vectors are shown in Figure 1-13.

FIGURE 1-13

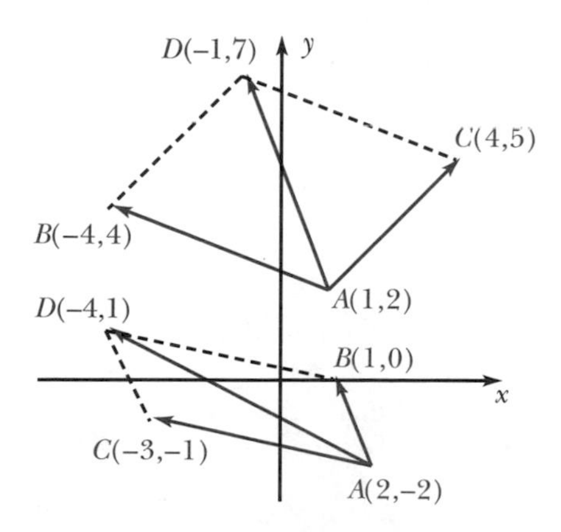

Notice that we cannot speak of a parallelogram when the vectors are collinear (i.e., lying in the same line), but the coordinate expression for D continues to make sense, so we can use it to define the sum in this case. However, to avoid this exception, the parallelogram rule for addition is sometimes replaced by the following equivalent "triangle" rule: move **AB** in a parallel fashion so that its initial point coincides with C; then **AD** is the vector from A to the new position (D) of B (Figure 1-14).

FIGURE 1-14

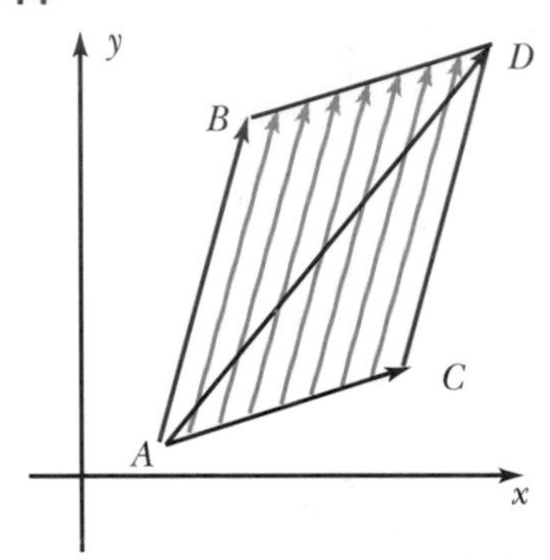

These same procedures can be applied to define the addition of vectors in R^3 (see Figure 1-15). Since any two noncollinear vectors **AB** and **AC** in R^3 lie in the plane containing the three points A, B, and C, the parallelogram or triangle rule for R^2 can be applied.

FIGURE 1-15

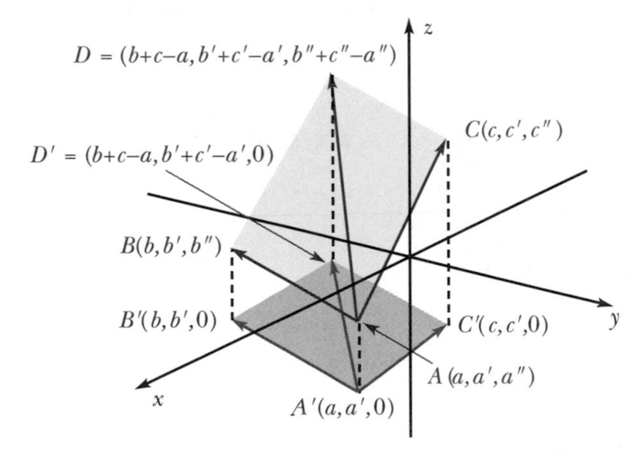

In terms of coordinates, let $A = (a,a',a'')$, $B = (b,b',b'')$, and $C = (c,c',c'')$ as shown. Then, the terminal point D of **AB** + **AC** is $(b + c - a, b' + c' - a', b'' + c'' - a'')$. To see this, we need only observe that the feet A', B', C', and D' of the perpendiculars from A, B, C, and D respectively to the xy plane define a parallelogram in R^2 (a fact of geometry that we will assume!). Then, using the result for R^2 (in the xy plane), D' has coordinates $(b + c - a, b' + c' - a', 0)$, since $A' = (a,a',0)$, $B' = (b,b',0)$, and $C' = (c,c',0)$. It follows that the first two coordinates of D must be $(b + c - a)$ and $(b' + c' - a')$ respectively. Similarly, using the feet of the perpendiculars to another coordinate plane, say, the xz plane, we see that D has the coordinates in R^3 as stated.

EXAMPLE 1-8

If $A = (3,6,2)$, $B = (4,1,5)$, and $C = (0,3,7)$, then **AD** has terminal point $D = (4 + 0 - 3, 1 + 3 - 6, 5 + 7 - 2) = (1,-2,10)$. ❑

The addition of vectors considered does not allow us to add vectors that have *different* initial points: what would the initial point of the "sum" be? This is one reason it is useful to introduce *a notion of vector which is free from initial points*. Furthermore, the idea of moving vectors around in R^3 is a natural one (we saw it in the triangle law). We can formalize these ideas with the following definition.

DEFINITION 1-7

Two vectors in R^3 (or R^2) are said to be **equal**, denoted **AB** = **CD**, if they have the same length ($|\mathbf{AB}| = |\mathbf{CD}|$), are parallel, and point in the same direction.

Figure 1-16 shows two groups of equal vectors in R^2. We have **AB** = **CD** = **OP**, and **EF** = **GH** = **OQ**, while **AB** is not equal to **EF** (we write **AB** ≠ **EF**).

FIGURE 1-16

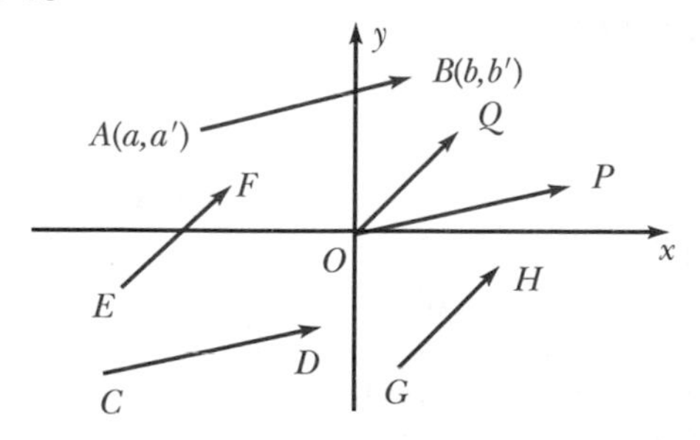

The next definition introduces a new meaning for the word "vector".

DEFINITION 1-8

A (free) **vector v** is the set of all (bound) vectors equal to some given vector **AB**. We write $\mathbf{v} = \{\mathbf{AB}\}$ to indicate this.

Thus, if **AB** and **CD** are any equal (bound) vectors ($\mathbf{AB} = \mathbf{CD}$), we say that they *represent* the same vector **v** and write $\mathbf{v} = \{\mathbf{AB}\} = \{\mathbf{CD}\}$. The vector **v** is called a free vector, because it stands for all vectors equal to **AB**, so is "free" of dependence on an initial point. However, we will usually omit the adjective "free" in referring to **v**. So, a vector $\mathbf{v} = \{\mathbf{AB}\}$ has infinitely many representatives, all of which are bound vectors having the same length and pointing in the same direction as **AB** (i.e., equal to **AB**). We write $\mathbf{v} = \{\mathbf{AB}\}$ to distinguish this collection of all vectors equal to **AB** from the vector **AB** itself. By convention, boldface letters such as **u, v, w**, etc., will stand for free vectors.

Clearly, we may speak of the length and direction of **v** as being the length and direction of any representative for **v**. We denote the length of **v** by $|\mathbf{v}|$. In Figure 1-16, $\mathbf{v} = \{\mathbf{AB}\} = \{\mathbf{CD}\} = \{\mathbf{OP}\}$, while **EF**, **GH**, and **OQ** are all representatives of a second vector, which we could denote by **w**. Thus, $|\mathbf{v}| = |\mathbf{AB}| = |\mathbf{CD}| = |\mathbf{OP}|$ and $|\mathbf{w}| = |\mathbf{EF}| = |\mathbf{GH}| = |\mathbf{OQ}|$.

While a vector as now defined has no initial point (it is a collection of equal vectors), we can nevertheless visualize it in terms of an *unique* bound vector, namely, its unique representative with initial point at the origin O.

In Figure 1-16, $\mathbf{v} = \{\mathbf{OP}\}$, where P has coordinates $(b - a, b' - a')$ when $A = (a,a')$ and $B = (b,b')$. To see this, suppose $P = (p,p')$. Since AB and OP are parallel and of the same length, $OABP$ is a parallelogram. Consequently, $\mathbf{OB} = \mathbf{OA} + \mathbf{OP}$, and by the result of Figure 1-12, $b = a + p$ and $b' = a' + p'$. Thus, $p = b - a$ and $p' = b' - a'$. We state this important observation as a theorem.

THEOREM 1-1

If **AB** is a vector in R^2 with $A = (a,a')$ and $B = (b,b')$, then $\mathbf{v} = \{\mathbf{AB}\}$ has an unique representative **OP** with initial point at the origin O and $P = (b - a, b' - a')$. ∎

For example, if $A = (-2,2)$ and $B = (1,3)$, then $\mathbf{v} = \{\mathbf{AB}\} = \{\mathbf{OP}\}$, with $O = (0,0)$ and $P = (1 - (-2), 3 - 2) = (3,1)$.

A similar result is true in R^3: if $\mathbf{v} = \{\mathbf{AB}\}$, where $A = (a,a',a'')$ and $B = (b,b',b'')$, then $\mathbf{v} = \{\mathbf{OP}\}$, where $P = (b - a, b' - a', b'' - a'')$. The proof is similar to the one for R^2 and is left for the reader.

As a result of these observations, we see that for every vector $\mathbf{v} = \{\mathbf{AB}\}$ in R^3, we have determined an unique point $P = (x,y,z) \in R^3$ such that $\mathbf{v} = \{\mathbf{AB}\} = \{\mathbf{OP}\}$. Conversely, given $P \in R^3$, $P \neq O$, the unique vector $\mathbf{v} = \{\mathbf{OP}\}$ in R^3 determines P. In order to accommodate the case $P = O$, we *define* the **zero vector**, denoted by **0**, to be the unique vector represented by any bound vectors **AB**, where $A = B$. In other words, $\mathbf{0} = \{\mathbf{AA}\}$, for any $A \in R^3$. Of course, $|\mathbf{0}| = 0$ and $\mathbf{0} = \{\mathbf{OO}\}$ determines the origin $O \in R^3$. We now have an example of the following important idea.

DEFINITION 1-9

Let X and Y be any two sets such that every $x \in X$ determines an unique $y \in Y$ and each $y \in Y$ is determined by an unique $x \in X$. Then we say that there is a **1-1 correspondence** between X and Y. We often write $X \leftrightarrow Y$ to denote this fact, and if $x \in X$ determines the unique element $y \in Y$, we write $x \leftrightarrow y$ and say that y **corresponds** to x.

When $X \leftrightarrow Y$ we may **identify** (regard as the same) corresponding elements of X and Y. This gives, in effect, two ways of thinking of the elements of X. Of course, this is essentially what we do in analytic geometry, where we identify each point of a plane with an ordered pair in R^2.

We may therefore summarize the preceding discussion as follows.

THEOREM 1-2

There is a 1-1 correspondence between the set of all vectors in R^3 and the set R^3 itself, with $\mathbf{v} \leftrightarrow (x,y,z)$, where $\mathbf{v} = \{\mathbf{OP}\}$ and $P = (x,y,z)$. ■

Note that $\mathbf{0} \leftrightarrow (0,0,0)$. A similar result holds for R^2.

DEFINITION 1-10

If $\mathbf{v} \leftrightarrow (x,y,z) \in R^3$, then x, y, and z are called the **components** of **v**. We denote this by writing $\mathbf{v} = (x,y,z)$. Similarly, if $\mathbf{u} \leftrightarrow (x,y) \in R^2$, x and y are the components of **u** and we write $\mathbf{u} = (x,y)$.

Of course, the components of the zero vector **0** in R^3 are 0, 0, and 0. Note that if **v** has components x, y, and z, then $|\mathbf{v}| = \sqrt{x^2 + y^2 + z^2}$.

EXAMPLE 1-9

(a) Let $\mathbf{v} = \{\mathbf{AB}\}$, where $A = (-2,5,-7)$ and $B = (4,0,3)$. Then, $\mathbf{v} = \{\mathbf{OP}\}$, where $P = (4 + 2, 0 - 5, 3 + 7) = (6,-5,10)$. Therefore, the components of **v** are 6, -5, and 10, and we write $\mathbf{v} = (6,-5,10)$.

(b) Suppose $\mathbf{u} = (-3,7)$. To find the representative $\mathbf{AB}$ of $\mathbf{u}$ having initial point $A = (-1,2)$, let $B = (b,b')$. Then, $b + 1 = -3$ and $b' - 2 = 7$, so $b = -4$, $b' = 9$, and $B = (-4,9)$. ❑

We now carry over the notion of addition of vectors to free vectors.

DEFINITION 1-11

If $\mathbf{v} = \{\mathbf{AB}\}$ and $\mathbf{w} = \{\mathbf{EF}\}$ are vectors in R^2 (or R^3), their **sum**, $\mathbf{v} + \mathbf{w}$, is the vector defined as follows: Choose representatives $\mathbf{RP}$ and $\mathbf{RQ}$ for $\mathbf{v}$ and $\mathbf{w}$ respectively having the same initial point R (Figure 1-17). Then $\mathbf{v} + \mathbf{w} = \{\mathbf{RP} + \mathbf{RQ}\}$, i.e., it is the vector whose representative is $\mathbf{RP} + \mathbf{RQ}$.

FIGURE 1-17

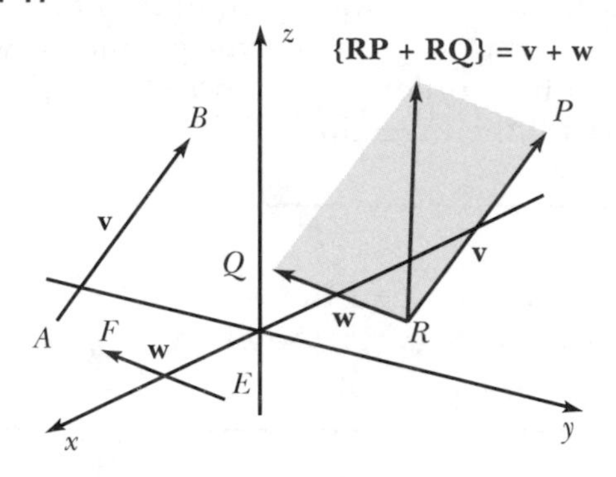

We leave as an exercise (question 10 in Exercises 1-2) the simple geometrical proof that this definition is independent of the representatives **AB** and **EF** chosen (and so makes sense!). Note that the construction above just amounts to "moving" **AB** and **EF** in a parallel fashion so that their initial points coincide.

FIGURE 1-18

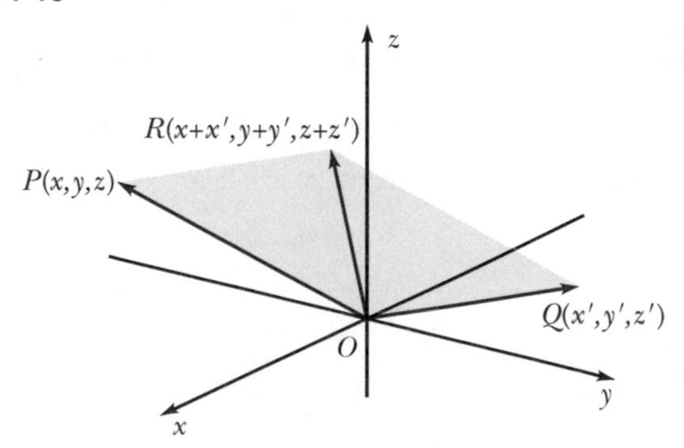

To simplify matters, we can always choose the representatives at the origin (since we can use any representatives we want). Suppose $\mathbf{v} = \{\mathbf{OP}\}$ and $\mathbf{w} = \{\mathbf{OQ}\}$ where $P = (x,y,z)$ and $Q = (x',y',z')$. Then the terminal point of $\mathbf{OR} = \mathbf{OP} + \mathbf{OQ}$ has coordinates $(x+x',y+y',z+z')$, and so $\mathbf{v} + \mathbf{w} = (x+x',y+y',z+z')$, as illustrated in Figure 1-18.

EXAMPLE 1-10

Suppose $\mathbf{v} = \{\mathbf{AB}\}$, $\mathbf{w} = \{\mathbf{EF}\}$, where $A = (-1,2,-3)$, $B = (2,4,-1)$, $E = (1,1,-2)$, and $F = (2,3,4)$. Using the same notation as above, **OP** has terminal point $P = (2 - (-1),4 - 2,-1 - (-3)) = (3,2,2,)$, and **OQ** has terminal point $Q = (2 - 1,3 - 1,4 - (-2)) = (1,2,6)$. Thus, $\mathbf{v} + \mathbf{w}$ is represented by the vector at the origin with terminal point $(3 + 1,2 + 2,2 + 6) = (4,4,8)$. In other words, $\mathbf{v} + \mathbf{w} = (4,4,8)$. ❑

The addition of vectors discussed above allows us to produce new vectors from two given vectors. There is another operation which produces a new vector from one given vector and a specified real number. This is called **scalar multiplication** and is defined as follows.

DEFINITION 1-12

If **AB** is a vector in R^2 (or R^3) and k is a number, the **scalar multiple** of k and **AB**, denoted $k\mathbf{AB}$, is the vector **AT** (with the same initial point) such that
(1) **AT** is collinear with **AB**,
(2) $|\mathbf{AT}| = |k||\mathbf{AB}|$,
(3) T is on the same side of A as B if $k > 0$, but is on the opposite side of A if $k < 0$ (if $k = 0$, $T = A$).

Numbers k used in this way to produce new vectors from given vectors are often called **scalars**. Some examples are shown in Figure 1-19.

FIGURE 1-19

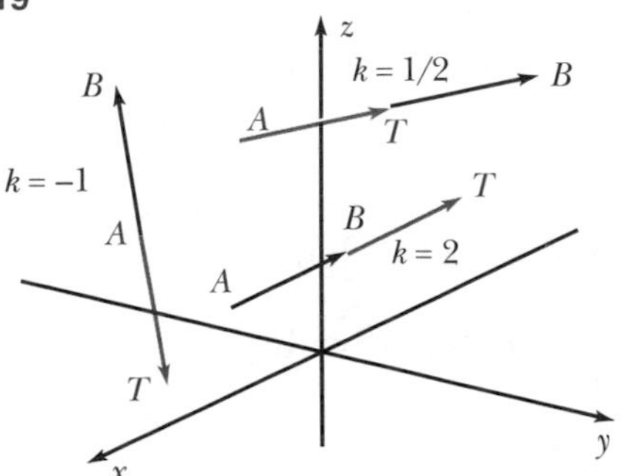

Naturally, we define scalar multiplication of (free) vectors in terms of the vectors representing them. If $\mathbf{v} = \{\mathbf{AB}\}$, $k\mathbf{v} = \{k\mathbf{AB}\}$. Once again, this definition is independent of the representative chosen. And, if $\mathbf{v} = (x,y,z)$, then $k\mathbf{v} = (kx,ky,kz)$. For, by definition, $k\mathbf{v} = \{k\mathbf{OP}\} = \{\mathbf{OT}\}$, where $P = (x,y,z)$. Since $\mathbf{OT}$ and $\mathbf{OP}$ are collinear and $|k\mathbf{OP}| = |k||\mathbf{OP}|$, it follows from the properties of similar triangles that $T = (kx,ky,kz) = k\mathbf{v}$. The reader is encouraged to make a sketch.

EXAMPLE 1-11

Suppose $\mathbf{v} = \{\mathbf{AB}\}$ where $A = (1,2,3)$ and $B = (3,5,9)$. Then, $\mathbf{v} = (2,3,6)$ and $|\mathbf{AB}| = (2^2 + 3^2 + 6^2)^{1/2} = 7$.

(i) For $k = 1.2$, $k\mathbf{v} = (2.4,3.6,7.2)$ and $k\mathbf{AB}$ has terminal point $(1+2.4,2+3.6,3+7.2) = (3.4,5.6,10.2)$.

(ii) For $k = -0.3$, $k\mathbf{v} = (-0.6,-0.9,-1.8)$ and $k\mathbf{AB}$ has terminal point $(1 - 0.6,2 - 0.9,3 - 1.8) = (0.4,1.1,1.2)$.

The reader should verify that in each case, $|k\mathbf{AB}| = |k||\mathbf{AB}|$ ($= 1.2\times 7 = 8.4$ and $0.3\times 7 = 2.1$ respectively). ❑

Note that the operations of addition and scalar multiplication as introduced are geometric in nature (the constructions can be carried out using ruler and compass) and do not require analytic geometry. However, the use of analytic geometry and the identification of a vector with its corresponding point in R^3 certainly simplify many computations and introduce new insights.

We next consider the basic "arithmetic" properties (i.e., pertaining to addition and scalar multiplication) of vectors in R^2 and R^3. They are important because together they imply that these sets of vectors constitute a *vector space*. The use of the word "vector" in this term is defined in Chapter 4.

THEOREM 1-3

Let $\mathbf{u}$, $\mathbf{v}$, and $\mathbf{w}$ be any vectors in R^2 (or R^3) and k and k' be any numbers.

(1) $\mathbf{u} + \mathbf{v} = \mathbf{v} + \mathbf{u}$ (**commutative law** of addition).
(2) $(\mathbf{u} + \mathbf{v}) + \mathbf{w} = \mathbf{u} + (\mathbf{v} + \mathbf{w})$ (**associative law** of addition).
(3) $\mathbf{u} + \mathbf{0} = \mathbf{u}$, where $\mathbf{0}$ is the zero vector; and $\mathbf{0}$ is the only vector with this property.
(4) Given any $\mathbf{u}$, there is an unique vector $\mathbf{v}$ such that $\mathbf{u} + \mathbf{v} = \mathbf{0}$. It is the **additive inverse** of $\mathbf{u}$ and is denoted $-\mathbf{u}$.
(5) $k(\mathbf{u} + \mathbf{v}) = k\mathbf{u} + k\mathbf{v}$.
(6) $(k + k')\mathbf{u} = k\mathbf{u} + k'\mathbf{u}$. (**distributive laws**)
(7) $(kk')\mathbf{u} = k(k'\mathbf{u})$.
(8) $1\mathbf{u} = \mathbf{u}$, where 1 is the number one.

PROOF AND REMARKS As noted earlier, the proofs of these properties may be either geometrical or analytic. We leave most of them as exercises for the reader, but comment on their significance as follows.

(1) This property means that the addition of two vectors is independent of the order in which we write them. A geometric proof consists simply of the observation that we get the same parallelogram (and hence the same diagonal) from the definition of $\mathbf{v} + \mathbf{u}$ as from that of $\mathbf{u} + \mathbf{v}$ (Figure 1-12). In an analytic proof, we let $\mathbf{u} = (x,y,z)$, $\mathbf{v} = (x',y',z')$ and observe that

$$\mathbf{u} + \mathbf{v} = (x+x',y+y',z+z') = (x'+x,y'+y,z'+z) = \mathbf{v} + \mathbf{u}$$

because of the commutativity of addition of numbers.

(2) This states that if we wish to form the sum of **three** vectors (which we must do successively in pairs, since addition is defined for pairs of vectors), the order in which we do so is immaterial. An important consequence is the fact that *we can write sums of any finite number of vectors unambiguously without brackets*:

$$\mathbf{u} + (\mathbf{v} + \mathbf{w}) = (\mathbf{u} + \mathbf{v}) + \mathbf{w} = \mathbf{u} + \mathbf{v} + \mathbf{w}.$$

See also the discussion following Definition 4-5. A geometric proof of this property (for R^2) is indicated in Figure 1-20. The analytic proof again results from the corresponding property of numbers. With **u, v** as above and $\mathbf{w} = (x'',y'',z'')$,

$$\begin{aligned}
&(\mathbf{u} + \mathbf{v}) + \mathbf{w} \\
&= ((x + x') + x'', (y + y') + y'', (z + z') + z'') \\
&= (x + x' + x'', y + y' + y'', z + z' + z'') \\
&= (x + (x' + x''), y + (y' + y''), z + (z' + z'')) \\
&= \mathbf{u} + (\mathbf{v} + \mathbf{w}).
\end{aligned}$$

FIGURE 1-20

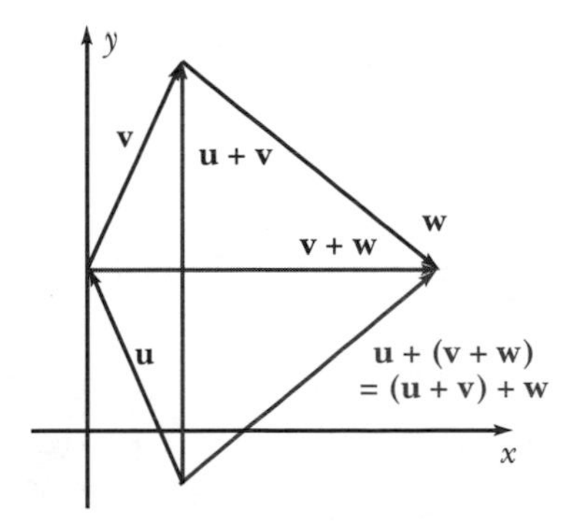

(3) The zero vector $\mathbf{0}$ is called the **additive identity** because adding it to any $\mathbf{u}$ leaves $\mathbf{u}$ identically the same. If $\mathbf{w}$ were another vector with this property, then $\mathbf{0} = \mathbf{w} + \mathbf{0} = \mathbf{w}$, which shows that $\mathbf{0}$ is unique.

(4) The vector $\mathbf{v}$ is just $(-1)\mathbf{u}$. For, if $\mathbf{u} = \{\mathbf{AB}\}$, $(-1)\mathbf{u} = \{(-1)\mathbf{AB}\}$, and $(-1)\mathbf{AB} = \mathbf{BA}$ is a vector of the same length pointing in the opposite direction. Since it is the only vector (why?) with the property $\mathbf{u} + \mathbf{v} = \mathbf{0}$, it is given a special symbol "$-\mathbf{u}$"; thus, $\mathbf{u} + (-\mathbf{u}) = \mathbf{0}$.

(5) A geometric proof is indicated in Figure 1-21 for R^2. It is based on the geometry of similar triangles.

FIGURE 1-21

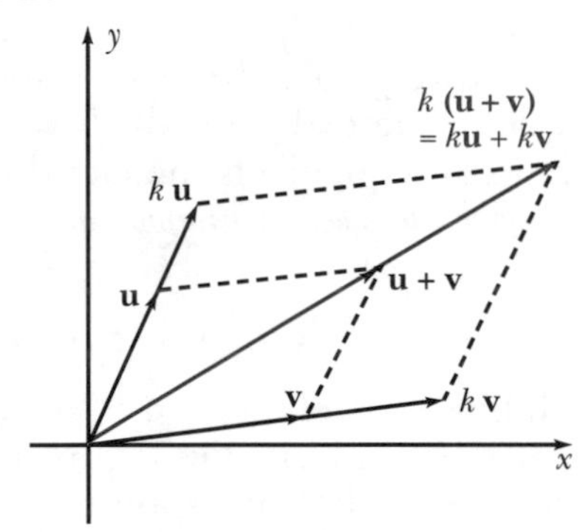

(6)–(8) The proofs of these properties are left for the reader. Properties (5)–(7) are called distributive laws because they show how scalar multiplication distributes the terms in sums of vectors or scalars. ■

EXAMPLE 1-12

(a) Let $\mathbf{u} = (-3,7)$, $\mathbf{v} = (4,5)$, and $\mathbf{w} = (-8,9)$. We find $|-2\mathbf{z}|$, where $\mathbf{z} = -3[2\mathbf{u} - 3(\mathbf{v} + 2\mathbf{w})]$.
Using Theorem 1-3, we have

$$\begin{aligned}\mathbf{z} &= -6\mathbf{u} + 9\mathbf{v} + 18\mathbf{w}\\ &= (18,-42) + (36,45) + (-144,162)\\ &= (-90,165).\end{aligned}$$

Therefore, $|-2\mathbf{z}| = |-2||\mathbf{z}| = 2(90^2 + 165^2)^{1/2} = 2\sqrt{35325} \approx 375.9$.

(b) Let $\mathbf{u} = (1,2,3)$, $\mathbf{v} = (-1,1,2)$, and $\mathbf{w} = (-1,1,-1)$. We find the vector $\mathbf{x}$ such that $2\mathbf{x} - 3\mathbf{u} + \mathbf{v} = 4\mathbf{w} - 2\mathbf{x}$.

Solving for **x**, we get

$$\begin{aligned} \mathbf{x} &= (1/4)(3\mathbf{u} - \mathbf{v} + 4\mathbf{w}) \\ &= (1/4)(0,9,3) \\ &= (0,9/4,3/4). \end{aligned}$$

The reader should identify those properties from Theorem 1-3 used in this calculation. ❑

APPLICATION (Optional)

We conclude this section with an application (from engineering) of the preceding ideas, and especially of the associative law of vector addition. Recall that the parallelogram law of addition is equivalent to the triangle law (Figure 1-14). It follows from the associative law that the sum of any finite number of vectors, $\mathbf{v}_1, \mathbf{v}_2, \ldots, \mathbf{v}_N$, is the vector **v** represented by the $(N + 1)^{\text{th}}$ side of the **polygon** whose sides are these vectors drawn in the order given, and where the initial point of each vector in the list is taken to be the terminal point of the preceding one. See Figure 1-22 for $N = 3$, 4, and 5.

FIGURE 1-22

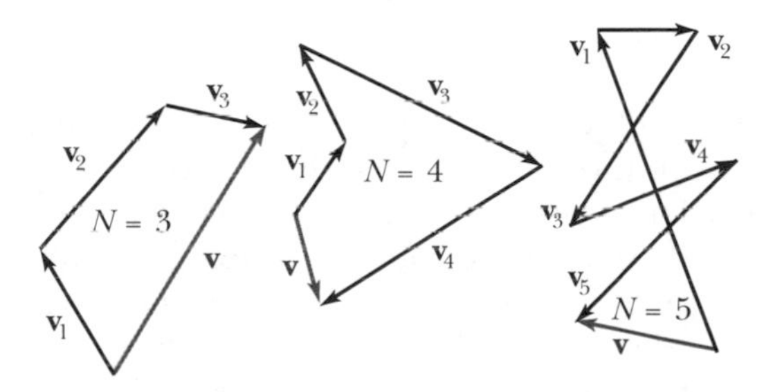

In this context, the sum **v** is often called the **resultant**. In Example 1-13 below, we apply this to a simple case where the vectors represent forces acting on a body. Roughly speaking, a *force* is something that attempts to change the state of motion (or of rest) of a body. For example, pushes or pulls on a body (i.e., an object) by agents such as people, machines, or engines are forces. The *weight* of a body is the gravitational force exerted by the earth on the body; it is directed from the body to the centre of the earth and is usually specified in units called kilograms-weight (kg-wt), where 1 kg-wt = 9.80 N (newtons). A force is a vector quantity: it has both magnitude and direction. In a force diagram, a force is drawn as a vector whose length represents its magnitude and whose direction represents its direction of application.

The following example shows how the polygon of vectors can be applied to force vectors. It involves Newton's Second Law (from physics) which states that if a body is at rest, the sum of the forces acting on it must be zero (i.e., the resultant is a **0** force vector).

EXAMPLE 1-13

A mountain climber is attempting to get from the top of a cliff face P to a higher cliff face Q by sliding down a smooth (frictionless) rope on a pulley to come to rest at a point R (and thereafter climbing the rope to get to Q!). Thus the rope is longer than the distance PQ, as shown in Figure 1-23. We are interested in finding where he comes to rest after his "slide", given his vertical heights h', h'' below the peaks P and Q respectively.

FIGURE 1-23

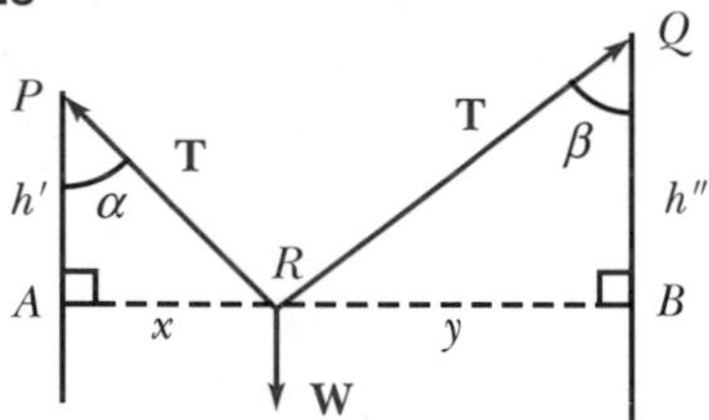

We solve this problem by recognizing that when he has reached his "equilibrium" position (i.e., is at rest), there are three forces acting on him: the tensions **T** in the rope section to each peak, and his weight **W**. The tensions in each part of the rope must be equal in magnitude since we assume the rope is smooth (the pulley does not exert any force on it). By Newton's Second Law, the sum of these forces must be zero; so if we draw the force (vector) polygon for the vectors $\mathbf{v}_1 = \mathbf{T}$, $\mathbf{v}_2 = \mathbf{T}$, and $\mathbf{v}_3 = \mathbf{W}$, the fourth side (their sum) must be **0**, the zero vector. We say that the polygon is **closed** by the three vectors. This is sketched in Figure 1-24.

FIGURE 1-24

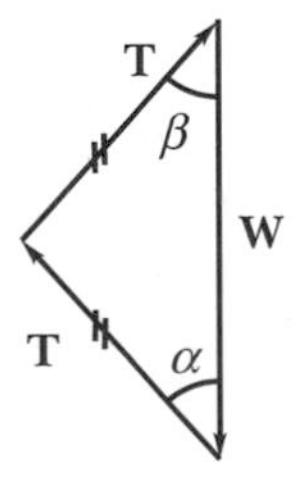

It follows that the angles α and β must be equal since the resulting triangle is isosceles. Thus, in Figure 1-23, triangles PAR and QBR are similar, and we conclude that $x/h' = y/h''$, i.e., $x = (h'/h'')y$. This locates the climber provided we know the horizontal distance, $x + y$, between the peaks. For example, the reader may check that if $x + y = 300'$, $h' = 50'$, and $h'' = 100'$, then $x = 100'$ and $y = 200'$. ❑.

EXERCISES 1-2

1. Draw the vectors in R^2 with initial and terminal points A and B respectively given as follows.

(a) $A = (-1,2), B = (3,4)$
(b) $A = (-4,-3), B = (0,-1)$
(c) $A = (3,0), B = (5,-2)$
(d) $A = (4,-2), B = (0,-4)$
(e) $A = (2,0), B = (0,2)$
(f) $A = (-2,0), B = (2,0)$
(g) $A = (0,0), B = (0,5)$
(h) $A = (0,-2), B = (4,-2)$

2. (a) Which of the vectors in question 1 are equal?

(b) Draw vectors bound at the origin and equal to each of the vectors in question 1. Determine the coordinates of the terminal point in each case.

3. Sketch the vectors **AB** in R^3 with initial and terminal points A and B respectively as specified below.

(a) $A = (6,7,1), B = (7,2,3)$
(b) $A = (-3,5,0), B = (-2,0,-4)$
(c) $A = (4,5,6), B = (2,3,5)$
(d) $A = (-5,3,7), B = (-2,1,5)$
(e) $A = (1,0,2), B = (2,-5,4)$

4. (a) Which of the vectors in question 3 are equal?

(b) Draw vectors bound at the origin and equal to each of the vectors in question 3. Determine the coordinates of the terminal point in each case.

5. Write down the components of the vectors $\mathbf{v}$ in R^2 and R^3 represented by each of the vectors in questions 1 and 3. Compute $|\mathbf{v}|$ in each case.

6. Draw vectors in R^2 representing the following points in the plane and having initial points A as indicated.

(a) $(2,3), A = (-1,0)$
(b) $(-1,-1), A = (2,4)$
(c) $(-2,0), A = (3,2)$
(d) $(0,1), A = (-2,-2)$

7. Draw a representative for the sum $\{\mathbf{AB}\} + \{\mathbf{CD}\}$ for each case below, with initial point at the origin.

(a) $A = (-1,2), B = (3,4), C = (-4,-3), D = (-3,0)$

(b) $A = (2,2)$, $B = (2,4)$, $C = (0,3)$, $D = (-2,-1)$

(c) $A = (2,-1)$, $B = (4,-2)$, $C = (-1,-1)$, $D = (-3,-3)$

8. For the vectors in question 7, verify that if $\mathbf{v} = \{\mathbf{AB}\} = (x,y)$ and $\mathbf{w} = \{\mathbf{CD}\} = (x',y')$, then $\mathbf{v} + \mathbf{w} = (x + x', y + y')$.

9. **(a)** Draw the scalar multiples $k\mathbf{AB}$ for each vector $\mathbf{AB}$ in question 1, for the following values of k: **(i)** $k = 2$; **(ii)** $k = -1.5$; **(iii)** $k = -0.5$.

(b) Verify for each case that if $\mathbf{v} = \{\mathbf{AB}\} = (x,y)$, then $k\mathbf{v} = (kx,ky)$.

10. Let $\mathbf{v}$ and $\mathbf{w}$ be vectors in R^3. Show that the definitions of $\mathbf{v} + \mathbf{w}$ and $k\mathbf{v}$ are independent of the vectors chosen to represent them (Definitions 1-11 and 1-12).

11. Prove properties (6)–(8) of Theorem 1-3 by **(a)** geometric methods, and **(b)** using components.

12. Let $\mathbf{u}$, $\mathbf{v}$, $\mathbf{w}$, $\mathbf{x}$, and $\mathbf{y}$ be the vectors in R^3 with representatives given in questions 3(a) through 3(e) respectively. Find the components of each of the following.

(a) $\mathbf{u} - 2\mathbf{v} + 3\mathbf{x}$ **(b)** $5\mathbf{y} - 3\mathbf{x} + \mathbf{u}$ **(c)** $-2(\mathbf{u} - 6\mathbf{y}) + \mathbf{v}$

(d) $22[\mathbf{u} - 3(\mathbf{x} + 4\mathbf{y})] - \mathbf{v} + 6\mathbf{w}$ **(e)** $-3[-4(\mathbf{u} + 6\mathbf{x})]$

13. **(a)** Find scalars a and b such that $\mathbf{w} = a\mathbf{u} + b\mathbf{v}$ where $\mathbf{u} = (-2,-1)$, $\mathbf{v} = (1,2)$, and $\mathbf{w} = (7,8)$.

(b) Show that there are no scalars a and b such that $\mathbf{w} = a\mathbf{x} + b\mathbf{y}$ where $\mathbf{x} = (-4,2)$, $\mathbf{y} = (2,-1)$, and $\mathbf{w} = (1,2)$.

14. Draw the polygon of vectors to determine the resultant of $\mathbf{u} + 2\mathbf{v} - 3\mathbf{w}$ for $\mathbf{u} = (2,1,-2)$, $\mathbf{v} = (2,3,-1)$, and $\mathbf{w} = (2,-1,0)$.

15. The supporting framework (called a truss) for the roofs of some buildings are made of isosceles triangles of beams, ABC, with A at the apex of the roof and BC horizontal and resting on the walls of the building; the equal sides AC and AB support the roof. Suppose that the lengths of BC and AC ($= AB$) are 10 m and 7.07 m respectively, that the weights of the beams can be neglected, and that a vertical load of 100 kg is placed at A (for example, by a heavy man on the roof). Use the polygon of forces for this situation to find the forces of compression in AC and AB, and the tension in BC (the unit of force will be kilogram-weight, 1 kg-wt = 9.80 N).

16. In Example 1-13 described above, suppose that the mountain climber proposes to get to the higher peak Q by pulling himself using an auxilliary rope to that peak, that h' and h'' are 10 m and 20 m respectively, that the horizontal distance between the peaks is 50 m, and that he weighs 80 kg. Find how hard he is pulling when he rests at a point 10 m vertically below the higher peak Q.

1-3 The Dot Product and Vector Geometry

In this section we introduce a concept for studying lengths of vectors and angles between them. Let us first consider the vectors in Figure 1-25 below. From the Pythagorean Theorem and its converse, we know that the triangle OPP' in R^2 defined by **u** and **u**$'$ will have a right angle at O iff (this means *if and only if*)

$$|OP|^2 + |OP'|^2 = |PP'|^2.$$

FIGURE 1-25

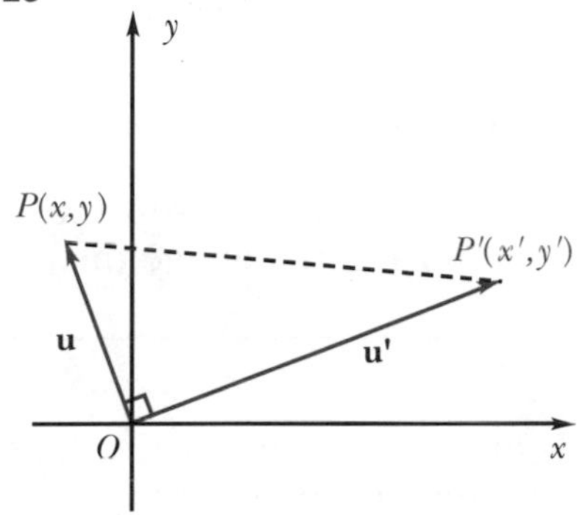

Replacing the lengths with the coordinate expressions, this equation becomes

$$(x^2 + y^2) + (x'^2 + y'^2) = (x' - x)^2 + (y' - y)^2,$$

or, after simplification,

$$xx' + yy' = 0.$$

Alternatively, this condition for **u** and **u**$'$ to be perpendicular (also described as **orthogonal**) may be obtained by recalling that the lines determined by (i.e., containing) **u** and **u**$'$ are perpendicular iff their slopes are negative reciprocals (if neither slope is 0): $y/x = -1/(y'/x')$. Simplification of this equation again leads to $xx' + yy' = 0$.

FIGURE 1-26

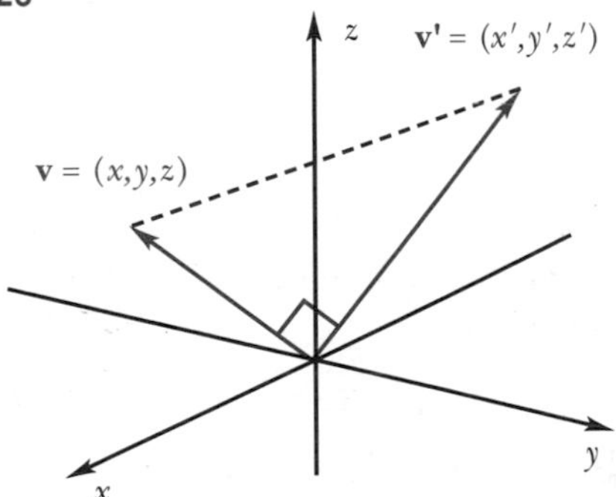

In Figure 1-26, we consider the similar situation in R^3. Again, the Pythagorean Theorem and its converse lead to the condition that the vectors $\mathbf{v} = (x,y,z)$ and $\mathbf{v}' = (x',y',z')$ are orthogonal iff

$$xx' + yy' + zz' = 0.$$

Thus we have proved the following theorem:

THEOREM 1-4

Two vectors $\mathbf{v} = (x,y,z)$ and $\mathbf{v}' = (x',y',z')$ different from $\mathbf{0}$ are orthogonal iff their components satisfy the equation

$$xx' + yy' + zz' = 0. \blacksquare$$

Note that $\mathbf{v}$ and $\mathbf{v}'$ are orthogonal iff any of their representatives are orthogonal. The following definition arises from the above theorem.

DEFINITION 1-13

If $\mathbf{v} = (x,y,z)$ and $\mathbf{v}' = (x',y',z')$ are two vectors in R^3, their **dot product** is the number denoted by $\mathbf{v}\cdot\mathbf{v}'$ and defined by

$$\mathbf{v}\cdot\mathbf{v}' = xx' + yy' + zz'.$$

Note that in R^2, the definition is similar except there is no zz' term.

It follows that *two vectors are orthogonal iff their dot product is 0*. Our interest in this numerical expression is increased by noting that if we take $\mathbf{v}' = \mathbf{v}$, we get

$$\mathbf{v}\cdot\mathbf{v} = x^2 + y^2 + z^2$$

which is the **square** of the length of $\mathbf{v}$ (see the remark following Definition 1-10). Therefore,

$$\mathbf{v}\cdot\mathbf{v} = |\mathbf{v}|^2.$$

EXAMPLE 1-14

Let $\mathbf{u} = (1,2,3)$, $\mathbf{v} = (-1,1,2)$, and $\mathbf{w} = (-1,1,-1)$ be the vectors in R^3 shown in Figure 1-27. We determine which of the vectors are orthogonal by taking their dot products in pairs.

$$\mathbf{u}\cdot\mathbf{v} = (1)(-1) + (2)(1) + (3)(2) = -1 + 2 + 6 = 7$$

$$\mathbf{u}\cdot\mathbf{w} = (1)(-1) + (2)(1) + (3)(-1) = -2$$

$$\mathbf{v}\cdot\mathbf{w} = (-1)(-1) + (1)(1) + (2)(-1) = 0$$

FIGURE 1-27

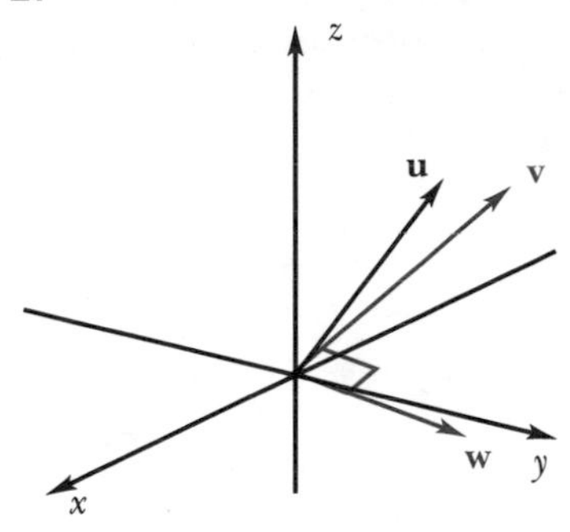

So, among the vectors **u**, **v**, and **w**, only **v** and **w** are orthogonal. ❑

To illustrate one of the properties of the dot product, we compare $(\mathbf{u} - \mathbf{v})\cdot\mathbf{w}$ and $\mathbf{u}\cdot\mathbf{w} - \mathbf{v}\cdot\mathbf{w}$:

$$\begin{aligned}(\mathbf{u} - \mathbf{v})\cdot\mathbf{w} &= (1 - (-1), 2 - 1, 3 - 2)\cdot(-1, 1, -1)\\ &= -2 + 1 + (-1)\\ &= -2.\end{aligned}$$

Using Example 1-14,

$$\mathbf{u}\cdot\mathbf{w} - \mathbf{v}\cdot\mathbf{w} = -2 - 0 = -2.$$

Therefore,

$$(\mathbf{u} - \mathbf{v})\cdot\mathbf{w} = \mathbf{u}\cdot\mathbf{w} - \mathbf{v}\cdot\mathbf{w}.$$

In fact, this is true for *any* vectors.

THEOREM 1-5 (Properties of the Dot Product)
Let **u**, **v**, and **w** be any vectors in R^2 (or R^3) and k be any number. Then,

(1) $\mathbf{u}\cdot\mathbf{v} = \mathbf{v}\cdot\mathbf{u}$,

(2) $(k\mathbf{u})\cdot\mathbf{v} = k(\mathbf{u}\cdot\mathbf{v})$,

(3) $(\mathbf{u} + \mathbf{v})\cdot\mathbf{w} = \mathbf{u}\cdot\mathbf{w} + \mathbf{v}\cdot\mathbf{w}$,

(4) $\mathbf{u}\cdot\mathbf{u} > 0$ unless $\mathbf{u} = \mathbf{0}$ (and $\mathbf{0}\cdot\mathbf{0} = 0$).

PROOF We leave the proofs of all but property (4) to the reader. As to the latter, if $\mathbf{u} = (x,y,z)$, then as we have seen, $\mathbf{u}\cdot\mathbf{u} = |\mathbf{u}|^2$, which is always positive unless $\mathbf{u} = \mathbf{0}$. ■

We now combine these properties with the law of cosines from trigonometry to obtain an expression for the angle between two vectors in terms of their dot product and lengths. In Figure 1-28, the vectors **u** and **v** in R^3 make a triangle whose third side may be represented by the vector $\mathbf{u} - \mathbf{v}$ as shown. (Note the triangle law in the figure, $\mathbf{v} + (\mathbf{u} - \mathbf{v}) = \mathbf{u}$.) The angle between **u** and **v** is represented there by α, which, by convention, means the angle α which lies between 0° and 180° inclusive. Then, by the law of cosines,

$$|\mathbf{u} - \mathbf{v}|^2 = |\mathbf{u}|^2 + |\mathbf{v}|^2 - 2|\mathbf{u}||\mathbf{v}|\cos\alpha.$$

FIGURE 1-28

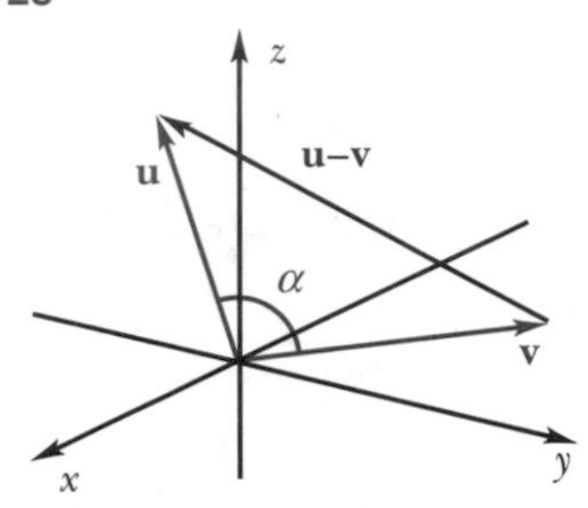

But, using the properties in Theorem 1-5,

$$|\mathbf{u} - \mathbf{v}|^2 = (\mathbf{u} - \mathbf{v})\cdot(\mathbf{u} - \mathbf{v}) = |\mathbf{u}|^2 - 2\mathbf{u}\cdot\mathbf{v} + |\mathbf{v}|^2.$$

Substitution into the first relation yields

$$-2\mathbf{u}\cdot\mathbf{v} = -2|\mathbf{u}||\mathbf{v}|\cos\alpha,$$

or,

$$\cos\alpha = \mathbf{u}\cdot\mathbf{v}/|\mathbf{u}||\mathbf{v}|. \tag{1-3}$$

Notice that $\cos\alpha = 0$ precisely when $\mathbf{u}\cdot\mathbf{v} = 0$, so we see again that two vectors are orthogonal ($\alpha = 90°$ and $\cos 90° = 0$) iff their dot product is 0. Also, observe that we can compute $\mathbf{u}\cdot\mathbf{v}$ from equation 1-3 given $|\mathbf{u}|$, $|\mathbf{v}|$, and $\cos\alpha$. Furthermore, recall the following:

(1) $0° < \alpha < 90°$ iff $\cos\alpha > 0$ (i.e., $\mathbf{u}\cdot\mathbf{v} > 0$).

(2) $90° < \alpha < 180°$ iff $\cos\alpha < 0$ (i.e., $\mathbf{u}\cdot\mathbf{v} < 0$).

(3) $\alpha = 0°$ when $\cos\alpha = 1$ and $\alpha = 180°$ when $\cos\alpha = -1$.

When the angle between **u** and **v** is 0° or 180°, we say that **u** and **v** are **parallel**. This means that the vectors that *represent* **u** and **v** are all parallel or collinear and holds true precisely when **u** and **v** are nonzero scalar multiples of one another: $\mathbf{u} = k\mathbf{v}$, $k \neq 0$. For, if $\mathbf{u} = k\mathbf{v}$, $k \neq 0$, with $\mathbf{u} = \{\mathbf{AB}\}$ and $\mathbf{v} = \{\mathbf{CD}\}$, then $\mathbf{AB} = k\mathbf{CD}$ are parallel or collinear by definition of the scalar product. Conversely, if **u** and **v** are parallel, suppose $\mathbf{u} = \{\mathbf{OP}\}$ and $\mathbf{v} = \{\mathbf{OQ}\}$. Then, **OP** is parallel to **OQ**, so O, P, and Q are collinear and $\mathbf{OP} = k\mathbf{OQ}$, for some $k \neq 0$. It follows that $\mathbf{u} = k\mathbf{v}$ for some $k \neq 0$.

When the angle between **u** and **v** is 0°, we say that **u** and **v** have the **same direction**. This holds iff $\mathbf{v} = k\mathbf{u}$, $k > 0$. When the angle between **u** and **v** is 180°, we say they have **opposite directions**. This holds iff $\mathbf{v} = k\mathbf{u}$, $k < 0$.

EXAMPLE 1-15

(a) Consider the vectors $\mathbf{u} = (-2,2,4)$ and $\mathbf{v} = (3,6,3)$ in R^3. We determine the angle, α, between $\mathbf{u}$ and $\mathbf{v}$.

$$\begin{aligned} |\mathbf{u}| &= [(-2)^2 + 2^2 + 4^2]^{1/2} = 2\sqrt{6} \\ |\mathbf{v}| &= (3^2 + 6^2 + 3^2)^{1/2} = 3\sqrt{6} \\ \mathbf{u}\cdot\mathbf{v} &= (-2)(3) + (2)(6) + (4)(3) = 18 \\ \cos\alpha &= \mathbf{u}\cdot\mathbf{v}/|\mathbf{u}||\mathbf{v}| \\ &= 18/(2\sqrt{6})(3\sqrt{6}) \\ &= 1/2 \end{aligned}$$

Therefore, $\alpha = 60°$.

On the other hand, $(-\mathbf{u})\cdot\mathbf{v} = -18$, so if β is the angle between $-\mathbf{u}$ and $\mathbf{v}$, we calculate that

$$\cos\beta = -18/(2\sqrt{6})(3\sqrt{6}) = -1/2,$$

making $\beta = 180° - 60° = 120°$, as was to be expected, since $\alpha + \beta = 180°$.

(b) Let $\mathbf{u} = (-1,0,3)$, $\mathbf{v} = (2,0,-6)$, and $\mathbf{w} = (-3,0,9)$.

(i) $\mathbf{v} = -2\mathbf{u}$, so $\mathbf{u}$ and $\mathbf{v}$ are parallel but have opposite directions.
(ii) $\mathbf{w} = 3\mathbf{u}$, so $\mathbf{u}$ and $\mathbf{w}$ are parallel and have the same direction. ❑

Frequently in applications we need to find families of mutually orthogonal vectors all having length 1. A systematic method for this is called the Gram-Schmidt process, which will be discussed in Chapter 7. Vectors of length 1 are called **unit vectors**. It follows from Theorem 1-5 that if $\mathbf{v} \neq \mathbf{0}$ is any vector, $\mathbf{v}/|\mathbf{v}|$ is a unit vector in the direction of $\mathbf{v}$. For, $|\mathbf{v}/|\mathbf{v}||^2 = (\mathbf{v}/|\mathbf{v}|)\cdot(\mathbf{v}/|\mathbf{v}|) = \mathbf{v}\cdot\mathbf{v}/|\mathbf{v}|^2 = 1$.

EXAMPLE 1-16

We find a unit vector in R^3 orthogonal to the vector $\mathbf{u} = (1,2,3)$. Suppose $\mathbf{v} = (a,b,c)$ is orthogonal to $\mathbf{u}$. Then, from $\mathbf{u}\cdot\mathbf{v} = 0$, we obtain the equation

$$a + 2b + 3c = 0.$$

Choosing arbitrarily $c = 0$ and $b = 1$, we find that $a = -2$, so the vector $\mathbf{v} = (-2,1,0)$ is one possible vector orthogonal to $\mathbf{u}$. Its length is $\sqrt{5}$, so the unit vector in the direction of $\mathbf{v}$ (orthogonal to $\mathbf{u}$) is

$$\begin{aligned} \mathbf{v}/|\mathbf{v}| &= (1/\sqrt{5})(-2,1,0) \\ &= (-2/\sqrt{5},1/\sqrt{5},0). \end{aligned}$$

Note that other choices are possible. ❑

EXAMPLE 1-17

Consider the triangle ABC shown in Figure 1-29. We find the smaller of the two angles between the medians BP and AQ as shown. Note that to solve this problem, we have positioned the triangle so that AB lies along the x axis with A at the origin. Now, by the geometry of similar triangles, the coordinates of P are (1,1), while those of Q are (3,1). The angle we seek is the angle, α, between the vectors

$$\mathbf{u} = \{\mathbf{PB}\} = (4 - 1, 0 - 1) = (3, -1)$$

and

$$\mathbf{v} = \{\mathbf{AQ}\} = (3 - 0, 1 - 0) = (3, 1).$$

We use the relation $\cos \alpha = \mathbf{u} \cdot \mathbf{v} / |\mathbf{u}||\mathbf{v}|$ to determine α.

$$\mathbf{u} \cdot \mathbf{v} = 3 \times 3 - 1 \times 1 = 8$$

$$|\mathbf{u}| = \sqrt{10}$$

$$|\mathbf{v}| = \sqrt{10}$$

So $\cos \alpha = 8/10 = 0.8$ and $\alpha \approx 36.9°$. ❑

FIGURE 1-29

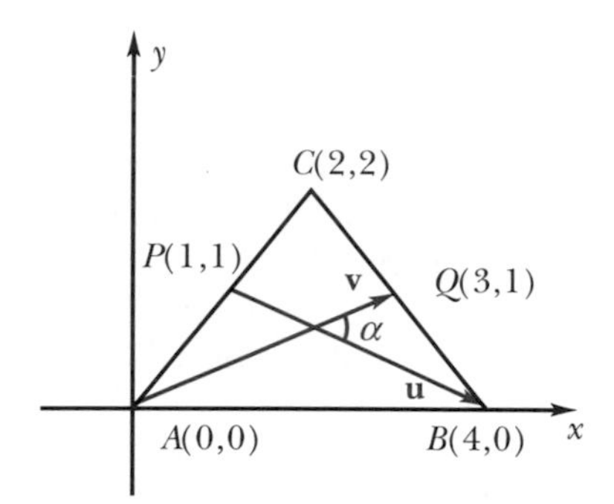

EXAMPLE 1-18

We use vector methods in R^2 to show the well-known proposition of Euclidean geometry that a diameter of a circle subtends a right angle at the circumference.

In Figure 1-30, P is an arbitrary point on the circumference of a circle which we have positioned as shown for convenience with diameter OQ along the positive x axis. We want to show that the angle OPQ is a right angle, or that the vectors $\mathbf{u} = \{\mathbf{OP}\}$ and $\mathbf{v} = \{\mathbf{QP}\}$ are perpendicular.

Let $\mathbf{r} = (1/2)\,\{\mathbf{OQ}\}$ be the vector from O to the center C of the circle, and $\mathbf{w} = \{\mathbf{CP}\}$. From the figure it is then clear that

$$\mathbf{u} = \mathbf{r} + \mathbf{w},$$
$$\mathbf{v} = \mathbf{w} - \mathbf{r},$$

so

$$\begin{aligned}\mathbf{u}\cdot\mathbf{v} &= (\mathbf{r} + \mathbf{w})\cdot(\mathbf{w} - \mathbf{r})\\ &= -\mathbf{r}\cdot\mathbf{r} + \mathbf{w}\cdot\mathbf{w} - \mathbf{w}\cdot\mathbf{r} + \mathbf{r}\cdot\mathbf{w}\\ &= \mathbf{w}\cdot\mathbf{w} - \mathbf{r}\cdot\mathbf{r}\\ &= |\mathbf{w}|^2 - |\mathbf{r}|^2.\end{aligned}$$

But both $\mathbf{w}$ and $\mathbf{r}$ have length equal to the radius of the circle, so $|\mathbf{w}| = |\mathbf{r}|$ and $\mathbf{u}\cdot\mathbf{v} = 0$. It follows that $\mathbf{u}$ and $\mathbf{v}$ are perpendicular. ❑

FIGURE 1-30

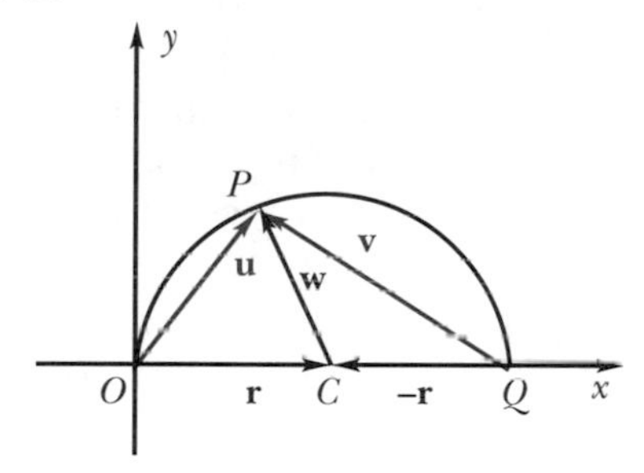

We next consider an idea which is central in linear algebra and which will be studied in detail in a more general setting in later chapters. This concept arises from the following problem (for simplicity, we discuss it for vectors in R^2). Given two vectors $\mathbf{u}$ and $\mathbf{v}$, we have seen that they determine an unique vector $\mathbf{u} + \mathbf{v} = \mathbf{w}$ (say) called their sum. Conversely, we may ask: given any three vectors $\mathbf{u}$, $\mathbf{v}$, and $\mathbf{w}$, can we "decompose" $\mathbf{w}$ as a sum of vectors $\mathbf{u}'$ and $\mathbf{v}'$ lying in the directions of $\mathbf{u}$ and $\mathbf{v}$ respectively?

Since vectors lying in the same direction must be scalar multiples of one another, we must have $\mathbf{u}' = k\mathbf{u}$ and $\mathbf{v}' = l\mathbf{v}$ for some numbers k and l, and therefore also $\mathbf{w} = k\mathbf{u} + l\mathbf{v}$, provided this decomposition is even possible. From the last equation it should be clear that this is not possible if $\mathbf{u}$ and $\mathbf{v}$ are collinear while $\mathbf{w}$ is not collinear with them, so we will assume that $\mathbf{u}$ and $\mathbf{v}$ are *not* collinear (i.e., not multiples of each other).

As we shall see in Chapter 4, this problem then always has a solution. For the moment, we shall consider the special case in which $\mathbf{u}$ and $\mathbf{v}$ are orthogonal (Figure 1-31). According to the parallelogram law of addition, we see from the figure that $\mathbf{u}'$ and $\mathbf{v}'$ are obtained for a given $\mathbf{w}$ as the vectors with terminal points at the feet of the perpendiculars to the lines containing $\mathbf{u}$ and $\mathbf{v}$. These vectors are called the **projections** of $\mathbf{w}$ on $\mathbf{u}$ and $\mathbf{v}$ respectively.

FIGURE 1-31

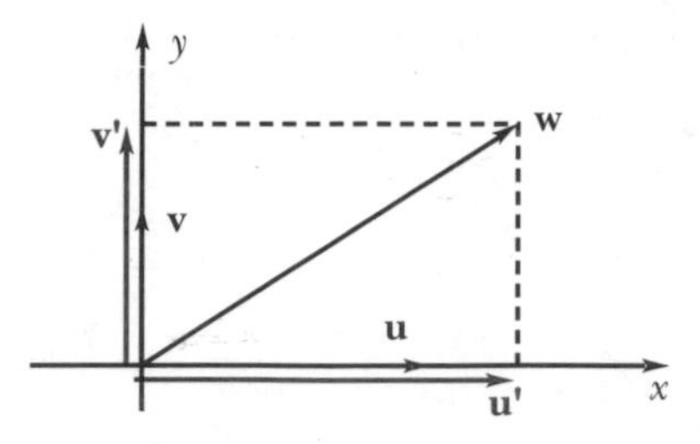

There is a practical problem analogous to the situation discussed above. Suppose you are pulling a sled or wagon over the ground and the vector **w** represents the force you are applying to the rope attached to it. How much of the force is actually used to move the sled horizontally, and how much of it goes into trying to lift the sled? A little reflection shows that the forces in question are precisely $\mathbf{u}'$ and $\mathbf{v}'$ respectively, since their sum produces **w**. In order to put as much of your effort as possible into moving the sled over the ground (for a fixed pull **w**), $\mathbf{v}'$ should be made as small as possible, i.e., the rope should be as nearly horizontal as possible. In physics, the determination of $\mathbf{u}'$ and $\mathbf{v}'$ is called the **resolution** of the force **w** into its horizontal and vertical components.

We can determine $\mathbf{u}'$ and $\mathbf{v}'$ analytically as follows using the dot product. Taking the dot product of both sides of the equation $\mathbf{w} = k\mathbf{u} + l\mathbf{v}$ with **u**, we get

$$\begin{aligned}\mathbf{w}\cdot\mathbf{u} &= (k\mathbf{u})\cdot\mathbf{u} + (l\mathbf{v})\cdot\mathbf{u}\\ &= k(\mathbf{u}\cdot\mathbf{u}) + 0\\ &= k(\mathbf{u}\cdot\mathbf{u})\end{aligned}$$

since $\mathbf{u}\cdot\mathbf{v} = 0$ (**u** and **v** are orthogonal). Thus,

$$k = \mathbf{w}\cdot\mathbf{u}/\mathbf{u}\cdot\mathbf{u} \text{ (provided } \mathbf{u} \neq \mathbf{0})$$

and therefore

$$\mathbf{u}' = k\mathbf{u} = (\mathbf{w}\cdot\mathbf{u}/\mathbf{u}\cdot\mathbf{u})\mathbf{u}.$$

Similarly, we can show that

$$\mathbf{v}' = (\mathbf{w}\cdot\mathbf{v}/\mathbf{v}\cdot\mathbf{v})\mathbf{v}.$$

DEFINITION 1-14

Given a vector **u** different from **0** and any vector **w**, the **vector projection** of **w** on **u** is the vector

$$\mathbf{u}' = (\mathbf{w}\cdot\mathbf{u}/\mathbf{u}\cdot\mathbf{u})\mathbf{u} = (\mathbf{w}\cdot\mathbf{u}/|\mathbf{u}|)\ \mathbf{u}/|\mathbf{u}|.$$

The scalar multiplying the **unit** vector $\mathbf{u}/|\mathbf{u}|$ to get $\mathbf{u}'$ is called the **scalar projection** of **w** on **u** and is given by $\mathbf{w}\cdot\mathbf{u}/|\mathbf{u}|$.

The same definition applies to vectors in R^3. Note that

$$\begin{aligned}|\mathbf{u}'| &= \{[\mathbf{w}\cdot\mathbf{u}/\mathbf{u}\cdot\mathbf{u}]\mathbf{u}\cdot[\mathbf{w}\cdot\mathbf{u}/\mathbf{u}\cdot\mathbf{u}]\mathbf{u}\}^{1/2}\\ &= |\mathbf{w}\cdot\mathbf{u}|/|\mathbf{u}|\\ &= |\text{the scalar projection of } \mathbf{w} \text{ on } \mathbf{u}|.\end{aligned}$$

In other words, the scalar projection of $\mathbf{w}$ on $\mathbf{u}$ is $\pm|\mathbf{u}'|$, depending on the sign of $\mathbf{w}\cdot\mathbf{u}$.

Alternatively, we can obtain these expressions from the geometry as follows. Let α be the angle between $\mathbf{w}$ and $\mathbf{u}$, as shown in Figure 1-32 for the case $0° < \alpha < 90°$. Then, as $\cos\alpha > 0$,

$$|\mathbf{u}'| = |\mathbf{w}|\cos\alpha.$$

FIGURE 1-32

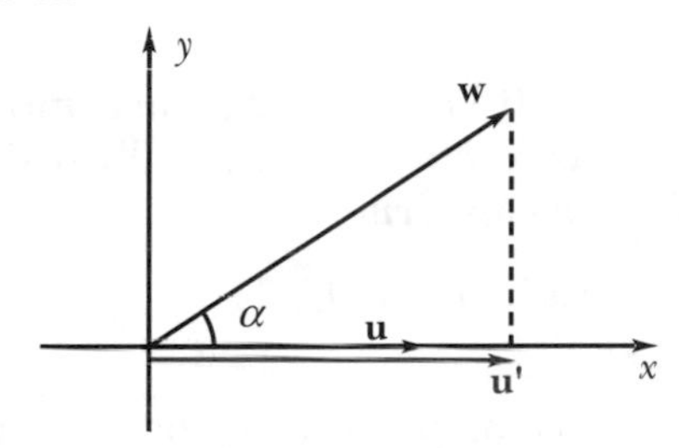

But, according to equation 1-3,

$$\cos\alpha = \mathbf{u}\cdot\mathbf{w}/|\mathbf{u}||\mathbf{w}|,$$

so that substitution yields

$$|\mathbf{u}'| = |\mathbf{w}|\,(\mathbf{u}\cdot\mathbf{w}/|\mathbf{u}||\mathbf{w}|) = \mathbf{w}\cdot\mathbf{u}/|\mathbf{u}|.$$

Therefore,

$$\begin{aligned}\mathbf{u}' &= |\mathbf{u}'| \times (\text{a unit vector in the direction of } \mathbf{u})\\ &= (\mathbf{w}\cdot\mathbf{u}/|\mathbf{u}|) \times (\mathbf{u}/|\mathbf{u}|)\\ &= (\mathbf{w}\cdot\mathbf{u}/\mathbf{u}\cdot\mathbf{u})\mathbf{u}\end{aligned}$$

as obtained earlier. The case $90° < \alpha < 180°$ is left for the reader.

EXAMPLE 1-19

(a) Consider the vectors $\mathbf{u} = (1,3)$ and $\mathbf{w} = (2,5)$ in R^2. The scalar projection of $\mathbf{w}$ on $\mathbf{u}$ is

$$\mathbf{w}\cdot\mathbf{u}/|\mathbf{u}| = (1\times2 + 3\times5)/(1^2 + 3^2)^{1/2} = 17/\sqrt{10}.$$

The vector projection of $\mathbf{w}$ on $\mathbf{u}$ is

$$(\mathbf{w}\cdot\mathbf{u}/|\mathbf{u}|)(\mathbf{u}/|\mathbf{u}|) = (17/\sqrt{10})(1/\sqrt{10})(1,3) = (17/10,51/10).$$

(b) Let $\mathbf{u} = (1,2,2)$ and $\mathbf{w} = (-3,0,-4)$. To find the vector projection $\mathbf{u}'$ of $\mathbf{w}$ on $\mathbf{u}$, we note that

$$\begin{aligned} \mathbf{w}\cdot\mathbf{u} &= -3 + 0 - 8 = -11, \\ |\mathbf{u}| &= 3, \end{aligned}$$

so

$$\mathbf{u}' = (-11/9)\mathbf{u} = (-11/9,-22/9,-22/9).$$

The scalar projection of $\mathbf{w}$ on $\mathbf{u}$ is $\mathbf{w}\cdot\mathbf{u}/|\mathbf{u}| = -11/3$. It is negative because the angle between $\mathbf{w}$ and $\mathbf{u}$ is greater than 90°. ❑

EXAMPLE 1-20

Let $\mathbf{i} = (1,0,0)$, $\mathbf{j} = (0,1,0)$, and $\mathbf{k} = (0,0,1)$. These are the **standard unit vectors** in R^3. Notice that these special vectors are mutually orthogonal. If $\mathbf{w} = (x,y,z)$ is any vector in R^3, we can write

$$\begin{aligned} \mathbf{w} &= x(1,0,0) + y(0,1,0) + z(0,0,1) \\ &= x\mathbf{i} + y\mathbf{j} + z\mathbf{k}. \end{aligned}$$

This expresses $\mathbf{w}$ as the sum of its vector projections on $\mathbf{i}$, $\mathbf{j}$, and $\mathbf{k}$. At the same time, x, y, and z are the scalar projections of $\mathbf{w}$ on $\mathbf{i}$, $\mathbf{j}$, and $\mathbf{k}$ respectively. ❑

EXERCISES 1-3

1. Find $\mathbf{u}\cdot\mathbf{v}$ for each of the following.
   (a) $\mathbf{u} = (3,-2,1)$, $\mathbf{v} = (2,1,-4)$
   (b) $\mathbf{u} = (-1.5,1,-0.5)$, $\mathbf{v} = (1,-3,5)$
   (c) $\mathbf{u} = (5,0,3)$, $\mathbf{v} = (-1,-0.5,2)$
   (d) $\mathbf{u} = (0,0,2)$, $\mathbf{v} = (2,3,0)$
   (e) $\mathbf{u} = (2,-7)$, $\mathbf{v} = (-4,2)$
2. For each pair of vectors in question 1, find $\cos\alpha$ where α is the angle between $\mathbf{u}$ and $\mathbf{v}$. Which pairs are orthogonal?
3. Find the angle between the vectors $\{\mathbf{AB}\}$ and $\{\mathbf{CD}\}$ where $A = (-2,3)$, $B = (2,-1)$, $C = (-3,-5)$, and $D = (-6,9)$.
4. Find the angle between the vectors $\{\mathbf{AB}\}$ and $\{\mathbf{CD}\}$ where $A = (1,2,2)$, $B = (3,4,1)$, $C = (-2,-3,7)$, and $D = (4,-6,9)$.

5. Find the value of k in each case such that the vectors are orthogonal.

(a) $\mathbf{u} = (3,-2k,4)$, $\mathbf{v} = (1,2,5)$

(b) $\mathbf{u} = (-k,-1,-1)$, $\mathbf{v} = (3,0,1)$

(c) $\mathbf{u} = (1,1,-k)$, $\mathbf{v} = (-2,k,-k)$

6. Find a unit vector in R^3 orthogonal to the vectors: **(a)** $(-2,3,7)$; **(b)** $(0,3,4)$.

7. Find a unit vector perpendicular to the plane containing the vectors $\mathbf{u} = (1,1,2)$ and $\mathbf{v} = (-1,1,3)$.

8. Let $\mathbf{u}$ and $\mathbf{v}$ be vectors in the plane where $\mathbf{u} = (\cos\alpha, \sin\alpha)$ and $\mathbf{v} = (\cos\beta, -\sin\beta)$.

(a) Find the angles between $\mathbf{u}$ and the x axis, and $\mathbf{v}$ and the x axis.

(b) Compute $\mathbf{u}\cdot\mathbf{v}$ and use it to find an expression for $\cos(\alpha + \beta)$.

9. Use appropriate vectors and dot products to find the interior angles of the triangle with vertices $(1,2)$, $(4,7)$, and $(2,-3)$.

10. Prove the so-called "polarization" identity below showing that the dot product is determined by lengths of vectors:

$$\mathbf{u}\cdot\mathbf{v} = (1/4)|\mathbf{u} + \mathbf{v}|^2 - (1/4)|\mathbf{u} - \mathbf{v}|^2,$$

where $\mathbf{u}$ and $\mathbf{v}$ are vectors in R^2 (or R^3).

11. Find the vector and scalar projections of each given vector $\mathbf{w}$ on the vector $\mathbf{u} = (-3,4)$: **(a)** $\mathbf{w} = (1,1)$; **(b)** $\mathbf{w} = (-5,7)$; **(c)** $\mathbf{w} = (0,4)$.

12. Find the vector and scalar projections of each given vector $\mathbf{w}$ on the vector $\mathbf{u} = (1,1,1)$: **(a)** $\mathbf{w} = (-2,3,7)$; **(b)** $\mathbf{w} = (4,3,2)$; **(c)** $\mathbf{w} = (-1,-2,3)$.

13. Determine a unit vector in R^3 orthogonal to $(1,-2,0)$ and $(2,3,-5)$ and making an acute angle with $\mathbf{k} = (0,0,1)$.

14. Recall that a rhombus is a parallelogram having sides of equal lengths. Use vector methods to show that the diagonals of a rhombus are orthogonal.

15. A man is pulling a sleigh by an attached rope with a force of 10 newtons, but the magnitude of the force actually tending to draw the sleigh forward horizontally is only 5 newtons. At what angle to the horizontal is the man holding the rope? What is the magnitude of the force tending to lift the sleigh vertically upward?

16. Prove properties (1)–(3) of Theorem 1-5.

17. Prove (using vector methods) that the line joining the midpoints of two sides of a triangle is parallel to the third side and has half the length.

18. Determine two mutually orthogonal vectors lying in the plane that contains the vectors $(1,1,1)$ and $(1,2,2)$.

1-4 Further Vector Geometry

Recall from the discussion at the end of section 1-1 that we cannot speak unambiguously of the angles between a line and the coordinate axes in R^3. However, for vectors in R^3 we have

> **DEFINITION 1-15**
>
> If $\mathbf{v}$ is a vector in R^3, its **direction angles** α, β, and γ are the angles between $\mathbf{v}$ and the standard unit vectors $\mathbf{i} = (1,0,0)$, $\mathbf{j} = (0,1,0)$, and $\mathbf{k} = (0,0,1)$.

These angles and vectors are illustrated in Figure 1-33.

FIGURE 1-33

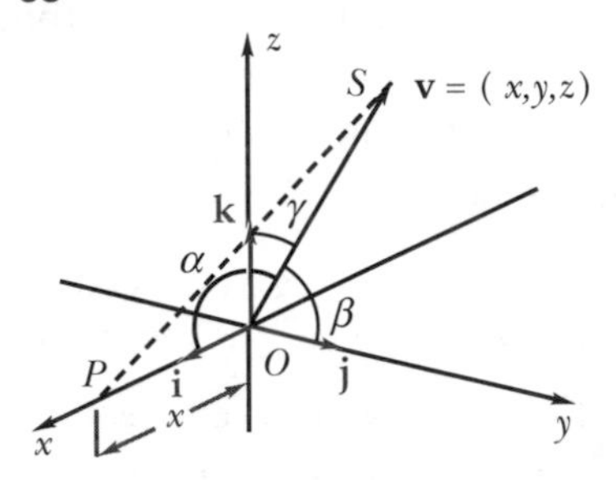

Notice that since **i**, **j**, and **k** are unit vectors in the direction of the positive coordinate axes, we may also think of α, β, *and* γ *as the angles between* $\mathbf{v}$ *and the positive coordinate axes*. Similarly, in R^2, a vector $\mathbf{v}$ makes two angles, α and β, with the standard unit vectors $\mathbf{i} = (1,0)$ and $\mathbf{j} = (0,1)$ respectively.

The direction angles of a vector are calculated from their cosines. Cos α, cos β, and cos γ are called the **direction cosines** of the vector.

> **THEOREM 1-6**
>
> If $\mathbf{v} = (x,y,z)$ is a vector in R^3, its **direction cosines** are given by $\cos \alpha = x/|\mathbf{v}|$, $\cos \beta = y/|\mathbf{v}|$, and $\cos \gamma = z/|\mathbf{v}|$.

PROOF From Figure 1-33, we see that $\cos \alpha = |OP|/|OS| = x/|\mathbf{v}|$ when α is acute; and $\cos \alpha = -|OP|/|OS| = x/|\mathbf{v}|$ when α is obtuse (i.e., $90° < \alpha < 180°$). Similarly, we obtain the results for $\cos \beta$ and $\cos \gamma$. ■

EXAMPLE 1-21

We find the direction cosines and direction angles of **(a)** $\mathbf{u} = (-2,-3,6)$, and **(b)** $\{\mathbf{PQ}\}$, where $P = (-3,0,2)$ and $Q = (-2,2,5)$.

(a) $|\mathbf{u}| = (4 + 9 + 36)^{1/2} = 7$. Thus,

$$\cos\alpha = -2/7 \approx -0.2857,$$
$$\cos\beta = -3/7 \approx -0.4286,$$
$$\cos\gamma = 6/7 \approx 0.8571.$$

From tables of cosines, we find

$$\alpha \approx 180° - 73.4° = 106.6°,$$
$$\beta \approx 180° - 64.6° = 115.4°,$$
$$\gamma \approx 31.0°.$$

(b) Let $\mathbf{v} = \{\mathbf{PQ}\} = (1,2,3)$. Then, $|\mathbf{v}| = \sqrt{14}$, so

$$\cos\alpha = 1/\sqrt{14} \approx 0.2674,$$
$$\cos\beta \approx 0.5348,$$
$$\cos\gamma \approx 0.8021.$$

Therefore, the direction angles are approximately 74.5°, 57.7°, and 36.5° respectively. ❑

COROLLARY

The direction cosines of a vector $\mathbf{v} = (x,y,z)$ satisfy the relation

$$(\cos\alpha)^2 + (\cos\beta)^2 + (\cos\gamma)^2 = 1.$$

PROOF The left-hand side of the equation is

$$(x/|\mathbf{v}|)^2 + (y/|\mathbf{v}|)^2 + (z/|\mathbf{v}|)^2 = (x^2 + y^2 + z^2)/|\mathbf{v}|^2 = 1,$$

which is the right-hand side of the equation. ■

The relation shows that if two of the direction angles of a vector are specified, there are only two possibilities for the third.

We can see a special case of this relation in R^2. Recall that R^2 may be thought of as the xy plane ($z = 0$) in R^3. Therefore R^2 consists of all vectors with $\gamma = 90°$, i.e., $\cos\gamma = 0$. The equation above becomes

$$(\cos\alpha)^2 + (\cos\beta)^2 = 1.$$

This last equation is equivalent to the well-known identity

$$(\sin\beta)^2 + (\cos\beta)^2 = 1,$$

since here $\alpha = 90° + \beta$ or $90° - \beta$, depending on the quadrant that $\mathbf{v}$ lies in. The reader should verify this using some sketches.

We now consider the alternative descriptions of a straight line which were mentioned at the end of section 1-1.

THEOREM 1-7

Let L be a line passing through a point $P' = (x',y',z')$ in R^3 and parallel to a vector $\mathbf{v} = (l,m,n)$. Then L is the set of points $P = (x,y,z)$ such that

$$\begin{aligned} x &= x' + lt \\ y &= y' + mt \\ z &= z' + nt \end{aligned}$$

for some number $t \in R$. These equations are called **parametric equations** for the line L.

FIGURE 1-34

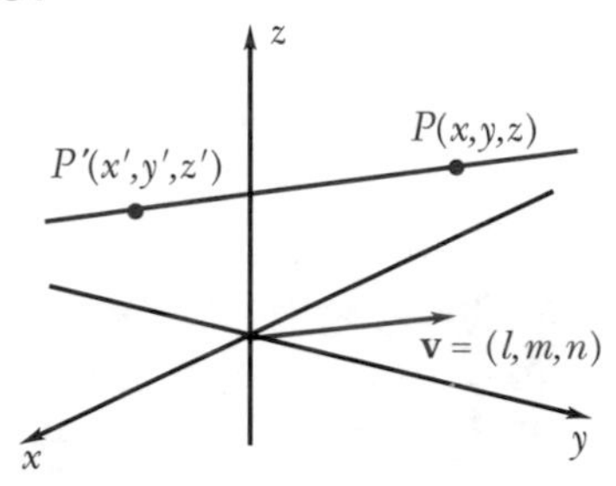

PROOF P (different from P') is on the line iff the vector $\{\mathbf{P'P}\}$ is parallel to the vector $\mathbf{v}$, as shown in Figure 1-34. By our earlier discussion, this is true iff $\{\mathbf{P'P}\} = t\mathbf{v}$ for some number $t \neq 0$. Since $\{\mathbf{P'P}\} = (x - x',y - y',z - z')$, it follows that P is on L iff

$$(x - x',y - y',z - z') = t(l,m,n),$$

or,

$$\begin{aligned} x - x' &= tl, \\ y - y' &= tm, \\ z - z' &= tn, \; t \in R, \; t \neq 0. \end{aligned}$$

Now, for $t = 0$, these equations simply yield the point $P = P' = (x',y',z')$. Therefore L is the set of points satisfying the equations as stated above. ■

In physics and engineering problems, it is often convenient to use Theorem 1-7 in vector form. Let $\mathbf{r} = (x,y,z)$ be the vector from the origin to an arbitrary point $P(x,y,z)$ on the line. In this context, $\mathbf{r}$ is called the **position vector** of a point on the line. Similarly, we may let $\mathbf{r}' = (x',y',z')$ be the position vector of the given point P' on the line, as in the theorem. Then we may state the theorem as follows:

COROLLARY 1

The **vector equation** of a line L through a point P' with position vector $\mathbf{r}' = (x',y',z')$ in R^3 and parallel to a vector $\mathbf{v} = (l,m,n)$ is

$$\mathbf{r} = \mathbf{r}' + t\mathbf{v},\ t \in R. \blacksquare$$

EXAMPLE 1-22

Let L be the line through the point $P'(1,2,2)$ and parallel to the vector $\mathbf{v} = (1,-2,3)$. We graph L by first finding its parametric equations. They are

$$x = 1 + t$$
$$y = 2 - 2t$$
$$z = 2 + 3t,\ t \in R.$$

Notice that when we put $t = 0$, we of course get the point $(1,2,2) = P'$. To graph this line we need another point on it. This we can obtain by setting t equal to some other convenient value. For example, with $t = 1$, we find that $P'' = (2,0,5)$ is on L. The line L is sketched in Figure 1-35.

FIGURE 1-35

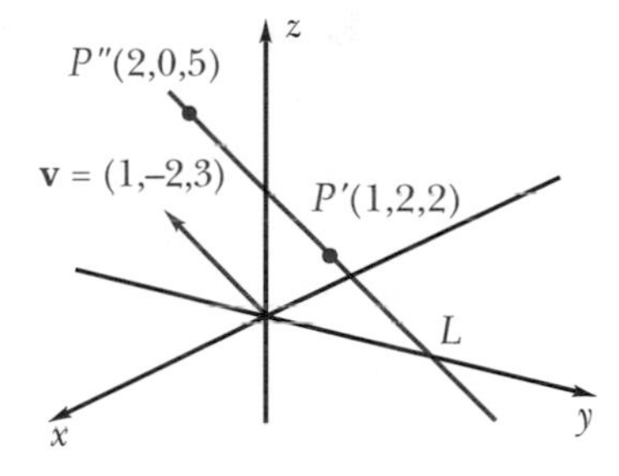

We can also easily find the intercepts with the coordinate planes. For example, the xy intercept ($z = 0$) is found by setting $z = 0$ in the last of the three parametric equations to get

$$0 = 2 + 3t,$$

or,

$$t = -2/3.$$

When we substitute this value for t into the equations for x and y, we get

$$x = 1 - 2/3 = 1/3,$$
$$y = 2 + 2(2/3) = 10/3.$$

These are the coordinates of the xy intercept. The other intercepts are found in a similar way. ❑

Another set of parametric equations can be obtained by noting that L is also parallel to the vector $c\mathbf{v}$ for any number $c \neq 0$. For example, with $c = -2$, $c\mathbf{v} = (-2,4,-6)$, so that we get parametric equations

$$x = 1 - 2t$$
$$y = 2 + 4t$$
$$z = 2 - 6t, t \in R.$$

It should be clear that these equations generate the *same* set of points in R^3 as the earlier set, namely, line L. Therefore L can be represented by infinitely many sets of parametric equations. For this reason, the components of *any* vector $\mathbf{v}$ parallel to L are called **direction numbers** for L.

COROLLARY 2

If a line L passes through the points $P'(x',y',z')$ and $P''(x'',y'',z'')$, it has parametric equations

$$x = x' + (x'' - x')t$$
$$y = y' + (y'' - y')t$$
$$z = z' + (z'' - z')t, t \in R.$$

PROOF This follows from the theorem because the line is parallel to $\mathbf{v} = \{\mathbf{P'P''}\} = (x'' - x', y'' - y', z'' - z')$. ■

The vector form of these equations is

$$\mathbf{r} = \mathbf{r}' + t(\mathbf{r}'' - \mathbf{r}'), t \in R,$$

where $\mathbf{r}'$ and $\mathbf{r}''$ are the position vectors of the given points on the line.

EXAMPLE 1-23

We find parametric equations for the line L with general equations

$$x + y + z = 3$$
$$2x - y + z = 5.$$

Using the results of Example 1-7, we know that L passes through the points $P'(0,-1,4)$ and $P''(2,0,1)$. In fact, these points are the intercepts of L with the yz and xz planes respectively. By Corollary 2 to Theorem 1-7, parametric equations for L are

$$\begin{aligned} x &= 0 + (2 - 0)t &&= 2t \\ y &= -1 + (0 - (-1))t &&= -1 + t \\ z &= 4 + (1 - 4)t &&= 4 - 3t, t \in R. \end{aligned}$$

Alternatively, we can solve the general equations for x and y in terms of z to get

$$\begin{aligned} x &= 8/3 - (2/3)z \\ y &= 1/3 - (1/3)z. \end{aligned}$$

So another set of parametric equations for L is

$$\begin{aligned} x &= 8/3 - (2/3)t \\ y &= 1/3 - (1/3)t \\ z &= t, t \in R. \ \square \end{aligned}$$

EXAMPLE 1-24

Let L be the line through the points $(-5,3,7)$ and $(-2,1,5)$, and L' be the line through $(0,6,-2)$ and $(-4,2,-4)$. Then L is parallel to the vector $\mathbf{v}$ with components $(-5 - (-2),3 - 1,7 - 5) = (-3,2,2)$. Similarly, L' is parallel to $\mathbf{v}' = (4,4,2)$. Since $\mathbf{v}\cdot\mathbf{v}' = -12 + 8 + 4 = 0$, we see that L and L' are orthogonal.

However, as shown below, L and L' do not lie in a common plane. Their parametric equations are

for L:

$$\begin{aligned} x &= -2 - 3t \\ y &= 1 + 2t \\ z &= 5 + 2t, t \in R; \end{aligned}$$

for L':

$$\begin{aligned} x &= 4t' \\ y &= 6 + 4t' \\ z &= -2 + 2t', t' \in R. \end{aligned}$$

Now if the lines were in the same plane, they would intersect and there would be a common point (x,y,z) on both lines. This is the case iff there are values for t and t' which produce the same x, y, and z values in the equations for L and L'. Equating the expressions for x gives

$$4t' = -2 - 3t,$$

while equating the expressions for y gives

$$6 + 4t' = 1 + 2t.$$

Substituting $4t'$ from the first equation into the second equation produces $t = 3/5$, so that $t' = -19/20$. Finally, substitution of these numbers into the z expressions yields $z = 31/5$ for L but $z = -39/10$ for L'. Since these are not the same, the lines do not intersect. The reader should confirm this conclusion with a sketch. ❑

We conclude this section with an important application of scalar projections.

EXAMPLE 1-25

We find the equation of the plane passing through the point (1,3,1) and perpendicular to the vector $\mathbf{N} = (1,2,3)$. We also find the perpendicular distance from the origin to the plane.

Let $P(x,y,z)$ be any point on the plane and Q be the point (1,3,1) as shown in Figure 1-36. If $\mathbf{v}$ is the vector represented by $\mathbf{QP}$, then $\mathbf{v}$ is perpendicular to $\mathbf{N}$ and therefore $\mathbf{v}\cdot\mathbf{N} = 0$. Since $\mathbf{v} = (x - 1, y - 3, z - 1)$, this yields

$$(x - 1)(1) + (y - 3)(2) + (z - 1)(3) = 0,$$

or, after simplifying,

$$x + 2y + 3z = 10.$$

This is therefore the equation of the plane.

FIGURE 1-36

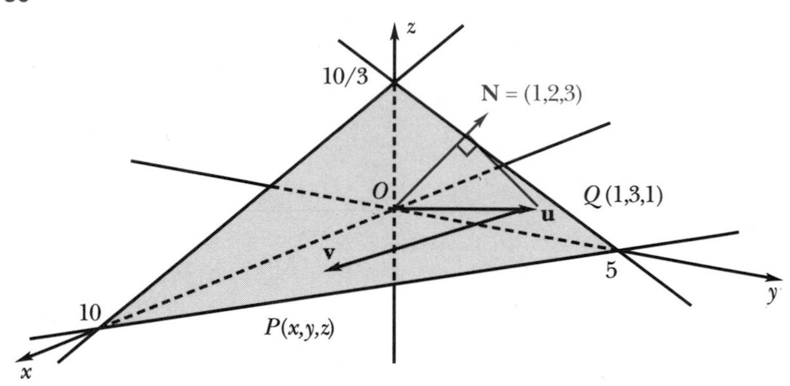

Letting O be the origin, we see from the figure that *the perpendicular distance of the plane from O is just the absolute value of the scalar projection of the vector $\mathbf{u} = \{\mathbf{OQ}\}$ on* $\mathbf{N}$ (since $\mathbf{N}$ is perpendicular to the plane). Thus, it is $|\mathbf{u}\cdot\mathbf{N}|/|\mathbf{N}|$. Now,

$$\mathbf{u}\cdot\mathbf{N} = 1\times 1 + 3\times 2 + 1\times 3 = 10,$$

and

$$|\mathbf{N}| = \{1^2 + 2^2 + 3^2\}^{1/2} = \sqrt{14}.$$

Therefore, the required distance is $10/\sqrt{14}$. ❑

More generally,

THEOREM 1-8

The equation of a plane in R^3 that is perpendicular to a unit vector $\mathbf{n}$ and has perpendicular distance d from the origin is

$$\mathbf{r}\cdot\mathbf{n} = \pm d \tag{1-4}$$

where $\mathbf{r} = (x,y,z)$ is the position vector of any point on the plane. The + sign applies if $\mathbf{r}\cdot\mathbf{n} > 0$ for some point on the plane, while the − sign applies if $\mathbf{r}\cdot\mathbf{n} < 0$.

Equation 1-4 is a *vector form* for the equation of a plane. Note that for $d \neq 0$, there are two planes perpendicular to a given vector $\mathbf{n}$, on opposite sides of the origin. Also, if $\mathbf{r}\cdot\mathbf{n} > 0$ (respectively, < 0) for some $\mathbf{r}$ to the plane, then $\mathbf{r}'\cdot\mathbf{n} > 0$ (respectively, < 0) for all $\mathbf{r}'$ to the plane, because $(\mathbf{r} - \mathbf{r}')\cdot\mathbf{n} = 0$, as $\mathbf{r} - \mathbf{r}'$ lies in the plane.

PROOF Since $\mathbf{n}$ is a unit vector, the left-hand side of the equation is simply the scalar projection of $\mathbf{r}$ on $\mathbf{n}$. If the angle between $\mathbf{r}$ and $\mathbf{n}$ is acute, so $\mathbf{r}\cdot\mathbf{n} > 0$, then, as in the preceding example, this equals the perpendicular distance d from the origin to the plane. If the angle is obtuse, then $\mathbf{r}\cdot\mathbf{n} < 0$, so $-\mathbf{r}\cdot\mathbf{n} = d$. Of course, $\mathbf{r}\cdot\mathbf{n} = 0$ iff $d = 0$. ■

EXAMPLE 1-26

The equations of the two planes perpendicular to the (unit) vector $\mathbf{n} = (1/3,2/3,2/3)$ and having a perpendicular distance of 3 units from the origin are

$$\mathbf{r}\cdot\mathbf{n} = \pm 3,$$

or,

$$x + 2y + 2z = \pm 9.$$

In order to single out either of these planes we must specify some point on the plane. For example, the one that intersects the positive x axis has equation

$$x + 2y + 2z = 9,$$

because, setting $y = z = 0$, we find $x = 9 > 0$. ❑

COROLLARY

The plane in R^3 passing through a given point $P'(x',y',z')$ and orthogonal to a vector $\mathbf{N} = (l,m,n)$ has equation

$$(x - x')l + (y - y')m + (z - z')n = 0.$$

PROOF Let $\mathbf{u} = \{\mathbf{OP'}\}$. Then, the scalar projection of $\mathbf{u}$ on the unit vector $\mathbf{N}/|\mathbf{N}|$ is $\pm d$, where d is the distance from O to the plane. By Theorem 1-8, the equation is therefore

$$\mathbf{r}\cdot\mathbf{N}/|\mathbf{N}| = \pm d = \mathbf{u}\cdot\mathbf{N}/|\mathbf{N}|,$$

or,

$$xl + ym + zn = x'l + y'm + z'n.$$

After rearranging, this produces the equation

$$(x - x')l + (y - y')m + (z - z')n = 0. \blacksquare$$

Note that the stated equation is linear in x, y, and z and has the general form of equation 1-2 for a plane as discussed in section 1-1:

$$ax + by + cz = d.$$

Conversely, this is the equation of the plane perpendicular to the vector (a,b,c) and passing through the point $P'(x',y',z')$, where x', y', and z' are any numbers satisfying equation 1-2. For, then we have

$$a(x - x') + b(y - y') + c(z - z') = d - d = 0.$$

This shows, in essence, that *linear equations in x, y, and z are equations of planes*, as stated in section 1-1. In this regard, readers should also look at question 12 below.

EXERCISES 1-4

1. **(a)** Find the direction cosines of the following vectors: **(i)** $(1,2,2)$; **(ii)** $(1,1,1)$; **(iii)** $(0,1,2)$; **(iv)** $(-3,2,-1)$; **(v)** $(4,-5,6)$.
   (b) Find the direction angles of the vectors in (a) above.
2. Find parametric equations for the lines specified below. In each case sketch the line by first finding its intercepts.
   (a) Through the point $(-2,4,5)$ and parallel to the vector $\mathbf{v} = (-1,-3,2)$.
   (b) Through the point $(-5,-3,-4)$ and parallel to the vector with direction angles $\alpha = 120°$, $\beta = 45°$, and $\gamma = 60°$.

(c) Through the point $(-4,2,-6)$ and parallel to the x axis.

(d) Through the points $(2,4,-3)$ and $(5,-1,3)$.

(e) Having general system of equations

$$\begin{aligned} x - 2y + 4z - 14 &= 0 \\ x + 20y - 18z + 30 &= 0. \end{aligned}$$

3. Suppose a line L has nonzero direction numbers k, l, and m and passes through the point $P'(x',y',z')$.

(a) Show that parametric equations for L may be written as

$$\begin{aligned} x &= x' + kt \\ y &= y' + lt \\ z &= z' + mt, t \in R. \end{aligned}$$

(b) Show that a system of general equations for L is given by

$$\begin{aligned} (x - x')/k &= (y - y')/l \\ (y - y')/l &= (z - z')/m. \end{aligned}$$

(c) The equations in (b) are called **symmetric equations** of the line through the point $P'(x',y',z')$ and having nonzero direction numbers. Find symmetric equations of the lines in question 2 where possible.

4. Find parametric equations of the line through the point $(1,2,3)$ and parallel to the line with equations

$$\begin{aligned} 2x + y - 5z + 1 &= 0 \\ 3x - 3y + z - 10 &= 0. \end{aligned}$$

5. **(a)** Show that the line through the points $(6,7,1)$ and $(7,2,-3)$ is parallel to the line through $(-3,5,0)$ and $(-2,0,-4)$.

(b) Show that the line through the points $(4,5,6)$ and $(2,3,5)$ is perpendicular to the line through the points $(-5,3,7)$ and $(-2,1,5)$.

(c) Determine whether or not the two lines in each of (a) and (b) lie in a common plane.

6. Find the equation of the plane perpendicular to the vector $(1,1,1)$ and passing through the point $(-1,3,-2)$. Sketch the plane by drawing its traces.

7. Find the equation of the plane perpendicular to the planes with equations $x + 2z - 7 = 0$ and $x - y - z - 5 = 0$, and passing through the origin $(0,0,0)$.

8. Find the cosine of the angle between the following pairs of planes. (Hint: The angle between two planes is the same as the angle between vectors that are orthogonal to the respective planes.)

(a) $y - 3z - 4 = 0$ and $2x + y - 5z - 7 = 0$

(b) $x + y - z - 2 = 0$ and $x = 0$

9. **(a)** Find the equation of the plane orthogonal to the vector $\mathbf{N} = (2,3,6)$ and passing through the point $(1,5,3)$.

(b) Find the perpendicular distance of the plane in (a) from the origin.

(c) Find the perpendicular distance of the plane in (a) from the point $(10,15,20)$. (Hint: First find the distance from the origin to the plane through this point and parallel to the original plane.)

10. Generalize the method used in question 9 to obtain a formula for the perpendicular distance from a point in R^3 to a plane $ax + by + cz = d$.

11. Use vector methods to obtain a formula for the distance from a point to a line in the plane.

12. Show that the equation of a plane perpendicular to a vector with direction cosines $\cos\alpha$, $\cos\beta$, and $\cos\gamma$ and whose perpendicular distance from the origin is d units is $x\cos\alpha + y\cos\beta + z\cos\gamma = +d$ or $-d$. Explain the two possible signs for d.

13. Consider the two planes $ax + by + cz = d$ and $a'x + b'y + c'z = d'$.

(a) Show that they are parallel iff $a = ka'$, $b = kb'$, and $c = kc'$ for some $k \neq 0$.

(b) Show that they are perpendicular iff $aa' + bb' + cc' = 0$.

14. **(a)** Find the equations of the two planes perpendicular to the vector $(-3,2,-1)$ and at a distance of 5 units from the origin.

(b) Find the equation of the plane in (a) that intersects the positive x axis.

15. The best "linear approximation" to the sphere $x^2 + y^2 + z^2 = d^2$ in R^3 at a point (x',y',z') on it is its *tangent plane* at that point, i.e., the plane touching the sphere only at (x',y',z'). This means that we may approximate points on the sphere near (x',y',z') by points on the tangent plane.

(a) Show that the equation of the tangent plane is

$$x'(x - x') + y'(y - y') + z'(z - z') = 0.$$

(b) Use this equation to compute the approximate change in z, $z - z'$, when a particle originally located at the point $(x',y',z') = (1,2,3)$ on the sphere given above with $d = \sqrt{14}$ moves to a new location on the sphere with $x = 1.01$ and $y = 2.01$.

(c) Compare the value obtained in (b) with the actual value obtained by using the equation of the sphere.

2 SYSTEMS OF LINEAR EQUATIONS AND MATRICES

2-1 Gauss-Jordan Elimination
2-2 Matrices and Gauss-Jordan Reduction
2-3 Matrices: Some Examples and Applications (Optional)
2-4 Matrix Algebra
2-5 Elementary Matrices and Matrix Inversion
2-6 Computational Considerations I (Optional)

2-1 Gauss-Jordan Elimination

In the previous chapter, we saw that systems of linear equations arise in many contexts in mathematics and its applications. We discussed some elementary methods for solving simple systems. To solve complex systems of many equations in many variables, we need systematic procedures that can be programmed for execution by a computer.

In this chapter, we discuss several such procedures, and some new concepts which naturally arise, notably, the idea of a matrix. These concepts are central to linear algebra. Furthermore, many computations involving vector spaces and linear maps (to be discussed later) boil down to solving systems of linear equations.

One of the principal methods for solving systems of linear equations in many variables is known as **Gauss-Jordan elimination**, after its inventors Carl Friedrich Gauss (1777–1855), the greatest mathematician of the first half of the 19th Century, and Camille Jordan (1838–1921), an eminent French mathematician. We begin our discussion by considering an important example of a system of equations used in economics. It is called a linear equilibrium model of supply and demand.

$$\begin{aligned} S - 8P &= -17 \\ D \quad + P &= 10 \\ D - S \quad &= 0 \end{aligned} \tag{2-1}$$

The variables D, S, and P are, respectively, the demand, supply, and price of some commodity. To say that this system of equations models the behaviour of the commodity means that the variables D, S, and P can only take on those values which, when substituted into the equations, make them true statements. We also say that such values *satisfy* the equations of the system and constitute a *solution*. For example, the first equation in the system indicates that for each dollar increase in the price P, the supply S (the number of units produced) will increase by 8 units. This reflects the fact that manufacturers will produce more when the price goes up. The second equation indicates that the demand D (the number of units consumers will buy) will decrease when the price increases. The third equation of the system, namely, $D = S$ (called the "equilibrium condition"), means that eventually the number of units produced will equal the number consumed. It allows us to eliminate one of the variables D and S (say, S) from the first two equations to produce a system of equations having the same solutions:

$$\begin{aligned} D \quad - 8P &= -17 \\ D \quad + P &= 10 \\ D - S \quad &= 0. \end{aligned} \tag{2-2}$$

The first two equations in this new system represent straight lines in the (P, D) plane whose intersection point (3,7) is the unique solution (together with $S = 7$) of the system. This graphical method of solution was discussed in section 1-1, and is illustrated in Figure 2-1.

FIGURE 2-1

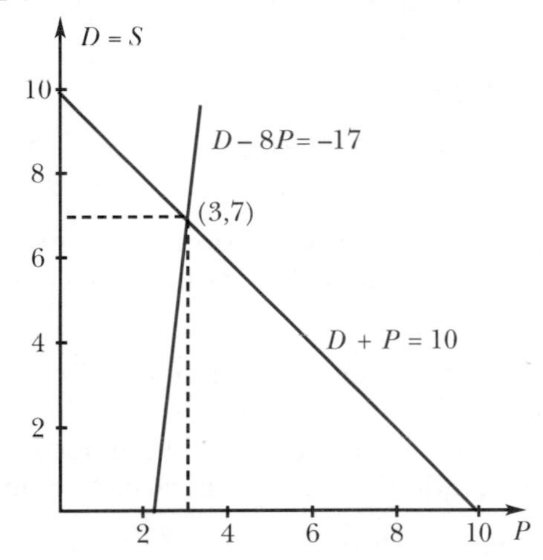

More generally,

> **DEFINITION 2-1**
>
> By a **system** of m linear equations in n variables (or unknowns) x_1, x_2, ..., x_n, we mean the set of m linear equations conventionally written as:
>
> $$\begin{array}{c} a_{11}x_1 + a_{12}x_2 + \dots + a_{1n}x_n = b_1 \\ a_{21}x_1 + a_{22}x_2 + \dots + a_{2n}x_n = b_2 \\ \dots\dots\dots\dots\dots\dots\dots\dots\dots \\ a_{m1}x_1 + a_{m2}x_2 + \dots + a_{mn}x_n = b_m \end{array} \qquad (2\text{-}3)$$
>
> where $a_{11}, a_{12}, \dots, a_{1n}; a_{21}, \dots, a_{2n}; \dots; a_{m1}, \dots, a_{mn}$; and $b_1, b_2, \dots, b_m$ arc constants.

We shall find it convenient to write this last condition as simply "a_{ij} and b_i are constants". Then, a_{ij} is called the **coefficient** of the variable x_j in equation i, and i, j are integers such that $1 \leqslant i \leqslant m$, $1 \leqslant j \leqslant n$.

A **solution** of the system 2-3 is then a specific ordered set (called an **n-tuple**) of n numbers which, when substituted for $x_1, \dots, x_n$ respectively, make all m equations true statements. The *set* of all solutions is called the **solution set**. In general, the system 2-3 could have infinitely many solutions, or exactly one solution, or no solution. It is called **consistent** if it has at least one solution, and **inconsistent** if it has no solution. The system is called **homogeneous** if $b_1 = \dots = b_m = 0$. In this case, there is always at least one solution, $x_1 = \dots = x_n = 0$, called the **trivial solution**.

EXAMPLE 2-1

We consider the system 2-1 discussed above. Here $a_{11} = 0$, $a_{12} = 1$, $a_{13} = -8$, etc., and the variables are $x_1 = D$, $x_2 = S$, and $x_3 = P$. The system is therefore

$$0D + 1S - 8P = -17 \qquad \text{(i)}$$
$$1D + 0S + 1P = 10 \qquad \text{(ii)}$$
$$1D - 1S + 0P = 0. \qquad \text{(iii)}$$

Of course, as shown in Figure 2-1, the only solution is the 3-tuple (7,7,3): $D = S = 7$ and $P = 3$. We now obtain this solution by Gauss-Jordan elimination as follows:

Step 1 We interchange equations (i) and (ii) so that the "first" (leftmost) variable appearing in the (new) first equation is the "first" variable $D = x_1$. Notice that its coefficient is 1, but that if it were not 1, we would make it so by multiplying equation (i) by its reciprocal. The new system is therefore

$$1D + 0S + 1P = 10 \qquad \text{(i)}$$
$$0D + 1S - 8P = -17 \qquad \text{(ii)}$$
$$1D - 1S + 0P = 0. \qquad \text{(iii)}$$

Step 2 Replace (iii) by (iii) − (i) to obtain a second new system,

$$1D + 0S + 1P = 10 \qquad \text{(i)}$$
$$0D + 1S - 8P = -17 \qquad \text{(ii)}$$
$$0D - 1S - 1P = -10. \qquad \text{(iii)}$$

In this system, D has been *eliminated* from (iii): it has coefficient 0. Notice that D has coefficient 0 in (ii), so we did not need to eliminate it in this case.

Step 3 Notice that in equation (ii), the coefficient of S, which is the leftmost variable with coefficient $\neq 0$, is 1. If it were not 1, we would make it so by multiplying equation (ii) by its reciprocal.

Step 4 Replace (iii) by (iii) + (ii) to obtain a third new system,

$$1D + 0S + 1P = 10 \qquad \text{(i)}$$
$$0D + 1S - 8P = -17 \qquad \text{(ii)}$$
$$0D + 0S - 9P = -27. \qquad \text{(iii)}$$

At this stage, the solution indicated earlier is clear since (iii) yields $P = 3$, which, when substituted back into (i) and (ii), gives $D = 7$ and $S = 7$ respectively. This method of eliminating variables and substituting back to get a solution is known as **Gaussian elimination**. It is important for solving large systems of equations, and is considered further in section 2-6. However, let us continue this "replacement" procedure a few steps further in order to illustrate the rules of Gauss-Jordan elimination and the special form of the system of equations which is produced.

Step 5 Since the coefficient of P in equation (iii) is not 1, we replace (iii) by $-1/9\times$(iii) to get

$$\begin{aligned} 1D + 0S + 1P &= 10 && \text{(i)} \\ 0D + 1S - 8P &= -17 && \text{(ii)} \\ 0D + 0S + 1P &= 3. && \text{(iii)} \end{aligned}$$

Step 6 Replace (i) by (i) $-$ (iii) to get

$$\begin{aligned} 1D + 0S + 0P &= 7 && \text{(i)} \\ 0D + 1S - 8P &= -17 && \text{(ii)} \\ 0D + 0S + 1P &= 3. && \text{(iii)} \end{aligned}$$

Step 7 Replace (ii) by (ii) $+\ 8\times$(iii) to get

$$\begin{aligned} 1D + 0S + 0P &= 7 && \text{(i)} \\ 0D + 1S + 0P &= 7 && \text{(ii)} \\ 0D + 0S + 1P &= 3 && \text{(iii)} \end{aligned}$$

which is the same as

$$\begin{aligned} D \qquad\qquad &= 7 && \text{(i)} \\ S \qquad &= 7 && \text{(ii)} \\ P &= 3. && \text{(iii)} \end{aligned}$$

❑

In the final form obtained in the above example, the system has these properties:

(1) The solution is obvious.
(2) The first variable, D ("x_1"), occurs in equation (i) only, the second, S ("x_2"), occurs in (ii) only, and the third, P("x_3"), occurs in (iii) only.

Notice that the procedure consists of *repetitions of only three types of operations applied to each system of equations:*

Type 1 interchanging two equations of a system.

Type 2 replacing an equation in a system by the same equation times a nonzero constant.

Type 3 replacing an equation in a system by the same equation plus a constant multiple of another equation.

We call these operations **elementary operations**. The steps of Gauss-Jordan elimination are summarized in the chart in Figure 2-2.

FIGURE 2-2

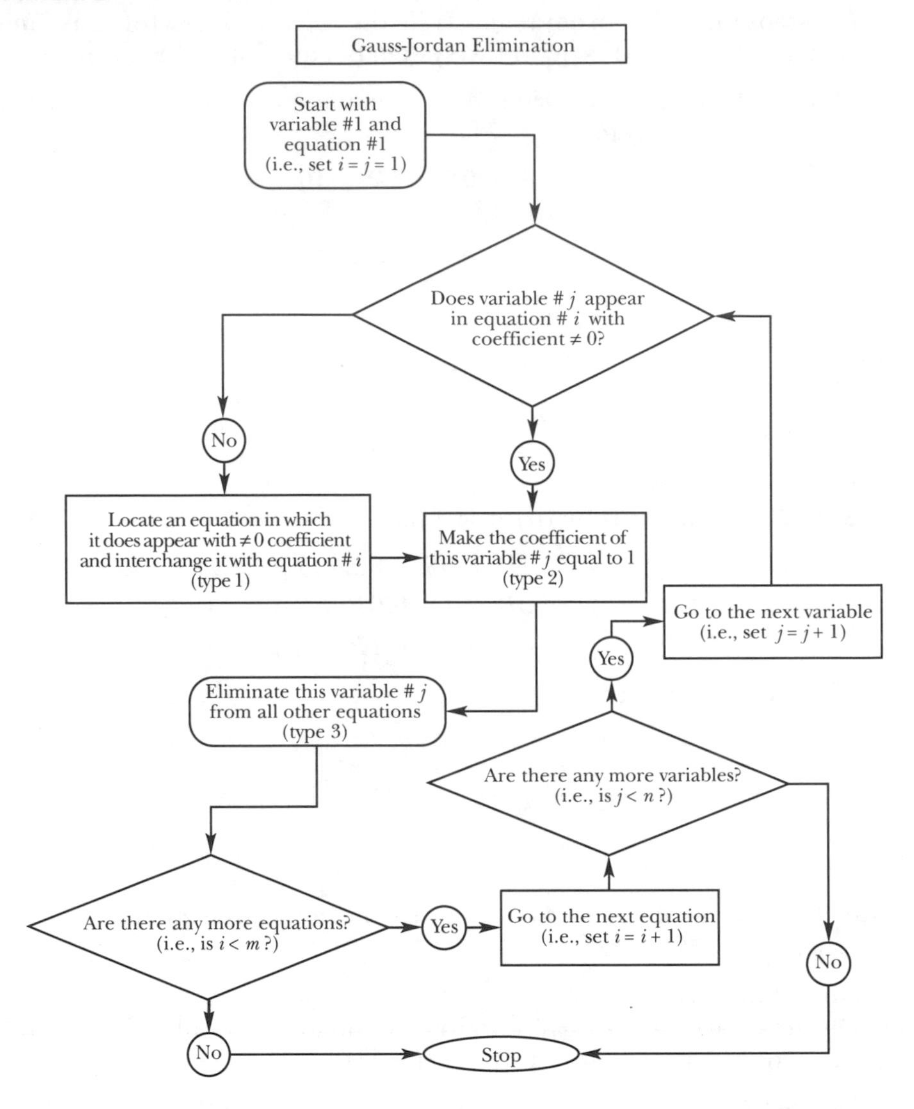

The procedure has the following properties:

(1) Each step in the procedure consists of applying exactly one elementary operation to the current system of equations.

(2) Each step replaces a system of equations by a new system having the *same* solutions. It is because of this fact that the solution of the final system is also the solution of the original one.

(3) The final system produced is in such a form that its solution is obvious by inspection.

Note that while property (2) may have seemed obvious in the example, we need to prove that it is true in general. To this end we first state the following:

DEFINITION 2-2

Two systems of linear equations are said to be **equivalent** if they have the same solutions.

THEOREM 2-1

If S' is a system of linear equations in n variables obtained from a system S by Gauss-Jordan elimination, then S and S' are equivalent.

PROOF (Optional) It is sufficient to prove the result for the case when S' has been obtained from S by a single elementary operation since Gauss-Jordan elimination simply consists of a finite repetition of such operations. However, we must consider each of the three types separately. We assume that S has the form of equations 2-3.

(1) If S' has been obtained by a type 1 operation, then S and S' consist of the same set of equations (but in a different order) and so have the same solutions.

(2) Next, suppose S' has been obtained by a type 2 operation. Then some equation of S', say the ith equation, is a nonzero multiple of equation i of S, and has the form

$$c(a_{i1}x_1 + \dots + a_{in}x_n) = cb_i,\ c \neq 0.$$

It follows that if $x_1, \dots, x_n$ is a solution of S, i.e., the n-tuple satisfies

$$a_{i1}x_1 + \dots + a_{in}x_n = b_i$$

in particular, it necessarily satisfies the above i^{th} equation of S', and so is a solution of S', as the other equations of S' are those of S. Conversely, if $x_1, \dots, x_n$ is a solution of S', it satisfies, in particular, the ith equation

$$c(a_{i1}x_1 + \dots + a_{in}x_n) = cb_i,$$

so that dividing by the nonzero constant c, we get

$$a_{i1}x_1 + \dots + a_{in}x_n = b_i.$$

This means that it is also a solution of the ith equation of S and, as noted earlier, since the other equations of S and S' are the same, it is a solution of the system S.

(3) Finally, suppose S' has been obtained from S by replacing the ith equation of S by the ith equation plus c times the jth equation $(i \neq j)$. Then the ith equation of S' is

$$(a_{i1} + ca_{j1})x_1 + \ldots + (a_{in} + ca_{jn})x_n = b_i + cb_j,$$

and as in (2) above, it is clear that any solution of S is a solution of this equation and thus of S'. Conversely, if $x_1, \ldots, x_n$ is a solution of S', it satisfies the above i^{th} equation and the jth equation of S' (= j^{th} equation of S):

$$a_{j1}x_1 + \ldots + a_{jn}x_n = b_j.$$

Multiplying this last equation by c and subtracting the result from equation i of S' above we get

$$a_{i1}x_1 + \ldots + a_{in}x_n = b_i.$$

This means that $x_1, \ldots, x_n$ is a solution of the ith equation of S and hence of system S. Thus, we have completed the proof as required. ■

EXAMPLE 2-2

Consider the system of three equations in three unknowns

$$\begin{aligned} 3y - 4z &= -6 && \text{(i)} \\ x + y + 5z &= 19 && \text{(ii)} \\ x + 4y + z &= 13. && \text{(iii)} \end{aligned}$$

The steps in the Gauss-Jordan elimination solution and the types of operations used are listed below.

Step 1 Since the variable $x_1 = x$ has nonzero coefficient in equation (ii) but not in (i), we interchange equations (i) and (ii), to place x in equation (i) (type 1).

$$\begin{aligned} x + y + 5z &= 19 && \text{(i)} \\ 3y - 4z &= -6 && \text{(ii)} \\ x + 4y + z &= 13 && \text{(iii)} \end{aligned}$$

Step 2 Now eliminate x from the other equations: replace (iii) with (iii) + $(-1)\times$(i) (type 3).

$$\begin{aligned} x + y + 5z &= 19 && \text{(i)} \\ 3y - 4z &= -6 && \text{(ii)} \\ 0x + 3y - 4z &= -6 && \text{(iii)} \end{aligned}$$

Step 3 Having eliminated x from all equations except equation (i), we work on y. We find that variable $x_2 = y$ has nonzero coefficient $(=3)$ in equation (ii). We make this coefficient equal to 1 by multiplying equation (ii) by 1/3 (type 2).

$$\begin{aligned} x + y + 5z &= 19 && \text{(i)} \\ y - (4/3)z &= -2 && \text{(ii)} \\ 3y - 4z &= -6 && \text{(iii)} \end{aligned}$$

Note that we are following the rules of Gauss-Jordan elimination to the letter. Here it would clearly have been quicker to observe that the last two equations in step 2 are the same, so that the last one can, in effect, be dropped, leading to the system

$$\begin{aligned} x + y + 5z &= 19 \\ 3y - 4z &= -6. \end{aligned}$$

Nevertheless, following the rules, we eliminate y from the other equations:

Step 4 Replace (iii) with (iii) + $(-3)\times$(ii) (type 3).

$$\begin{aligned} x + y + 5z &= 19 \\ y - (4/3)z &= -2 \\ 0y - 0z &= 0 \end{aligned}$$

Then, dropping the last (zero) equation, we have

$$\begin{aligned} x + y + 5z &= 19 && \text{(i)} \\ y - (4/3)z &= -2 && \text{(ii)} \end{aligned}$$

Step 5 Replace (i) with (i) + $(-1)\times$(ii) (type 3).

$$\begin{aligned} x + 0y + (19/3)z &= 21 && \text{(i)} \\ y - (4/3)z &= -2 && \text{(ii)} \end{aligned}$$

Notice that now the procedure stops since there are no remaining nonzero variables to which the rules can be applied (we have already used equations (i) and (ii)). The system has infinitely many solutions, since there are more variables than nonzero equations. For example, we may assign z any value, and then solve the two nonzero equations for x and y. Thus, if we put $z = 0$, we get $x = 21$ and $y = -2$, while if we put $z = 3$, we get $x = 2$ and $y = 2$. Note that since Gauss-Jordan elimination does not change the solution sets of systems of equations, these are also solutions of the original system. ❑

EXAMPLE 2-3

This example illustrates the way in which inconsistency of a system is revealed through a *contradiction*. We try to solve the system

$$\begin{aligned} x_1 + 2x_2 - 3x_3 &= -2 && \text{(i)} \\ 3x_1 - x_2 + 2x_3 &= 7 && \text{(ii)} \\ 5x_1 + 3x_2 - 4x_3 &= 2. && \text{(iii)} \end{aligned}$$

The steps are as follows.
(ii) + (−3) × (i):

$$\begin{aligned} x_1 + 2x_2 - 3x_3 &= -2 && \text{(i)} \\ -7x_2 + 11x_3 &= 13 && \text{(ii)} \\ 5x_1 + 3x_2 - 4x_3 &= 2 && \text{(iii)} \end{aligned}$$

(iii) + (−5) × (i):

$$\begin{aligned} x_1 + 2x_2 - 3x_3 &= -2 && \text{(i)} \\ -7x_2 + 11x_3 &= 13 && \text{(ii)} \\ -7x_2 + 11x_3 &= 12 && \text{(iii)} \end{aligned}$$

At this stage we recognize that, as a pair, equations (ii) and (iii) constitute a contradiction, unless there is no solution for the system. For, if there were a solution x_1, x_2, x_3, then by (ii) and (iii), we would conclude $13 = 12$. Thus the original system can have no solution either. ❑

EXERCISES 2-1

1. Solve the following systems of equations, if possible, using Gauss-Jordan elimination.

(a) $$\begin{aligned} x - y &= 1 \\ 2x &= 4 \end{aligned}$$

(b) $$\begin{aligned} 2x - z &= -1 \\ x + y - z &= 0 \\ 2x - y + 2z &= 3 \end{aligned}$$

(c) $$\begin{aligned} 2x + 3y - z &= 19 \\ 3x - 2y + 3z &= 7 \end{aligned}$$

(d) $$\begin{aligned} x_1 + 2x_2 - x_3 &= 0 \\ 2x_1 + 5x_2 + 5x_3 &= 0 \\ x_1 + 4x_2 + 7x_3 &= 0 \\ x_1 + 3x_2 + 3x_3 &= 0 \end{aligned}$$

2. Use Gauss-Jordan elimination to solve the systems of equations in question 2, Exercises 1-1.

3. Use Gauss-Jordan elimination to solve the system consisting of the three equations in question 8, Exercises 1-1.

4. Use Gauss-Jordan elimination to show that the system

$$\begin{aligned} 4x + 4y + 3z &= 12 \\ x + y + (3/4)z &= 1 \end{aligned}$$

is inconsistent. What is the geometric interpretation of this fact?

5. Show that the two systems below are equivalent.

$$\begin{aligned} x + y \quad\quad &= 1 \\ 2x + y \quad\quad &= 0 \\ x \quad\quad + z &= 0 \end{aligned} \qquad\qquad \begin{aligned} -2x + y + z &= 5 \\ 3x + 2y + 3z &= 4 \\ x + 2y - z &= 2 \end{aligned}$$

6. A company manufacturing television sets finds that the cost of producing x sets is given by $C(x) = 120\,000 + 60x$ dollars. The company sells the sets at $200 each. Find the number of sets it should manufacture in order to break even, assuming that all the sets made are sold.

7. An electronics firm produces two types of printed circuit board; both types have to be processed in its etching and soldering departments. The times required for the two types in each department are given below (in minutes):

	Type 1	*Type 2*
Etching	4	3
Soldering	1	2

There are three men working in the etching department, but only one in the soldering department. Find the hourly production of the two types of board that exactly uses the full time available in the two departments.

8. A man has invested $20 000 in two ways: one part in a savings account at 11% per annum, and the other part in a mortgage fund yielding 16% per annum. If the former yields $400 less than the latter, how much has been invested in each?

9. Use Gauss-Jordan elimination to show the following for the system

$$\begin{aligned} ax + by &= e \\ cx + dy &= f \end{aligned}$$

where a, b, c, d, e, and f are constants.

(a) That it has an unique solution

$$x = (de - bf)/(ad - bc) \text{ and } y = (af - ce)/(ad - bc),$$

if $ad - bc \neq 0$.

(b) That it has no solution if $a/c = b/d$ but these fractions are not equal to e/f.

(c) That it has infinitely many solutions if $a/c = b/d = e/f$.

10. A pneumatic switch turns off the cooling system which it controls when the pressure on its left side becomes equal to half the pressure on its right side. The pressure $L(t)$ on the left side is given by $L(t) = 0.6t - 1.2$, where t is the elapsed time (in hours) since midnight ($t = 0$). The pressure $R(t')$ on the right side is given by $R(t') = 0.4t' + 6.4$ where t' is the elapsed time (in hours) since 6 a.m. ($t' = 0$). Find the time at which the cooling system turns off.

2-2 Matrices and Gauss-Jordan Reduction

A re-examination of the examples in the previous section will show that the variables in a system of equations play *no* role in the computations of Gauss-Jordan elimination. The reason is that the procedure works with their coefficients, while the variables only serve to tell us where the various coefficients belong. Thus, omitting the variables (and equality signs too) would not affect the computations and would save considerable writing. When we omit these symbols in writing a system of equations, we get, in effect, a **matrix**. Consider again the system S of m equations in n unknowns (equation 2-3):

$$\begin{array}{l} a_{11}x_1 + a_{12}x_2 + \ldots + a_{1n}x_n = b_1 \\ a_{21}x_1 + a_{22}x_2 + \ldots + a_{2n}x_n = b_2 \\ \cdots\cdots\cdots\cdots\cdots\cdots\cdots\cdots \\ a_{m1}x_1 + a_{m2}x_2 + \ldots + a_{mn}x_n = b_m. \end{array}$$

DEFINITION 2-3

The $m \times n$ array of numbers given by

$$A = \begin{bmatrix} a_{11} & a_{12} & \cdots & a_{1n} \\ a_{21} & a_{22} & \cdots & a_{2n} \\ \cdots & \cdots & \cdots & \cdots \\ a_{m1} & a_{m2} & \cdots & a_{mn} \end{bmatrix}$$

is called the **coefficient matrix** A of the system S.

More generally, *any* $m \times n$ array A of numbers as given above is called an $m \times n$ matrix A (regardless of its origin). It has m rows and n columns. It will frequently be convenient to abbreviate the notation to $A = (a_{ij})$. The number a_{ij} (or A_{ij}) is then called the (i,j) **entry** of the matrix A. It is the number in row i and column j and we say that A is a matrix of **order** (or size) $m \times n$.

The $m \times 1$ matrix B given by

$$B = \begin{bmatrix} b_1 \\ b_2 \\ \cdots \\ b_m \end{bmatrix}$$

is called the **matrix of constants** of the system S.

The $m \times (n+1)$ matrix $(A{:}B)$ defined by

$$(A{:}B) = \begin{bmatrix} a_{11} & \cdots & a_{1n} & b_1 \\ a_{21} & \cdots & a_{2n} & b_2 \\ \cdots & \cdots & \cdots & \cdots \\ a_{m1} & \cdots & a_{mn} & b_m \end{bmatrix}$$

is called the **augmented matrix** of the system. Note that $(A{:}B)$ *is just obtained by writing A and B side-by-side as shown, and contains all the information relevant to the system of equations S.*

EXAMPLE 2-4

Consider the system

$$\begin{aligned} x_1 \qquad + \ x_3 &= 10 \\ x_2 - 8x_3 &= -17 \\ x_1 - x_2 \qquad &= 0, \end{aligned}$$

which is the system of Example 2-1 rewritten. The coefficient matrix is

$$A = \begin{bmatrix} 1 & 0 & 1 \\ 0 & 1 & -8 \\ 1 & -1 & 0 \end{bmatrix},$$

the matrix of constants is

$$B = \begin{bmatrix} 10 \\ -17 \\ 0 \end{bmatrix},$$

and the augmented matrix is

$$(A{:}B) = \begin{bmatrix} 1 & 0 & 1 & 10 \\ 0 & 1 & -8 & -17 \\ 1 & -1 & 0 & 0 \end{bmatrix}. \ \square$$

EXAMPLE 2-5

Let C be the 2×4 matrix given by

$$C = \begin{bmatrix} 2 & 3 & -1 & 4 \\ -1 & 0 & 2 & -1 \end{bmatrix}.$$

Then the system of equations having C as augmented matrix is

$$\begin{aligned} 2x_1 + 3x_2 - \ x_3 &= 4 \\ -x_1 \qquad + 2x_3 &= -1. \ \square \end{aligned}$$

These examples show that *we can pass freely from a system of linear equations to a matrix, and vice-versa*, by removing or inserting variables and "=" signs respectively. It should be noted, however, that matrices arise in mathematics in many other contexts, as we shall later see.

We now take up the notion suggested at the beginning of this section, that is, we solve a system of equations without writing variables and "=" signs. In other words, we work with the augmented matrix of the system. The corresponding operations applied to the matrices are called **elementary row operations**, and we call the procedure itself **Gauss-Jordan reduction** of the augmented matrix, to distinguish it from the earlier procedure dealing with equations.

EXAMPLE 2-6

We apply Gauss-Jordan reduction to the matrix of the system

$$\begin{aligned} y - 8z &= -17 \\ x \quad + z &= 10 \\ x - y \quad &= 0. \end{aligned}$$

The sequence of steps in Gauss-Jordan reduction appears below in the right column, while the corresponding operations on the equations appear in the left column (i.e., Gauss-Jordan elimination). The reader should compare this example with Example 2-1, and with the flow chart in Figure 2-2, replacing the words "variable" and "equation" by "entry" and "row", respectively.

Equations *Matrix*

$$\begin{aligned} y - 8z &= -17 \\ x \quad + z &= 10 \\ x - y \quad &= 0 \end{aligned} \qquad \begin{bmatrix} 0 & 1 & -8 & -17 \\ 1 & 0 & 1 & 10 \\ 1 & -1 & 0 & 0 \end{bmatrix}$$

We start the flow chart with $i = j = 1$. To bring a "1" into row $i = 1$ and column $j = 1$, we interchange rows 1 and 2 leading to

$$\begin{aligned} x \quad + z &= 10 \\ y - 8z &= -17 \\ x - y \quad &= 0 \end{aligned} \qquad \begin{bmatrix} 1 & 0 & 1 & 10 \\ 0 & 1 & -8 & -17 \\ 1 & -1 & 0 & 0 \end{bmatrix}$$

Now there is a 1 in row $i = 1$ and column $j = 1$, so we use it to eliminate the other entries in column 1.

$$\begin{aligned} x \quad + z &= 10 \\ y - 8z &= -17 \\ -y - z &= -10 \end{aligned} \qquad \begin{bmatrix} 1 & 0 & 1 & 10 \\ 0 & 1 & -8 & -17 \\ 0 & -1 & -1 & -10 \end{bmatrix}$$

We next increase i, j to 2 as directed by the flow chart. There is a 1 in row $i = 2$ and column $j = 2$, which we use to eliminate the other entries in that column.

$$\begin{aligned} x \quad + \quad z &= 10 \\ y - 8z &= -17 \\ -9z &= -27 \end{aligned} \qquad \begin{bmatrix} 1 & 0 & 1 & 10 \\ 0 & 1 & -8 & -17 \\ 0 & 0 & -9 & -27 \end{bmatrix}$$

Increase i, j to 3; there is a -9 in row $i = 3$ and column $j = 3$. We make this entry 1, and then use it to eliminate all other entries in column 3.

$$\begin{aligned} x \quad + \quad z &= 10 \\ y - 8z &= -17 \\ z &= 3 \end{aligned} \qquad \begin{bmatrix} 1 & 0 & 1 & 10 \\ 0 & 1 & -8 & -17 \\ 0 & 0 & 1 & 3 \end{bmatrix}$$

$$\begin{aligned} x \qquad &= 7 \\ y - 8z &= -17 \\ z &= 3 \end{aligned} \qquad \begin{bmatrix} 1 & 0 & 0 & 7 \\ 0 & 1 & -8 & -17 \\ 0 & 0 & 1 & 3 \end{bmatrix}$$

$$\begin{aligned} x &= 7 \\ y &= 7 \\ z &= 3 \end{aligned} \qquad \begin{bmatrix} 1 & 0 & 0 & 7 \\ 0 & 1 & 0 & 7 \\ 0 & 0 & 1 & 3 \end{bmatrix}$$

At this point, we have completed the procedure since $i = m = 3$ (there are no more variables and equations to work with). It is important to note that the final matrix is of course the augmented matrix of the system of equations at its left, and that conversely, given only that matrix, we can reconstruct the corresponding system of equations. The last matrix has a special form called **row-reduced echelon form**, corresponding to the simplicity of the associated system of equations. ❑

DEFINITION 2-4

An $m \times n$ matrix A is in **row echelon** form if it has the following properties:

(1) All rows of A consisting entirely of zeros (called **zero rows**) appear below all rows with at least one nonzero entry.

(2) The left-most nonzero entry (called the **leading entry**) in any nonzero row is a 1.

(3) The leading entry in any nonzero row appears to the right of the leading entry in any row above it.

A matrix that also satisfies (4) below is said to be in **row-reduced echelon form**:

(4) If a column contains a leading row entry 1, then all other entries in that column are zero.

EXAMPLE 2-7

(a) The following matrices are in row echelon form:

$$\begin{bmatrix} 1 & 2 \\ 0 & 1 \\ 0 & 0 \\ 0 & 0 \end{bmatrix}, \quad \begin{bmatrix} 0 & 1 & 6 & -2 \\ 0 & 0 & 1 & 3 \\ 0 & 0 & 0 & 1 \\ 0 & 0 & 0 & 0 \end{bmatrix}.$$

(Note that they are *not* in row-reduced echelon form.)

(b) The following matrices are in row-reduced echelon form:

$$\begin{bmatrix} 0 & 1 & 0 & 4 \\ 0 & 0 & 1 & -1 \\ 0 & 0 & 0 & 0 \end{bmatrix}, \quad \begin{bmatrix} 1 & 3 & 0 & 0 & -1 \\ 0 & 0 & 1 & 0 & 2 \\ 0 & 0 & 0 & 1 & -4 \end{bmatrix},$$

$$\begin{bmatrix} 1 & 0 & 0 & 0 & 0 \\ 0 & 1 & 0 & 0 & 0 \\ 0 & 0 & 0 & 0 & 0 \end{bmatrix}.$$

(c) The following matrices are neither in row echelon form nor in row-reduced echelon form:

$$\begin{bmatrix} 1 & 0 & 2 & 0 \\ 0 & 0 & 0 & 0 \\ 0 & 0 & 0 & 1 \end{bmatrix}, \quad \begin{bmatrix} 0 & 0 & 1 & 3 & 0 \\ 0 & 0 & 0 & 0 & 1 \\ 0 & 1 & 0 & -2 & 0 \end{bmatrix}, \quad \begin{bmatrix} 0 & -1 & 0 \\ 0 & 0 & 1 \\ 0 & 0 & 0 \end{bmatrix}. \quad \square$$

As we have seen in Example 2-6, for each of the three operations on equations used in Gauss-Jordan elimination, there is a corresponding operation on matrices used in Gauss-Jordan reduction. They are called elementary row operations of types 1, 2, and 3 respectively. Their effects are summarized below (compare with the earlier discussion of Gauss-Jordan elimination):

Type 1 interchanging two rows of a matrix.

Type 2 replacing a row by the same row multiplied by a nonzero constant.

Type 3 replacing a row by the same row plus a constant multiple of another row.

DEFINITION 2-5

If an $m \times n$ matrix A' can be obtained from another $m \times n$ matrix A by a finite number of elementary row operations, then we say A' is **row-equivalent** to A.

By this definition, then, if A' is obtained from A by a *single* elementary row operation, it is certainly row-equivalent to A. Thus, for example, each matrix in the sequence of matrices in Example 2-6 is row-equivalent to every other matrix in that sequence.

The next theorem states the **symmetry** of row-equivalence and has important consequences for later work with matrices.

THEOREM 2-2

If A' is row-equivalent to A, then A is row-equivalent to A'.

PROOF This follows from the fact that to every elementary row operation e, there is a corresponding elementary row operation, denoted e^{-1}, which "undoes" the work of e when e and then e^{-1} are applied to a matrix in succession (in either order). For the three possibilities for e, the corresponding inverse operations e^{-1} are listed in the table below.

Operation	e	e^{-1}
Type 1	interchange rows i and j	interchange rows i and j
Type 2	replace row i by $k \times$ row i, $k \neq 0$	replace row i by $(1/k) \times$ row i, $k \neq 0$
Type 3	replace row i by row $i + k \times$ row j, $(i \neq j)$	replace row i by row $i - k \times$ row j, $(i \neq j)$

For example,

$$A' = \begin{bmatrix} 1 & 2 & 3 \\ 6 & 9 & 12 \end{bmatrix}$$

is obtained from

$$A = \begin{bmatrix} 1 & 2 & 3 \\ 4 & 5 & 6 \end{bmatrix}$$

by replacing row 2 by row 2 + 2 × row 1, a type 3 operation e. Conversely, if we replace row 2 of A' by row 2 − 2 × row 1 (i.e., apply e^{-1} to A') we get back A. Similar results are obtained when we use type 1 and type 2 operations. We may summarize these facts by writing

$$e^{-1}(e(A)) = e(e^{-1}(A)) = A, \text{ for any matrix } A,$$

where $e(A)$ is the matrix obtained from A by applying an elementary row operation e to it.

Now, to prove the theorem, as noted in the example above, it is sufficient to suppose that A' has been obtained from A by a single elementary row operation. Then we can write $A' = e(A)$, so it follows that applying e^{-1} to both sides yields

$$A = e^{-1}(e(A)) = e^{-1}(A'),$$

showing that A is row-equivalent to A', as required, since e^{-1} is an elementary row operation. ■

This result shows that the relation of row-equivalence is **symmetric**; in fact, it is an equivalence relation (see question 5, Exercises 2-2).

EXAMPLE 2-8

Each matrix in Example 2-6 is row-equivalent to every other matrix in that example, as already noted. In particular, they are all row-equivalent to the last one,

$$\begin{bmatrix} 1 & 0 & 0 & 7 \\ 0 & 1 & 0 & 7 \\ 0 & 0 & 1 & 3 \end{bmatrix}$$

which is in row-reduced echelon form. ❑

Example 2-8 illustrates the next theorem.

> **THEOREM 2-3**
>
> Every $m \times n$ matrix is row-equivalent to an unique $m \times n$ matrix in row-reduced echelon form.

PROOF We do not provide a detailed general proof, but it should be intuitively plausible from our experience with Gauss-Jordan reduction that every matrix A is row-equivalent to at least one matrix A' in row-reduced echelon form. It is important, although the proof is not obvious, that there is only *one* such A'. For, there are many possible sequences of elementary row operations which lead to a row-reduced echelon matrix. The theorem says they must all lead to the same A'. ■

The corollary that follows allows us to work with augmented matrices rather than with the corresponding equations.

COROLLARY

Two systems of m equations in n variables are equivalent iff their corresponding augmented matrices are row-equivalent to the same row-reduced echelon matrix (and hence also to each other). ■

EXAMPLE 2-9

The system of equations S with augmented matrix

$$(A{:}B) = \begin{bmatrix} 1 & 2 & -3 & -2 \\ 3 & -1 & 2 & 7 \\ 5 & 3 & -4 & 2 \end{bmatrix}$$

is inconsistent (compare Example 2-3!). To see this, we apply Gauss-Jordan reduction to $(A{:}B)$ to get the following matrices:

$$(A{:}B) - \begin{bmatrix} 1 & 2 & -3 & -2 \\ 3 & -1 & 2 & 7 \\ 5 & 3 & -4 & 2 \end{bmatrix}$$

$$\begin{bmatrix} 1 & 2 & -3 & -2 \\ 0 & -7 & 11 & 13 \\ 5 & 3 & -4 & 2 \end{bmatrix} \quad \text{(row 2 to row } 2-3 \times \text{row 1)}$$

$$\begin{bmatrix} 1 & 2 & -3 & -2 \\ 0 & -7 & 11 & 13 \\ 0 & -7 & 11 & 12 \end{bmatrix} \quad \text{(row 3 to row } 3-5 \times \text{row 1)}$$

$$\begin{bmatrix} 1 & 2 & -3 & -2 \\ 0 & -7 & 11 & 13 \\ 0 & 0 & 0 & -1 \end{bmatrix} \quad \text{(row 3 to row 3 } - \text{ row 2)}$$

$$= (A'{:}B').$$

Notice that we have not followed Gauss-Jordan reduction exactly here, and that $(A'{:}B')$ is not a row echelon matrix. Nevertheless, the system S' corresponding to $(A'{:}B')$ has as its last equation

$$0x_1 + 0x_2 + 0x_3 = -1,$$

which has no solution. Therefore, the original system S is also inconsistent. ❑

EXAMPLE 2-10

We find the 3×1 matrices B such that the system of equations with augmented matrix $(A{:}B)$ is consistent, where A is the matrix in Example 2-9 above. The augmented matrix for the system is

$$(A{:}B) = \begin{bmatrix} 1 & 2 & -3 & b_1 \\ 3 & -1 & 2 & b_2 \\ 5 & 3 & -4 & b_3 \end{bmatrix}.$$

We find the conditions that the b's must satisfy so that the system has a solution. Applying Gauss-Jordan reduction, we find that $(A{:}B)$ is row equivalent to

$$\begin{bmatrix} 1 & 2 & -3 & b_1 \\ 0 & -7 & 11 & b_2 - 3b_1 \\ 0 & -7 & 11 & b_3 - 5b_1 \end{bmatrix}$$

(compare Example 2-9). Finally, subtracting row 2 from row 3 yields

$$\begin{bmatrix} 1 & 2 & -3 & b_1 \\ 0 & -7 & 11 & b_2 - 3b_1 \\ 0 & 0 & 0 & b_3 - b_2 - 2b_1 \end{bmatrix}$$

from which we see that the system will be consistent if B is such that $b_3 - b_2 - 2b_1 = 0$. ❑

EXAMPLE 2-11

Solve (if possible) the following system of equations using Gauss-Jordan reduction.

$$\begin{aligned} 2x_1 - x_2 + x_3 &= 1 \\ x_1 + 2x_2 - x_3 &= 3 \\ x_1 + 7x_2 - 4x_3 &= 8 \end{aligned}$$

Solution The system has augmented matrix

$$\begin{bmatrix} 2 & -1 & 1 & 1 \\ 1 & 2 & -1 & 3 \\ 1 & 7 & -4 & 8 \end{bmatrix}.$$

Applying Gauss-Jordan reduction to it yields (the reader should note the types of the operations used):

$$\begin{bmatrix} 1 & 2 & -1 & 3 \\ 2 & -1 & 1 & 1 \\ 1 & 7 & -4 & 8 \end{bmatrix}$$

$$\begin{bmatrix} 1 & 2 & -1 & 3 \\ 0 & -5 & 3 & -5 \\ 0 & 5 & -3 & 5 \end{bmatrix}$$

$$\begin{bmatrix} 1 & 2 & -1 & 3 \\ 0 & 1 & -3/5 & 1 \\ 0 & 0 & 0 & 0 \end{bmatrix}$$

$$\begin{bmatrix} 1 & 0 & 1/5 & 1 \\ 0 & 1 & -3/5 & 1 \\ 0 & 0 & 0 & 0 \end{bmatrix}$$

which is a row-reduced echelon matrix. The corresponding system of equations is

$$\begin{aligned} x_1 \quad + (1/5)x_3 &= 1 \\ x_2 - (3/5)x_3 &= 1. \end{aligned}$$

There are infinitely many solutions which can be obtained by assigning arbitrary values to the variable x_3. For example,

(i) $x_3 = 0$ yields $x_1 = 1, x_2 = 1$;
(ii) $x_3 = 5$ yields $x_1 = 0, x_2 = 4$;
(iii) $x_3 = -5$ yields $x_1 = 2, x_2 = -2$. ❑

EXAMPLE 2-12

Solve (if possible) the system of equations below.

$$\begin{aligned} y + z + 5s + 6t &= -1 \\ -2x - y \quad + 2s + t &= -2 \\ x + 2y + 2z \quad + 4t &= 2 \\ x \quad - z + 3s + t &= -1 \\ 2y + 3z - 3s + 3t &= 3 \end{aligned}$$

Solution The reduction of the augmented matrix is shown below.

$$\begin{bmatrix} 0 & 1 & 1 & 5 & 6 & -1 \\ -2 & -1 & 0 & 2 & 1 & -2 \\ 1 & 2 & 2 & 0 & 4 & 2 \\ 1 & 0 & -1 & 3 & 1 & -1 \\ 0 & 2 & 3 & -3 & 3 & 3 \end{bmatrix}$$

$$\begin{bmatrix} 1 & 2 & 2 & 0 & 4 & 2 \\ -2 & -1 & 0 & 2 & 1 & -2 \\ 0 & 1 & 1 & 5 & 6 & -1 \\ 1 & 0 & -1 & 3 & 1 & -1 \\ 0 & 2 & 3 & -3 & 3 & 3 \end{bmatrix}$$

$$\begin{bmatrix} 1 & 2 & 2 & 0 & 4 & 2 \\ 0 & 3 & 4 & 2 & 9 & 2 \\ 0 & 1 & 1 & 5 & 6 & -1 \\ 0 & -2 & -3 & 3 & -3 & -3 \\ 0 & 2 & 3 & -3 & 3 & 3 \end{bmatrix}$$

$$\begin{bmatrix} 1 & 2 & 2 & 0 & 4 & 2 \\ 0 & 1 & 1 & 5 & 6 & -1 \\ 0 & 3 & 4 & 2 & 9 & 2 \\ 0 & -2 & -3 & 3 & -3 & -3 \\ 0 & 0 & 0 & 0 & 0 & 0 \end{bmatrix}$$

$$\begin{bmatrix} 1 & 0 & 0 & -10 & -8 & 4 \\ 0 & 1 & 1 & 5 & 6 & -1 \\ 0 & 0 & 1 & -13 & -9 & 5 \\ 0 & 0 & -1 & 13 & 9 & -5 \\ 0 & 0 & 0 & 0 & 0 & 0 \end{bmatrix}$$

$$\begin{bmatrix} 1 & 0 & 0 & -10 & -8 & 4 \\ 0 & 1 & 0 & 18 & 15 & -6 \\ 0 & 0 & 1 & -13 & -9 & 5 \\ 0 & 0 & 0 & 0 & 0 & 0 \\ 0 & 0 & 0 & 0 & 0 & 0 \end{bmatrix}$$

The corresponding system of equations is

$$\begin{aligned} x \qquad - 10s - \ \ 8t &= 4 \\ y \quad + 18s + 15t &= -6 \\ z - 13s - \ \ 9t &= 5 \end{aligned}$$

or,

$$\begin{aligned} x &= \ \ 4 + 10s + 8t \\ y &= -6 - 18s - 15t \\ z &= \ \ 5 + 13s + 9t. \end{aligned}$$

The system therefore has infinitely many solutions, which can be obtained by assigning arbitrary values to s and t. ❑

EXERCISES 2-2

1. **(a)** Write the coefficient, constant, and augmented matrices of the following systems of equations.

(i)
$$\begin{aligned} a - 2b + c \qquad &= 0 \\ a - b \qquad &= -3 \\ 2a \qquad + 3d &= 4 \\ -b + c - 2d &= 2 \\ a + b + c + d &= 0 \end{aligned}$$

(ii)
$$\begin{aligned} -x_1 + (1/2)x_2 \qquad + x_4 &= -1 \\ x_3 + x_4 &= 0 \end{aligned}$$

(b) Write the systems of equations with augmented matrices given below.

(i) $\begin{bmatrix} -1 & 0 & 3 & 2 \\ 1 & 1 & 2 & 4 \\ -2 & 0 & 3 & 1 \end{bmatrix}$ **(ii)** $\begin{bmatrix} -1 & 2 & 3 \\ 4 & -2 & 1 \\ 6 & -1 & 0 \\ 3 & 2 & 1 \end{bmatrix}$

2. **(a)** Which of the following matrices are in row-reduced echelon form?

(i) $\begin{bmatrix} 1 & 0 & 0 \\ 0 & 0 & 0 \\ 0 & 1 & 0 \end{bmatrix}$ **(ii)** $\begin{bmatrix} 0 & 1 & 0 & 0 & 3 \\ 0 & 0 & 0 & 0 & 1 \end{bmatrix}$

(iii) $\begin{bmatrix} 0 & 1 & 0 & 3 & 2 & 6 \\ 0 & 0 & 1 & -2 & 3 & 4 \\ 0 & 0 & 0 & 0 & 0 & 0 \end{bmatrix}$ **(iv)** $\begin{bmatrix} 0 & 1 & 2 \\ 1 & 0 & -3 \\ 0 & 0 & 0 \end{bmatrix}$

(b) Which of the following matrices are in row echelon form but not in row-reduced echelon form?

(i) $\begin{bmatrix} 1 & -1 & 3 \\ 0 & 1 & 2 \\ 0 & 0 & 1 \end{bmatrix}$ **(ii)** $\begin{bmatrix} 0 & 1 & -1 & 0 \\ 0 & 0 & 1 & 0 \\ 0 & 0 & 0 & 0 \end{bmatrix}$

(iii) $\begin{bmatrix} 0 & 1 & 0 & 0 \\ 0 & 0 & 1 & 0 \end{bmatrix}$ **(iv)** $\begin{bmatrix} 0 & 2 & -1 & 0 & 3 \\ 0 & 0 & 0 & 1 & 0 \\ 0 & 0 & 0 & 0 & 1 \end{bmatrix}$

3. In each of the following, the matrix A' has been obtained from a matrix A of the same order by a single elementary row operation of the type indicated. Find A in each case.

(a) $A' = \begin{bmatrix} -1 & 2 \\ 3 & 4 \\ 0 & 1 \end{bmatrix}$ (type 1, to rows 1 and 3)

(b) $A' = \begin{bmatrix} 2 & -1 & 3 & 2 \\ 1 & 0 & -2 & -1 \\ 1 & 0 & 0 & 1 \end{bmatrix}$ (type 2, row 2 multiplied by 1/2)

(c) $A' = \begin{bmatrix} 1 & -2 & 0 & 3 \\ 0 & 1 & 2 & 6 \\ 0 & 0 & 0 & 0 \end{bmatrix}$ (type 3, row 3 replaced by row $3 - 2 \times$ row 1)

4. In each of the following cases, transform the given matrix to row-reduced echelon form, and solve (if possible) the corresponding system of equations having it as augmented matrix.

(a) $\begin{bmatrix} 4 & -1 & 14 \\ 2 & 1 & 4 \end{bmatrix}$ **(b)** $\begin{bmatrix} 1 & -1 & -1 & 1 \\ 1 & 1 & 1 & -1 \\ 2 & 2 & -1 & 6 \end{bmatrix}$

(c) $\begin{bmatrix} 1 & -7 & 17 \\ 2 & 3 & 0 \\ 3 & -4 & 16 \end{bmatrix}$ **(d)** $\begin{bmatrix} 3 & 4 & 1 & 8 \\ 5 & -2 & 3 & 10 \\ -1 & 3 & 4 & 19 \end{bmatrix}$

(e) $\begin{bmatrix} 2 & 0 & 10 \\ 1 & 1/2 & 2 \\ 3 & 1/2 & 12 \\ 1 & -1/2 & -2 \end{bmatrix}$ **(f)** $\begin{bmatrix} 2 & 1 & 0 & 0 \\ -1 & 0 & 1 & 0 \\ 0 & 1 & 1 & 0 \end{bmatrix}$

5. Show that the relation of row-equivalence is an equivalence relation on the set of $m \times n$ matrices as follows:

(a) It is reflexive: any matrix A is row-equivalent to itself.

(b) It is symmetric: if A is row-equivalent to A'', then A'' is row-equivalent to A.

(c) It is transitive: if A is row-equivalent to A' and A' is row-equivalent to A'', then A is row-equivalent to A''.

6. Use Gauss-Jordan reduction to solve the systems of equations in question 1, Exercises 2-1.

7. Use Gauss-Jordan reduction to show that the systems of equations in question 5, Exercises 2-1, are equivalent.

8. **(a)** Show that the homogeneous system of two equations

$$\begin{aligned} x_1 + 2x_2 + x_3 &= 0 \\ x_1 - x_2 - x_3 &= 0 \end{aligned}$$

in three variables has infinitely many solutions and, in particular, at least one solution different from the trivial solution $x_1 = x_2 = x_3 = 0$.

(b) Generalize (a) for an arbitrary homogeneous system of m equations in n unknowns with $m < n$ (more variables than equations) as follows:

(i) Show that if the augmented matrix $(A{:}B)$ of the system (with B an $m \times 1$ matrix of 0's) is row-equivalent to a row-reduced echelon matrix $(A'{:}B')$, then $(A'{:}B')$ has fewer than n nonzero rows (note that the entries of the $m \times 1$ matrix B' are all 0 too).

(ii) Show that because of (i), the system with augmented matrix $(A'{:}B')$ has at least one "free" or "undetermined" variable which may be assigned arbitrary real values.

(iii) Conclude that the original system has infinitely many solutions. These facts will be proved in Chapter 4 as a particular case of a more general result.

9. The static equilibrium of a horizontal beam loaded with vertical weights and supported at various points as shown in Figure 2-3 is determined by two principles:

(i) The sum of the forces up equals the sum of the forces down (see the discussion on forces before Example 1-13).

(ii) The sum of the counterclockwise moments of the forces about any point P equals the sum of the clockwise moments of the forces about that point. Here the *moment* of a force of F kilograms-weight at a perpendicular distance of d meters from P is the product $d \times F$. It is considered counterclockwise if the force would tend to cause a counterclockwise rotation of the beam about P, and clockwise otherwise. For example, in Figure 2-3, the weight at C exerts a counterclockwise moment about point B (a support) but a clockwise moment about A (the other support).

FIGURE 2-3

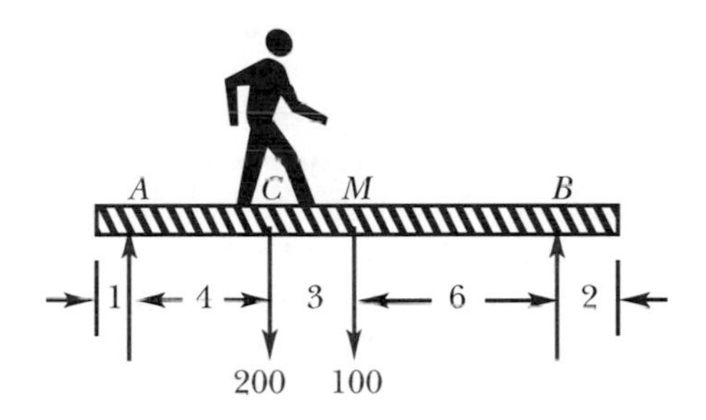

The figure represents a 16-m wooden beam weighing 100 kg; the weight may be assumed to act at the midpoint M. The beam rests on two blocks at A and B, and is loaded with a 200-kg man standing at point C. Use the principles stated above to find the weight supported by the blocks A and B.

10. In a two-commodity market (as distinct from the one-commodity market discussed in section 2-2), the supply and demand for each commodity depend on the prices of *both* commodities. Adopting the more conventional notation from economics, let $q^d{}_1$, $q^d{}_2$ be the demands for products 1 and 2 respectively, and $q^s{}_1$, $q^s{}_2$ be the respective supplies. Find the equilibrium solution (where supply equals demand for all commodities) for the market described by the linear equations

$$\begin{aligned} q^d{}_1 &= -p_1 + 2p_2 + 2 \qquad & q^d{}_2 &= 3p_1 - 2p_2 - 4 \\ q^s{}_1 &= 2p_1 - p_2 + 1 \qquad & q^s{}_2 &= -p_1 + p_2 - 1. \end{aligned}$$

Here the p's are the prices for the commodities.

11. If A is an $m \times n$ matrix, its *transpose* A^t is the $n \times m$ matrix whose rows are the columns of A in the same order: $A^t{}_{ij} = A_{ji}$. Find the transposes of the matrices in question 4 above. (The notion of transpose will be discussed in section 2-5.)

2-3 Matrices: Some Examples and Applications (Optional)

As noted in the previous section, although the notion of matrices arises in the theory of linear equations, there are many other contexts in mathematics and its applications where this concept is useful. Matrices are considered to have been invented by Arthur Cayley (1821–1895), an eminent English mathematician, although the term matrix was first used by his close friend and collaborator, James Joseph Sylvester (1814–1897). Both men made numerous contributions to mathematics and its applications.

Since a matrix consists of an array of numbers, an immediate use of this notion is to summarize information. This is essentially what we did in defining the augmented matrix of a system of equations. Some other situations are described in the following examples.

EXAMPLE 2-13

A mining company owns mines in three locations, Sudbury, Falconbridge, and Rouyn, where the company also has smelters. Owing to the different ores mined at the three locations, and to the different smelting facilities at the refineries, it may be cheaper to transport ore from a mine in one location to a smelter at another. Thus, the total cost of refining an ore may be considered to consist of a transportation component and a smelting component. These costs for a ton of ore can be conveniently summarized in two 3×3 matrices as follows. Let us denote the three locations above by the integers 1,2, and 3 respectively. Then we may define the transportation cost matrix T by T_{ij} = the cost (in dollars) of transporting one ton of ore from mine i to smelter j. For example, if T is given by

$$T = \begin{bmatrix} 0 & 2 & 6 \\ 2 & 0 & 3 \\ 6 & 3 & 0 \end{bmatrix},$$

then $T_{23} = 3$ means the cost of shipping one ton of ore from Falconbridge to Rouyn is \$3. Notice that $T_{ij} = T_{ji}$, which reflects that the cost of transporting ore from mine i to smelter j is the same as from mine j to smelter i. However, this is an assumption and is not necessarily a fact in reality. Because $T_{ij} = T_{ji}$, the matrix T is said to be **symmetric**. It also has the special property that $T_{ii} = 0$, which reflects that there is no cost in transporting ore from a mine at site i to a smelter at the same site i. Again, this is an assumption. We say that the **diagonal entries** of T are 0.

Similarly, we define the smelting cost matrix S by S_{ij} = the cost in dollars of smelting a ton of ore from mine i at smelter j. For example, if

$$S = \begin{bmatrix} 50 & 75 & 60 \\ 30 & 60 & 50 \\ 120 & 80 & 10 \end{bmatrix},$$

then we conclude, for example, that it costs \$80 to smelt a ton of ore from the mine at Rouyn in the smelter at Falconbridge. Clearly, the 3×3 matrix whose entries are the *sums* of the corresponding entries of T and S is the matrix of *total* refining costs per ton. We shall see in the next section that there are good reasons for calling this matrix $T + S$. ❑

This type of example is of interest because it belongs to a large class of problems known as **transportation problems**. We illustrate with the same example by further supposing that the mines have daily productions and the smelters have daily capacities as given in the table below.

Location	*Mine Production* (tons)	*Smelter Capacity* (tons)
Sudbury	1000	2000
Falconbridge	3000	1500
Rouyn	2000	2500

Then the transportation problem for the mining company is to minimize the total cost of refining the production of its mines subject to the conditions imposed by the mine productions and smelter capacities as given in the table.

Special methods (known as transportation algorithms) are available to solve this type of problem, but they are outside the scope of this book. It is of interest to note that transportation problems themselves belong to an even larger class called **linear programming** problems (LP for short). These problems are concerned with the *maximization* or *minimization* of a linear function of several variables subject to a set of linear equations and/or inequalities in those variables. The equations and/or inequalities are often called **constraints**. (A discussion of LP problems and their solution is provided in Appendix C.)

In the mining example, we denote by x_{ij} the number of tons (per day) that the company should ship from mine i to smelter j. The reader should verify that the total cost function is

$$\begin{aligned} x_0 = {} & (0 + 50)x_{11} + (2 + 75)x_{12} + (6 + 60)x_{13} + (2 + 30)x_{21} + (0 + 60)x_{22} + \\ & (3 + 50)x_{23} + (6 + 120)x_{31} + (3 + 80)x_{32} + (0 + 10)x_{33}. \end{aligned}$$

The problem for the company then is to minimize x_0 subject to the mine productions and smelter capacities. These result in six constraints as follows. First, there are three relating to the production of the mines. For example, the mine at Sudbury produces 1000 tons per day, and this must equal $x_{11} + x_{12} + x_{13}$, which is the total amount shipped from Sudbury each day. Similarly, there are three inequalities relating to the capacities of the three smelters. We leave the completion of the formulation as an exercise for the reader (question 4, Exercises 2-3).

Matrices may also be associated with **graphs**. Here a graph is a collection of line segments (in R^2 or R^3) called **edges**. Every edge must have two distinct endpoints called **vertices**, with the property that if two edges meet, they do so only at their vertices. (This is usually called a geometrical graph in the literature.) The graph is called a **directed graph** or **digraph** if each edge has been labelled with a direction by means of an arrowhead, as in Figures 2-4 and 2-5.

There are numerous situations where it is useful to represent relationships among objects by a graph. For example, the graph in Figure 2-4 could represent the flow of information from a sender at point 1 to a receiver at point 7. The arrows (edges) represent communication links via the various subsidiary points 2, 3, 4, 5, and 6; one could think of a system of microwave transmitters and receivers. In this context, the graph is called a **communications network**. We shall return to this term later in Example 2-14.

FIGURE 2-4

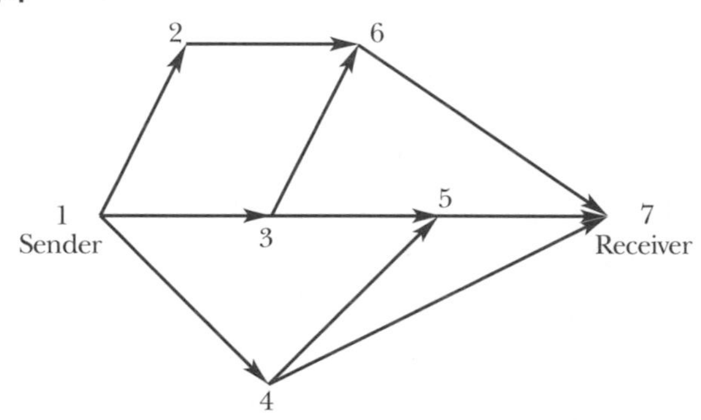

Alternatively, the arrows could represent activities in a complex project and the points of intersection their starting or finishing times. Then the graph displays the hierarchy of activities in the project between start (1) and completion (7). As such, the graph is an example of a **PERT** network, which stands for "**P**roject **E**valuation and **R**eview **T**echnique".

There are several ways of associating matrices with graphs (both directed and undirected). We consider two of them in the next example.

EXAMPLE 2-14

Suppose a directed graph has v vertices and e edges. It is common to label the vertices with positive integers and the edges with letters as in Figure 2-5. Then the **incidence matrix** of the graph is the $v \times e$ matrix with entries defined by

$$a_{ij} = \begin{cases} +1, \text{ if edge } j \text{ flows out of vertex } i, \\ -1, \text{ if edge } j \text{ flows into vertex } i, \\ 0, \text{ otherwise.} \end{cases}$$

FIGURE 2-5

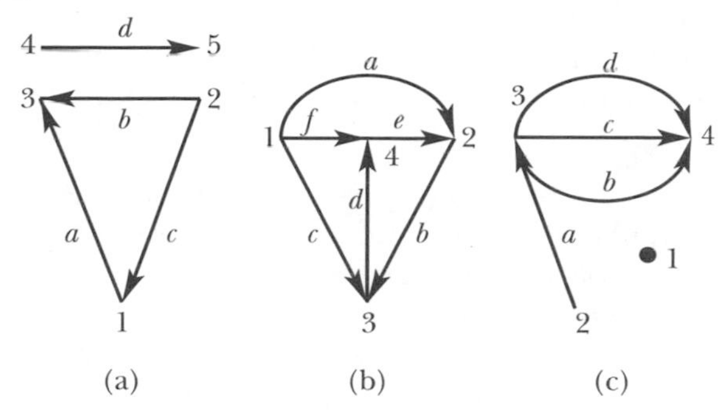

Thus, the matrices for the graphs in Figure 2-5 are (columns enumerate edges and rows enumerate vertices):

$$\textbf{(a)}\begin{bmatrix} 1 & 0 & -1 & 0 \\ 0 & 1 & 1 & 0 \\ -1 & -1 & 0 & 0 \\ 0 & 0 & 0 & 1 \\ 0 & 0 & 0 & -1 \end{bmatrix} \quad \textbf{(b)}\begin{bmatrix} 1 & 0 & 1 & 0 & 0 & 1 \\ -1 & 1 & 0 & 0 & -1 & 0 \\ 0 & -1 & -1 & 1 & 0 & 0 \\ 0 & 0 & 0 & -1 & 1 & -1 \end{bmatrix},$$

and

$$\textbf{(c)}\begin{bmatrix} 0 & 0 & 0 & 0 \\ 1 & 0 & 0 & 0 \\ -1 & 1 & 1 & 1 \\ 0 & -1 & -1 & -1 \end{bmatrix}.$$

Observe that since each edge has two vertices and flows out of one into the other, every column in the matrix of each graph has exactly two nonzero entries, namely, $+1$ and -1. Thus, we can see that certain properties of a directed graph are reflected in corresponding properties of its incidence matrix. Some of these are explored further in later sections.

Another matrix that can be associated with a directed graph is its **vertex** or **adjacency matrix**. This is the $v \times v$ matrix with entries v_{ij} defined by $v_{ij} = k$ if there are k edges *from i to j*. Therefore, the vertex matrices of the graphs in Figure 2-5 are

$$\textbf{(a)}\begin{bmatrix} 0 & 0 & 1 & 0 & 0 \\ 1 & 0 & 1 & 0 & 0 \\ 0 & 0 & 0 & 0 & 0 \\ 0 & 0 & 0 & 0 & 1 \\ 0 & 0 & 0 & 0 & 0 \end{bmatrix}, \quad \textbf{(b)}\begin{bmatrix} 0 & 1 & 1 & 1 \\ 0 & 0 & 1 & 0 \\ 0 & 0 & 0 & 1 \\ 0 & 1 & 0 & 0 \end{bmatrix}, \text{ and } \textbf{(c)}\begin{bmatrix} 0 & 0 & 0 & 0 \\ 0 & 0 & 1 & 0 \\ 0 & 0 & 0 & 3 \\ 0 & 0 & 0 & 0 \end{bmatrix}.$$

Notice that the diagonal entries of a vertex matrix are necessarily zero since we defined the edge of a graph as having two distinct vertices (there are no "loops").

Graphs (a) and (b) in Figure 2-5 are examples of communications networks because there is at most one edge from any vertex i to another vertex j. Graph (c) is not, as there are three edges from vertex 3 to vertex 4 (see the entry 3 in row 3, column 4 of its vertex matrix). ❑

One of the many applications of digraphs and their matrices is in the analysis of group relationships in sociology and biology. For example, suppose the leadership of a political party consists of four individuals whom we represent by the vertices 1, 2, 3, and 4 of a graph. Then, if we represent the fact that an individual i in this leadership group influences the opinion of another j by an arrow from i to j, we obtain a directed graph. By inspection of the vertex matrix for this graph, we can determine the "leader" i of the group from the row i having the most nonzero entries, since the "leader" presumably influences the most people. In the vertex matrix for a leadership group modelled by Figure 2-5(b), for example, we see that individual 1 is the leader, as there are 3 entries $= 1$ in row 1.

Digraphs and vertex matrices can be used also to represent the history of the spread of a contagious disease among the individuals in a group: a directed edge from vertex i to vertex j indicates that individual j was exposed to disease-carrying individual i. Thus the number of times an individual was in contact with the disease is represented in the vertex matrix by the sum of the entries in the column for that individual. For example, using the graph of Figure 2-5(b) again to model this situation in a group of four individuals, we see that individual 1 was in contact with no one having the disease (because column 1 of the vertex matrix contains only zeros). But, he spread it to individuals 2, 3, and 4. Also, for example, these latter individuals were each in contact with two people having the disease.

Other applications of graphs and matrices are found in many different disciplines: electrical network theory, communications, logic, and the study of language, to name a few.

EXERCISES 2-3

1. Write the incidence matrices of the directed graphs in Figure 2-6.

FIGURE 2-6

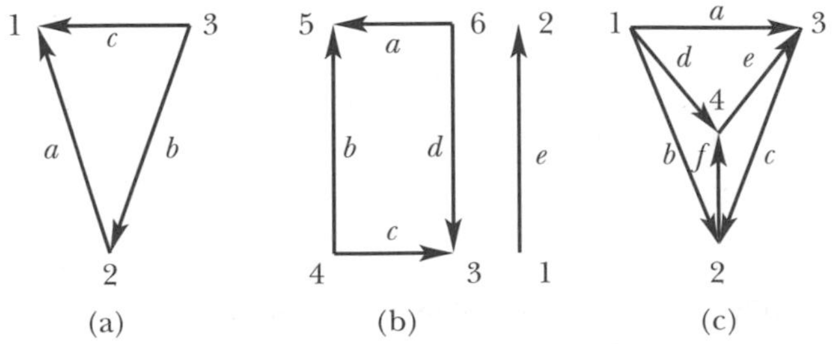

2. Draw the directed graphs whose incidence matrices are given below:

(a) $$\begin{bmatrix} 1 & 0 & 1 & -1 \\ 0 & -1 & -1 & 1 \\ -1 & 1 & 0 & 0 \end{bmatrix}$$

(b) $$\begin{bmatrix} 0 & 1 & 0 & 0 & -1 & 0 \\ -1 & -1 & 1 & 1 & 0 & 0 \\ 1 & 0 & 0 & 0 & 0 & -1 \\ 0 & 0 & -1 & -1 & 1 & 1 \end{bmatrix}$$

3. A manufacturer of glass products converts sand into three qualities of glass by a melting process. Two types of sand are available, type 1 and type 2. The production costs of the different glass types are as follows: One ton of optical quality glass costs \$6 and \$10 to produce from type 1 and type 2 sand respectively. The corresponding costs for plate and bottle glass are \$4 and \$5, and \$2 and \$3 respectively.

(a) Write a 2×3 matrix whose (i,j) entry is the cost per ton in dollars of producing glass j from sand i, where the glass types are numbered in the order described above.

(b) Suppose the firm has available 60 and 40 tons of type 1 and type 2 sand respectively, and that it wants to convert all of it to glass to fill orders for 25 tons of optical glass, 40 tons of plate glass, and 35 tons of bottle glass. Formulate the problem of minimizing the production cost as an LP problem.

(c) Show that the production of 25 tons of optical glass from type 1 sand, 35 and 5 tons of plate glass from types 1 and 2 respectively, and 35 tons of bottle glass from type 2 sand is a solution in the sense that it satisfies the constraints, but that it does not yield the least possible production cost. (Hint: use trial and error to produce another solution satisfying the constraints and yielding a lesser total cost.)

4. Complete the formulation of the transportation problem discussed in Example 2-13.

5. In the game called "Guess a Bill", one player either (i) places a 1, 2, 5, or 10 dollar bill in an envelope, or (ii) places nothing in the envelope. The second player must then write 1, 2, 5, or 10 on the envelope. If the number matches the denomination of the bill, he wins that bill, but if it doesn't, he must pay the first player the value of that bill. If there was no bill in the envelope, the first player must pay the second (the one who is guessing) half the value of the number written on the envelope. Write the 4×5 matrix whose (i,j) entry is the payoff to the player who is guessing when he writes the ith denomination on the envelope and it actually contains the jth (represent 1 dollar by 1, 2 dollars by 2, 5 dollars by 3, and so on, and denote the case of an empty envelope by $j = 5$). (This is an example of a *two-person-zero-sum* game and the matrix defined above is called the *payoff matrix*. The players want to determine strategies of play that, over many games, maximize their average payoffs. It can be shown that the maximization of this problem can also be solved by LP methods.)

6. The incidence matrix of an undirected graph with v vertices and e edges is the $v \times e$ matrix with

$$a_{ij} = \begin{cases} 1, & \text{if vertex } i \text{ lies on edge } j, \\ 0, & \text{otherwise.} \end{cases}$$

(a) Write the incidence matrices for the graphs in Figures 2-5 and 2-6 regarded as undirected graphs.

(b) What can be said about the sum of the entries in any column of the incidence matrix of an undirected graph?

7. Find the vertex matrices of the directed graphs in Figure 2-6.

8. **(a)** Which of the following matrices could be vertex matrices for a communications network?

(i) $\begin{bmatrix} 0 & 0 & 0 & 0 \\ 1 & 0 & 1 & 0 \\ 0 & 1 & 0 & 0 \\ 0 & 0 & 1 & 0 \end{bmatrix}$ **(ii)** $\begin{bmatrix} 0 & 2 & 1 \\ 0 & 0 & 1 \\ 0 & 0 & 0 \end{bmatrix}$

(b) Without drawing the corresponding graph, answer the following questions about the communications network with vertex matrix V given by

$$V = \begin{bmatrix} 0 & 1 & 1 & 0 & 0 \\ 0 & 0 & 1 & 0 & 0 \\ 1 & 0 & 0 & 1 & 0 \\ 0 & 0 & 1 & 0 & 0 \\ 0 & 1 & 0 & 0 & 0 \end{bmatrix}.$$

(i) How many individuals communicate directly to more than one individual?

(ii) Can every individual communicate directly or indirectly (i.e., through one or more individuals) to every other individual?

(c) Draw a graph for the matrix V above to verify your answers to part (b).

9. The directed graph in Figure 2-7 pertains to traffic flow from Toronto to Montreal through various intervening cities. The edges of the graph are labelled with the capacities (in thousands of cars per hour) for the directions shown. Find the maximum possible flow of cars along the network from Toronto to Montreal, using the "Max Flow-Min Cut Theorem": for any straight line C separating the sender and receiver, consider the sum of the capacities of edges cut by C; the maximum flow is the minimum of such sums for all possible "cuts" C.

FIGURE 2-7

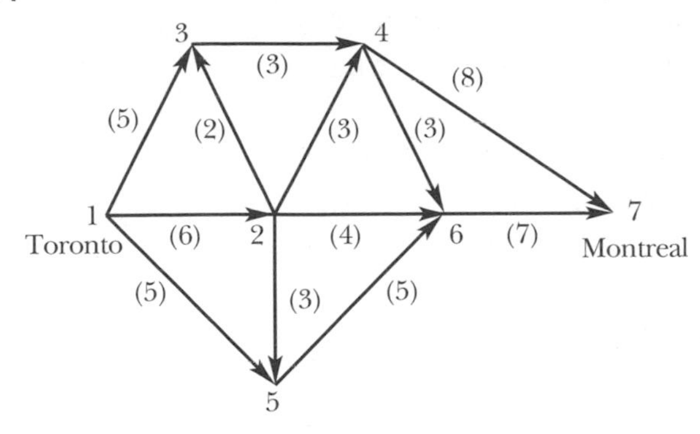

2-4 Matrix Algebra

Matrices possess algebraic properties analogous (but not identical) to those of the set of real numbers: we shall define multiplication and addition of matrices, as well as an operation called scalar multiplication. This is possible, essentially, because of related properties of systems of linear equations. We shall use these properties to motivate the definitions.

DEFINITION 2-6

If A is an $m \times n$ matrix and B is an $m \times 1$ matrix, we denote the system of linear equations S with augmented matrix $(A{:}B)$ by $AX = B$. The symbol X, for the moment, is there to remind us of the variables $x_1, \ldots, x_n$. This notation suggests that we may interpret a *system* of linear equations as a *single* equation relating two $m \times 1$ matrices AX and B, where AX is the $m \times 1$ matrix

$$AX = \begin{bmatrix} a_{11}x_1 & + \ldots + & a_{1n}x_n \\ a_{21}x_1 & + \ldots + & a_{2n}x_n \\ \ldots & \ldots & \ldots \\ a_{m1}x_1 & + \ldots + & a_{mn}x_n \end{bmatrix} \tag{2-4}$$

Of course, the entries of AX are not numbers, so AX may be thought of as a "matrix variable".

To complete this idea, we must make the next definition.

DEFINITION 2-7

Two $m \times n$ matrices C and D are equal, denoted $C = D$, iff $c_{ij} = d_{ij}$ for all i and j, i.e., they have equal entries in corresponding positions. If C and D are of different orders, then $C \neq D$.

So, to say $AX = B$ (as a matrix equation) means precisely that all equations in the system are true statements: the equality of the ith entries of these matrices is exactly the ith equation of the system S. In other words, *$AX = B$ is a matrix equation equivalent to the system S.*

EXAMPLE 2-15

Consider the system of equations

$$\begin{aligned} 2x_1 - 3x_2 &= -1 \\ x_1 + \ \ x_2 &= 2. \end{aligned}$$

The equivalent matrix equation is

$$\begin{bmatrix} 2x_1 & - & 3x_2 \\ x_1 & + & x_2 \end{bmatrix} = \begin{bmatrix} -1 \\ 2 \end{bmatrix}$$

which is an equation stating the equality of two matrices of order 2×1. It is true precisely for those x_1 and x_2 that are solutions of the system. ❑

In fact, more is implicit in the notation $AX = B$. For, AX suggests that this $m \times 1$ matrix may be regarded as a "product" of A and X, provided we can identify X as a matrix. The most natural choice for representing X is

$$X = \begin{bmatrix} x_1 \\ x_2 \\ \cdots \\ x_n \end{bmatrix}$$

which is an $n \times 1$ matrix of variables. We then see that the "product" of the $m \times n$ matrix A and the $n \times 1$ matrix X is to be the $m \times 1$ matrix AX with entries as defined by equation 2-4. It is important to note that the entry in row i of AX is computed using *only* row i of A, as well as X. It is, from 2-4, $a_{i1}x_1 + a_{i2}x_2 + \ldots + a_{in}x_n$. This suggests the next definition.

DEFINITION 2-8

If $A = (a_i)$ is a $1 \times n$ matrix (called a row vector) and $B = (b_i)$ is $n \times 1$ (called a column vector), their **product** is the number $AB = a_1b_1 + \ldots + a_nb_n$.

EXAMPLE 2-16

If

$$A = \begin{bmatrix} 1 & 2 & -3 \end{bmatrix} \text{ and } B = \begin{bmatrix} 4 \\ -1 \\ 2 \end{bmatrix},$$

$AB = (1)(4) + (2)(-1) + (-3)(2) = -4.$
Note that BA (in that order) is not defined as yet, but will be shortly. ❑

With Definition 2-8, we can now interpret the product AX as the $m \times 1$ matrix whose entries are the products of row vectors (the rows of A) and the column vector X. AX is called the **product** of A and X. And, in regard to the equation $AX = B$, we can say that the matrix

$$X = \begin{bmatrix} x_1 \\ x_2 \\ \cdots \\ x_n \end{bmatrix}$$

is a **solution** of this matrix equation iff the n-tuple $(x_1, ..., x_n)$ is a solution of the original system of linear equations (2-3). This makes it natural to *identify* (regard as the same) an n-tuple $(x_1, ..., x_n)$ and its corresponding matrix above and we will use both notations interchangeably as appropriate.

The product AX so defined involves a matrix X having only one column, but it suggests the following definition.

DEFINITION 2-9

Let A be an $m \times n$ matrix and C be an $n \times p$ matrix. Their **product** AC is the $m \times p$ matrix whose (i,j) entry, $(AC)_{ij}$, is the product of the ith row, $A_{[i]}$, of A and the jth column, $C^{[j]}$, of C:

$$\begin{aligned}(AC)_{ij} &= A_{[i]}C^{[j]} = a_{i1}c_{1j} + a_{i2}c_{2j} + \ldots + a_{in}c_{nj} \\ &= \sum_{k=1}^{n} a_{ik}c_{kj}.\end{aligned}$$

Observe that this definition is possible because *the number of columns of A equals the number of rows of C*. For, this fact means that the rows, $A_{[i]}$, of A are row vectors with the same number of entries as the columns, $C^{[j]}$ (column vectors), of C, so that we are merely applying Definition 2-8 to obtain the entries of AC.

EXAMPLE 2-17

(a) If

$$A = \begin{bmatrix} 1 & 2 \\ 3 & 4 \end{bmatrix}, 2 \times 2, \text{ and } C = \begin{bmatrix} 5 \\ 6 \end{bmatrix}, 2 \times 1,$$

then AC is 2×1:

$$AC = \begin{bmatrix} [1 \quad 2]\begin{bmatrix} 5 \\ 6 \end{bmatrix} \\ \\ [3 \quad 4]\begin{bmatrix} 5 \\ 6 \end{bmatrix} \end{bmatrix} = \begin{bmatrix} 5+12 \\ 15+24 \end{bmatrix} = \begin{bmatrix} 17 \\ 39 \end{bmatrix}.$$

Notice that CA is not defined: the number of columns of C is 1, different from the number of rows of A. However, if

$$D = [-3 \quad 2],$$

then both DC and CD are defined (but not equal!):

$$DC = [-3 \quad 2]\begin{bmatrix} 5 \\ 6 \end{bmatrix} = -3 \quad \text{(a number, or } 1 \times 1 \text{ matrix)}$$

$$CD = \begin{bmatrix} 5 \\ 6 \end{bmatrix}[-3 \quad 2] = \begin{bmatrix} (5)(-3) & (5)(2) \\ (6)(-3) & (6)(2) \end{bmatrix}$$

$$= \begin{bmatrix} -15 & 10 \\ -18 & 12 \end{bmatrix} \quad \text{(a } 2 \times 2 \text{ matrix)}$$

(b) If

$$A = \begin{bmatrix} 2 & -1 \\ 1 & 0 \\ -3 & 4 \end{bmatrix} \text{ and } C = \begin{bmatrix} 1 & -5 & -2 \\ 3 & 0 & 4 \end{bmatrix},$$

then,

$$AC = \begin{bmatrix} 2 & -1 \\ 1 & 0 \\ -3 & 4 \end{bmatrix}\begin{bmatrix} 1 & -5 & -2 \\ 3 & 0 & 4 \end{bmatrix}$$

$$= \begin{bmatrix} [2 \quad -1]\begin{bmatrix} 1 \\ 3 \end{bmatrix} & [2 \quad -1]\begin{bmatrix} -5 \\ 0 \end{bmatrix} & [2 \quad -1]\begin{bmatrix} -2 \\ 4 \end{bmatrix} \\ [1 \quad 0]\begin{bmatrix} 1 \\ 3 \end{bmatrix} & [1 \quad 0]\begin{bmatrix} -5 \\ 0 \end{bmatrix} & [1 \quad 0]\begin{bmatrix} -2 \\ 4 \end{bmatrix} \\ [-3 \quad 4]\begin{bmatrix} 1 \\ 3 \end{bmatrix} & [-3 \quad 4]\begin{bmatrix} -5 \\ 0 \end{bmatrix} & [-3 \quad 4]\begin{bmatrix} -2 \\ 4 \end{bmatrix} \end{bmatrix}$$

$$= \begin{bmatrix} 2 - 3 & -10 + 0 & -4 - 4 \\ 1 + 0 & -5 + 0 & -2 + 0 \\ -3 + 12 & 15 + 0 & 6 + 16 \end{bmatrix} = \begin{bmatrix} -1 & -10 & -8 \\ 1 & -5 & -2 \\ 9 & 15 & 22 \end{bmatrix}.$$

In this case, the product CA is also defined and equals

$$\begin{bmatrix} 1 & -5 & -2 \\ 3 & 0 & 4 \end{bmatrix}\begin{bmatrix} 2 & -1 \\ 1 & 0 \\ -3 & 4 \end{bmatrix} = \begin{bmatrix} 2 - 5 + 6 & -1 + 0 - 8 \\ 6 + 0 - 12 & -3 + 0 + 16 \end{bmatrix}$$

$$= \begin{bmatrix} 3 & -9 \\ -6 & 13 \end{bmatrix}.$$

This example shows that $AC \neq CA$ even when both products are defined. We say that multiplication of matrices is *not* a commutative operation. In this case, AC and CA are not even matrices of the same order.

(c) Let

$$C = \begin{bmatrix} 1 & 0 \\ -1 & 1 \end{bmatrix} \text{ and } D = \begin{bmatrix} -1 & 1 \\ 0 & 0 \end{bmatrix}.$$

Then CD and DC are both defined and have the same order, but

$$CD = \begin{bmatrix} -1 & 1 \\ 1 & -1 \end{bmatrix}, \text{ while } DC = \begin{bmatrix} -2 & 1 \\ 0 & 0 \end{bmatrix}.$$

Obviously $CD \neq DC$. ❑

In the light of these examples, it should be clear that when A and B are **square** (i.e., $n \times n$) **matrices** of the same order, the products AB and BA are always defined, but not necessarily equal. In particular, the products $AA = A^2$, $AAA = A^3$, etc., are defined. In this regard, the reader should look at questions 3 and 4 at the end of this section.

EXAMPLE 2-18

The cost of refining a ton of ore from a certain mine in three available smelters may be summarized in a 1×3 cost matrix C given by

$$C = [50 \quad 75 \quad 60]$$

This means, for example, that the cost of refining a ton of the ore at smelter 2 is \$75 (see Example 2-13). So, if the numbers of tons smelted at each refinery are 200, 400, and 300 respectively, the total cost of refining the ore is

$$50\times 200 + 75\times 400 + 60\times 300 = \begin{bmatrix}50 & 75 & 60\end{bmatrix}\begin{bmatrix}200\\400\\300\end{bmatrix}$$

$= CQ$ (= \$58 000), where Q denotes the 3×1 matrix of quantities of ore on the right. Now, if S is the smelting cost matrix

$$S = \begin{bmatrix}50 & 75 & 60\\30 & 60 & 50\\120 & 80 & 10\end{bmatrix}$$

(as in Example 2-13) whose (i,j) entry is the cost in dollars per ton of smelting ore from mine i at smelter j, then the matrix product

$$SQ = \begin{bmatrix}50 & 75 & 60\\30 & 60 & 50\\120 & 80 & 10\end{bmatrix}\begin{bmatrix}200\\400\\300\end{bmatrix} = \begin{bmatrix}58\,000\\45\,000\\59\,000\end{bmatrix}$$

(a 3×1 matrix) contains as its entries the costs for the three mines of refining identical shipments of 200, 400, and 300 tons of ore at the smelters. ❑

Having introduced an operation on matrices which is reminiscent of multiplication of numbers, it is natural to ask whether there are further operations on matrices analogous to other algebraic properties of numbers. In fact, we saw in Example 2-13 that it might be useful and natural to speak of the sum of matrices. The next example provides further motivation in terms of systems of linear equations.

EXAMPLE 2-19

Consider the two linear equations

$$a_1x_1 + b_1x_2 + c_1x_3 = d_1 \tag{2-5}$$

and

$$a_2x_1 + b_2x_2 + c_2x_3 = d_2. \tag{2-6}$$

We may regard *each* of these as a system of *one* linear equation in three variables, with augmented matrices

$$\begin{bmatrix}a_1 & b_1 & c_1 & d_1\end{bmatrix} \text{ and } \begin{bmatrix}a_2 & b_2 & c_2 & d_2\end{bmatrix}$$

respectively. Now, as we saw in the proof of Theorem 2-1, if

$$X = \begin{bmatrix}x_1\\x_2\\x_3\end{bmatrix}$$

is a solution of *both* equations, it is also a solution of their *sum*, which is the equation with 1×4 augmented matrix

$$[a_1 + a_2 \quad b_1 + b_2 \quad c_1 + c_2 \quad d_1 + d_2].$$

It is natural to call this matrix the **sum** of the matrices above.

Similarly, X is also a solution of any **multiple**

$$ka_1x_1 + kb_1x_2 + kc_1x_3 = kd_1$$

of equation 2-5 when it is a solution of 2-5 itself. The augmented matrix of the multiple of equation 2-5 is

$$[ka_1 \quad kb_1 \quad kc_1 \quad kd_1],$$

which is a **scalar multiple** of the augmented matrix of 2-5. ❑

More generally,

DEFINITION 2-10

If A and B are $m \times n$ matrices, their **sum** $A + B$ is the $m \times n$ matrix with entries $a_{ij} + b_{ij}$ obtained by adding corresponding entries of A and B. If k is any number, the **scalar multiple** kA is the $m \times n$ matrix obtained by multiplying each entry of A by k. When a number k is used in this way, it is referred to as a **scalar**.

EXAMPLE 2-20

(a) $$\begin{bmatrix} -1 & 0 \\ 2 & 3 \end{bmatrix} + \begin{bmatrix} 2 & -2 \\ 4 & -1 \end{bmatrix} = \begin{bmatrix} 1 & -2 \\ 6 & 2 \end{bmatrix}$$

(b) $$(-3)\begin{bmatrix} -1 & 0 \\ 2 & 3 \end{bmatrix} = \begin{bmatrix} 3 & 0 \\ -6 & -9 \end{bmatrix}$$

(c) $$\begin{bmatrix} 2 & -4 \\ 5 & -1 \\ 0 & 3 \end{bmatrix} + \begin{bmatrix} 1 & 2 \\ 3 & 4 \\ 5 & 6 \end{bmatrix} = \begin{bmatrix} 3 & -2 \\ 8 & 3 \\ 5 & 9 \end{bmatrix}$$

(d) $$\begin{bmatrix} 4 & 5 \\ -1 & 2 \end{bmatrix} + \begin{bmatrix} 1 & 2 \\ 3 & 4 \\ 5 & 6 \end{bmatrix}$$

is not defined since the matrices in question don't have the same order. ❑

EXAMPLE 2-21

The transportation costs of shipping certain manufactured goods from two warehouses to four retail outlets are given as the entries in the following 2×4 matrix T (in dollars):

$$T = \begin{bmatrix} 200 & 390 & 410 & 136 \\ 180 & 90 & 250 & 88 \end{bmatrix}.$$

If the cost of transportation is increased by 10%, the cost matrix becomes the scalar multiple

$$(1.1)T = \begin{bmatrix} 220 & 429 & 451 & 149.6 \\ 198 & 99 & 275 & 96.8 \end{bmatrix}. \square$$

We summarize the essential properties of **matrix algebra** in the following "Omnibus" theorem. (Compare this with Theorem 1-3.)

THEOREM 2-4

Let $A, B, C, \ldots$ be matrices and $a, b, c, \ldots$ be numbers. Then, provided that the matrices in question are defined, the following hold:

(1) $A + B = B + A$ (**commutative law** of addition).

(2) $(A + B) + C = A + (B + C)$ (**associative law** of addition).

(3) For all m, n positive integers, there is an $m \times n$ matrix O defined as having all entries 0 and the property that $A + O = A$ for any matrix A. Furthermore, O is the only matrix with this property and is called the **zero matrix**.

(4) Given an $m \times n$ matrix A, there is an unique matrix denoted by $-A = (-a_{ij})$ with the property $A + (-A) = O$. It is called the **additive inverse** of A.

(5) $a(A + B) = aA + aB$.
(6) $(a + b)A = aA + bA$.
(7) $(ab)A = a(bA)$.
(5)–(7): (**distributive laws**)

(8) $1A = A$ where 1 is the number one.

(9) $(AB)C = A(BC)$ (**associative law** of multiplication).

(10) $A(B + C) = AB + AC$ and $(A + B)C = AC + BC$ (**distributive laws** of multiplication).

(11) $a(AB) = (aA)B = A(aB)$ (**distributive law** of scalar multiplication over matrix multiplication).

(12) For each positive integer n, there is an $n \times n$ matrix I called the $n \times n$ (multiplicative) **identity** such that $AI = IA = A$ for any $n \times n$ matrix A. I is the matrix with 1's on the main diagonal and 0's elsewhere.

Many of the proofs are left to the reader. In any case, the reader may wish to skip them for the time being and proceed to the examples.

PROOF Properties (1), (2), (3), (5), (6), (7), (8), and (11) are consequences of analogous properties of numbers, and of the fact that addition and scalar multiplication of matrices (when defined) are defined in terms of addition and multiplication of individual entries of the matrices in question. We refer the reader to Example 2-22 which illustrates these properties. The proofs for (4), (9), (10), and (12) are provided below.

(4) By definition,

$$A + (-A) = (a_{ij} + (-a_{ij})) = (0) = O.$$

Notice that if B were another matrix such that $A + B = O$, then, adding $-A$ to both sides yields

$$-A + (A + B) = -A.$$

Then, using (2), (1), (4), and (3), the left side becomes

$$(-A + A) + B = (A + (-A)) + B = O + B - B.$$

It follows that $B = -A$ and $-A$ is unique.

(9) Let A be $m \times n$, B be $n \times p$, C be $p \times q$ so that the products $(AB)C$ and $A(BC)$ are defined and are of the order $m \times q$. Then

$$[(AB)C]_{ij} = \sum_{k=1}^{p} (AB)_{ik}C_{kj} = \sum_{k=1}^{p} (\sum_{r=1}^{n} A_{ir}B_{rk})C_{kj} = \sum_{r=1}^{n} A_{ir} (\sum_{k=1}^{p} B_{rk}C_{kj})$$

$$= \sum_{r=1}^{n} A_{ir}(BC)_{rj} = [A(BC)]_{ij},$$

that is, $(AB)C = A(BC)$ as required. Readers who are not comfortable with the summation notation should write out the sums in the proof in full for, say, $n = p = 2$.

(10) We prove the first of these and leave the other for the reader. Let A, B be $m \times n$ and C be $n \times p$. Then, using the corresponding property of numbers, we obtain

$$[(A + B)C]_{ij} = \sum_{k=1}^{n} (A + B)_{ik}C_{kj} = \sum_{k=1}^{n} (A_{ij}C_{kj} + B_{ik}C_{kj})$$

$$= \sum_{k=1}^{n} A_{ik}C_{kj} + \sum_{k=1}^{n} B_{ik}C_{kj} = [AC]_{ij} + [BC]_{ij} = [AC + BC]_{ij}.$$

(12) This property applies only to $n \times n$ matrices. As indicated, the matrix I is defined to have the form

$$I = \begin{bmatrix} 1 & 0 & 0 & \dots & 0 \\ 0 & 1 & 0 & \dots & 0 \\ \dots & \dots & \dots & \dots & \dots \\ 0 & 0 & \dots & 1 & 0 \\ 0 & 0 & \dots & 0 & 1 \end{bmatrix}$$

For $n = 2$ and 3, the matrix I is therefore

$$I = \begin{bmatrix} 1 & 0 \\ 0 & 1 \end{bmatrix}, \quad I = \begin{bmatrix} 1 & 0 & 0 \\ 0 & 1 & 0 \\ 0 & 0 & 1 \end{bmatrix}.$$

The entries of I are frequently denoted by δ_{ij}, where

$$\delta_{ij} = \begin{cases} 1, \text{if } i = j, \\ 0, \text{if } i \neq j. \end{cases}$$

To establish the property that I is a multiplicative identity, consider AI:

$$[AI]_{ij} = \sum_{k=1}^{n} A_{ik} I_{kj} = \sum_{k=1}^{n} A_{ik} \delta_{kj} = A_{ij},$$

since $\delta_{kj} = 0$ when $k \neq j$. Thus $AI = A$. Similarly, we can show $IA = A$. In fact, it can be shown that I is the only matrix with this property. This is left for the reader. ■

EXAMPLE 2-22 (Matrix Algebra)

(a) Commutative law of addition: $A + B = B + A$

$$\begin{bmatrix} 1 & -1 & 0 \\ 2 & 3 & 1 \end{bmatrix} + \begin{bmatrix} 2 & 2 & 4 \\ -2 & 0 & 1 \end{bmatrix} = \begin{bmatrix} 3 & 1 & 4 \\ 0 & 3 & 2 \end{bmatrix}$$

$$\begin{bmatrix} 2 & 2 & 4 \\ -2 & 0 & 1 \end{bmatrix} + \begin{bmatrix} 1 & -1 & 0 \\ 2 & 3 & 1 \end{bmatrix} = \begin{bmatrix} 3 & 1 & 4 \\ 0 & 3 & 2 \end{bmatrix}$$

(b) Associative law of addition: $(A+B) + C = A + (B+C)$

$$\left\{\begin{bmatrix} 1 & 2 \\ 3 & 4 \end{bmatrix} + \begin{bmatrix} 5 & 6 \\ 7 & 8 \end{bmatrix}\right\} + \begin{bmatrix} 9 & 10 \\ 11 & 12 \end{bmatrix}$$

$$= \begin{bmatrix} 6 & 8 \\ 10 & 12 \end{bmatrix} + \begin{bmatrix} 9 & 10 \\ 11 & 12 \end{bmatrix} = \begin{bmatrix} 15 & 18 \\ 21 & 24 \end{bmatrix}$$

$$\begin{bmatrix} 1 & 2 \\ 3 & 4 \end{bmatrix} + \{\begin{bmatrix} 5 & 6 \\ 7 & 8 \end{bmatrix} + \begin{bmatrix} 9 & 10 \\ 11 & 12 \end{bmatrix}\}$$

$$= \begin{bmatrix} 1 & 2 \\ 3 & 4 \end{bmatrix} + \begin{bmatrix} 14 & 16 \\ 18 & 20 \end{bmatrix} = \begin{bmatrix} 15 & 18 \\ 21 & 24 \end{bmatrix}$$

(c) Additive inverse: $A + (-A) = O$

$$\text{If } A = \begin{bmatrix} 1 & 2 \\ 3 & 4 \\ 5 & 6 \end{bmatrix}, -A = \begin{bmatrix} -1 & -2 \\ -3 & -4 \\ -5 & -6 \end{bmatrix},$$

$$\text{then } A + (-A) = \begin{bmatrix} 1-1 & 2-2 \\ 3-3 & 4-4 \\ 5-5 & 6-6 \end{bmatrix} = O.$$

(d) Distributive law: $(a+b)A = aA + bA$

$$(2+3)\begin{bmatrix} 1 & -1 & 2 \\ 0 & 3 & 4 \end{bmatrix} = 5\begin{bmatrix} 1 & -1 & 2 \\ 0 & 3 & 4 \end{bmatrix} = \begin{bmatrix} 5 & -5 & 10 \\ 0 & 15 & 20 \end{bmatrix}$$

$$2\begin{bmatrix} 1 & -1 & 2 \\ 0 & 3 & 4 \end{bmatrix} + 3\begin{bmatrix} 1 & -1 & 2 \\ 0 & 3 & 4 \end{bmatrix}$$

$$= \begin{bmatrix} 2 & -2 & 4 \\ 0 & 6 & 8 \end{bmatrix} + \begin{bmatrix} 3 & -3 & 6 \\ 0 & 9 & 12 \end{bmatrix} = \begin{bmatrix} 5 & -5 & 10 \\ 0 & 15 & 20 \end{bmatrix}$$

(e) Distributive law: $(ab)A = a(bA)$

$$(-2 \times 3)\begin{bmatrix} 1 & 2 \\ 3 & 4 \end{bmatrix} = (-6)\begin{bmatrix} 1 & 2 \\ 3 & 4 \end{bmatrix} = \begin{bmatrix} -6 & -12 \\ -18 & -24 \end{bmatrix}$$

$$(-2)\{3\begin{bmatrix} 1 & 2 \\ 3 & 4 \end{bmatrix}\} = (-2)\begin{bmatrix} 3 & 6 \\ 9 & 12 \end{bmatrix} = \begin{bmatrix} -6 & -12 \\ -18 & -24 \end{bmatrix}$$

(f) Associative law of multiplication: $(AB)C = A(BC)$

$$\{\begin{bmatrix} 1 & 2 \\ 3 & 4 \end{bmatrix}\begin{bmatrix} 5 & -2 & 3 \\ 1 & 0 & 2 \end{bmatrix}\}\begin{bmatrix} 1 & -1 \\ 2 & 3 \\ 0 & 1 \end{bmatrix}$$

$$= \begin{bmatrix} 7 & -2 & 7 \\ 19 & -6 & 17 \end{bmatrix}\begin{bmatrix} 1 & -1 \\ 2 & 3 \\ 0 & 1 \end{bmatrix} = \begin{bmatrix} 3 & -6 \\ 7 & -20 \end{bmatrix}$$

$$\begin{bmatrix} 1 & 2 \\ 3 & 4 \end{bmatrix}\{\begin{bmatrix} 5 & -2 & 3 \\ 1 & 0 & 2 \end{bmatrix}\begin{bmatrix} 1 & -1 \\ 2 & 3 \\ 0 & 1 \end{bmatrix}\} = \begin{bmatrix} 1 & 2 \\ 3 & 4 \end{bmatrix}\begin{bmatrix} 1 & -8 \\ 1 & 1 \end{bmatrix} = \begin{bmatrix} 3 & -6 \\ 7 & -20 \end{bmatrix}$$

(g) Distributive law of multiplication: $A(B \pm C) = AB \pm AC$

$$\begin{bmatrix} 1 & -1 \\ 2 & -3 \\ 0 & 1 \end{bmatrix} \left\{ \begin{bmatrix} 1 & 0 \\ 1 & 1 \end{bmatrix} - \begin{bmatrix} 2 & 3 \\ -3 & 2 \end{bmatrix} \right\}$$

$$= \begin{bmatrix} 1 & -1 \\ 2 & -3 \\ 0 & 1 \end{bmatrix} \begin{bmatrix} -1 & -3 \\ 4 & -1 \end{bmatrix} = \begin{bmatrix} -5 & -2 \\ -14 & -3 \\ 4 & -1 \end{bmatrix}$$

$$\begin{bmatrix} 1 & -1 \\ 2 & -3 \\ 0 & 1 \end{bmatrix} \begin{bmatrix} 1 & 0 \\ 1 & 1 \end{bmatrix} - \begin{bmatrix} 1 & -1 \\ 2 & -3 \\ 0 & 1 \end{bmatrix} \begin{bmatrix} 2 & 3 \\ -3 & 2 \end{bmatrix}$$

$$= \begin{bmatrix} 0 & -1 \\ -1 & -3 \\ 1 & 1 \end{bmatrix} - \begin{bmatrix} 5 & 1 \\ 13 & 0 \\ -3 & 2 \end{bmatrix} = \begin{bmatrix} -5 & -2 \\ -14 & -3 \\ 4 & -1 \end{bmatrix} \square$$

As we shall see in Chapter 4, properties (1) – (8) imply that the set of $m \times n$ matrices is a **vector space**, which is the principal object of interest in this book. The set of $n \times n$ matrices, having the full set of properties listed in Theorem 2-4, is an example of a "linear algebra".

We conclude this section with an operation on matrices which has no analogue in the set of real numbers.

DEFINITION 2-11

If A is an $m \times n$ matrix, its **transpose** is the $n \times m$ matrix A^t with entries given by $A^t_{ij} = A_{ji}$. Therefore, the rows of A^t are precisely the columns of A and vice-versa.

EXAMPLE 2-23

Let A, B, and C be given by

$$A = \begin{bmatrix} 0 & 2 \\ 3 & 1 \end{bmatrix}, B = \begin{bmatrix} 1 & 2 & -1 \\ 2 & 0 & 5 \end{bmatrix}, C = \begin{bmatrix} 4 \\ 3 \\ 5 \end{bmatrix}.$$

Then,

$$A^t = \begin{bmatrix} 0 & 3 \\ 2 & 1 \end{bmatrix}, B^t = \begin{bmatrix} 1 & 2 \\ 2 & 0 \\ -1 & 5 \end{bmatrix}, C^t = [4 \quad 3 \quad 5]. \square$$

The transpose operation can naturally be applied to sums, differences, products, and scalar products of matrices. Its properties in this regard are summarized in the next theorem.

THEOREM 2-5 (Properties of the Transpose Operation)

(1) $(A + B)^t = A^t + B^t$ where A and B are matrices of the same order.

(2) $(kA)^t = k(A^t)$ where A is any matrix and k is a scalar.

(3) $(A^t)^t = A$ for any matrix A.

(4) $(AB)^t = B^t A^t$ provided the product AB is defined.

PROOF We leave (1), (2), and (3) as exercises for the reader, and prove (4) using the summation notation. Let A be $m \times n$ and B be $n \times p$. Then,

$$(AB)^t_{ij} = (AB)_{ji} = \sum_{k=1}^{n} A_{jk}B_{ki} = \sum_{k=1}^{n} B_{ki}A_{jk} = \sum_{k=1}^{n} B^t_{ik}A^t_{kj} = (B^tA^t)_{ij}. \blacksquare$$

Notice that property (3) simply says that "the transpose of the transpose is the original matrix", and (4) says "the transpose of a product is the product of the transposes in the *reverse* order".

EXAMPLE 2-24

Let A and B be given by

$$A = \begin{bmatrix} 1 & 2 \\ -1 & 0 \end{bmatrix}, B = \begin{bmatrix} 2 & -1 & 0 \\ 0 & 3 & 1 \end{bmatrix}.$$

Then,

$$AB = \begin{bmatrix} 2 & 5 & 2 \\ -2 & 1 & 0 \end{bmatrix}, (AB)^t = \begin{bmatrix} 2 & -2 \\ 5 & 1 \\ 2 & 0 \end{bmatrix}, B^t = \begin{bmatrix} 2 & 0 \\ -1 & 3 \\ 0 & 1 \end{bmatrix}, A^t = \begin{bmatrix} 1 & -1 \\ 2 & 0 \end{bmatrix},$$

so we see that $(AB)^t = B^tA^t$. ❑

EXAMPLE 2-25

In Example 2-13, we saw an example of a symmetric matrix T: it had the property that $T^t = T$, or, T is equal to its own transpose. Such matrices arise frequently and are important in mathematics and its applications. For example, it can be shown that the graph of a quadratic equation in x and y such as $11x^2 - 6xy + 3y^2 = 18$ is a curve called a *conic section*. In this case, it is an ellipse with centre at the origin. We can represent this equation as a matrix product involving a symmetric matrix by writing it as

$$[x \quad y]\begin{bmatrix} 11 & -3 \\ -3 & 3 \end{bmatrix}\begin{bmatrix} x \\ y \end{bmatrix} = 18.$$

This application is discussed further in Chapter 7. ❑

EXERCISES 2-4

1. Compute the following products of row and column vectors.

(a) $[-1 \quad 2]\begin{bmatrix} 3 \\ -2 \end{bmatrix}$ **(b)** $[-2 \quad 0 \quad 3]\begin{bmatrix} -4 \\ 5 \\ -2 \end{bmatrix}$

(c) $[1/2 \quad -1 \quad 2/3 \quad 1/4]\begin{bmatrix} -1 \\ 1/3 \\ -6 \\ 4/5 \end{bmatrix}$

2. Let

$$A = \begin{bmatrix} 2 & -1 \\ 3 & 2 \end{bmatrix}, B = \begin{bmatrix} -1 & 2 & 3 \\ 4 & -2 & 0 \end{bmatrix}, C = \begin{bmatrix} 2 \\ -1 \\ 3 \end{bmatrix},$$

$$D = [-1 \quad 0 \quad 3], E = \begin{bmatrix} 1 & 3 & -1 \\ 0 & 2 & 4 \\ -2 & -1 & 1 \end{bmatrix}.$$

Evaluate, if possible, each of the following expressions, or state why it is not defined.

(a) DC **(b)** AB **(c)** BC **(d)** CD
(e) BE **(f)** $AB + 3BE$ **(g)** $A(BC)$ **(h)** $(AB)C - B(EC)$
(i) $C^t - 2DE$ **(j)** $2DD^t$ **(k)** EB **(l)** $AB - 2E$

3. If A is an $n \times n$ matrix, it is natural to define $A^2 = AA$, $A^3 = A(AA) = (AA)A$, and, in general, $A^n = AA^{n-1}$, $n > 1$. We put $A^0 = I$, the $n \times n$ identity matrix; and $A^1 = A$.

(a) Compute A^2, A^3, and $2A^3 - A^2 + A$, when

$$A = \begin{bmatrix} 2 & -1 \\ 3 & 2 \end{bmatrix}.$$

(b) Show that if

$$A = \begin{bmatrix} 0 & 0 \\ 1 & 0 \end{bmatrix},$$

then $A^2 = O$, the 2×2 zero matrix. (This illustrates another difference between matrix algebra and the arithmetic of real numbers.)

(c) Show that $A^2 - 3A - 4I = O$ when

$$A = \begin{bmatrix} 2 & 3 \\ 2 & 1 \end{bmatrix}.$$

4. Let

$$A = \begin{bmatrix} 0 & 1 \\ 0 & 2 \end{bmatrix}, B = \begin{bmatrix} 1 & 2 \\ 3 & 4 \end{bmatrix}, C = \begin{bmatrix} -2 & 1 \\ 3 & 4 \end{bmatrix}.$$

(a) Show that $AB = AC$. (Since $B \neq C$, this shows that we cannot "cancel" A as in arithmetic.)

(b) Use (a) above to find a matrix $D \neq O$ such that $AD = O$.

5. Let

$$E = \begin{bmatrix} 1 & 0 & 0 \\ 0 & 0 & 1 \\ 0 & 0 & 1 \end{bmatrix}.$$

Show that $E^2 = E$.

6. If

$$A = \begin{bmatrix} 1 & 2 \\ -5 & 0 \end{bmatrix} \text{ and } B = \begin{bmatrix} 2 & -1 & 3 \\ 0 & 3 & 1 \end{bmatrix},$$

show that $(AB)^t = B^t A^t$.

7. Write the following systems of equations in matrix form.

(a)
$$\begin{aligned} 2x_1 - 3x_2 &= 1 \\ -x_1 + 2x_2 &= 0 \\ 3x_1 - x_2 &= 2 \end{aligned}$$

(b)
$$\begin{aligned} 2a + b + 3c + 4d &= 0 \\ 3a - b + 2c &= 3 \\ -2a + b - 4c + 3d &= 2 \end{aligned}$$

8. Given

$$A = \begin{bmatrix} 0 & 1 \\ 0 & 2 \end{bmatrix}, B = \begin{bmatrix} 1 & 2 \\ 3 & 4 \end{bmatrix}, C = \begin{bmatrix} -2 & 1 \\ 3 & 4 \end{bmatrix}, a = 2, b = -3,$$

verify parts (1) through (8) of Theorem 2-4.

9. Consider the two systems of equations

$$\begin{aligned} z_1 &= b_{11}y_1 + b_{12}y_2 \\ z_2 &= b_{21}y_1 + b_{22}y_2 \end{aligned} \quad \text{and} \quad \begin{aligned} y_1 &= a_{11}x_1 + a_{12}x_2 \\ y_2 &= a_{21}x_1 + a_{22}x_2. \end{aligned}$$

By substituting y_1 and y_2 from the second system into the first system, show that $Z = BAX$, where

$$Z = \begin{bmatrix} z_1 \\ z_2 \end{bmatrix}, A = (a_{ij}), B = (b_{ij}), \text{ and } X = \begin{bmatrix} x_1 \\ x_2 \end{bmatrix}.$$

10. In a certain city, the total population remains constant, but each year 20% of the people living in the center move to the suburbs, while 10% of those living in the suburbs move to the center.

(a) Denote by c and s the populations of the center and suburbs respectively at any given time, and by c' and s' the corresponding populations one year later. Obtain a system of two linear equations relating c' and s' to c and s.

(b) Write the system of equations obtained in (a) as a matrix equation $P' = MP$ where M is an appropriate 2×2 matrix and

$$P' = \begin{bmatrix} c' \\ s' \end{bmatrix}, P = \begin{bmatrix} c \\ s \end{bmatrix}.$$

(c) Obtain a matrix equation for the populations

$$P'' = \begin{bmatrix} c'' \\ s'' \end{bmatrix}$$

in the center and suburbs after two years.

(d) Use the result of (c) to obtain the fraction of the population still living in the center of the city after two years.

11. A communications network has vertex matrix A given by

$$A = \begin{bmatrix} 0 & 1 & 1 & 1 \\ 1 & 0 & 0 & 1 \\ 0 & 1 & 0 & 0 \\ 1 & 0 & 1 & 0 \end{bmatrix}.$$

(a) Draw a directed graph for this matrix.

(b) Compute A^2 and verify from the graph that $(A^2)_{ij}$ is the number of two-stage communication links (i.e., passing through exactly *one* other individual) from individual i to individual j.

(c) Prove the observation made in (b) in general using the fact that $(A^2)_{ij} = a_{i1}a_{1j} + a_{i2}a_{2j} + \ldots + a_{iv}a_{vj}$, where v is the number of vertices of the graph.

(d) Compute A^3 and interpret its entries as in (b).

(e) Generalize to an interpretation of A^n, where n is a positive integer.

12. If A is the vertex matrix of a communications network, and $k > 0$ is the least number so that every individual has a communication link with at most k stages to every other individual, show that k is the smallest positive integer such that the matrix $A + A^2 + \ldots + A^k$ has no zero off-diagonal entries.

13. In the analysis of electrical networks, it is frequently convenient to regard a larger network as composed of smaller networks called "black boxes" and joined (cascaded) by wires as shown in Figure 2-8 (other connections are also possible). Each black box is completely characterized (for the purposes of the analysis) by a system of two linear equations relating its two input variables to its two output variables (usually these variables are electrical current and voltage).

FIGURE 2-8

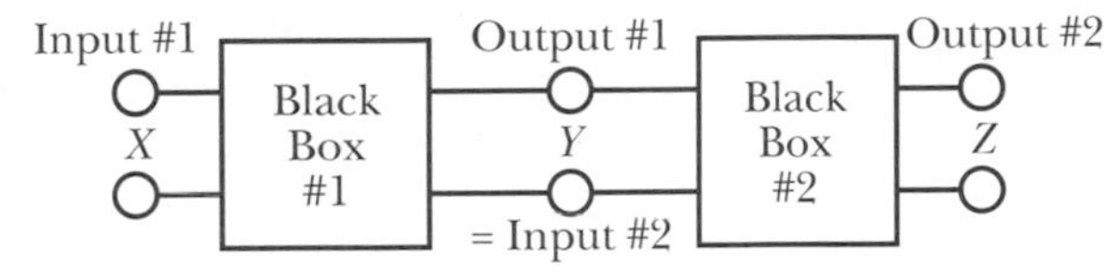

(a) Use the ideas and notation of question 9 to obtain a matrix equation for the output Z of box 2 in terms of the input X to box 1, where

$$Z = \begin{bmatrix} z_1 \\ z_2 \end{bmatrix}, X = \begin{bmatrix} x_1 \\ x_2 \end{bmatrix}.$$

(b) If the black boxes 1 and 2 are characterized by matrices A and B respectively, given by

$$B = \begin{bmatrix} 1 & -2 \\ 3 & 1/2 \end{bmatrix}, A = \begin{bmatrix} 1/2 & -1 \\ 1/2 & 2 \end{bmatrix},$$

find the matrix of the black box "equivalent" to boxes 1 and 2 cascaded as shown.

2-5 Elementary Matrices and Matrix Inversion

We consider again the system of m equations in n unknowns

$$AX = B \tag{2-7}$$

which we now regard as a single matrix equation. It bears a great similarity to the simple linear equation $ax = b$, where a and b are numbers. We can easily solve this linear equation by multiplying both sides by $1/a = a^{-1}$, the reciprocal of a, provided that $a \neq 0$. Then $x = a^{-1} b$.

Suppose we try to approach the matrix equation in the same way. We ask: Is there an $m \times m$ matrix P such that when we multiply both sides of 2-7 on the left by P (recall that order is important in multiplication of matrices!), the resulting equation $PAX = PB$ becomes $X = PB$? If there were, then, we must have

$$X = PAX \text{ for } any\ n\times 1 \text{ matrix } X. \tag{2-8}$$

Note that P here has the same role as $1/a$ in the equation with numbers; we don't denote P by $1/A$ or A^{-1} because we don't as yet have a meaning for this symbol for matrices.

As we shall see, this is *not* in general possible. To begin with, the matrix PAX is $m\times 1$, while X is $n\times 1$, so in order that $PAX = X$, we must first have $m = n$. Suppose then that A is $n\times n$, so $AX = B$ is a system of n equations in the same number, n, of unknowns. Then, by equation 2-8, we must have $PAX = X$ for *any* $n\times 1$ matrix X as a necessary condition for solving $AX = B$ as $X = PB$. The circumstances in which this happens are given in the next result.

THEOREM 2-6

Let C be an $n\times n$ matrix. Then $CX = X$ for all $n\times 1$ matrices X iff $C = I$, the $n\times n$ identity.

PROOF If $C = I$, we have already seen that $CX = IX = X$ for any X. Conversely, suppose $CX = X$ for all such X. We can then make n different "nice" choices for X which will reveal that C must in fact be I:

Choice 1 Put X equal to the 1st column of I,

$$X = \begin{bmatrix} 1 \\ 0 \\ \cdot\cdot \\ \cdot\cdot \\ 0 \end{bmatrix}, CX = \begin{bmatrix} c_{11} & c_{12} & \cdots & c_{1n} \\ c_{21} & c_{22} & \cdots & c_{2n} \\ \cdots & \cdots & \cdots & \cdots \\ c_{n1} & c_{n2} & \cdots & c_{nn} \end{bmatrix} \begin{bmatrix} 1 \\ 0 \\ \cdot\cdot \\ \cdot\cdot \\ 0 \end{bmatrix} = \begin{bmatrix} c_{11} \\ c_{21} \\ \cdots \\ c_{n1} \end{bmatrix},$$

the first column of C. Then, since $CX = X$ for this X, we conclude that the 1st column of C is the 1st column of I.

Choice 2 Put X equal to the 2nd column of I,

$$X = \begin{bmatrix} 0 \\ 1 \\ \cdot\cdot \\ \cdot\cdot \\ 0 \end{bmatrix}, CX = \begin{bmatrix} c_{11} & c_{12} & \cdots & c_{1n} \\ c_{21} & c_{22} & \cdots & c_{2n} \\ \cdots & \cdots & \cdots & \cdots \\ c_{n1} & c_{n2} & \cdots & c_{nn} \end{bmatrix} \begin{bmatrix} 0 \\ 1 \\ \cdot\cdot \\ \cdot\cdot \\ 0 \end{bmatrix} = \begin{bmatrix} c_{12} \\ c_{22} \\ \cdots \\ c_{n2} \end{bmatrix},$$

the 2nd column of C. Again, since $CX = X$, we conclude that the 2nd column of C coincides with the 2nd column of I. The reader should now see the pattern, and see that the nth (last) choice for X is the nth column of I. Thus all the columns of C are equal to the corresponding columns of I, so $C = I$. ■

EXAMPLE 2-26

This example illustrates the fact that $CX = i^{\text{th}}$ column of C when $X = i^{\text{th}}$ column of I. Let

$$C = \begin{bmatrix} 1 & 2 \\ 3 & 4 \end{bmatrix}.$$

(a) When $X = \begin{bmatrix} 1 \\ 0 \end{bmatrix}$,

$$CX = \begin{bmatrix} 1 \\ 3 \end{bmatrix},$$

which is the first column of C.

(b) When $X = \begin{bmatrix} 0 \\ 1 \end{bmatrix}$,

$$CX = \begin{bmatrix} 2 \\ 4 \end{bmatrix},$$

which is the second column of C. ❑

> **DEFINITION 2-12**
> An $n \times n$ matrix A is **nonsingular** or **invertible** if there is an $n \times n$ matrix P such that $PA = I = AP$.

We shall see later that $PA = I$ implies $AP = I$ and vice-versa, so in effect A is invertible if we can find an $n \times n$ matrix P such that *one* of AP or $PA = I$. Now if P exists, (but it need not!) then it is unique. For, if also $QA = I$, then $Q = QI = Q(AP) = (QA)P = IP = P$. We call this unique matrix P such that $PA = I$ the **inverse** of A and denote it A^{-1}; it is completely characterized by $A^{-1}A = AA^{-1} = I$.

We shall shortly develop a technique for finding inverses when they exist. However, *to check that a suspected inverse of A actually is an inverse, we only have to multiply.*

EXAMPLE 2-27

(a) Let

$$A = \begin{bmatrix} 2 & 5 \\ 1 & 3 \end{bmatrix}.$$

Then,

$$A^{-1} = \begin{bmatrix} 3 & -5 \\ -1 & 2 \end{bmatrix},$$

because

$$AA^{-1} = \begin{bmatrix} 2 & 5 \\ 1 & 3 \end{bmatrix}\begin{bmatrix} 3 & -5 \\ -1 & 2 \end{bmatrix} = \begin{bmatrix} 1 & 0 \\ 0 & 1 \end{bmatrix} = I.$$

(b) Let A be an invertible $n \times n$ matrix, so A^{-1} exists. Then, the transpose A^t is also invertible and $(A^t)^{-1} = (A^{-1})^t$. To see this, we multiply A^t and $(\mathrm{A}^{-1})^t$ and, using Theorem 2-5, we get the identity matrix:

$$A^t(A^{-1})^t = (A^{-1}A)^t = I^t = I.$$

(c) Consider

$$B = \begin{bmatrix} 1 & -2 \\ -2 & 4 \end{bmatrix}.$$

B is not invertible. For, if it were, the homogeneous system of equations $BX = O$ could be solved as suggested earlier as $X = B^{-1}BX = B^{-1}O = O$, which has to be the only solution. But it is easy to see that

$$X = \begin{bmatrix} 2 \\ 1 \end{bmatrix}$$

is a solution different from O, so B can't be invertible. ❑

Part (c) above is an illustration of a result which is important in its own right.

THEOREM 2-7

If A is an $n \times n$ invertible matrix, the system of equations $AX = B$ has exactly one solution: $X = A^{-1}B$.

PROOF It remains only to show that the solution stated is unique. If Y were another solution, then $AY = B$, so $Y = A^{-1}B$ also. Thus $Y = X$ and there is only one solution. ■

EXAMPLE 2-28

Consider the system of linear equations

$$\begin{aligned} 3x + 4y + z &= 16 \\ 5x - 2y + 3z &= 20 \\ -x + 3y + 4z &= 38. \end{aligned}$$

The coefficient matrix is

$$A = \begin{bmatrix} 3 & 4 & 1 \\ 5 & -2 & 3 \\ -1 & 3 & 4 \end{bmatrix}.$$

Using a method for computing the inverse of A which we shall discuss later, we find that

$$A^{-1} = \begin{bmatrix} 17/130 & 1/10 & -7/65 \\ 23/130 & -1/10 & 2/65 \\ -1/10 & 1/10 & 1/5 \end{bmatrix}.$$

Then, by the preceding theorem, the only solution of the system is

$$X = A^{-1}B = A^{-1}\begin{bmatrix} 16 \\ 20 \\ 38 \end{bmatrix} = \begin{bmatrix} 0 \\ 2 \\ 8 \end{bmatrix}. \square$$

In order to gain some insight into the method of finding the inverse of a matrix A, we consider first a class of matrices for which this is easy. These are the elementary matrices, defined below. Readers will notice that in the definition, the steps of Gauss-Jordan reduction (section 2-2) appear once again. Indeed, we shall see that *the problem of finding the inverse of an* $n \times n$ *matrix is essentially a problem of solving* n *(special) systems of* n *linear equations*. (See Example 2-33.)

The reader who is primarily interested in computing inverses can skip to Theorem 2-12 and the examples following it.

DEFINITION 2-13

An $n \times n$ matrix E is called an **elementary matrix** if it can be obtained from the $n \times n$ identity by the application of a single elementary row operation.

Since there are three types of elementary row operations, we will often refer to the corresponding elementary matrices as being of the first, second, or third *kind* (or *type*) respectively.

EXAMPLE 2-29

(a) First consider 2×2 matrices. There is only one possible elementary row operation of type 1 (since there are only 2 rows). It follows that the only elementary matrix of the first kind is

$$\begin{bmatrix} 0 & 1 \\ 1 & 0 \end{bmatrix},$$

obtained by interchanging the rows of

$$\begin{bmatrix} 1 & 0 \\ 0 & 1 \end{bmatrix}.$$

The only 2×2 elementary matrices of the second kind are

$$\begin{bmatrix} k & 0 \\ 0 & 1 \end{bmatrix} \text{and} \begin{bmatrix} 1 & 0 \\ 0 & k \end{bmatrix},$$

obtained by multiplying a row by k, where k is a nonzero constant. Note that with $k = 1$, we get I, so I is elementary. Finally, the only 2×2 elementary matrices of the third kind are

$$\begin{bmatrix} 1 & 0 \\ k & 1 \end{bmatrix} \text{and} \begin{bmatrix} 1 & k \\ 0 & 1 \end{bmatrix},$$

obtained by adding k times one row to the other.

(b) The reader should identify the kind of elementary matrix for each of the following:

$$\begin{bmatrix} 1 & 0 & 0 \\ 0 & 1 & 0 \\ 1 & 0 & 1 \end{bmatrix}, \begin{bmatrix} 1 & -2 & 0 \\ 0 & 1 & 0 \\ 0 & 0 & 1 \end{bmatrix}, \begin{bmatrix} 0 & 1 & 0 \\ 1 & 0 & 0 \\ 0 & 0 & 1 \end{bmatrix}.$$

(c) The following are not elementary matrices:

$$\begin{bmatrix} -1 & 0 \\ 1 & 1 \end{bmatrix}, \begin{bmatrix} 1 & 1 & 0 \\ 0 & 0 & 1 \\ 0 & 1 & 0 \end{bmatrix}, \begin{bmatrix} 1 & 0 \\ 0 & 1 \\ 1 & 0 \end{bmatrix}. \square$$

It is convenient to recall the notation used in the proof of Theorem 2-2 in connection with row equivalence of matrices. Note that *any elementary matrix is row equivalent to the identity matrix of the same order*. If E is an elementary matrix obtained from I by the elementary row operation e, then we shall write $E = e(I)$. Now in Theorem 2-2, we saw that any e has an "inverse" operation e^{-1} with the property that if A is any matrix, then $e^{-1}(e(A)) = A$. This suggests the fact that if $E = e(I)$, it is nonsingular with inverse $E^{-1} = e^{-1}(I)$.

EXAMPLE 2-30

Let

$$E = \begin{bmatrix} k & 0 \\ 0 & 1 \end{bmatrix}.$$

Then according to the above suggestion,

$$E^{-1} = \begin{bmatrix} 1/k & 0 \\ 0 & 1 \end{bmatrix}$$

since the inverse operation of multiplying row 1 by k is multiplying row 1 by $1/k$ ($k \neq 0$). The reader should check by multiplication that this is indeed the inverse of E. ❑

The next result is needed to establish the truth of the suggestion made earlier.

THEOREM 2-8

If e is an elementary row operation and A, B are matrices such that AB is defined, then $e(AB) = e(A)B$.

PROOF (Outline) The idea of the proof stems from the fact that the ith row of the product AB is obtained only from the ith row of A together with the columns of B. Thus, the effect of applying e to AB is captured in $e(A)$ as e affects rows only. A detailed proof would consider the three possibilities for e and is left for the reader. ■

COROLLARY

If e is an elementary row operation with corresponding $m \times m$ elementary matrix $E = e(I)$, then for any $m \times n$ matrix A, $e(A) = EA$.

PROOF By the theorem $e(A) = e(IA) = e(I)A = EA$. ■

The corollary "translates" elementary row operations into multiplication on the left by an elementary matrix.

EXAMPLE 2-31

Let e interchange the rows of

$$A = \begin{bmatrix} 1 & 2 \\ 3 & 4 \end{bmatrix}.$$

Then,

$$e(A) = \begin{bmatrix} 3 & 4 \\ 1 & 2 \end{bmatrix} = \begin{bmatrix} 0 & 1 \\ 1 & 0 \end{bmatrix}\begin{bmatrix} 1 & 2 \\ 3 & 4 \end{bmatrix} = EA. \square$$

THEOREM 2-9

If $E = e(I)$ is an elementary matrix, then it is nonsingular with inverse $E^{-1} = e^{-1}(I)$, where e^{-1} is the inverse of operation e.

PROOF By Theorem 2-8,

$$E^{-1}E = e^{-1}(I)E = e^{-1}(IE) = e^{-1}(E) = e^{-1}(e(I)) = I. \blacksquare$$

EXAMPLE 2-32

If

$$E = \begin{bmatrix} 1 & 0 & 2 \\ 0 & 1 & 0 \\ 0 & 0 & 1 \end{bmatrix},$$

then

$$E^{-1} = \begin{bmatrix} 1 & 0 & -2 \\ 0 & 1 & 0 \\ 0 & 0 & 1 \end{bmatrix},$$

since E is obtained from I by adding $2 \times \text{row } 3$ to row 1 and the inverse of this operation is subtracting $2 \times \text{row } 3$ from row 1. $\square$

Having established the existence of a class of nonsingular matrices, the elementary matrices, it is a surprising fact that *any other* nonsingular matrix is necessarily a product of these. We shall arrive at a proof of this fact in a number of steps. In so doing, we shall also develop an effective technique for finding inverses.

THEOREM 2-10

(1) If A and B are nonsingular $n \times n$ matrices, then so is AB and $(AB)^{-1} = B^{-1}A^{-1}$. (Note the reversed order!)

(2) If $A_1, A_2, \ldots, A_k$ are nonsingular $n \times n$ matrices, then so is their product $A_1A_2 \ldots A_k$ and $(A_1A_2 \ldots A_k)^{-1} = A_k^{-1}A_{k-1}^{-1} \ldots A_2^{-1}A_1^{-1}$.

PROOF (1) We use the associative law of multiplication:

$$(AB)(B^{-1}A^{-1}) = A(BB^{-1})A^{-1} = AIA^{-1} = AA^{-1} = I.$$

(2)
$$\begin{aligned}
&(A_k^{-1}A_{k-1}^{-1} \ldots A_2^{-1}A_1^{-1})(A_1A_2 \ldots A_k) \\
&= (A_k^{-1}A_{k-1}^{-1} \ldots A_2^{-1})(A_1^{-1}A_1)(A_2 \ldots A_k) \\
&= (A_k^{-1}A_{k-1}^{-1} \ldots A_2^{-1})I(A_2 \ldots A_k) \\
&= (A_k^{-1}A_{k-1}^{-1} \ldots A_2^{-1})(A_2 \ldots A_k) \\
&= \ldots = I,
\end{aligned}$$

after k such steps of combining two inverse matrices to yield the identity I. ■

COROLLARY

Any product of elementary matrices is nonsingular. ■

Notice that we have now proved half of the result mentioned just before Theorem 2-10. Suppose now that A is an $n \times n$ nonsingular matrix. Then the system of homogeneous equations $AX = O$ has only the trivial solution $X = O$. Now we know how to solve such a system: we apply Gauss-Jordan reduction to the augmented matrix $(A{:}O)$ to obtain a row-reduced echelon matrix $(A'{:}O)$ such that the system $A'X = O$ has the same solutions as $AX = O$, namely, $X = O$; i.e., $x_1 = x_2 = \ldots = x_n = 0$. This means that the system $A'X = O$ has the form

$$\begin{aligned}
x_1 \qquad\qquad\qquad &= 0 \\
x_2 \qquad\qquad &= 0 \\
\cdots\cdots\cdots\cdots\cdots\cdots & \\
x_n &= 0.
\end{aligned}$$

From this, it should be clear that the coefficient matrix $A' = I$. Now suppose that k elementary row operations $e_1, e_2, \ldots, e_k$ were used in succession to change A to A'. Then we may write

$$I = A' = e_k\, e_{k-1} \ldots e_2e_1A. \tag{2-9}$$

Using the corollary to Theorem 2-8, we translate this into matrix notation as

$$I = E_k E_{k-1} \dots E_2 E_1 A,$$

where E_i is the matrix corresponding to e_i. Multiplying both sides by $(E_k E_{k-1} \dots E_2 E_1)^{-1}$ we get

$$A = (E_k E_{k-1} \dots E_2 E_1)^{-1} = E_1^{-1} E_2^{-1} \dots E_k^{-1}, \qquad (2\text{-}10)$$

which shows that A is a product of elementary matrices. Thus, we have proved the next theorem.

THEOREM 2-11

An $n \times n$ matrix A is nonsingular iff it is a product of elementary matrices. ■

It is important to note that we have also proved that

(1) A is nonsingular iff it is row-equivalent to I;

(2) the elementary matrices entering into the expression for A^{-1} correspond to elementary row operations which, when applied to A, will transform it to I (equation 2-9).

This last observation leads to the **principal method** for finding A^{-1} when it exists, while the first observation tells us when it does not exist. Specifically, let us take the inverse of both sides of equation 2-10 to get

$$A^{-1} = E_k E_{k-1} \dots E_2 E_1 = E_k E_{k-1} \dots E_2 E_1 I = e_k e_{k-1} \dots e_2 e_1(I). \qquad (2\text{-}11)$$

The next theorem follows from equation 2-11.

THEOREM 2-12

If A is a nonsingular matrix, then the sequence of elementary row operations that transforms A to I (equation 2-9) will transform I to A^{-1} (equation 2-11). ■

In practice, *to find A^{-1}, we proceed by placing A and I side-by-side in an augmented matrix (A:I) and then applying elementary row operations to transform it to (I:A^{-1}).* Thus, in effect, we are solving n systems of equations with coefficient matrix A and with right-hand sides equal to the n columns of I.

EXAMPLE 2-33

Find the inverse of

$$A = \begin{bmatrix} 1 & 3 & 1 \\ 1 & 0 & 0 \\ 0 & 1 & 1 \end{bmatrix}.$$

Solution The steps are as follows:

$$(A{:}I) = \begin{bmatrix} 1 & 3 & 1 & 1 & 0 & 0 \\ 1 & 0 & 0 & 0 & 1 & 0 \\ 0 & 1 & 1 & 0 & 0 & 1 \end{bmatrix}$$

row 2 to row 2 − row 1:

$$\begin{bmatrix} 1 & 3 & 1 & 1 & 0 & 0 \\ 0 & -3 & -1 & -1 & 1 & 0 \\ 0 & 1 & 1 & 0 & 0 & 1 \end{bmatrix}$$

interchange rows 2 and 3:

$$\begin{bmatrix} 1 & 3 & 1 & 1 & 0 & 0 \\ 0 & 1 & 1 & 0 & 0 & 1 \\ 0 & -3 & -1 & -1 & 1 & 0 \end{bmatrix}$$

row 3 to row 3 + 3×row 2:

$$\begin{bmatrix} 1 & 3 & 1 & 1 & 0 & 0 \\ 0 & 1 & 1 & 0 & 0 & 1 \\ 0 & 0 & 2 & -1 & 1 & 3 \end{bmatrix}$$

row 1 to row 1 − 3×row 2:

$$\begin{bmatrix} 1 & 0 & -2 & 1 & 0 & -3 \\ 0 & 1 & 1 & 0 & 0 & 1 \\ 0 & 0 & 2 & -1 & 1 & 3 \end{bmatrix}$$

row 1 to row 1 + row 3:

$$\begin{bmatrix} 1 & 0 & 0 & 0 & 1 & 0 \\ 0 & 1 & 1 & 0 & 0 & 1 \\ 0 & 0 & 2 & -1 & 1 & 3 \end{bmatrix}$$

(Note that once again we are not following Gauss-Jordan reduction literally, in order to take advantage of obvious time-saving operations.)

row 3 to (1/2)×row 3:

$$\begin{bmatrix} 1 & 0 & 0 & 0 & 1 & 0 \\ 0 & 1 & 1 & 0 & 0 & 1 \\ 0 & 0 & 1 & -1/2 & 1/2 & 3/2 \end{bmatrix}$$

row 2 to row 2 − row 3:

$$\begin{bmatrix} 1 & 0 & 0 & 0 & 1 & 0 \\ 0 & 1 & 0 & 1/2 & -1/2 & -1/2 \\ 0 & 0 & 1 & -1/2 & 1/2 & 3/2 \end{bmatrix} = (I{:}A^{-1}).$$

So, the inverse of A is

$$A^{-1} = \begin{bmatrix} 0 & 1 & 0 \\ 1/2 & -1/2 & -1/2 \\ -1/2 & 1/2 & 3/2 \end{bmatrix}. \square$$

EXAMPLE 2-34

Find, if possible, the inverse of

$$B = \begin{bmatrix} 1 & 2 & 3 \\ 2 & 3 & 4 \\ 0 & 1 & 2 \end{bmatrix}.$$

Solution The steps are:

$$(B{:}I) = \begin{bmatrix} 1 & 2 & 3 & 1 & 0 & 0 \\ 2 & 3 & 4 & 0 & 1 & 0 \\ 0 & 1 & 2 & 0 & 0 & 1 \end{bmatrix}$$

$$\begin{bmatrix} 1 & 2 & 3 & 1 & 0 & 0 \\ 0 & -1 & -2 & -2 & 1 & 0 \\ 0 & 1 & 2 & 0 & 0 & 1 \end{bmatrix}$$

$$\begin{bmatrix} 1 & 2 & 3 & 1 & 0 & 0 \\ 0 & 1 & 2 & 2 & -1 & 0 \\ 0 & 0 & 0 & -2 & 1 & 1 \end{bmatrix}$$

$= (B'{:}C)$, say. It is now apparent that B has no inverse since B', the left matrix, cannot be row-equivalent to I on account of the row of zeros. $\square$

We return now to the general problem stated at the beginning of this section. Given a system of m equations in n unknowns, $AX = B$, can we find a matrix P such that $PAX = PB$ becomes $X = PB$? We have seen that this is possible when $m = n$ *and* A is invertible, in which case $P = A^{-1}$. Furthermore, we have seen that

$$P = E_k E_{k-1} \ldots E_2 E_1,$$

the product of elementary matrices corresponding to the elementary row operations that will transform A to I: $e_k e_{k-1} \ldots e_2 e_1(A) = I$.

Now when $m \neq n$, so that A is not square, we obviously can't transform A to the identity I, but we can transform A to a row-reduced echelon matrix A', by Theorem 2-3. In other words,

$$e_k e_{k-1} \ldots e_2 e_1(A) = A'$$

for some elementary row operations $e_1, \ldots, e_k$. Translating into matrix language, we get

$$E_k E_{k-1} \ldots E_2 E_1 A = A'.$$

Now, putting $P = E_k E_{k-1} \ldots E_2 E_1$, an $m \times m$ invertible matrix (A is $m \times n$) and multiplying $AX = B$ by P, we get $PAX = PB$, or, since $PA = A'$,

$$A'X = PB \tag{2-12}$$

where, as stated, A' is a row-reduced echelon matrix. Equation 2-12 is the analogue of $X = PB$ for the case $m \neq n$; its solution is immediate since A' is in row-reduced echelon form. We have thus proved:

> **THEOREM 2-13**
>
> If A is an $m \times n$ matrix and $e_1, \ldots, e_k$ are elementary row operations that transform A to a row-reduced echelon matrix A', then $A' = PA$, where P is an $m \times m$ invertible matrix: $P = E_k E_{k-1} \ldots E_2 E_1 = e_k e_{k-1} \ldots e_2 e_1(I)$. Furthermore, the system of equations $AX = B$ is equivalent to $A'X = PB$. ■

EXAMPLE 2-35

Consider the matrix

$$A = \begin{bmatrix} 2 & -1 & 3 \\ 1 & 0 & 4 \end{bmatrix}.$$

We obtain a matrix P such that $PA = A'$ is in row-reduced echelon form by applying Gauss-Jordan reduction to $(A{:}I)$ where I is the 2×2 identity. The steps are:

$$(A{:}I) = \begin{bmatrix} 2 & -1 & 3 & 1 & 0 \\ 1 & 0 & 4 & 0 & 1 \end{bmatrix}$$

$$\begin{bmatrix} 1 & 0 & 4 & 0 & 1 \\ 2 & -1 & 3 & 1 & 0 \end{bmatrix}$$

$$\begin{bmatrix} 1 & 0 & 4 & 0 & 1 \\ 0 & -1 & -5 & 1 & -2 \end{bmatrix}$$

$$\begin{bmatrix} 1 & 0 & 4 & 0 & 1 \\ 0 & 1 & 5 & -1 & 2 \end{bmatrix} = (A'{:}P).$$

Therefore, the required matrix is

$$P = \begin{bmatrix} 0 & 1 \\ -1 & 2 \end{bmatrix}.$$

Furthermore, given

$$B = \begin{bmatrix} 3 \\ -2 \end{bmatrix},$$

the system $AX = B$ is equivalent to $A'X = PB$, which is

$$\begin{bmatrix} 1 & 0 & 4 \\ 0 & 1 & 5 \end{bmatrix} \begin{bmatrix} x_1 \\ x_2 \\ x_3 \end{bmatrix} = \begin{bmatrix} 0 & 1 \\ -1 & 2 \end{bmatrix} \begin{bmatrix} 3 \\ -2 \end{bmatrix} = \begin{bmatrix} -2 \\ -7 \end{bmatrix}.$$

By inspection, its solution is

$$\begin{aligned} x_1 &= -2 - 4x_3 \\ x_2 &= -7 - 5x_3. \end{aligned}$$

There are infinitely many solutions which can be obtained by assigning arbitrary values to x_3. ❑

EXERCISES 2-5

1. Let

$$A = \begin{bmatrix} 1 & 2 & 3 \\ 4 & 5 & 6 \\ 7 & 8 & 9 \end{bmatrix}, X_1 = \begin{bmatrix} 1 \\ 0 \\ 0 \end{bmatrix}, X_2 = \begin{bmatrix} 0 \\ 1 \\ 0 \end{bmatrix}, X_3 = \begin{bmatrix} 0 \\ 0 \\ 1 \end{bmatrix}.$$

Verify that the products AX_1, AX_2, and AX_3 are the columns 1, 2, and 3 respectively of A.

2. By computing AA^{-1}, verify that each of the matrices A given below has the inverse indicated.

(a) $A = \begin{bmatrix} 2 & -1 \\ -3 & 2 \end{bmatrix}; A^{-1} = \begin{bmatrix} 2 & 1 \\ 3 & 2 \end{bmatrix}.$

(b) $A = \begin{bmatrix} 4 & 3 & 2 \\ 3 & 5 & 2 \\ 2 & 2 & 1 \end{bmatrix}; A^{-1} = \begin{bmatrix} -1 & -1 & 4 \\ -1 & 0 & 2 \\ 4 & 2 & -11 \end{bmatrix}.$

3. By considering the associated homogeneous system of equations, show that

$$B = \begin{bmatrix} -3 & -6 \\ 1/2 & 1 \end{bmatrix}$$

cannot be invertible.

4. Use the results of question 2 above to find the unique solutions of the following systems of equations.

(a)
$$\begin{aligned} 2x_1 - x_2 &= -1 \\ -3x_1 + 2x_2 &= 2 \end{aligned}$$

(b)
$$\begin{aligned} 4a + 3b + 2c &= 2 \\ 3a + 5b + 2c &= -3 \\ 2a + 2b + c &= 1 \end{aligned}$$

5. Which of the following matrices are elementary?

(a) $\begin{bmatrix} 2 & 0 \\ 0 & 0 \end{bmatrix}$ **(b)** $\begin{bmatrix} -1 & 0 \\ 0 & 1 \end{bmatrix}$ **(c)** $\begin{bmatrix} 0 & 1 \\ 1 & 0 \end{bmatrix}$

(d) $\begin{bmatrix} 0 & 1 & 0 \\ 0 & 0 & 1 \\ 1 & 0 & 0 \end{bmatrix}$ **(e)** $\begin{bmatrix} 1 & 0 & 0 \\ 0 & 1 & 0 \\ -2 & 0 & 1 \end{bmatrix}$ **(f)** $\begin{bmatrix} 0 & -1 & 0 \\ 1 & 0 & 0 \\ 0 & 0 & 1 \end{bmatrix}$

(g) $\begin{bmatrix} 1 & 0 & 0 \\ 0 & 1 & 0 \\ 0 & 0 & -2 \end{bmatrix}$ **(h)** $\begin{bmatrix} 0 & 0 \\ 0 & 0 \end{bmatrix}$ **(i)** $\begin{bmatrix} 1 & 0 & 0 & 0 \\ 0 & 1 & 0 & 0 \\ 0 & 0 & 1 & 0 \end{bmatrix}$

6. For those matrices in question 5 that are elementary, identify in each case the elementary row operation used to obtain it from the identity matrix.

7. For the matrices A and B in each of the following, find an elementary matrix E such that $B = EA$.

(a) $A = \begin{bmatrix} -2 & 1 \\ 0 & 3 \end{bmatrix}, B = \begin{bmatrix} -2 & 1 \\ 4 & 1 \end{bmatrix}$

(b) $A = \begin{bmatrix} -1 & 3 \\ 8 & -2 \end{bmatrix}, B = \begin{bmatrix} 8 & -2 \\ -1 & 3 \end{bmatrix}$

(c) $A = \begin{bmatrix} 4 & -3 & 2 \\ 1 & 0 & 2 \\ 2 & -3 & 2 \end{bmatrix}, B = \begin{bmatrix} 4 & -3 & 2 \\ -1 & 3 & 0 \\ 2 & -3 & 2 \end{bmatrix}$

8. Find the inverse of each of the following elementary matrices.

(a) $\begin{bmatrix} 0 & 1 \\ 1 & 0 \end{bmatrix}$ **(b)** $\begin{bmatrix} 1 & 0 \\ 0 & -3 \end{bmatrix}$

(c) $\begin{bmatrix} 1 & 0 & 0 \\ 0 & 1 & 2 \\ 0 & 0 & 1 \end{bmatrix}$ **(d)** $\begin{bmatrix} 1 & -2 & 0 & 0 \\ 0 & 1 & 0 & 0 \\ 0 & 0 & 1 & 0 \\ 0 & 0 & 0 & 1 \end{bmatrix}$

9. (a) Give an example of 2×2 matrices A and B such that $A + B$ is not invertible.

(b) Give an example of 2×2 matrices A and B such that $A + B$ is invertible.

10. A square matrix is called a *diagonal matrix* if all its off-diagonal entries are zero: $a_{ij} = 0$ when $i \neq j$. Prove that such a matrix A is invertible provided no diagonal entry is zero, and find A^{-1}.

11. Show that A below is invertible iff $ad - bc \neq 0$.

$$A = \begin{bmatrix} a & b \\ c & d \end{bmatrix}$$

12. Find the inverse (if possible) of the following matrices using Gauss-Jordan reduction of an appropriate augmented matrix.

(a) $\begin{bmatrix} 2 & -1 \\ 0 & 4 \end{bmatrix}$ **(b)** $\begin{bmatrix} 2 & 1 \\ 6 & 3 \end{bmatrix}$ **(c)** $\begin{bmatrix} 1 & 2 & 3 \\ 0 & -1 & 2 \\ 0 & 0 & 3 \end{bmatrix}$

(d) $\begin{bmatrix} 3 & -1 \\ -5 & 2 \end{bmatrix}$ **(e)** $\begin{bmatrix} 2 & -5 & 5 \\ 2 & -3 & 8 \\ 3 & 8 & 7 \end{bmatrix}$ **(f)** $\begin{bmatrix} 1 & 2 & -3 \\ 1 & -2 & 1 \\ 5 & -2 & -3 \end{bmatrix}$

(g) $\begin{bmatrix} 0 & 5 & -7 & 1 \\ -2 & 0 & 0 & 1 \\ 1 & -2 & 3 & -1 \\ 1 & -1 & 1 & 0 \end{bmatrix}$ **(h)** $\begin{bmatrix} 1 & 1 & 0 \\ 1 & 2 & 5 \\ 2 & 4 & 3 \end{bmatrix}$

13. Show that the following matrices are singular.

(a) $\begin{bmatrix} 2 & 3 \\ 4 & 6 \end{bmatrix}$ **(b)** $\begin{bmatrix} 3 & 1 & 5 \\ 2 & 4 & 1 \\ -4 & 2 & -9 \end{bmatrix}$ **(c)** $\begin{bmatrix} -1 & 0 & 2 & 3 \\ 1 & 6 & -7 & 3 \\ 2 & 2 & -4 & 1 \\ -3 & 2 & 1 & 1 \end{bmatrix}$

14. Solve the following systems of equations by finding the inverse of the coefficient matrix.

(a) $2a + 5b = 8$
$6a - 2b = 10$

(b) $2x + 6y + 4z = 1$
$x + 4y + 6z = -3$
$x + 3y + 3z = 2$

(c) $a + 2b + 3c = 18$
$2a + 3b + 4c = -20$
$3a + 4b + 6c = -9$

15. (a) Show that the matrices below have no inverse.

(i) $A = \begin{bmatrix} 1 & 2 & 3 \\ 0 & 0 & 0 \\ -4 & 2 & 3 \end{bmatrix}$ **(ii)** $B = \begin{bmatrix} 0 & -1 & 2 \\ 0 & 2 & -3 \\ 0 & 3 & 4 \end{bmatrix}$

(b) Generalize (a) by proving that a square matrix is not invertible if it has a row or column of zeros.

16. A square matrix A is called *upper triangular* if the entries below the main diagonal (respectively, *lower triangular*, if above the main diagonal) are zero. Show that upper and lower triangular matrices are nonsingular provided none of the diagonal entries is zero.

17. **(a)** A matrix $A \neq O$ is called *nilpotent* if $A^k = O$ for some positive integer k. Show that any nilpotent matrix is singular.

(b) Show that the following matrices are nilpotent.

(i) $\begin{bmatrix} 0 & 0 \\ 2 & 0 \end{bmatrix}$ **(ii)** $\begin{bmatrix} 0 & 1 & 2 \\ 0 & 0 & 3 \\ 0 & 0 & 0 \end{bmatrix}$

18. Use Theorem 2-13 to obtain a nonsingular matrix P such that the matrix PA is in row-reduced echelon form for each of the matrices A given below.

(a) $\begin{bmatrix} 2 & 0 \\ -1 & 1 \end{bmatrix}$ **(b)** $\begin{bmatrix} 1 & 0 & -2 \\ 0 & 0 & -1 \end{bmatrix}$

(c) $\begin{bmatrix} 1 & 1 & 1 \\ 1 & 1 & 1 \\ 1 & 1 & 0 \\ 1 & 0 & 0 \end{bmatrix}$ **(d)** $\begin{bmatrix} -3 & 2 \\ 0 & 1 \\ -2 & 0 \end{bmatrix}$

19. Let

$$A = \begin{bmatrix} 1 & 1 & -1 \\ 4 & 1 & -2 \end{bmatrix}.$$

By finding a nonsingular 2×2 matrix P such that PA is in row-reduced echelon form, solve the system of equations $AX = B$, where B is given as follows.

(a) $\begin{bmatrix} -1 \\ 2 \end{bmatrix}$ **(b)** $\begin{bmatrix} 1 \\ 2 \end{bmatrix}$ **(c)** $\begin{bmatrix} 18 \\ -9 \end{bmatrix}$ **(d)** $\begin{bmatrix} 0 \\ 0 \end{bmatrix}$

20. **(a)** Express the following matrices as products of elementary matrices.

(i) $\begin{bmatrix} -2 & 1 \\ -1 & 3 \end{bmatrix}$ **(ii)** $\begin{bmatrix} 1 & 1 & 1 \\ 1 & 2 & 2 \\ 1 & 2 & 1 \end{bmatrix}$

(b) Explain why the matrix below is not a product of elementary matrices.

$$\begin{bmatrix} 2 & -1 & 3 \\ -1 & 0 & 1 \\ 0 & -1 & 5 \end{bmatrix}$$

21. **(a)** Find the equation of the plane passing through the points (1,0,0), (0,1,0), and (0,0,1) by solving an appropriate system of three linear equations in three variables.

(b) Find the equation of the plane passing through the points $(-1,-2,-3)$, $(-1,-1,2)$, and $(2,2,4)$.

2-6 Computational Considerations I (Optional)

In our discussion of the solution of systems of equations so far, we have concentrated almost entirely on the method of Gauss-Jordan elimination or reduction. While this method is particularly well suited for theoretical discussions, it is not, in fact, a practical one for solving the large systems of equations which arise in everyday problems in business and the sciences. Such problems are usually solved using computers, as the number of calculations involved would make it impossible to solve them "by hand".

For example, it can be shown that the number of arithmetic operations (i.e., additions, subtractions, multiplications, and divisions) required to solve a system of n linear equations in n unknowns by Gauss-Jordan elimination is approximately n^3 for large n. Thus, for a system of 1000 equations in 1000 unknowns (which is quite common in practice), the number of operations required is about one billion. Readers may wish to estimate how long it would take them to do this! While individual computers vary considerably, let us suppose that a computer can perform about one million arithmetic operations per second. Thus, it could perform the whole job in a matter of minutes (there are operations besides arithmetic ones which a computer must perform to solve a system). On the other hand, computer time is expensive, so there is nevertheless a real need for methods of solution that are more efficient than Gauss-Jordan elimination.

It can be shown that the number of arithmetic operations required for a solution by Gaussian elimination is about 50% less (i.e., of the order of $(2/3)n^3$ for large n). Consequently, Gaussian elimination is one of several important methods used in practice, although often some variations are used to make it more effective. The phrase *Gaussian elimination* was introduced in section 2-1. It refers to the procedure in which the augmented matrix for the system is reduced to row echelon form *only* (not to row-reduced echelon form); the resulting system is then solved by "back-substitution", starting from the last equation.

EXAMPLE 2-36

We solve the system

$$\begin{aligned} x + 2y + 3z &= 6 \\ 2x + 3y + z &= 6 \\ 3x + y + 2z &= 6 \end{aligned}$$

by Gaussian elimination. The reduction of the augmented matrix to row echelon form appears below. The number of arithmetic operations at each stage is compared with the number required for a Gauss-Jordan solution.

$$(A{:}B) = \begin{bmatrix} 1 & 2 & 3 & 6 \\ 2 & 3 & 1 & 6 \\ 3 & 1 & 2 & 6 \end{bmatrix}$$

$\begin{bmatrix} 1 & 2 & 3 & 6 \\ 0 & -1 & -5 & -6 \\ 0 & -5 & -7 & -12 \end{bmatrix}$ (8 multiplications, 8 subtractions, the same as for Gauss-Jordan reduction at this point)

$\begin{bmatrix} 1 & 2 & 3 & 6 \\ 0 & 1 & 5 & 6 \\ 0 & -5 & -7 & -12 \end{bmatrix}$ (3 multiplications, as for Gauss-Jordan)

$\begin{bmatrix} 1 & 2 & 3 & 6 \\ 0 & 1 & 5 & 6 \\ 0 & 0 & 18 & 18 \end{bmatrix}$ (3 multiplications, 3 subtractions; Gauss-Jordan would have required 6 of each to eliminate the 2 in row 1 as well)

$\begin{bmatrix} 1 & 2 & 3 & 6 \\ 0 & 1 & 5 & 6 \\ 0 & 0 & 1 & 1 \end{bmatrix}$ (2 divisions; at this point, Gauss-Jordan would have required 10 operations)

The resulting system of equations is

$$\begin{aligned} x + 2y + 3z &= 6 \\ y + 5z &= 6 \\ z &= 1. \end{aligned}$$

Using "back-substitution", we now find

$$\begin{aligned} y &= 6 - 5z = 6 - 5(1) = 1 \quad \text{(2 operations)} \\ x &= 6 - 2y - 3z = 6 - 2 - 3 = 1 \quad \text{(4 operations).} \end{aligned}$$

The total number of operations for Gaussian elimination is found to be 33, as compared with 41 for Gauss-Jordan elimination. In fact, these numbers can be reduced somewhat by a different organization of the computation, but we will not pursue this. ❑

EXAMPLE 2-37

We solve the system of equations from Example 2-12 by Gaussian elimination. As we saw in that example, the augmented matrix is transformed from

$$\begin{bmatrix} 0 & 1 & 1 & 5 & 6 & -1 \\ -2 & -1 & 0 & 2 & 1 & -2 \\ 1 & 2 & 2 & 0 & 4 & 2 \\ 1 & 0 & -1 & 3 & 1 & -1 \\ 0 & 2 & 3 & -3 & 3 & 3 \end{bmatrix}$$

to

$$\begin{bmatrix} 1 & 2 & 2 & 0 & 4 & 2 \\ 0 & 1 & 1 & 5 & 6 & -1 \\ 0 & 3 & 4 & 2 & 9 & 2 \\ 0 & -2 & -3 & 3 & -3 & -3 \\ 0 & 0 & 0 & 0 & 0 & 0 \end{bmatrix}$$

The remaining steps in the Gaussian reduction are

$$\begin{bmatrix} 1 & 2 & 2 & 0 & 4 & 2 \\ 0 & 1 & 1 & 5 & 6 & -1 \\ 0 & 0 & 1 & -13 & -9 & 5 \\ 0 & 0 & -1 & 13 & 9 & -5 \\ 0 & 0 & 0 & 0 & 0 & 0 \end{bmatrix}$$

and

$$\begin{bmatrix} 1 & 2 & 2 & 0 & 4 & 2 \\ 0 & 1 & 1 & 5 & 6 & -1 \\ 0 & 0 & 1 & -13 & -9 & 5 \\ 0 & 0 & 0 & 0 & 0 & 0 \\ 0 & 0 & 0 & 0 & 0 & 0 \end{bmatrix}.$$

The corresponding system of equations is (compare Example 2-12)

$$\begin{aligned} x + 2y + 2z \qquad\quad + 4t &= 2 \\ y + \;\; z + \;\; 5s + 6t &= -1 \\ z - 13s - 9t &= 5. \end{aligned}$$

We may assign arbitrary values to s and t and solve by back-substitution. For example, with $s = t = 0$, we find $z = 5$, $y = -1 - 5 = -6$, and $x = 2 - 10 + 12 = 4$. ❑

A second problem which arises in solving realistic systems of equations is that the coefficients in them are not usually the "nice" numbers we have used in our examples. Rather than small integers, they are usually numbers expressed to several decimal places. The number of decimal places appearing might be a result, for example, of the measurements that produced the data. In any case, in order to solve the equations, whether by hand or by computer, we are usually forced to limit the number of digits we carry by rounding. For example, the number $7/6 = 1.166...$ might be rounded to 1.167 if it were necessary to do computations with 4-digit numbers.

Clearly, the more digits we carry, the more work we must do; but the fewer digits we carry, the less accurate the answer will be. In the case of computers, a number N is usually expressed in **floating point** form. This means that N is written as $N = \pm M \times 10^k$ for some integer k and $M \in R$ with $0.1 \leq M < 1$. Then a number N is said to be expressed to n **significant digits** if, in this floating point form, M has n digits.

EXAMPLE 2-38

The table below lists some numbers, their floating point forms rounded to 7 (significant) digits, and their floating point forms rounded to 3 digits.

Number	*7-digit*	*3-digit*
$7/6$	$0.116\,666\,7 \times 10^1$	0.117×10^1
$-\sqrt{200}$	$-0.141\,421\,4 \times 10^2$	-0.141×10^2
$0.002\,344$	$0.234\,400\,0 \times 10^{-2}$	0.234×10^{-2}
-2325	$-0.232\,500\,0 \times 10^4$	-0.232×10^4
-2315	$-0.231\,500\,0 \times 10^4$	-0.232×10^4 ❑

The last two numbers in Example 2-38 illustrate a common convention for rounding a number to d digits:

(1) If the $(d + 1)^{th}$ digit is greater than 5, increase the dth by 1.

(2) If the $(d + 1)^{th}$ digit is less than 5, leave the dth unaltered.

(3) If the $(d + 1)^{th}$ digit is 5, increase the dth by 1 if that would make it even, and otherwise leave it as it is (i.e., even).

Computers vary in the number of digits they can handle in the fractional part M of a number N. Typically, this ranges from 7 to 11 digits. Now, whenever a computer performs an arithmetic operation, the result is rounded to the number of digits the computer in question can handle. This may introduce an error in the result which is called **rounding error**.

EXAMPLE 2-39

Let us suppose that a computer works in 3-digit floating point arithmetic. This means that the results of all computations are rounded to 3 significant figures (3 digits in floating point form).

(a) For 28.5×5.99 ($= 170.715$), it would give 0.171×10^3.

(b) For $(28.5 \times 5.99) \times 0.322$ ($= 54.970\,23$), it would use the result of (a) to get 55.1, rather than the true value 54.970 23.

This example illustrates the **cumulative effect** of rounding errors in computations to a fixed finite number of digits. ❑

As we have seen, Gaussian elimination involves a great many arithmetic operations. Consequently, because of rounding errors, solving realistic systems of equations by this method may lead to results which are seriously wrong.

EXAMPLE 2-40

Consider the system of equations

$$\begin{aligned} 0.001x + y &= 0.999 \\ x - y &= 0.002. \end{aligned}$$

By adding the equations, we see immediately that the exact solution is $x = 1, y = 0.998$. Suppose, however, that the system were solved using 3-digit floating point arithmetic. The steps in the usual Gaussian elimination would be:

$$\begin{bmatrix} 0.001 & 1 & 0.999 \\ 1 & -1 & 0.002 \end{bmatrix} \rightarrow \begin{bmatrix} 1 & 1000 & 999 \\ 1 & -1 & 0.002 \end{bmatrix} \rightarrow \begin{bmatrix} 1 & 1000 & 999 \\ 0 & -1000 & -999 \end{bmatrix}$$

(because, to 3 significant figures, $-1 - 1000 = -1000$ and $0.002 - 999 = -999$). Therefore, the final matrix is

$$\begin{bmatrix} 1 & 1000 & 999 \\ 0 & 1 & 0.999 \end{bmatrix}$$

The corresponding equations are

$$\begin{aligned} x + 1000y &= 999 \\ y &= 0.999 \end{aligned}$$

from which we obtain $x = 999 - 999 = 0$, a very poor approximation to $x = 1$ indeed (the value for y is good). ❑

A complete discussion of this surprising result is beyond the scope of this text (see, for example, *Applied Linear Algebra, 2nd edition*, by Noble and Daniel, Prentice-Hall, Inc., 1977). However, we note that in this example, the very small coefficient 0.001 of x caused the large numbers 1000 and 999 in the usual Gaussian elimination to be subtracted from the relatively much smaller numbers -1 and 0.002 respectively, with rounding errors as noted.

This problem may usually be avoided by using a modification of Gaussian elimination called **pivoting**. In fact, there are several variations, of which we describe only one, called *partial pivoting*. The steps follow Gaussian elimination, except that rather than using the first available nonzero entry in a column to eliminate the entries below it, we choose the one whose absolute value is *largest*. This entry is then called the *pivot*. We say that we "pivot on it" when we use it to eliminate other column entries.

EXAMPLE 2-41

We solve the system in the preceding example by pivoting on the second entry (shown bracketed, []) in column 1, since it has absolute value 1 >

0.001. We first bring it to the (1,1) position by a row interchange. The steps in pivoting are therefore

$$\begin{bmatrix} 0.001 & 1 & 0.999 \\ [1] & -1 & 0.002 \end{bmatrix} \to \begin{bmatrix} [1] & -1 & 0.002 \\ 0.001 & 1 & 0.999 \end{bmatrix} \to \begin{bmatrix} 1 & -1 & 0.002 \\ 0 & 1 & 0.999 \end{bmatrix}$$

($1 + 0.001 = 1$ and $0.001 \times 0.002 = 0$ to 3 digits). The corresponding equations are now $y = 0.999$ (as before), $x = 0.002 + 0.999 = 1.001 \approx 1$ (to 3 digits), which is actually the exact solution. ❑

In practice, the solution of large systems of equations by pivoting is usually preceded by **scaling**, another operation intended to improve the accuracy and reliability of the solutions. One of many possible scaling strategies is to ensure that the coefficients of the system do not differ too much from 1 in absolute value. This is achieved by (i) multiplying one or more equations by nonzero constants, and (ii) replacing one or more variables x by $x' = cx$, where c is an appropriate nonzero number.

EXAMPLE 2-42

Consider the system

$$\begin{aligned} 0.01x + 0.5y - z &= 1 \\ 0.02x + 1.01y + 1.01z &= -3 \\ 0.001x + 0.26y + 0.01z &= 0.11 \end{aligned}$$

We scale by (i) multiplying the third equation by 10 and (ii) letting $x' = 0.01x$. The resulting scaled system is

$$\begin{aligned} x' + 0.5y - z &= 1 \\ 2x' + 1.01y + 1.01z &= -3 \\ x' + 2.6y + 0.1z &= 1.1 \end{aligned}$$

Next, we solve by partial pivoting using 3-digit arithmetic. The sequence of augmented matrices is listed below, with pivots at each stage bracketed.

$$\begin{bmatrix} 1 & 0.5 & -1 & 1 \\ [2] & 1.01 & 1.01 & -3 \\ 1 & 2.6 & 0.1 & 1.1 \end{bmatrix} \to \begin{bmatrix} [2] & 1.01 & 1.01 & -3 \\ 1 & 0.5 & -1 & 1 \\ 1 & 2.6 & 0.1 & 1.1 \end{bmatrix}$$

$$\to \begin{bmatrix} [1] & 0.505 & 0.505 & -1.5 \\ 1 & 0.5 & -1 & 1 \\ 1 & 2.6 & 0.1 & 1.1 \end{bmatrix} \to \begin{bmatrix} 1 & 0.505 & 0.505 & -1.5 \\ 0 & -0.005 & -1.51 & 2.5 \\ 0 & [2.10] & -0.405 & 2.6 \end{bmatrix}$$

(here we have rounded -1.505 to -1.51 and 2.095 to 2.10)

$$\to \begin{bmatrix} 1 & 0.505 & 0.505 & -1.5 \\ 0 & [2.10] & -0.405 & 2.6 \\ 0 & -0.005 & -1.51 & 2.5 \end{bmatrix} \to \begin{bmatrix} 1 & 0.505 & 0.505 & -1.5 \\ 0 & [1] & -0.193 & 1.24 \\ 0 & -0.005 & -1.51 & 2.5 \end{bmatrix}$$

$$\to \begin{bmatrix} 1 & 0.505 & 0.505 & -1.5 \\ 0 & [1] & -0.193 & 1.24 \\ 0 & 0 & -1.51 & 2.51 \end{bmatrix}$$

(again we have rounded in obtaining the last row)

By back-substitution with rounding to 3 digits, we obtain

$$\begin{aligned} z &= -1.662 \approx -1.66, \\ y &= 1.24 + 0.193(-1.66) \approx 0.920, \\ x' &= -1.5 - 0.505(0.92) + 0.505(1.66) \approx -1.127 \approx -1.13. \end{aligned}$$

Thus, $x = 100x' = -113$. This compares well with the actual solution, which, to 5 significant figures, is $x = -112.38$, $y = 0.919\,32$, and $z = -1.6642$. On the other hand, using ordinary Gaussian elimination and 3-digit floating point arithmetic, we would get the very poor approximation $x = -66.0$, $y = 0.000$ and $z = -1.66$. We leave the details for the reader. ❑

The above discussion and examples are intended only to alert the reader to some of the problems in the very large and active field of numerical solution of systems of linear equations. For more detail, the reader should consult one of the many texts on numerical methods in linear algebra.

We conclude with a brief discussion of another family of methods which are faster for certain systems of equations. These are called **iterative methods** because they consist of repeated applications ("iterations") of a procedure which produces an approximation to the solution. When the method works, the approximations get better and better; they are said to *converge* to the actual solution. Otherwise, they are said to *diverge*, and the method fails. Consider the system

$$\begin{aligned} 10x + 2y - z &= 11 \\ x - 15y + 2z &= -12 \\ -2x + y + 20z &= 19. \end{aligned}$$

It is easy to see by substitution that the exact solution is $x = y = z = 1$, and it is unique. We solve the ith equation for the ith variable to get

$$\begin{aligned} x &= (1/10)(11 - 2y + z) \quad (i = 1) \\ y &= (-1/15)(-12 - x - 2z) \quad (i = 2) \\ z &= (1/20)(19 + 2x - y) \quad (i = 3) \end{aligned} \tag{2-13}$$

(Clearly this only works for systems $AX = B$ where A is square and has nonzero diagonal elements.)

In the **Jacobi method** (after Carl Gustav Jacob Jacobi, 1804–1851, eminent German mathematician who made highly original contributions to both mathematics and physics), we use these equations to obtain a new approximation for the solutions by substituting the preceding approximation for x, y, and z into the right-hand sides. This procedure is then repeated until an approximation of desired accuracy is obtained, provided the approximations converge.

In the absence of a better first approximation, obtained, for example, by guessing, the procedure is started by putting $x = y = z = 0$ into the right-hand sides of 2-13. We get

$$\begin{aligned} x &= (1/10)(11) = 1.1 \\ y &= (-1/15)(-12) = 0.8 \\ z &= (1/20)(19) = 0.95 \end{aligned}$$

For the second iteration, we put these values into the right-hand sides of 2-13 to get

$$\begin{aligned} x &= (1/10)[11 - 2(0.8) + 0.95] = 1.035 \\ y &= (-1/15)[-12 - 1.1 - 2(0.95)] = 1 \\ z &= (1/20)[19 + 2(1.1) - 0.8] = 1.02. \end{aligned}$$

The results of the first five iterations are shown in the table below.

Iteration	1	2	3	4	5
x	1.1	1.035	1.002	0.999 35	0.999 875
y	0.8	1	1.005	1.000 6	1.000 03
z	0.95	1.02	1.0035	0.999 95	0.999 932

We see that after only four iterations, we are very close to the actual answer.

In fact, for some large systems of equations, it is faster to approximate the solutions in this way. The actual number of arithmetic operations may be fewer than for solution by pivoting. This is particularly the case when the coefficient matrix A contains many zeros, in which case it is said to be **sparse**.

A variation of the Jacobi method which is often faster arises from the observation that since we compute the new values of x, y, and z in sequence for each iteration, we might as well use the new value of x to compute the new value of y, the new values of x and y to compute the new value of z, and so on (when there are more than 3 variables). This is called the **Gauss-Seidel method** (Philipp Ludwig Seidel, 1821–1896, German mathematician).

For the example above, we would therefore get, for the first iteration (starting with $x = y = z = 0$ as before),

$$\begin{aligned} x &= (1/10)(11 + 0 + 0) = 1.1 \\ y &= (-1/15)[-12 - 1.1 - 2(0)] = 0.873 \\ z &= (1/20)[19 + 2(1.1) - 0.873] = 1.0163. \end{aligned}$$

The results of the first three iterations are shown below.

Iteration	1	2	3
x	1.1	1.026 97	0.999 456
y	0.873	1.003 97	1.000 300
z	1.0163	1.002 50	0.999 931

In this example, it appears that the Gauss-Seidel method converges faster. While this is usually true, it is *not* always true.

As noted earlier, these methods *cannot* be used for all systems of n equations in n unknowns. However, when the coefficient matrix A of $AX = B$ has the property

$$|a_{ii}| > \sum_{j=1}^{n} |a_{ij}|$$

for all $i = 1, 2, \ldots, n$, then it can be shown that the iterative methods described above converge to the unique solution of the system. Such matrices A are called (strictly) **diagonally dominant**. The matrix in the example above is clearly of this type. The reader should note that such a matrix is necessarily nonsingular.

EXERCISES 2-6

1. Use Gaussian elimination to solve the following systems of equations.

(a)
$$\begin{aligned} x - 2y + 3z &= 6 \\ -2x + y - z &= -1 \\ 5x - 3y + z &= 2 \end{aligned}$$

(b)
$$\begin{aligned} 3w + 4x - y + z &= -3 \\ 2x - z &= -1 \\ 5w - 6x + 2z &= 9 \\ w + x + y - z &= 2 \end{aligned}$$

2. **(a)** Solve the systems of equations in question 1, Exercises 2-1 by Gaussian elimination.

(b) Solve the systems of equations in Examples 2-11 and 2-12 by Gaussian elimination.

3. **(a)** Express the following numbers in floating point form. Where possible, state the number of significant figures to which the number is expressed.

(i) 1234.0 **(ii)** $-26/4$ **(iii)** $-0.004\,27$ **(iv)** 2425

(v) -23.364 **(vi)** 0.0004 **(vi)** -2435

(b) Round each of the numbers in (a) to **(i)** 3 significant digits, **(ii)** 2 significant digits.

(c) Use 3-digit floating point arithmetic to compute **(i)** $(199 + 0.3) + 1.4$, **(ii)** $199 + (0.3 + 1.4)$.

(d) The fact that the answers are not the same illustrates that addition is not associative when using rounding to a fixed number of digits. Make an example using three numbers to see whether this is the case for multiplication.

4. **(a)** Use partial pivoting to solve the systems of equations in question 1 above.

(b) Use partial pivoting to find a row echelon form for the matrices in question 4, Exercises 2-2.

5. Use partial pivoting to solve the systems in Examples 2-11 and 2-12.

6. Solve the following systems of equations by **(i)** Gaussian elimination, and **(ii)** partial pivoting. In both cases, round to 3 significant digits after each arithmetic operation. Compare the results.

(a)
$$\begin{aligned} 0.002x + 0.5y &= 2.1 \\ x - 0.3y &= 4 \end{aligned}$$

(b)
$$\begin{aligned} 0.130x - 2.10y + z &= -1.07 \\ -1.20x + 0.500y + 3.30z &= 9.70 \\ -3x + 0.110y - 0.900z &= -5.48 \end{aligned}$$

7. Solve the following system of equations by scaling and partial pivoting. Use 3-digit arithmetic.

$$\begin{aligned} 0.002x - 0.03y - 0.005z &= -0.102 \\ 2x - 3y + 4z &= -3 \\ 0.1x + 0.05y + 0.1z &= 0.25 \end{aligned}$$

8. Solve the system in Example 2-42 by partial pivoting without scaling, and using 3-digit arithmetic. Compare with the earlier results.

9. A system of equations $AX = B$ where A is $n \times n$ is called *ill-conditioned* if a small change in one or more of the entries of $(A{:}B)$ causes a large change in the solutions of the system. This is a highly undesirable situation, since if the data for $(A{:}B)$ result from experimental measurement, they are necessarily subject to "experimental error". Solve the system

$$\begin{aligned} 2.001x + 5y &= 7.001 \\ 2x + 5y &= 7 \end{aligned}$$

and also the slightly changed system

$$\begin{aligned} 2.001x + 5y &= 7 \\ 2x + 5y &= 7, \end{aligned}$$

to show that these systems are ill-conditioned.

10. Solve the following systems of equations using **(i)** the Jacobi method, and **(ii)** the Gauss-Seidel method. Use 3-digit floating point arithmetic in both cases.

(a)
$$\begin{aligned} 15x - 3y + z &= 13 \\ -2x + 20y - z &= 17 \\ 4x + 3y - 18z &= -11 \end{aligned}$$

(b)
$$\begin{aligned} -10.1x + 3y - z &= -2.1 \\ 2.1x - 15y + 3.2z &= -39.7 \\ -2y + 10z &= 4 \end{aligned}$$

(c)
$$\begin{aligned} x - 20y + 2z &= 0 \\ 10x + z &= 10 \\ -x - y + 10z &= 99 \end{aligned}$$

(Hint: rearrange the equations to make the coefficient matrix strictly diagonally dominant.)

11. An $n \times n$ matrix A is called *tridiagonal* if its only nonzero entries are those on the main diagonal and those immediately above or below the main diagonal. For example, the matrix A given below is tridiagonal. Tridiagonal matrices are important examples of sparse matrices. It can be shown that for such matrices, the system $AX = B$ is solved faster using the Gauss-Seidel method than the Jacobi method (if they converge).

$$A = \begin{bmatrix} 10 & -1 & 0 & 0 & 0 & 0 \\ 2 & 10 & 1 & 0 & 0 & 0 \\ 0 & -1 & -10 & 2 & 0 & 0 \\ 0 & 0 & 1 & 10 & 1 & 0 \\ 0 & 0 & 0 & -1 & 10 & 1 \\ 0 & 0 & 0 & 0 & -1 & -20 \end{bmatrix}$$

(a) Solve the system $AX = B$ where A is the matrix above and B is the 6×1 matrix with entries $11, -7, -11, -8, 10$, and 19 using **(i)** the Jacobi method, and **(ii)** the Gauss-Seidel method.

(b) Solve the system of (a) above using partial pivoting.

3

DETERMINANTS

3-1 Introduction
3-2 Properties of Determinants
3-3 Cramer's Rule
3-4 The Cross Product (Optional)

3-1 Introduction

The question of whether or not an $n \times n$ matrix A is invertible is an important one in linear algebra. We know how to find the inverse of A when it exists. And, we already have several criteria for invertibility. For example, A is invertible iff it is row-equivalent to the $n \times n$ identity matrix I. We shall develop other criteria in Chapter 4.

In this chapter, we define a "numerical" criterion for invertibility in the following sense: we want to assign a number to A (calculated from its entries, of course) so that we will know whether A is invertible or not by mere inspection of that number. We call this number the **determinant** of A since it will "determine" whether or not A is invertible. We denote it by *det* A.

Thus, "*det*" may be regarded as a function which assigns a number (*det* A) to any square matrix A. We shall find that A will be invertible precisely when *det* $A \neq 0$. The theory of determinants was first considered by the French mathematician Alexandre-Théophile Vandermonde (1735–1796).

The reader may find the following analogous situation useful for comparison. Recall that the **discriminant** $D = b^2 - 4ac$ of a quadratic equation $ax^2 + bx + c = 0$ allows us to decide whether or not the equation has real roots: it does when $D \geq 0$, and it doesn't when $D < 0$.

In the following discussion, we will show how our existing criteria for invertibility lead to formulae for *det* A when A is 2×2 and 3×3. We will then use these as motivation for the general formula for *det* A when A is $n \times n$.

First, let us use Gauss-Jordan reduction to find the inverse of an arbitrary 2×2 matrix A:

$$A = \begin{bmatrix} a_{11} & a_{12} \\ a_{21} & a_{22} \end{bmatrix}.$$

We suppose $a_{11} \neq 0$. Then, the steps to find the inverse are

$$\begin{bmatrix} a_{11} & a_{12} & 1 & 0 \\ a_{21} & a_{22} & 0 & 1 \end{bmatrix} \rightarrow \begin{bmatrix} 1 & a_{12}/a_{11} & 1/a_{11} & 0 \\ a_{21} & a_{22} & 0 & 1 \end{bmatrix}$$

$$\rightarrow \begin{bmatrix} 1 & a_{12}/a_{11} & 1/a_{11} & 0 \\ 0 & a_{22} - a_{12}a_{21}/a_{11} & -a_{21}/a_{11} & 1 \end{bmatrix}.$$

From the last matrix, it is clear that A^{-1} will exist iff

$$a_{22} - a_{12}a_{21}/a_{11} = (a_{11}a_{22} - a_{12}a_{21})/a_{11} \neq 0,$$

i.e., iff

$$a_{11}a_{22} - a_{12}a_{21} \neq 0.$$

For this reason, we will define the number *det* A by

$$\mathit{det}\, A = a_{11}a_{22} - a_{12}a_{21} \tag{3-1}$$

when A is the general 2×2 matrix above. The reader should carry out the analogous calculation for the case $a_{11} = 0$. Also, question 2, Exercises 3-1, provides a geometrical interpretation of *det* A for 2×2 matrices.

EXAMPLE 3-1

(a) $det\begin{bmatrix} 1 & 2 \\ 3 & 4 \end{bmatrix} = 4 - 6 = -2;$ **(b)** $det\begin{bmatrix} 1 & -2 \\ -2 & 4 \end{bmatrix} = 4 - 4 = 0.$

The reader should check that the matrix in (a) is invertible, while the one in (b) is not. ❑

Alternatively, let us make use of the fact that if a 2×2 matrix A as above is *not* invertible, then neither is A^t, since $(A^t)^{-1} = (A^{-1})^t$. (See Example 2-27(b).) It follows that A^t is row-equivalent to a matrix with a zero row, as otherwise A^t would be row-equivalent to I. By the steps of Gauss-Jordan reduction, this means that the rows of A^t must be multiples of each other, or, since the columns of A are the rows of A^t, that the columns of A are multiples of each other. Therefore, there is a scalar c such that

$$\begin{bmatrix} a_{11} \\ a_{21} \end{bmatrix} = c\begin{bmatrix} a_{12} \\ a_{22} \end{bmatrix}.$$

This yields the system of equations

$$\begin{aligned} a_{11} &= ca_{12} && \text{(i)} \\ a_{21} &= ca_{22}. && \text{(ii)} \end{aligned}$$

Now, we can eliminate c by subtracting $a_{12}\times$(ii) from $a_{22}\times$(i) to get

$$a_{11}a_{22} - a_{12}a_{21} = ca_{12}a_{22} - ca_{22}a_{12} = 0.$$

Thus, we see that if A is singular, *det* $A = 0$.

Next, we apply the preceding argument to a 3×3 matrix

$$A = \begin{bmatrix} a_{11} & a_{12} & a_{13} \\ a_{21} & a_{22} & a_{23} \\ a_{31} & a_{32} & a_{33} \end{bmatrix}$$

to obtain an expression for *det* A. Suppose that A is not invertible, so that A^t is row equivalent to a matrix having a zero row. Again, by the steps of Gauss-Jordan reduction, this means that one of the columns of A must be a sum of multiples of the other two. For simplicity, let us suppose that the first column is a sum of multiples of the others. Then there are scalars c_1 and c_2 such that

$$\begin{bmatrix} a_{11} \\ a_{21} \\ a_{31} \end{bmatrix} = c_1\begin{bmatrix} a_{12} \\ a_{22} \\ a_{32} \end{bmatrix} + c_2\begin{bmatrix} a_{13} \\ a_{23} \\ a_{33} \end{bmatrix}.$$

This leads to the following system of three equations for c_1 and c_2:

$$a_{11} = c_1 a_{12} + c_2 a_{13} \qquad \text{(i)}$$
$$a_{21} = c_1 a_{22} + c_2 a_{23} \qquad \text{(ii)}$$
$$a_{31} = c_1 a_{32} + c_2 a_{33}. \qquad \text{(iii)}$$

In a manner similar to the 2×2 case discussed earlier, we can eliminate c_1 and c_2 from these equations by adding suitable multiples of them as follows:

Multiply (i) by $a_{22}a_{33} - a_{32}a_{23}$;
(ii) by $a_{12}a_{33} - a_{32}a_{12}$;
(iii) by $a_{12}a_{23} - a_{22}a_{13}$.

The results are

$$a_{11}a_{22}a_{33} - a_{11}a_{32}a_{23} = \underset{\text{(A)}}{c_1a_{12}a_{22}a_{33}} - \underset{\text{(B)}}{c_1a_{12}a_{32}a_{23}} + \underset{\text{(C)}}{c_2a_{13}a_{22}a_{33}} - \underset{\text{(D)}}{c_2a_{13}a_{32}a_{23}} \qquad \text{(i}'\text{)}$$

$$a_{21}a_{12}a_{33} - a_{21}a_{32}a_{13} = \underset{\text{(A)}}{c_1a_{22}a_{12}a_{33}} - \underset{\text{(E)}}{c_1a_{22}a_{32}a_{13}} + \underset{\text{(F)}}{c_2a_{23}a_{12}a_{33}} - \underset{\text{(D)}}{c_2a_{23}a_{32}a_{13}} \qquad \text{(ii}'\text{)}$$

$$a_{31}a_{12}a_{23} - a_{31}a_{22}a_{13} = \underset{\text{(B)}}{c_1a_{32}a_{12}a_{23}} - \underset{\text{(E)}}{c_1a_{32}a_{22}a_{13}} + \underset{\text{(F)}}{c_2a_{33}a_{12}a_{23}} - \underset{\text{(C)}}{c_2a_{33}a_{22}a_{13}}. \qquad \text{(iii}'\text{)}$$

Now, the right-hand sides of these equations contain six pairs of equal terms as indicated ((A) through (F)). They can be made to cancel completely if we take (i′) − (ii′) + (iii′) to get

$$a_{11}a_{22}a_{33} - a_{11}a_{32}a_{23} - a_{21}a_{12}a_{33} + a_{21}a_{32}a_{13} + a_{31}a_{12}a_{23} - a_{31}a_{22}a_{13} = 0. \qquad \text{(3-2)}$$

Thus, we shall call the left-hand side of equation 3-2 *det A*, when A is 3×3 as above.

The reader may have noticed that the multiplying factors we used were, in fact, determinants of 2×2 matrices obtained from A by deleting a row and a column. For example,

$$a_{22}a_{33} - a_{32}a_{23} = det \begin{bmatrix} a_{22} & a_{23} \\ a_{32} & a_{33} \end{bmatrix},$$

which is the determinant of the 2×2 matrix formed from A by deleting row 1 and column 1. This motivates the following definition.

DEFINITION 3-1

If A is an $n \times n$ matrix, we denote by $A(i/j)$ the $(n-1) \times (n-1)$ matrix obtained from A by deleting row i and column j. There are n^2 matrices of this type. The number $det\, A(i/j)$ is called the (i,j) **minor** of A.

With this notation, and by grouping terms in equation 3-2, we see that for the 3×3 matrix A,

$$det\, A = a_{11} det\, A(1/1) - a_{21} det\, A(2/1) + a_{31} det\, A(3/1) \qquad (3\text{-}3)$$

In fact, there are five other expressions for $det\, A$ using different minors; the interested reader may wish to write some of them down. However, we shall be doing this in a systematic way for a general $n \times n$ matrix later.

EXAMPLE 3-2

(a) Let us use formula 3-3 to compute $det\, A$ where

$$A = \begin{bmatrix} 2 & 3 & -1 \\ -1 & 0 & 2 \\ 3 & -1 & 1 \end{bmatrix}.$$

$$A(1/1) = \begin{bmatrix} 0 & 2 \\ -1 & 1 \end{bmatrix}, A(2/1) = \begin{bmatrix} 3 & -1 \\ -1 & 1 \end{bmatrix}, A(3/1) = \begin{bmatrix} 3 & -1 \\ 0 & 2 \end{bmatrix},$$

$a_{11} = 2$, $a_{21} = -1$, and $a_{31} = 3$, so

$$\begin{aligned} det\, A &= 2[(0)(1) - (-1)(2)] - (-1)[(3)(1) - (-1)(-1)] \\ &\quad + 3[(3)(2) - (0)(-1)] \\ &= 2(2) + 1(2) + 3(6) = 24. \end{aligned}$$

The reader should check that A is invertible.

(b) Let B be given by

$$B = \begin{bmatrix} 1 & -1 & 1 \\ -2 & 3 & 0 \\ -1 & 2 & 1 \end{bmatrix}.$$

Using the same formula, we find

$$det\, B = 1(3 - 0) - (-2)(-1 - 2) + (-1)(0 - 3) = 3 - 6 + 3 = 0.$$

The reader should check that B is singular. ❑

Now it should be plausible that we could, in principle, employ the same method (or Gauss-Jordan reduction) to obtain an expression for a number we would call *det A* when A is of any size (but, the algebra would be formidable!). In order to see what this expression should look like, let us look again at the *form* of *det A* for the 3×3 case. By rearranging the factors in the terms of equation 3-2, we can write

$$\begin{aligned} det\ A = {} & a_{11}a_{22}a_{33} - a_{11}a_{23}a_{32} + a_{13}a_{21}a_{32} - \\ & a_{12}a_{21}a_{33} + a_{12}a_{23}a_{31} - a_{13}a_{22}a_{31}. \end{aligned} \qquad (3\text{-}4)$$

Each of the six terms in this expression has the form $a_{1\_}\,a_{2\_}\,a_{3\_}$, where the blanks are filled by all six possible arrangements of the integers 1, 2, 3. These are (1,2,3), (1,3,2), (3,1,2), (2,1,3), (2,3,1), and (3,2,1), which are the **permutations** of the set of integers {1,2,3,}. In general,

DEFINITION 3-2

A **permutation** σ ("sigma") of the set of integers $\{1,2,...,n\}$ is an arrangement of these n integers denoted by $(\sigma(1),\sigma(2),...,\sigma(n))$. In this notation, each $\sigma(i)$ denotes one of the integers from 1 to n such that (a) when $i \neq j$, $\sigma(i) \neq \sigma(j)$; (b) every integer in $\{1,2,...,n\}$ is $\sigma(i)$ for some $i = 1, 2, ..., n$.

For example, if we denote the permutation (3,1,2) of {1,2,3} by σ, then $\sigma(1) = 3$, $\sigma(2) = 1$, and $\sigma(3) = 2$. So, a permutation σ may be regarded as a function σ: $\{1,2,...,n\} \rightarrow \{1,2,...,n\}$ with the property that its values include *all* elements of the set $\{1,2,...,n\}$. And the notation $(\sigma(1),\sigma(2),...,\sigma(n))$ for the permutation is simply a list of the *values* of σ at the numbers 1, 2, ..., n.

With this point of view, the permutation $(1,2,...,n)$ (i.e., the "natural arrangement" of $\{1,2,...,n\}$) is the identity function on the set $\{1,2,...,n\}$; we denote this permutation by ι ("iota"). We have already listed all the permutations of {1,2,3} and noted that there were 6. In general,

LEMMA 3-1

There are $n! = n\times(n-1)\times(n-2)...3\times 2\times 1$ permutations of $\{1,2,...,n\}$, where n is a positive integer.

PROOF Each permutation $(\sigma(1),...,\sigma(n))$ is formed by selecting exactly one of the integers in $\{1,2,...,n\}$ for the value $\sigma(i)$, $i = 1, 2, ..., n$, with no duplicates. Thus $\sigma(1)$ can be chosen in n ways, and, for each of these ways, $\sigma(2)$ can be chosen in $n-1$ ways, as we can't use the value assigned to $\sigma(1)$. We continue in this way, and find that the total number of ways of choosing values is $n\times(n-1)...3\times 2\times 1 = n!$, as stated. ∎

EXAMPLE 3-3

(a) There are $4! = 24$ permutations of the set $\{1,2,3,4\}$.

(b) There are $5! = 120$ permutations of $\{1,2,3,4,5\}$. ❑

The set of all $n!$ permutations of $\{1,2,\ldots,n\}$ is denoted $S(n)$, and is called the **symmetric group of degree n**. It plays an important role in many areas of mathematics, especially algebra. The term "group" arises because if σ and τ ("tau") are elements of $S(n)$, we can form their **composition** $\sigma\tau$ (as functions), which is also an element of $S(n)$. In this context, $\sigma\tau$ is called the "product" of σ and τ.

For example, in $S(3)$, let $\sigma = (3,2,1)$ and $\tau = (2,3,1)$. This means that $\sigma(1) = 3$, $\sigma(2) = 2$, $\sigma(3) = 1$, $\tau(1) = 2$, $\tau(2) = 3$, and $\tau(3) = 1$. Thus, using the definition of composition of functions, $\sigma\tau(1) = \sigma(2) = 2$, $\sigma\tau(2) = \sigma(3) = 1$, and $\sigma\tau(3) = \sigma(1) = 3$. Therefore, we see that $\sigma\tau = (2,1,3)$, which is an element of $S(3)$. These algebraic properties of $S(n)$ are explored in the exercises.

Returning to equation 3-4 for *det A*, we can now fill in the blanks to represent each of the six terms as $a_{1\sigma(1)}a_{2\sigma(2)}a_{3\sigma(3)}$, where σ is one of the six elements of $S(3)$. The equation for *det A* is then the sum over all permutations in $S(3)$ with an appropriate $+$ or $-$ sign for each.

Our next task is to obtain a rule to determine the appropriate sign ($+$ or $-$) in each of those terms in *det A*. To this end, notice that each term is obtained from the previous one (in the order written in 3-4) by an interchange of exactly two subscripts. For example, $a_{11}a_{2\mathbf{3}}a_{3\mathbf{2}}$ is obtained from $a_{11}a_{2\mathbf{2}}a_{\mathbf{3}3}$ by interchanging the **2** and **3** (indicated by boldface). And after each such interchange, the sign of the term changes.

Alternatively, we can determine the sign of a term as $+$ or $-$ according to whether the *total* number of interchanges that has taken place (from the first one, $a_{11}a_{22}a_{33}$) is even or odd respectively. For example, the third term, $a_{13}a_{21}a_{32}$, has a $+$ sign, since two interchanges have taken place in obtaining it from $a_{11}a_{22}a_{33}$. This phenomenon is described by the notion of "sign" of a permutation.

DEFINITION 3-3

A permutation σ is called **even** or **odd** according to whether the number of interchanges of pairs of adjacent integers used to obtain $(\sigma(1),\ldots,\sigma(n))$ from $\iota = (1,2,\ldots,n)$ (or vice-versa) is even or odd respectively. We say that the **sign** of σ, denoted *sgn* σ (or $\varepsilon(\sigma)$), is $+1$ when σ is even, and -1 when σ is odd.

EXAMPLE 3-4

(a) The permutation $\sigma = (3,2,1)$ in $S(3)$ is odd, because

$$(3,2,1) \to (3,1,2) \to (1,3,2) \to (1,2,3)$$

is a sequence of three interchanges which changes σ to the natural ordering $\iota = (1,2,3)$. Note that the reverse steps will produce σ from ι. So, $sgn\ \sigma = -1$.

(b) The permutation $\tau = (2,3,1)$ is even because

$$(2,3,1) \to (2,1,3) \to (1,2,3)$$

is an even number of interchanges. We write $sgn\ \tau = +1$ (or 1).

(c) The identity permutation is always even: $sgn\ \iota = 1$. ❑

Remarks

(1) There are in general many sequences of interchanges of adjacent pairs which will change a given permutation to the identity. For example,

(i) $(3,2,1) \to (2,3,1) \to (2,1,3) \to (1,2,3)$

and

(ii) $(3,2,1) \to (2,3,1) \to (2,1,3) \to (2,3,1) \to (2,1,3) \to (1,2,3)$

are different sequences from the one given for σ in Example 3-4(a) above. However, it can be shown that *the number of interchanges in all such sequences is always even, or always odd, depending only on the permutation.* This is certainly the case for the three sequences given for (3,2,1): they all have an odd number of interchanges.

(2) We state without proof the following equivalent method for finding the sign of a permutation σ. For each $\sigma(i)$, count the number of integers to its right that are less than $\sigma(i)$, and add these for all i to obtain a total $p(\sigma)$. Then σ is even iff $p(\sigma)$ is even, and odd otherwise.

EXAMPLE 3-5

(a) If $\sigma = (3,2,1) \in S(3)$, $p(\sigma) = 2 + 1 + 0 = 3$, so σ is odd.

(b) If $\tau = (2,3,1) \in S(3)$, $p(\tau) = 1 + 1 + 0 = 2$, so τ is even.

(c) If μ ("mu") $= (1,4,3,2) \in S(4)$, $p(\mu) = 0 + 2 + 1 + 0 = 3$, so μ is odd. ❑

With these preliminaries, we can now rewrite equation 3-4 for *det A* (where *A* is 3×3) as

$$det\,A = \sum_{\sigma} sgn\,\sigma\, a_{1\sigma(1)} a_{2\sigma(2)} a_{3\sigma(3)}$$

where the summation notation means to sum over all six permutations in $S(3)$. We generalize this form to obtain the next definition.

DEFINITION 3-4

If A is an $n \times n$ matrix, its **determinant** $det\,A$ is defined by

$$det\,A = \sum_{\sigma} sgn\,\sigma\, a_{1\sigma(1)} a_{2\sigma(2)} \dots a_{n\sigma(n)}. \qquad (3\text{-}5)$$

Note the following:

(1) There are $n!$ terms in this sum: the summation notation here indicates summation over all $n!$ of the $\sigma \in S(n)$.
(2) Each term is a product of n entries from A.
(3) Each term contains a single entry from every row and every column of A, since $\sigma(1), \dots, \sigma(n)$ are distinct integers in $\{1,2,\dots,n\}$.

EXAMPLE 3-6

If I is the $n \times n$ identity matrix, $det\,I = 1$. For, by definition 3-4,

$$det\,I = \sum_{\sigma} sgn\,\sigma\, I_{1\sigma(1)} \dots I_{n\sigma(n)}.$$

Now, since the off-diagonal entries of I are 0, it follows that for any j, $I_{j\sigma(j)} = 0$ unless $\sigma(j) = j$, in which case $I_{jj} = 1$. Thus, all terms in the sum are zero except when $\sigma(j) = j$ for all j, which means $\sigma = \iota$. So

$$det\,I = sgn\,\iota\, I_{11} I_{22} \dots I_{nn} = 1. \;\square$$

While the definition of $det\,A$ above is useful for theoretical purposes (as we shall see), it is less useful for computing determinants of specific matrices. Instead, we can compute $det\,A$ in terms of determinants of $(n-1) \times (n-1)$ submatrices $A(i/j)$ as discussed earlier for the case $n = 3$ (equation 3-3).

DEFINITION 3-5

If A is $n \times n$, the (i,j) **cofactor** of A is $(-1)^{i+j}\, det\,A(i/j)$. This is often denoted C_{ij} and called the **cofactor of the entry** a_{ij} because it is calculated using the submatrix obtained from A by crossing out the row and column containing a_{ij}. Notice that there are n^2 cofactors when A is an $n \times n$ matrix.

The formulae that generalize equation 3-3 for an $n \times n$ matrix are given in the theorem below, the proof of which we omit.

THEOREM 3-1

(1) For any $i = 1, 2, \ldots, n$,

$$det\,A = \sum_{j=1}^{n} (-1)^{i+j} a_{ij} det\,A(i/j) = \sum_{j=1}^{n} a_{ij} C_{ij}.$$

This is called the **expansion of *det A* by the ith row** because it is the sum of the products of entries from row i with their cofactors (note that we sum on *columns* j).

(2) For any $j = 1, 2, \ldots, n$,

$$det\,A = \sum_{i=1}^{n} (-1)^{i+j} a_{ij} det\,A(i/j) = \sum_{i=1}^{n} a_{ij} C_{ij}.$$

This is called the **expansion of *det A* by the jth column** (note that we sum over *rows* i). ■

These formulae are attributed to Pierre-Simon de Laplace (1749–1827), one of the most eminent French scientists of all time. Laplace made important contributions to celestial mechanics and other areas of mathematical physics.

EXAMPLE 3-7

(a) Let A be the matrix of Example 3-2 (a):

$$A = \begin{bmatrix} 2 & 3 & -1 \\ -1 & 0 & 2 \\ 3 & -1 & 1 \end{bmatrix}.$$

In that example, we have actually computed *det A* by its expansion along column $j = 1$. For,

$$C_{11} = (-1)^{1+1} det\,A(1/1) = det \begin{bmatrix} 0 & 2 \\ -1 & 1 \end{bmatrix} = 2,$$

$$C_{21} = (-1)^{2+1} det\,A(2/1) = -det \begin{bmatrix} 3 & -1 \\ -1 & 1 \end{bmatrix} = -2,$$

$$C_{31} = (-1)^{3+1} det\,A(3/1) = det \begin{bmatrix} 3 & -1 \\ 0 & 2 \end{bmatrix} = 6.$$

So, using Theorem 3-1 (part (b)),

$$det\, A = a_{11}C_{11} + a_{21}C_{21} + a_{31}C_{31} = (2)(2) + (-1)(-2) + (3)(6) = 24,$$

as before.

(b) Next, let us evaluate $det\, A$ by expansion along row 2. The relevant cofactors are

$$C_{21} = -2 \text{ (as in (a))},$$

$$C_{22} = (-1)^{2+2} det\, A(2/2) = det\begin{bmatrix} 2 & -1 \\ 3 & 1 \end{bmatrix} = 5,$$

$$C_{23} = (-1)^{2+3} det\, A(2/3) = -det\begin{bmatrix} 2 & 3 \\ 3 & -1 \end{bmatrix} = 11.$$

So, by Theorem 3-1 (part (a)),

$$det\, A = a_{21}C_{21} + a_{22}C_{22} + a_{23}C_{23} = (-1)(-2) + (0)(5) + (2)(11) = 24,$$

as before. Notice that, on account of the 0 in row 2, we did not really need to calculate the cofactor C_{22}. ❑

In general, since the theorem gives $2n$ ways to compute $det\, A$, we should choose an expansion along a row or column that has the most 0 entries.

EXAMPLE 3-8

Let B be the matrix

$$B = \begin{bmatrix} -1 & -2 & 4 & 0 \\ 2 & 3 & 0 & 1 \\ 0 & 1 & 1 & 2 \\ 3 & 1 & 0 & -2 \end{bmatrix}.$$

Column 3 contains the most 0's, so we expand by that column. We need only the cofactors C_{13} and C_{33}. They are

$$C_{13} = (-1)^{1+3} det\begin{bmatrix} 2 & 3 & 1 \\ 0 & 1 & 2 \\ 3 & 1 & -2 \end{bmatrix} = 2\, det\begin{bmatrix} 1 & 2 \\ 1 & -2 \end{bmatrix} - 0 + 3\, det\begin{bmatrix} 3 & 1 \\ 1 & 2 \end{bmatrix}$$

(by expansion along column 1 of the 3×3 matrix)

$$= 2(-4) + 3(5) = 9,$$

$$C_{33} = (-1)^{3+3} det\begin{bmatrix} -1 & -2 & 0 \\ 2 & 3 & 1 \\ 3 & 1 & -2 \end{bmatrix}$$

$$= -det\begin{bmatrix} 3 & 1 \\ 1 & -2 \end{bmatrix} - (-2)\, det\begin{bmatrix} 2 & 1 \\ 3 & -2 \end{bmatrix} + 0$$

(by expansion along row 1 of the 3×3 matrix)

$$= (-1)(-7) + 2(-7) = -7.$$

Thus,

$$det\ B = b_{13}C_{13} + b_{33}C_{33} = 4(9) + 1(-7) = 29. \square$$

In this section and the next, we will study several properties of determinants to obtain computational procedures that are more efficient than the use of row and column expansions. One of the most important of these is illustrated in the next example.

EXAMPLE 3-9

Let A be the general 2×2 matrix

$$A = \begin{bmatrix} a_{11} & a_{12} \\ a_{21} & a_{22} \end{bmatrix}.$$

The cofactors of A are then the 1×1 matrices

$$\begin{aligned} C_{11} &= (-1)^{1+1} det\,A(1/1) = a_{22}, \\ C_{12} &= (-1)^{1+2} det\,A(1/2) = -a_{21}, \\ C_{21} &= -a_{12}, \\ C_{22} &= a_{11}. \end{aligned}$$

So, for example, expanding by the first row, we find

$$det\ A = a_{11}C_{11} + a_{12}C_{12} = a_{11}a_{22} - a_{12}a_{21},$$

as in equation 3-1.

Now consider the matrix A' obtained from A by interchanging columns 1 and 2:

$$A' = \begin{bmatrix} a_{12} & a_{11} \\ a_{22} & a_{21} \end{bmatrix}.$$

Then we see that

$$det\ A' = a_{12}a_{21} - a_{11}a_{22} = -det\ A.$$

A similar result holds if we interchange the rows of A. $\square$

In general, we have

LEMMA 3-2

If A' is obtained from an $n \times n$ matrix A by interchanging two *adjacent* rows (or columns), then $det\ A' = -det\ A$.

PROOF (Optional) This is included primarily to illustrate the "group" properties of permutations. A simpler proof is possible using Theorem 3-1 and is indicated in question 8, Exercises 3-1, for the case in which adjacent columns are interchanged.

We consider the case in which A' has been obtained from A by interchanging rows i and $i+1$. Now,

$$det\, A' = \sum_{\sigma} sgn\, \sigma\, a'_{1\sigma(1)} \ldots a'_{n\sigma(n)}.$$

But $a'_{ij} = a_{i+1\,j}$ and $a'_{i+1\,j} = a_{ij}$ for any j since we have interchanged rows i and $i+1$. Consequently, $a'_{i\sigma(i)} = a_{i+1\ \sigma(i)}$ and $a'_{i+1\ \sigma(i+1)} = a_{i\ \sigma(i+1)}$. It follows that

$$det\, A' = \sum_{\sigma} sgn\, \sigma\, a_{1\sigma(1)} \ldots a_{i+1\ \sigma(i)} a_{i\ \sigma(i+1)} \ldots a_{n\ \sigma(n)}. \tag{3-6}$$

Next, let τ be the permutation in $S(n)$ that interchanges only i and $i+1$, and leaves all other integers fixed:

$$\tau(i) = i+1,\ \tau(i+1) = i \text{ and } \tau(j) = j, j \neq i, i+1;$$

that is,

$$\tau = (1,2,\ldots,i-1,i+1,i,i+2,\ldots,n).$$

Then, $a_{j\sigma(j)} = a_{j\ \sigma\tau(j)}$ for $j \neq i,\ i+1$, while $a_{i+1\ \sigma(i)} = a_{i+1\ \sigma\tau(i+1)}$ and $a_{i\sigma(i+1)} = a_{i\sigma\tau(i)}$. Substituting into equation 3-6 and interchanging the ith and $(i+1)^{\text{th}}$ factors yield

$$det\, A' = \sum_{\sigma} sgn\, \sigma\, a_{1\sigma\tau(1)} \ldots a_{i\sigma\tau(i)} a_{i+1\ \sigma\tau(i+1)} \ldots a_{n\sigma\tau(n)}. \tag{3-7}$$

The right-hand side looks almost like $det\, A$ (see equation 3-5) except that $\sigma\tau$ appears rather than σ by itself. We can bring equation 3-7 to the required form by recalling that $\sigma\tau$ is, in fact, a permutation in $S(n)$. Furthermore, $sgn\ \sigma\tau = -sgn\ \sigma$ because $\sigma\tau$ involves one more interchange than σ, owing to τ (see also question 5, Exercises 3-1).

Finally, *every* permutation π ("pi") in $S(n)$ can be expressed as $\pi = \sigma\tau$ for exactly one σ in $S(n)$, namely, $\sigma = \pi\tau$, for then

$$\sigma\tau = (\pi\tau)\tau = \pi(\tau\tau) = \pi\iota = \pi$$

(because $\tau\tau = \iota$: the composition $\tau\tau$ changes i to $i+1$ and back to i again). Replacing $\sigma\tau$ by π and using $sgn\ \sigma = (-1)\ sgn\ \pi$, the sum in 3-7 can be rewritten as

$$det\, A' = \sum_{\pi} (-1) sgn\ \pi\, a_{1\pi(1)} \ldots a_{n\pi(n)} = -det\, A,$$

thus completing the proof (note that both the sum over σ and the sum over π involve all $n!$ permutations of $S(n)$). ■

EXAMPLE 3-10

Consider the matrix A with

$$A = \begin{bmatrix} 0 & 1 & 0 \\ 0 & 0 & 1 \\ 1 & 0 & 0 \end{bmatrix}.$$

By Lemma 3-2,

$$det\,A = -det \begin{bmatrix} 0 & 1 & 0 \\ 1 & 0 & 0 \\ 0 & 0 & 1 \end{bmatrix} = (-1)^2\, det \begin{bmatrix} 1 & 0 & 0 \\ 0 & 1 & 0 \\ 0 & 0 & 1 \end{bmatrix} = det\, I = 1. \square$$

EXERCISES 3-1

1. Use equation 3-1 to compute the determinants of the following 2×2 matrices.

(a) $\begin{bmatrix} -1 & 2 \\ 3/2 & -2 \end{bmatrix}$ **(b)** $\begin{bmatrix} 4 & 2 \\ 6 & 3 \end{bmatrix}$ **(c)** $\begin{bmatrix} 6 & 0 \\ -10 & 1 \end{bmatrix}$

(d) $\begin{bmatrix} 1 & 1 \\ 1 & 1 \end{bmatrix}$ **(e)** $\begin{bmatrix} -1/2 & 1 \\ 4 & 2 \end{bmatrix}$

2. Let A be the matrix

$$\begin{bmatrix} 2 & 1 \\ 1 & 3 \end{bmatrix}.$$

(a) The *parallelogram formed by its rows* is the parallelogram with vertices (0,0), (2,1), (2,1) + (1,3) = (3,4), and (1,3). Show that $det\,A$ is the area of this parallelogram.

(b) Let A be any 2×2 matrix. Show that $|det\,A|$ is the area of the parallelogram formed by the rows of A. What is the geometrical interpretation of $det\,A = 0$?

(c) Show that if

$$A = \begin{bmatrix} 1 & 0 & 0 \\ 0 & 2 & 1 \\ 0 & 1 & 3 \end{bmatrix},$$

then $det\,A$ is the volume of the parallelepiped having the rows of A as its edges.

3. Use equation 3-2 to compute the determinant of the following 3×3 matrices.

(a) $\begin{bmatrix} -1 & 1 & 0 \\ 2 & 0 & 1 \\ 1 & -1 & 1 \end{bmatrix}$ **(b)** $\begin{bmatrix} 1 & 2 & 3 \\ 4 & 5 & 6 \\ 7 & 8 & 9 \end{bmatrix}$ **(c)** $\begin{bmatrix} 1 & 2 & 1 \\ 0 & 3 & 3 \\ 0 & -1 & -1 \end{bmatrix}$

Check your answers by using equation 3-3.

4. Make a list of all 24 permutations in $S(4)$ together with their signs.

5. For the following pairs of permutations σ and τ in $S(3)$, find their product (composition) $\sigma\tau$, and verify that in every case, $sgn\ \sigma\tau = sgn\ \sigma\ sgn\ \tau$.

   (a) $\sigma = (1,2,3)$, $\tau = (2,3,1)$

   (b) $\sigma = (3,1,2)$, $\tau = (2,1,3)$

   (c) $\sigma = (1,3,2)$, $\tau = (1,3,2)$

6. If σ is a permutation in $S(n)$, its *inverse*, denoted σ^{-1}, is the permutation in $S(n)$ with the property that $\sigma^{-1}\sigma = \sigma\sigma^{-1} = \iota$. ($\sigma^{-1}$ is simply the inverse function of σ; $\sigma^{-1}: \{1,2,...,n\} \rightarrow \{1,2,...,n\}$. For example, if $\sigma = (2,3,1)$ in $S(3)$, then $\sigma^{-1} = (3,1,2)$.) Find the inverse of each permutation in $S(3)$ and verify that $sgn\ \sigma^{-1} = sgn\ \sigma$.

7. Find the determinant of the following matrices by **(a)** expanding along row 2; **(b)** expanding along column 2.

   (i) $\begin{bmatrix} 2 & -1 \\ 1 & 3 \end{bmatrix}$ **(ii)** $\begin{bmatrix} -2 & 3 & 0 \\ 1 & 1 & 1 \\ -3 & 4 & 2 \end{bmatrix}$

   (iii) $\begin{bmatrix} 2 & 0 & 3 & 4 \\ 1 & -2 & 1 & 3 \\ 0 & 1 & 2 & 2 \\ -4 & 1 & 0 & 1 \end{bmatrix}$ **(iv)** $\begin{bmatrix} 0 & 0 & 1 & -3 \\ 2 & -4 & 1 & 0 \\ 3 & 2 & 8 & -2 \\ 1 & -5 & 3 & 5 \end{bmatrix}$

   (v) $\begin{bmatrix} 0 & 1 & 2 & 3 & 0 \\ -1 & 0 & 0 & 0 & 0 \\ 0 & 2 & 0 & 1 & 1 \\ 0 & -3 & 0 & 0 & 0 \\ 1 & 1 & 1 & 1 & 1 \end{bmatrix}$

8. Use the expansion along column j of A' to show that if A' has been obtained from an $n \times n$ matrix A by interchanging columns j and $j+1$, then $det\ A' = -det\ A$.

9. **(a)** Use Definition 3-4 to show that if a matrix A has a row or column of 0's, then $det\ A = 0$.

   (b) Use Theorem 3-1 to prove the statement in (a).

10. Use Definition 3-4 to show that if A is a diagonal matrix (i.e., $a_{ij} = 0$ if $i \neq j$), then $det\ A = a_{11}a_{22}...a_{nn}$.

3-2 Properties of Determinants

We have seen that the formula for the determinant of a square matrix arises naturally from a relation that its entries must satisfy when it is singular. In this section, we explore several properties of the determinant function. For example, we consider how the determinant function behaves with respect to elementary row operations, transposes, and products. These properties are important for essentially two reasons. First, they will allow us to reduce the work in computing determinants of large matrices. And second, certain of these properties completely characterize the determinant function among all functions which assign a number to a square matrix. This fact may be of some comfort to readers who are bewildered by the complexity of the determinant formulae.

To begin with, we will show that $det\, A^t = det\, A$. This will have the consequence that *any results we state about the behaviour of det A in regard to its rows will also be true in regard to its columns*, as the columns of A are the rows of A^t.

We expand $det\, A^t$ by its ith row to get

$$det\, A^t = \sum_{j=1}^{n} (-1)^{i+j} a_{ji} det\, A^t(i/j),$$

where we have used the fact that $(A^t)_{ij} = a_{ji}$. Now, $A^t(i/j)$ is the matrix obtained from A^t by removing row i and column j. Some reflection will show that this is $A(j/i)^t$. To illustrate this, suppose

$$A = \begin{bmatrix} 1 & 2 & 3 \\ 4 & 5 & 6 \\ 7 & 8 & 9 \end{bmatrix}, A^t = \begin{bmatrix} 1 & 4 & 7 \\ 2 & 5 & 8 \\ 3 & 6 & 9 \end{bmatrix}.$$

Then we have, for example,

$$A^t(3/2) = \begin{bmatrix} 1 & 7 \\ 2 & 8 \end{bmatrix} = \begin{bmatrix} 1 & 2 \\ 7 & 8 \end{bmatrix}^t = A(2/3)^t.$$

Therefore, after substitution, we see that

$$det\, A^t = \sum_{j=1}^{n} (-1)^{i+j} a_{ji} det\, A(j/i)^t.$$

Now $A(j/i)$ is an $(n-1)\times(n-1)$ matrix, so, using induction on n, we may assume the result we wish to prove is true for this case: $det\, A(j/i)^t = det\, A(j/i)$. Substituting into the last equation, we get

$$det\, A^t = \sum_{j=1}^{n} (-1)^{i+j} a_{ji} det\, A(j/i) = det\, A$$

(by expansion along column i of A). And, for $n = 1$, the result is certainly true as 1×1 matrices are single numbers. By induction, the result is therefore true for all n. We have proved

THEOREM 3-2

If A is square, $\det A^t = \det A$. ■

EXAMPLE 3-11

Let A be given by

$$A = \begin{bmatrix} 1 & -1 & 0 & 3 \\ 2 & 0 & -1 & 2 \\ 0 & 2 & 2 & 0 \\ -1 & 0 & 3 & 1 \end{bmatrix}.$$

Then, expanding along column 2,

$$\det A = (-1)^{1+2}(-1) \det \begin{bmatrix} 2 & -1 & 2 \\ 0 & 2 & 0 \\ -1 & 3 & 1 \end{bmatrix} +$$

$$(-1)^{(3+2)}(2) \det \begin{bmatrix} 1 & 0 & 3 \\ 2 & -1 & 2 \\ -1 & 3 & 1 \end{bmatrix}$$

$$= 8 + 14 - 30 = -8.$$

And,

$$A^t = \begin{bmatrix} 1 & 2 & 0 & -1 \\ -1 & 0 & 2 & 0 \\ 0 & -1 & 2 & 3 \\ 3 & 2 & 0 & 1 \end{bmatrix},$$

so expanding by row 2,

$$\det A^t = (-1)^{2+1}(-1) \det \begin{bmatrix} 2 & 0 & -1 \\ -1 & 2 & 3 \\ 2 & 0 & 1 \end{bmatrix} + (-1)^{2+3}(2) \det \begin{bmatrix} 1 & 2 & -1 \\ 0 & -1 & 3 \\ 3 & 2 & 1 \end{bmatrix}.$$

Note that this expression contains the transposes of the matrices in the expansion of $\det A$. The reader can check that it therefore leads to the same value, so $\det A^t = -8$. ❑

The next property is a generalization of Lemma 3-2 proved in the preceding section.

THEOREM 3-3

If A' is obtained from an $n \times n$ matrix A by interchanging *any* two rows (or columns), then $det\, A' = -det\, A$.

PROOF We show that an interchange of rows (or columns) always requires an *odd* number of interchanges of *adjacent* rows (respectively, columns). To see this, suppose A' is obtained from A by interchanging rows i and $i+k$, $k > 0$. This can be done using k interchanges of adjacent rows to bring row i to the $(i+k)^{\text{th}}$ position, and then $k-1$ interchanges of adjacent rows to bring row $i+k$ to the ith position:

$$(1,\ldots,i,i+1,\ldots,i+k-1,i+k,\ldots,n)$$
$$\rightarrow (1,\ldots,i-1,i+1,\ldots,i+k-1,i+k,i,\ldots,n)$$

(k interchanges to move i to the right of $i+k$)

$$(1,\ldots,i-1,i+1,\ldots,i+k-1,i+k,i,\ldots,n)$$
$$\rightarrow (1,\ldots,i-1,i+k,i+1,\ldots,i+k-1,i,\ldots,n)$$

($k-1$ interchanges to move $i+k$ to the left of $i+1$)

So, the total number of interchanges of adjacent rows is $k + (k-1) = 2k-1$, which is odd. By $2k-1$ applications of Lemma 3-2, we find $det\, A' = (-1)^{2k-1}\, det\, A = -det\, A$, thus completing the proof. (The argument also applies to columns. The reader will be asked to provide another proof using cofactors in question 6, Exercises 3-2.) ■

Note that Theorem 3-3 tells us how $det\, A$ changes when A is changed by an elementary row operation of type 1.

EXAMPLE 3-12

Consider the matrix A with

$$A = \begin{bmatrix} 1 & 0 & 0 \\ 1 & 2 & 3 \\ 4 & 5 & 6 \end{bmatrix}.$$

Let A' be obtained by interchanging columns 1 and 3:

$$A' = \begin{bmatrix} 0 & 0 & 1 \\ 3 & 2 & 1 \\ 6 & 5 & 4 \end{bmatrix}.$$

Then, $det\, A = -3$, while $det\, A' = 3$, so $det\, A' = -det\, A$. ❑

COROLLARY

If A has two identical rows (or columns), then $det\, A = 0$.

PROOF Let A' be obtained from A by interchanging the two identical rows (columns). Then by the theorem, $det\, A' = -det\, A$. But, $A' = A$, so $det\, A' = det\, A$. It follows that $2det\, A = 0$, so $det\, A = 0$. ■

EXAMPLE 3-13

Consider the matrix

$$A = \begin{bmatrix} 1 & 2 & 3 \\ 0 & -2 & 1 \\ 1 & 2 & 3 \end{bmatrix}$$

which has identical first and third rows. Expanding by column 1, we find

$$det\, A = det \begin{bmatrix} -2 & 1 \\ 2 & 3 \end{bmatrix} + det \begin{bmatrix} 2 & 3 \\ -2 & 1 \end{bmatrix}$$

$$= -8 + 8 = 0. \square$$

Next, suppose we multiply every entry of the ith row of a matrix A by a constant $c \neq 0$. This is an elementary row operation of type 2. Let us denote the result by A'. Then, $a'_{kj} = a_{kj}$ for $k \neq i$ while $a'_{ij} = ca_{ij}$. If we now expand A' by the ith row, we get

$$det\, A' = \sum_{j=1}^{n} (-1)^{i+j} a'_{ij} det\, A'(i/j)$$

$$= \sum_{j=1}^{n} (-1)^{i+j} c a_{ij} det\, A(i/j) = c\, det\, A,$$

because $A'(i/j) = A(i/j)$. Thus we have proved

THEOREM 3-4

If A' is obtained from A by multiplying a row (or column) by $c \in R$, then

$$det\, A' = c\,det\, A. \blacksquare$$

EXAMPLE 3-14

With

$$A = \begin{bmatrix} -1 & 2 & 3 \\ 4 & 0 & 1 \\ -2 & 1 & 3 \end{bmatrix},$$

$$det\, A = (-4)\, det \begin{bmatrix} 2 & 3 \\ 1 & 3 \end{bmatrix} - det \begin{bmatrix} -1 & 2 \\ -2 & 1 \end{bmatrix} = -15.$$

Let A' be obtained from A by multiplying column 3 by -2:

$$A' = \begin{bmatrix} -1 & 2 & -6 \\ 4 & 0 & -2 \\ -2 & 1 & -6 \end{bmatrix}.$$

Then,

$$det\, A' = (-4)\, det \begin{bmatrix} 2 & -6 \\ 1 & -6 \end{bmatrix} + 2\, det \begin{bmatrix} -1 & 2 \\ -2 & 1 \end{bmatrix} = 30 = (-2) det\, A. \square$$

Remark If A' is obtained from an $n \times n$ matrix A by multiplying *every* entry of A by c, then $det\, A' = c^n\, det\, A$.

Next, an elementary row operation of type 3 replaces some row i of a matrix by row i + $c \times$ row k, where $k \neq i$. We will obtain the behaviour of $det\, A$ under such an operation as a consequence of the next result, which is important in its own right.

THEOREM 3-5

Suppose each entry of the ith row of a matrix A is expressed as a sum $(a'_{ij} + a''_{ij})$. Then

$$det\, A = det\, A' + det\, A'',$$

where A' and A'' have the same entries as A except in row i, in which they have entries a'_{ij} and a''_{ij} respectively, $j = 1, \ldots, n$.

In other words, the theorem says

$$det \begin{bmatrix} a_{11} & \cdots & \cdots & \cdots & a_{1n} \\ \cdots & \cdots & \cdots & \cdots & \cdots \\ (a'_{i1}+a''_{i1}) & \cdots & \cdots & \cdots & (a'_{in}+a''_{in}) \\ \cdots & \cdots & \cdots & \cdots & \cdots \\ a_{n1} & \cdots & \cdots & \cdots & a_{nn} \end{bmatrix}$$

$$= det \begin{bmatrix} a_{11} & \cdots & \cdots & a_{1n} \\ \cdots & \cdots & \cdots & \cdots \\ a'_{i1} & \cdots & \cdots & a'_{in} \\ \cdots & \cdots & \cdots & \cdots \\ a_{n1} & \cdots & \cdots & a_{nn} \end{bmatrix} + det \begin{bmatrix} a_{11} & \cdots & \cdots & a_{1n} \\ \cdots & \cdots & \cdots & \cdots \\ a''_{i1} & \cdots & \cdots & a''_{in} \\ \cdots & \cdots & \cdots & \cdots \\ a_{n1} & \cdots & \cdots & a_{nn} \end{bmatrix}$$

PROOF (Optional)

$$\begin{aligned} det\ A &= \sum_{\sigma} sgn\ \sigma\, a_{1\sigma(1)} \cdots a_{i\sigma(i)} \cdots a_{n\sigma(n)} \\ &= \sum_{\sigma} sgn\ \sigma\, a_{1\sigma(1)} \cdots (a'_{i\sigma(i)} + a''_{i\sigma(i)}) \cdots a_{n\sigma(n)} \\ &= \sum_{\sigma} sgn\ \sigma\, a_{1\sigma(1)} \cdots a'_{i\sigma(i)} \cdots a_{n\sigma(n)} + \sum_{\sigma} sgn\ \sigma\, a_{1\sigma(1)} \cdots a''_{i\sigma(i)} \quad a_{n\sigma(n)} \\ &= det\ A' + det\ A''. \ \blacksquare \end{aligned}$$

COROLLARY

If A' is obtained from A by adding a scalar multiple of row k to row i (or column k to column i) with $k \neq i$, then $det\ A' = det\ A$.

PROOF By Theorems 3-5 and 3-4 (where $c \in R$),

$$det\ A' = det \begin{bmatrix} a_{11} & \cdots & \cdots & \cdots & a_{1n} \\ \cdots & \cdots & \cdots & \cdots & \cdots \\ a_{i1}+ca_{k1} & \cdots & \cdots & \cdots & a_{in}+ca_{kn} \\ \cdots & \cdots & \cdots & \cdots & \cdots \\ a_{n1} & \cdots & \cdots & \cdots & a_{nn} \end{bmatrix}$$

$$= det \begin{bmatrix} a_{11} & \cdots & \cdots & a_{1n} \\ \cdots & \cdots & \cdots & \cdots \\ a_{i1} & \cdots & \cdots & a_{in} \\ \cdots & \cdots & \cdots & \cdots \\ a_{n1} & \cdots & \cdots & a_{nn} \end{bmatrix} + c\, det \begin{bmatrix} a_{11} & \cdots & \cdots & a_{1n} \\ \cdots & \cdots & \cdots & \cdots \\ a_{k1} & \cdots & \cdots & a_{kn} \\ \cdots & \cdots & \cdots & \cdots \\ a_{n1} & \cdots & \cdots & a_{nn} \end{bmatrix}$$

$$= det\ A + 0 = det\ A,$$

since the second matrix above has rows i and k identical. ■

This result is particularly useful for simplifying the computation of determinants. We use it to *introduce as many zeros as possible in a row (or column) and then expand along that row (respectively, column).*

EXAMPLE 3-15

Compute $det\, A$, where

$$A = \begin{bmatrix} 1 & 2 & 3 \\ 2 & 3 & 1 \\ 3 & 1 & 2 \end{bmatrix}.$$

Add $(-2) \times$ row 1 to row 2 and $(-3) \times$ row 1 to row 3 to get

$$det\, A = det \begin{bmatrix} 1 & 2 & 3 \\ 0 & -1 & -5 \\ 0 & -5 & -7 \end{bmatrix} = det \begin{bmatrix} -1 & -5 \\ -5 & -7 \end{bmatrix} = 7 - 25 = -18. \square$$

In fact, to compute the determinant of larger matrices, *it is often more efficient to use elementary row operations to change the matrix to an upper triangular matrix and then use Theorem 3-6 below.* (For definitions of upper and lower triangular matrices, see question 16, Exercises 2-5.)

THEOREM 3-6

If a matrix A is triangular, $det\, A = a_{11}a_{22}...a_{nn}$, the product of its diagonal entries.

PROOF Let us suppose that A is upper triangular so that it has the form

$$A = \begin{bmatrix} a_{11} & a_{12} & \cdots & a_{1n} \\ 0 & a_{22} & \cdots & a_{2n} \\ \cdots & \cdots & \cdots & \cdots \\ 0 & \cdots & 0 & a_{nn} \end{bmatrix}.$$

Note that $A(n/n)$ is again an $(n-1)\times(n-1)$ upper triangular matrix:

$$A(n/n) = \begin{bmatrix} a_{11} & a_{12} & \cdots & a_{1\,n-1} \\ 0 & a_{22} & \cdots & a_{2\,n-1} \\ \cdots & \cdots & \cdots & \cdots \\ 0 & \cdots & 0 & a_{n-1\,n-1} \end{bmatrix}.$$

So, expanding $det\, A$ by the last row, we see that $det\, A = a_{nn} det\, A(n/n)$. Continuing in this way (i.e., using induction on n to compute $det\, A(n/n)$), we see that $det\, A = a_{11}a_{22}...a_{nn}$, as stated. If A is lower triangular, then its transpose is upper triangular, so the result is also true in that case. ■

EXAMPLE 3-16

Let A be the 4×4 matrix

$$A = \begin{bmatrix} 2 & -4 & 3 & 1 \\ 1 & -2 & 0 & -1 \\ 3 & -5 & -2 & 1 \\ -2 & 2 & 3 & 3 \end{bmatrix}.$$

We find $det\, A$ by first transforming A to upper triangular form using elementary row operations of type 3 and type 1. This process is similar to Gauss reduction to row echelon form except that we don't make leading entries equal to 1. Recall that type 1 elementary row operations change the sign of the determinant. The steps are:

$$det\, A = det\begin{bmatrix} 2 & -4 & 3 & 1 \\ 1 & -2 & 0 & -1 \\ 3 & -5 & -2 & 1 \\ -2 & 2 & 3 & 3 \end{bmatrix}$$

$$= det\begin{bmatrix} 2 & -4 & 3 & 1 \\ 0 & 0 & -3/2 & -3/2 \\ 0 & 1 & -13/2 & -1/2 \\ 0 & -2 & 6 & 4 \end{bmatrix}$$

$$= -det\begin{bmatrix} 2 & -4 & 3 & 1 \\ 0 & 1 & -13/2 & -1/2 \\ 0 & 0 & -3/2 & -3/2 \\ 0 & -2 & 6 & 4 \end{bmatrix}$$

$$= -det\begin{bmatrix} 2 & -4 & 3 & 1 \\ 0 & 1 & -13/2 & -1/2 \\ 0 & 0 & -3/2 & -3/2 \\ 0 & 0 & -7 & 3 \end{bmatrix}$$

$$= -det\begin{bmatrix} 2 & -4 & 3 & 1 \\ 0 & 1 & -13/2 & -1/2 \\ 0 & 0 & -3/2 & -3/2 \\ 0 & 0 & 0 & 10 \end{bmatrix}$$

$$= -(2)(1)(-3/2)(10) = 30.$$ □

When A is $n \times n$, it is not hard to see that the evaluation of *det* A, either by the definition or by cofactors, requires approximately $n(n!)$ arithmetic operations. On the other hand, the number of arithmetic operations required to transform A to upper triangular form can be shown to be approximately $2n^3/3$ for large n. The calculation of *det* A then requires an additional $n - 1$ multiplications of the n diagonal elements, yielding an approximate total of $2n^3/3 + n$ arithmetic operations. When $n = 3$, the two methods therefore require about the same work, but $n(n!)$ grows much faster than n^3, so direct evaluation very quickly becomes unattractive: for example, for $n = 4$, $n(n!) = 96$, while $2n^3/3 + n = 47$; for $n = 6$, $n(n!) = 4320$, while $2n^3/3 + n = 150$.

We have now obtained the behaviour of *det* A under all three types of elementary row operations on A. These properties apply, in particular, when $A = I$, the $n \times n$ identity matrix. Thus we have

THEOREM 3-7

If E is an $n \times n$ elementary matrix,

(1) *det* $E = -1$ if E is of type 1;

(2) *det* $E = c$ if E is of type 2 (obtained from I by multiplication of a row by $c \neq 0$);

(3) *det* $E = 1$ if E is of type 3.

PROOF Since *det* $I = 1$ (Example 3-6), these facts are immediate consequences of Theorems 3-3, 3-4, and 3-5 (Corollary) respectively. ■

Now, we know that the result $e(A)$ of applying an elementary row operation e to a matrix A is $e(A) = EA$ where $E = e(I)$. We therefore have

COROLLARY

If A is $n \times n$ and E is an $n \times n$ elementary matrix, *det* EA = *det* E *det* A.

PROOF Using the three possibilities for *det* E as itemized in the theorem, the right-hand side of the equation above becomes $-$*det* A, *cdet* A, and *det* A when E is of types 1, 2, and 3 respectively. But our earlier theorems show that each of these is exactly *det* $(e(A))$ = *det* EA, thus completing the proof. ■

Theorem 3-7 itself also shows that *det* $E \neq 0$ for any elementary matrix E. When we further recall that any square matrix A is invertible iff it is a product of a finite number of elementary matrices we get

THEOREM 3-8

A matrix A is invertible iff $det\ A \neq 0$.

PROOF (Optional) If A is invertible, $A = E_1E_2...E_k$, a product of elementary matrices. Then, applying the corollary to Theorem 3-7,

$$\begin{aligned} det\ A &= det\ (E_1E_2...E_k) \\ &= det\ (E_1(E_2...E_k)) \\ &= det\ E_1 det\ (E_2...E_k) \\ &= ... = ... = ... \\ &= det\ E_1 det\ E_2...det\ E_k \quad \text{(by induction on } k\text{)} \end{aligned}$$

and this is nonzero since each factor is nonzero.

Conversely, if A is not invertible, then it is row-equivalent to a matrix A' in row-reduced echelon form having at least one zero row. So, by expansion along that zero row, we see that $det\ A' = 0$. Then, as above,

$$det\ A = det\ (E_1...E_kA') = det\ E_1...det\ E_k\ det\ A' = 0,$$

as required. ■

This result is fundamental because it shows that *invertibility is indeed characterized by the number det A.*

EXAMPLE 3-17

(a) If A is given by

$$A = \begin{bmatrix} 1 & 2 & 3 & 1 \\ 0 & 3 & 4 & 3 \\ 0 & 0 & 0 & -2 \\ 0 & 0 & 0 & 5 \end{bmatrix}$$

then $det\ A = 0$ (why?), so A is singular.

(b) If A is a triangular matrix with no diagonal entries equal to 0, then A is invertible.

(c) If A is an $n \times n$ **skew-symmetric matrix**, then $A = -A^t$. Therefore,

$$det\ A = det\ (-A^t) = (-1)^n det\ A^t = (-1)^n det\ A,$$

since $-A^t$ is the matrix obtained from A^t by multiplying all n rows by -1. Thus, if n is odd, $det\ A = -det\ A$, so $det\ A = 0$. This shows that a skew-symmetric matrix of *odd* order is not invertible. The result is *not* true when n is even. For example.

$$\begin{bmatrix} 0 & 1 \\ -1 & 0 \end{bmatrix}$$

is invertible and skew-symmetric. ❑

> **THEOREM 3-9**
>
> If A and B are $n \times n$, $\det AB = \det A \det B$.

PROOF If A is invertible, we can write $A = E_1E_2...E_k$, a product of elementary matrices.

Then,

$$\det AB = \det (E_1...E_kB) = \det (E_1...E_k) \det B = \det A \det B,$$

using the same argument as in Theorem 3-8.

On the other hand, if A is singular, there are elementary matrices E_i such that $E_1E_2...E_kA$ has at least one zero row. So,

$$(E_1E_2...E_k)AB = (E_1E_2...E_kA)B$$

has at least one zero row, because the left factor of the last product does. This means that AB is row equivalent to a matrix having at least one zero row, so AB is singular too. Therefore,

$$\det AB = 0 = 0\det B = \det A \det B,$$

thus completing the proof. ■

We apply this result to obtain a formula for the determinant of **block triangular matrices**. These are square matrices A such that A (or the transpose of A) has the form

$$\begin{bmatrix} B_{11} & B_{12} & \dots & B_{1k} \\ O & B_{22} & \dots & B_{2k} \\ \dots & \dots & \dots & \dots \\ O & \dots & O & B_{kk} \end{bmatrix} \tag{3-8}$$

where the B_{ij} and O's are matrices of sizes such that (i) A is square and (ii) the matrices B_{ii} "on the diagonal" are square (O is a zero matrix).

In fact, we shall mostly be interested in the case where $k = 2$, so that we can write

$$A = \begin{bmatrix} B & C \\ O & D \end{bmatrix} \text{ or } \begin{bmatrix} B & O \\ C & D \end{bmatrix}. \tag{3-9}$$

EXAMPLE 3-18

Let A be the 5×5 matrix

$$A = \begin{bmatrix} 1 & 2 & -1 & 2 & 0 \\ 2 & 2 & 3 & -1 & 1 \\ 0 & 0 & 1 & -2 & 0 \\ 0 & 0 & -1 & 1 & 1 \\ 0 & 0 & 2 & -1 & 0 \end{bmatrix}.$$

Then A has the form of 3-9 with

$$B = \begin{bmatrix} 1 & 2 \\ 2 & 2 \end{bmatrix}, \quad D = \begin{bmatrix} 1 & -2 & 0 \\ -1 & 1 & 1 \\ 2 & -1 & 0 \end{bmatrix}.$$

Note that C is 2×3 and O is 3×2. We shall show that $det\ A = det\ B\ det\ D$. Now, $det\ B = 2 - 4 = -2$ and

$$det\ D = (-1)^{2+3}\ det \begin{bmatrix} 1 & -2 \\ 2 & -1 \end{bmatrix} = -3.$$

So $det\ B\ det\ D = 6$.

We can evaluate $det\ A$ by expanding along column 5 after subtracting row 4 from row 2 to introduce another 0. We get

$$det\ A = det \begin{bmatrix} 1 & 2 & -1 & 2 & 0 \\ 2 & 2 & 4 & -2 & 0 \\ 0 & 0 & 1 & -2 & 0 \\ 0 & 0 & -1 & 1 & 1 \\ 0 & 0 & 2 & -1 & 0 \end{bmatrix}$$

$$= (-1)^{4+5}\ det \begin{bmatrix} 1 & 2 & -1 & 2 \\ 2 & 2 & 4 & -2 \\ 0 & 0 & 1 & -2 \\ 0 & 0 & 2 & -1 \end{bmatrix} = -det \begin{bmatrix} 1 & 2 & 3 & 0 \\ 2 & 2 & 0 & 0 \\ 0 & 0 & -3 & 0 \\ 0 & 0 & 2 & -1 \end{bmatrix}$$

$$= det \begin{bmatrix} 1 & 2 & 3 \\ 2 & 2 & 0 \\ 0 & 0 & -3 \end{bmatrix} = (-2)(-3) = 6.$$

Therefore, we see that $det\ A = det\ B\ det\ D$. ❑

THEOREM 3-10

If an $n \times n$ matrix has the block form of equation 3-9, then $det\ A = det\ B\ det\ D$.

PROOF (Outline — Optional) First, if B or D is singular, we see that the same is true for A, so that $det\ A = 0 = det\ B\ det\ D$.

Next, if B and D are both invertible and A has the first form in equation 3-9, we may "factorize" A as

$$A = \begin{bmatrix} B & O \\ O & I \end{bmatrix} \begin{bmatrix} I & B^{-1}CD^{-1} \\ O & I \end{bmatrix} \begin{bmatrix} I & O \\ O & D \end{bmatrix} \tag{3-10}$$

where the I's and O's are identity and zero matrices of appropriate sizes. Cofactor expansion and an induction argument will show that the determinants of the first and third matrices are $det\ B$ and $det\ D$ respectively, while the middle matrix has determinant 1 (it is upper triangular). The result then follows by Theorem 3-9. ■

Note that, by induction on k, it also follows that *the determinant of a matrix having the form of equation 3-8 is the product of the determinants of the matrices on the diagonal.*

EXAMPLE 3-19

Let A be the 6×6 matrix

$$A = \begin{bmatrix} 0 & 0 & 0 & 0 & -1 & 1 \\ 0 & 0 & 1 & -1 & 2 & 4 \\ 2 & -1 & 3 & 1 & 2 & 1 \\ 0 & 0 & 0 & 0 & 1 & 1 \\ -1 & -1 & 2 & 2 & 3 & 5 \\ 0 & 0 & 2 & 2 & 2 & 2 \end{bmatrix}.$$

A is not itself block triangular, but by 4 row interchanges we find

$$det\ A = (-1)^4\ det \begin{bmatrix} 2 & -1 & 3 & 1 & 2 & 1 \\ -1 & -1 & 2 & 2 & 3 & 5 \\ 0 & 0 & 1 & -1 & 2 & 4 \\ 0 & 0 & 2 & 2 & 2 & 2 \\ 0 & 0 & 0 & 0 & -1 & 1 \\ 0 & 0 & 0 & 0 & 1 & 1 \end{bmatrix}$$

$$= det \begin{bmatrix} 2 & -1 \\ -1 & -1 \end{bmatrix} det \begin{bmatrix} 1 & -1 \\ 2 & 2 \end{bmatrix} det \begin{bmatrix} -1 & 1 \\ 1 & 1 \end{bmatrix}$$

$$= (-3)(4)(-2) = 24.\ \square$$

We conclude this section by describing (without proofs) those properties that uniquely characterize the determinant function.

THEOREM 3-11

The determinant function *det* is the *only* function that assigns a number to an $n \times n$ matrix and has the following properties:

(1) It satisfies Theorems 3-4 and 3-5.

(2) It satisfies Theorem 3-3, Corollary.

(3) It assigns value 1 to the identity matrix I (by Example 3-6). ■

As a result of property (1), we say that the determinant is a **linear function** of each row of an $n \times n$ matrix when the other $n - 1$ rows are kept fixed. This type of function is studied in detail in Chapter 5. As a result of property (2), we say that the determinant is an **alternating function** of the rows.

Remark For an important *non-property* of *det*, the reader should look at question 5 below.

EXERCISES 3-2

1. For the matrices A in question 1, Exercises 3-1, verify by computation that $det\ A^t = det\ A$. Prove this for any 2×2 matrix using equation 3-1 for $det\ A$.

2. Show that the area of the parallelogram in the plane having sides determined by the vectors (a,b) and (c,d) is the same as the one determined by the vectors (a,c) and (b,d). (See question 2, Exercises 3-1.)

3. Compute the determinants of the following matrices by introducing 0's in a row or column and expanding using cofactors.

(a) $\begin{bmatrix} -1 & 2 & 3 \\ 2 & -1 & 2 \\ 3 & 2 & 1 \end{bmatrix}$ (b) $\begin{bmatrix} 1 & -1 & 1 \\ -2 & 2 & 1 \\ 3 & -3 & 1 \end{bmatrix}$

(c) $\begin{bmatrix} 4 & -1 & 2 & 1 \\ 1 & 1 & 1 & 1 \\ -2 & 0 & 3 & 3 \\ 1 & -1 & -2 & 1 \end{bmatrix}$

4. Show that

(a) $det \begin{bmatrix} 1 & a & b+c \\ 1 & b & a+c \\ 1 & c & a+b \end{bmatrix} = 0,$

(b) $det \begin{bmatrix} 1 & a & a^2 \\ 1 & b & b^2 \\ 1 & c & c^2 \end{bmatrix} = (b-a)(c-a)(c-b),$

(c) $det \begin{bmatrix} a & b & c \\ b & c & a \\ c & a & b \end{bmatrix} = 3abc - (a^3 + b^3 + c^3).$

5. Construct an example to show that, in general, $det\ (A + B) \neq det\ A + det\ B$.

6. Use cofactor expansions to show that if two rows of a matrix A are interchanged to yield a new matrix A', then $det\ A' = -det\ A$. (Hint: If rows i and k are interchanged with $i < k$, expand A' by cofactors on row i and use Lemma 3-2 to assert that $det\ A'(i/j) = (-1)^{k-i+1} det\ A(k/j)$).

7. Evaluate the determinants of the matrices below by first reducing them to upper triangular form.

(a) $\begin{bmatrix} -1 & 2 & 3 & -2 \\ 0 & 2 & -3 & 1 \\ -2 & 2 & 2 & 2 \\ 1 & -1 & 1 & -1 \end{bmatrix}$

(b) $\begin{bmatrix} 2 & -1 & 0 & 3 & 2 \\ 3 & -3 & 2 & 1 & 4 \\ 0 & 1 & -1 & -1 & -1 \\ -1 & 2 & 3 & -1 & 2 \\ 4 & 0 & 2 & 3 & 1 \end{bmatrix}$

8. Let A and B be $n \times n$. Use determinants to show that if A is singular, then so is AB. Is the converse true?

9. **(a)** Use determinants to show that if AB is invertible, then so is BA, where A, B are $n \times n$. Must A or B themselves be invertible?

(b) Show that if A is invertible, then so is A^{-1} and $det\ A^{-1} = (det\ A)^{-1}$.

10. For what values of c are the following matrices singular?

(a) $\begin{bmatrix} c+1 & -2 \\ -1 & c \end{bmatrix}$

(b) $\begin{bmatrix} c & -1 & 1 \\ 0 & c & 2 \\ c & 0 & 1 \end{bmatrix}$

11. Compute the determinants of the following matrices.

(a) $\begin{bmatrix} 2 & 2 & 3 & 0 & 0 \\ 1 & -1 & 1 & 0 & 0 \\ 0 & 1 & 1 & 0 & 0 \\ 0 & 0 & 0 & -2 & 1 \\ 0 & 0 & 0 & 2 & 1 \end{bmatrix}$

(b) $\begin{bmatrix} 0 & 0 & 4 & -1 & 2 & 3 \\ 0 & 0 & 0 & 0 & 4 & 6 \\ -2 & 2 & 6 & 5 & 1 & -2 \\ 0 & 0 & 3 & 2 & 1 & 1 \\ 0 & 0 & 0 & 0 & 3 & 4 \\ 1 & 1 & 4 & -1 & 2 & 0 \end{bmatrix}$

12. **(a)** Show that (x_1,y_1), (x_2,y_2), and (x_3,y_3) are collinear points in the plane iff

$$\text{(i) } det \begin{bmatrix} x_2 - x_1 & x_3 - x_1 \\ y_2 - y_1 & y_3 - y_1 \end{bmatrix} = 0, \quad \text{or (ii) } det \begin{bmatrix} 1 & x_1 & y_1 \\ 1 & x_2 & y_2 \\ 1 & x_3 & y_3 \end{bmatrix} = 0.$$

(b) Show that if the points in (a) above are not collinear, then the absolute value of the determinant in (ii) is twice the area of the triangle formed by the three points. (Hint: First assume the points all lie in the first quadrant, and express the area in terms of areas of trapezoids.)

(c) Use (b) to compute the area of the triangle formed by the points $(-1,1)$, $(3,2)$, and $(4,-2)$.

13. A *transposition* is a permutation in $S(n)$ that interchanges exactly two integers and leaves all the others fixed. For example, in $S(4)$, $\sigma = (4,2,3,1)$ and $\tau = (1,4,3,2)$ are transpositions interchanging 1 and 4, and 2 and 4, respectively. We denote the transposition in $S(n)$ that interchanges i and j $(i \neq j)$ by $<ij>$. Thus the transpositions in $S(4)$ noted above are $<14>$ and $<24>$ respectively.

(a) Show that any permutation in $S(n)$ may be expressed as a product of transpositions. (Hint: For example, in $S(4)$ we may write $(3,4,2,1) = <14><12><13>$.)

(b) Show that if τ is a transposition, $sgn\ \tau = -1$. (Hint: Use the argument in Theorem 3-3.)

(c) Show that $sgn\ \sigma\tau = sgn\ \sigma\ sgn\ \tau$ for any σ, τ in $S(n)$. (Hint: Use the results of (a) and (b).)

14. Let σ be a permutation in $S(n)$, and let P be the matrix obtained from the $n \times n$ identity matrix by permuting its rows by σ; i.e., the ith row of P is the $\sigma(i)^{\text{th}}$ row of I. Show that $det\ P = sgn\ \sigma$.

3-3 Cramer's Rule

In this section, we show that the inverse of a matrix A can actually be expressed in terms of its cofactors and $det\ A$. As well, we obtain a formula for the solutions of a system $AX = B$ of n linear equations in n unknowns. While these results do not replace our earlier methods for computing inverses or solutions of systems of equations (because they are much less efficient), they are important for theoretical reasons: they provide explicit formulae for the inverse and for the solutions of $AX = B$.

Let us begin our discussion of the formula for A^{-1} with the 2×2 case:

$$A = \begin{bmatrix} a_{11} & a_{12} \\ a_{21} & a_{22} \end{bmatrix}.$$

Continuing the usual row reduction of $(A{:}I)$ started in section 3-1 (assuming $det\ A \neq 0$ and $a_{11} \neq 0$) we find

$$\begin{bmatrix} a_{11} & a_{12} & 1 & 0 \\ a_{21} & a_{22} & 0 & 1 \end{bmatrix} \to \begin{bmatrix} 1 & a_{12}/a_{11} & 1/a_{11} & 0 \\ 0 & a_{22} - a_{12}a_{21}/a_{11} & -a_{21}/a_{11} & 1 \end{bmatrix}$$

$$\to \begin{bmatrix} 1 & a_{12}/a_{11} & 1/a_{11} & 0 \\ 0 & det\ A & -a_{21} & a_{11} \end{bmatrix} \to \begin{bmatrix} 1 & a_{12}/a_{11} & 1/a_{11} & 0 \\ 0 & 1 & -a_{21}/det\ A & a_{11}/det\ A \end{bmatrix}$$

$$\to \begin{bmatrix} 1 & 0 & 1/a_{11} + a_{12}a_{21}/a_{11}det\ A & -a_{12}/det\ A \\ 0 & 1 & -a_{21}/det\ A & a_{11}/det\ A \end{bmatrix}$$

$$= \begin{bmatrix} 1 & 0 & a_{22}/det\ A & -a_{12}/det\ A \\ 0 & 1 & -a_{21}/det\ A & a_{11}/det\ A \end{bmatrix}$$

(because $1/a_{11} + a_{12}a_{21}/a_{11}det\ A = a_{11}a_{22}/a_{11}det\ A = a_{22}/det\ A$). So, we see that if A is invertible, its inverse is

$$A^{-1} = (1/det\ A)\begin{bmatrix} a_{22} & -a_{12} \\ -a_{21} & a_{11} \end{bmatrix}. \tag{3-11}$$

Now, in Example 3-9, we saw that the cofactors of a 2×2 matrix A are $C_{11} = a_{22}$, $C_{12} = -a_{21}$, $C_{21} = -a_{12}$, and $C_{22} = a_{11}$. If we let C be the matrix whose entries are the cofactors, then 3-11 can be written as $A^{-1} = (1/det\ A)C^t$. This leads to

DEFINITION 3-6

The (classical) **adjoint matrix** of an $n \times n$ matrix A is the transpose of the matrix C of cofactors of A. It is denoted $adj\ A$, so $adj\ A = C^t$.

EXAMPLE 3-20

Let

$$A = \begin{bmatrix} 2 & 3 & -1 \\ -1 & 0 & 2 \\ 3 & -1 & 1 \end{bmatrix}.$$

This is the same matrix used in Examples 3-2(a) and 3-7. We have seen that $C_{11} = 2$, $C_{21} = -2$, $C_{31} = 6$, $C_{22} = 5$, and $C_{23} = 11$. The other cofactors are easily seen to be $C_{12} = 7$, $C_{13} = 1$, $C_{32} = -3$, and $C_{33} = 3$. Therefore,

$$C = \begin{bmatrix} 2 & 7 & 1 \\ -2 & 5 & 11 \\ 6 & -3 & 3 \end{bmatrix}, \quad adj\, A = \begin{bmatrix} 2 & -2 & 6 \\ 7 & 5 & -3 \\ 1 & 11 & 3 \end{bmatrix}. \square$$

The next result will allow us to generalize our observation about 2×2 matrices.

THEOREM 3-12

If A is $n \times n$, $A\, adj\, A = (det\, A)I$.

PROOF (Optional) Since both sides are $n \times n$ matrices, we have to show that they have the same (i,j) entry for every $i, j = 1, 2, ..., n$, that is, $(A\, adj\, A)_{ij} = (det\, A)I_{ij}$.

(a) Consider first the case $i = j$ (the diagonal entries, $I_{ii} = 1$).

$$(A\, adj\, A)_{ii} = \sum_{k=1}^{n} a_{ik}\, (adj\, A)_{ki} = \sum_{k=1}^{n} a_{ik} C_{ik} = det\, A = (det\, A)I_{ii},$$

by Theorem 3-1, since the last sum is just the expansion of $det\, A$ by the ith row.

(b) Suppose next that $i \neq j$ (the off-diagonal entries). Then certainly $I_{ij} = 0$. To show that $(A\, adj\, A)_{ij} = 0$, let B be the $n \times n$ matrix obtained from A by replacing the jth row of A by the ith row:

$$B = \begin{bmatrix} a_{11} & a_{12} & \cdots & a_{1n} \\ \cdots & \cdots & \cdots & \cdots \\ a_{i1} & a_{i2} & \cdots & a_{in} \\ \cdots & \cdots & \cdots & \cdots \\ a_{i1} & a_{i2} & \cdots & a_{in} \\ \cdots & \cdots & \cdots & \cdots \\ a_{n1} & a_{n2} & \cdots & a_{nn} \end{bmatrix} \begin{matrix} \\ \\ (i) \\ \\ (j) \\ \\ \\ \end{matrix} \qquad (i < j \text{ here}).$$

Then $b_{jk} = a_{ik}$ and $B(j/k) = A(j/k)$ for every k. Since B has two identical rows, $det\ B = 0$. We expand $det\ B$ along the jth row to get

$$0 = det\ B = \sum_{k=1}^{n} (-1)^{j+k} b_{jk} det\ B(j/k)$$

$$= \sum_{k=1}^{n} (-1)^{j+k} a_{ik} det\ A(j/k) = \sum_{k=1}^{n} a_{ik} C_{jk}$$

$$= \sum_{k=1}^{n} a_{ik} (adj\ A)_{kj} = (A\ adj\ A)_{ij},$$

as required. ■

COROLLARY

If A is invertible, $A^{-1} = (1/det\ A) adj\ A$.

PROOF If A is invertible, $det\ A \neq 0$, so Theorem 3-12 tells us that $A[(1/det\ A) adj\ A] = I$. Since $AA^{-1} = I$, $(1/det\ A) adj\ A$ has the property uniquely characterizing A^{-1}. ■

EXAMPLE 3-21

(a) If $A = \begin{bmatrix} 1 & 2 \\ 3 & 4 \end{bmatrix}$, then $A^{-1} = (-1/2) \begin{bmatrix} 4 & -2 \\ -3 & 1 \end{bmatrix}$.

(b) Continuing Example 3-20, we find A^{-1} for the 3×3 matrix given there. Since $det\ A = 24$ (from Example 3-2(a)),

$$A^{-1} = (1/24) \begin{bmatrix} 2 & -2 & 6 \\ 7 & 5 & -3 \\ 1 & 11 & 3 \end{bmatrix} = \begin{bmatrix} 2/24 & -2/24 & 6/24 \\ 7/24 & 5/24 & -3/24 \\ 1/24 & 11/24 & 3/24 \end{bmatrix}.$$

The reader may also check that

$$A\ adj\ A = \begin{bmatrix} 24 & 0 & 0 \\ 0 & 24 & 0 \\ 0 & 0 & 24 \end{bmatrix} = (det\ A)I. \ \square$$

As noted earlier, the formula given in the corollary of Theorem 3-12 is *not* an efficient method for finding A^{-1} when n is large (greater than 2). This is because the number of steps involved increases as $n!$, as opposed to n^3 for

Gauss-Jordan reduction. However, the formula is important since it provides a "closed expression" (formula) for the inverse.

Next, we consider a system of n linear equations $AX = B$ where B is $n \times 1$. If A is invertible, we know that it has an unique solution $X = A^{-1}B$. Using the formula for A^{-1} just obtained, we can write

$$X = (1/det\, A)(adj\, A)B.$$

Both sides of this equation are $n \times 1$ matrices. With $X = (x_i)$ and $B = (b_k)$, we take the $i^{th} = (i,1)$ entry of both sides to get

$$x_i = X_{i1} = [(1/det\, A)(adj\, A)B]_{i1} = (1/det\, A) \sum_{k=1}^{n} (adj\, A)_{ik}B_{k1}$$

$$= (1/det\, A) \sum_{k=1}^{n} C_{ki}b_k = (1/det\, A) \sum_{k=1}^{n} (-1)^{k+i} det\, A(k/i)b_k. \qquad (3\text{-}12)$$

Now the sum in equation 3-12 is precisely the expansion along the ith column of the determinant of the $n \times n$ matrix denoted $A(i)$ obtained from A by replacing the ith column with B:

$$A(i) = \begin{bmatrix} a_{11} & \cdots & a_{1\,i-1} & b_1 & a_{1\,i+1} & \cdots & a_{1n} \\ \cdots & \cdots & \cdots & \cdots & \cdots & \cdots & \cdots \\ a_{n1} & \cdots & a_{n\,i-1} & b_n & a_{n\,i+1} & \cdots & a_{nn} \end{bmatrix}$$

(columns $i-1$, i, $i+1$ marked above the matrix)

Consequently, 3-12 may be written $x_i = det\, A(i)/det\, A$ and we have proved

THEOREM 3-13 (Cramer's Rule)

Let $AX = B$ be a system of n equations in n unknowns such that A is nonsingular. Then the solutions are given by the formula

$$x_i = det\, A(i)/det\, A,\ i = 1, 2, \ldots, n,$$

where $A(i)$ is the $n \times n$ matrix obtained from A by replacing its ith column by B. ■

The formula is named after Gabriel Cramer (1704–1752), a Swiss mathematician who also studied the theory of conics.

EXAMPLE 3-22

We use Cramer's rule to solve the system $AX = B$ where

$$A = \begin{bmatrix} 2 & 3 & -1 \\ -1 & 0 & 2 \\ 3 & -1 & 1 \end{bmatrix} \text{ and } B = \begin{bmatrix} 24 \\ -48 \\ 72 \end{bmatrix}.$$

(A is the matrix in Examples 3-20 and 3-21(b).)
The matrices $A(i)$, $i = 1, 2, 3$, are

$$A(1) = \begin{bmatrix} 24 & 3 & -1 \\ -48 & 0 & 2 \\ 72 & -1 & 1 \end{bmatrix}, A(2) = \begin{bmatrix} 2 & 24 & -1 \\ -1 & -48 & 2 \\ 3 & 72 & 1 \end{bmatrix},$$

$$A(3) = \begin{bmatrix} 2 & 3 & 24 \\ -1 & 0 & -48 \\ 3 & -1 & 72 \end{bmatrix}.$$

The reader may verify that

$$\det A(1) = 576, \det A(2) = -288, \text{ and } \det A(3) = -288.$$

Since $\det A = 24$ (from Example 3-21(b)), the solution of the system is

$$x_1 = 576/24 = 24, x_2 = -288/24 = -12, \text{ and } x_3 = -12.$$

The reader may check this solution using the inverse of A found in Example 3-21(b). ❑

Just as the formula for A^{-1} is not an efficient method for computing A^{-1}, Cramer's rule is *not* an efficient method for solving systems of equations in more than two unknowns. However, the explicit formulae for the solutions are useful in many areas of mathematics and its applications.

EXERCISES 3-3

1. Use the corollary of Theorem 3-12 to compute the inverse of the following matrices.

(a) $\begin{bmatrix} 1 & 2 \\ 2 & -1 \end{bmatrix}$ **(b)** $\begin{bmatrix} -1 & 2 \\ 3 & -5 \end{bmatrix}$

(c) $\begin{bmatrix} -1 & 1 & 2 \\ 0 & 1 & 1 \\ 1 & 1 & 1 \end{bmatrix}$ **(d)** $\begin{bmatrix} -2 & 3 & 1 \\ 3 & -1 & 2 \\ 1 & -2 & 3 \end{bmatrix}$

2. Compute the inverse of each of the matrices below by **(i)** using the corollary of Theorem 3-12, and **(ii)** using Gauss-Jordan reduction. Compare the number of arithmetic operations used in the two methods.

(a) $\begin{bmatrix} 1 & -1 & 1 \\ 1 & 1 & 1 \\ -1 & 1 & 1 \end{bmatrix}$ **(b)** $\begin{bmatrix} 1 & -2 & 1 & 1 \\ -1 & 0 & 1 & 1 \\ 1 & 1 & 0 & 1 \\ -2 & 1 & 1 & 0 \end{bmatrix}$

3. For the matrices A in question 1, verify that $(A\ adj\ A)_{12} = 0$ and $(A\ adj\ A)_{22} = det\ A$.

4. Use Cramer's rule to solve the following systems of equations, where possible.

(a)
$$\begin{aligned} x_1 + 2x_2 &= 3 \\ 2x_1 - x_2 &= -1 \end{aligned}$$

(b)
$$\begin{aligned} x_1 + x_2 + x_3 &= 1 \\ x_2 + x_3 &= 1 \\ x_3 &= 1 \end{aligned}$$

(c)
$$\begin{aligned} x_1 + x_2 - x_3 &= -1 \\ 2x_1 - x_2 - x_3 &= 2 \\ -x_1 - x_2 + 2x_3 &= -1 \end{aligned}$$

(d)
$$\begin{aligned} -x_1 + x_2 + x_3 &= 1 \\ 2x_1 - x_2 + 2x_1 &= -2 \\ -x_1 + 2x_2 + 5x_3 &= 4 \end{aligned}$$

5. Use Cramer's rule to prove that if $AX = B$ is a system of n equations in n unknowns such that A and B have integral entries and $|det\ A| = 1$, then the solutions X are also integers.

6. Let A be an upper (respectively, lower) triangular matrix.

(a) Show that $adj\ A$ is also upper (respectively, lower) triangular.

(b) Use (a) to show that A^{-1} is upper (respectively, lower) triangular.

7. Show that for any 2×2 matrix A, $adj\ (adj\ A) = A$.

8. **(a)** Prove that $det(adj\ A) = (det\ A)^{n-1}$ when A is $n \times n$.

(b) Show that if A is invertible, then so is $adj\ A$, and $(adj\ A)^{-1} = (1/det\ A)A = adj(A^{-1})$.

9. Use Cramer's rule to solve for x_1 (only) in the following systems of equations, if possible.

(a)
$$\begin{aligned} 2x_1 + 2x_2 + 3x_3 &= 2 \\ x_1 - x_2 + x_3 &= 0 \\ x_2 + x_3 &= -3 \\ - 2x_4 + x_5 &= 1 \\ 2x_4 + x_5 &= 5 \end{aligned}$$

(b)
$$\begin{aligned} x_1 + 2x_2 + x_3 + x_4 &= -5 \\ x_1 - x_2 + x_3 - x_5 &= 3 \\ 2x_2 + 2x_3 &= -1 \\ 3x_1 - 2x_4 + x_5 &= 0 \\ -x_1 + x_2 - x_3 + 2x_4 + x_5 &= 2 \end{aligned}$$

10. Let A be an invertible $n \times n$ matrix and B be $n \times 1$. Consider the system $AX = B$.

(a) Use Cramer's rule to show that if B is a nonzero multiple of a column j of A, then $x_i = 0$, $i \neq j$.

(b) Use Cramer's rule to show that if B is a sum of nonzero multiples of columns $j_1, j_2, \ldots, j_k$ of A, then $x_i = 0$ for $i \neq j_1, j_2, \ldots, j_k$.

3-4 The Cross Product (Optional)

Our last application of determinants shows how the determinant, together with the dot product in R^3 (Chapter 1), allows us to define a new "product" of vectors in R^3. Let $\mathbf{u}$ and $\mathbf{v}$ be vectors in R^3. The next result defines a vector in R^3 called the **cross product** of $\mathbf{u}$ and $\mathbf{v}$, and denoted $\mathbf{u} \times \mathbf{v}$. This construction is important in many problems in physics and engineering, because, as shown later, $\mathbf{u} \times \mathbf{v}$ is orthogonal to both $\mathbf{u}$ and $\mathbf{v}$. We defer the proof to Chapter 7 because it depends on ideas from Chapters 5 and 7. Readers who do not intend to cover Chapter 7, may regard Theorem 3-14 or its corollary as the *definition* of $\mathbf{u} \times \mathbf{v}$. The formula for calculating $\mathbf{u} \times \mathbf{v}$ is given in the corollary.

THEOREM 3-14 (Cross Product in R^3)
Let $\mathbf{u} = (u_1, u_2, u_3)$ and $\mathbf{v} = (v_1, v_2, v_3)$ be arbitrary vectors in R^3. Then there is an unique vector $\mathbf{u} \times \mathbf{v}$ in R^3 which is completely characterized by the equation

$$(\mathbf{u} \times \mathbf{v}) \cdot \mathbf{w} = det \begin{bmatrix} u_1 & u_2 & u_3 \\ v_1 & v_2 & v_3 \\ w_1 & w_2 & w_3 \end{bmatrix} \qquad (3\text{-}13)$$

for any vector $\mathbf{w} = (w_1, w_2, w_3)$ in R^3. ■

Note that the theorem defines the vector $\mathbf{u} \times \mathbf{v}$ by indicating its dot product with any vector $\mathbf{w} \in R^3$. In fact, this allows us to compute the components of $\mathbf{u} \times \mathbf{v}$ as given in the corollary.

COROLLARY

$$\mathbf{u} \times \mathbf{v} = (u_2v_3 - v_2u_3)\mathbf{i} - (u_1v_3 - v_1u_3)\mathbf{j} + (u_1v_2 - v_1u_2)\mathbf{k}.$$

To make this easier to remember, we write

$$\mathbf{u} \times \mathbf{v} = det \begin{bmatrix} \mathbf{i} & \mathbf{j} & \mathbf{k} \\ u_1 & u_2 & u_3 \\ v_1 & v_2 & v_3 \end{bmatrix}.$$

Note that the above determinant is *symbolic* only — the array is not a matrix in the usual sense since the first row consists of vectors and the other two rows contain scalars.

Since $\mathbf{i} = (1,0,0)$, $\mathbf{j} = (0,1,0)$, and $\mathbf{k} = (0,0,1)$, the **components** of $\mathbf{u} \times \mathbf{v}$ are

$$(u_2v_3 - v_2u_3),\ (u_3v_1 - v_3u_1),\ \text{and } (u_1v_2 - v_1u_2)$$

respectively.

PROOF (Optional) By Theorem 3-14,

$$(\mathbf{u}\times\mathbf{v})\cdot\mathbf{i} = det\begin{bmatrix} u_1 & u_2 & u_3 \\ v_1 & v_2 & v_3 \\ 1 & 0 & 0 \end{bmatrix} = u_2v_3 - v_2u_3.$$

Similarly, we see that

$$(\mathbf{u}\times\mathbf{v})\cdot\mathbf{j} = u_3v_1 - v_3u_1$$

and

$$(\mathbf{u}\times\mathbf{v})\cdot\mathbf{k} = u_1v_2 - v_1u_2.$$

Now suppose that

$$\mathbf{u}\times\mathbf{v} = (x,y,z) = x\mathbf{i} + y\mathbf{j} + z\mathbf{k}.$$

Taking the dot product of both sides with **i** yields

$$(\mathbf{u}\times\mathbf{v})\cdot\mathbf{i} = x\mathbf{i}\cdot\mathbf{i} + y\mathbf{j}\cdot\mathbf{i} + z\mathbf{k}\cdot\mathbf{i} = x,$$

since $\mathbf{i}\cdot\mathbf{i} = 1$, $\mathbf{j}\cdot\mathbf{i} = 0 = \mathbf{k}\cdot\mathbf{i}$. Similarly, $(\mathbf{u}\times\mathbf{v})\cdot\mathbf{j} = y$ and $(\mathbf{u}\times\mathbf{v})\cdot\mathbf{k} = z$. Therefore, the components of $\mathbf{u}\times\mathbf{v}$ are as stated. ■

EXAMPLE 3-23

Let $\mathbf{u} = (-2,1,3)$ and $\mathbf{v} = (-1,2,1)$. Then,

$$\mathbf{u}\times\mathbf{v} = det\begin{bmatrix} \mathbf{i} & \mathbf{j} & \mathbf{k} \\ -2 & 1 & 3 \\ -1 & 2 & 1 \end{bmatrix} = (1 - 6)\mathbf{i} - (-2 + 3)\mathbf{j} + (-4 + 1)\mathbf{k}$$

$$= -5\mathbf{i} - \mathbf{j} - 3\mathbf{k} = (-5,-1,-3).$$

Notice that

$$(\mathbf{u}\times\mathbf{v})\cdot\mathbf{u} = (-5)(-2) + (-1)(1) + (-3)(3) = 0,$$

showing that $\mathbf{u}\times\mathbf{v}$ is orthogonal to **u**. Similarly, we can show that it is orthogonal to **v**. ❑

THEOREM 3-15 (Properties of the Cross Product)
Let **u**, **u**′, **v**, and **v**′ be vectors in R^3 and c be a scalar. Then,

(1) $(\mathbf{u}\times\mathbf{v})\cdot\mathbf{u} = (\mathbf{u}\times\mathbf{v})\cdot\mathbf{v} = 0$ (so, $\mathbf{u}\times\mathbf{v}$ is orthogonal to both **u** and **v**),

(2) $\mathbf{u}\times\mathbf{v} = -\mathbf{v}\times\mathbf{u}$,

(3) $(c\mathbf{u}+\mathbf{u}')\times\mathbf{v} = c(\mathbf{u}\times\mathbf{v}) + \mathbf{u}'\times\mathbf{v}$ and $\mathbf{u}\times(c\mathbf{v}+\mathbf{v}') = c(\mathbf{u}\times\mathbf{v}) + \mathbf{u}\times\mathbf{v}'$ (linearity),

(4) $\mathbf{u}\times\mathbf{u} = \mathbf{0}$,

(5) $\mathbf{u}\times\mathbf{0} = \mathbf{0}\times\mathbf{u} = \mathbf{0}$,

(6) $|\mathbf{u}\times\mathbf{v}|^2 = |\mathbf{u}|^2\,|\mathbf{v}|^2 - (\mathbf{u}\cdot\mathbf{v})^2$.

PROOF All except (6) follow at once from the properties of the determinant. For example, by equation 3-13,

$$(\mathbf{u}\times\mathbf{v})\cdot\mathbf{u} = det\begin{bmatrix} u_1 & u_2 & u_3 \\ v_1 & v_2 & v_3 \\ u_1 & u_2 & u_3 \end{bmatrix} = 0,$$

since the matrix has two identical rows. This proves the first part of (1). We leave the proofs of (2) through (5) for the reader. To see (6), we have

$$\begin{aligned} |\mathbf{u}\times\mathbf{v}|^2 &= (u_2v_3 - v_2u_3)^2 + (u_3v_1 - v_3u_1)^2 + (u_1v_2 - v_1u_2)^2 \\ &= (u_1{}^2 + u_2{}^2 + u_3{}^2)(v_1{}^2 + v_2{}^2 + v_3{}^2) - (u_1v_1 + u_2v_2 + u_3v_3)^2 \\ &\quad \text{(as the reader may verify by expansion)} \\ &= |\mathbf{u}|^2\,|\mathbf{v}|^2 - (\mathbf{u}\cdot\mathbf{v})^2. \blacksquare \end{aligned}$$

EXAMPLE 3-24

(a) We compute all possible cross products of **i**, **j**, and **k**. First,

$$\mathbf{i}\times\mathbf{j} = det\begin{bmatrix} \mathbf{i} & \mathbf{j} & \mathbf{k} \\ 1 & 0 & 0 \\ 0 & 1 & 0 \end{bmatrix} = \mathbf{k} = -\mathbf{j}\times\mathbf{i},$$

by property (2) of Theorem 3-15. Similarly, we can show that

$$\mathbf{i}\times\mathbf{k} = -\mathbf{k}\times\mathbf{i} = -\mathbf{j} \text{ and } \mathbf{j}\times\mathbf{k} = -\mathbf{k}\times\mathbf{j} = \mathbf{i}.$$

Of course,

$$\mathbf{i}\times\mathbf{i} = \mathbf{j}\times\mathbf{j} = \mathbf{k}\times\mathbf{k} = \mathbf{0},$$

by property (4) of the theorem.

It is interesting to note that, as a result of the above,

$$(\mathbf{i}\times\mathbf{j})\times\mathbf{j} = \mathbf{k}\times\mathbf{j} = -\mathbf{i},$$

while

$$\mathbf{i}\times(\mathbf{j}\times\mathbf{j}) = \mathbf{i}\times\mathbf{0} = \mathbf{0}.$$

This illustrates that, in general, $(\mathbf{u}\times\mathbf{v})\times\mathbf{w} \neq \mathbf{u}\times(\mathbf{v}\times\mathbf{w})$. The *associative law does not hold for the cross products* (as opposed to products of numbers or matrices).

(b) The relations in (a) above are often used to simplify the algebra in expressions involving cross products. For example, if $\mathbf{u} = (2,-3,1)$ and $\mathbf{v} = (-1,2,-2)$ then

$$\begin{aligned}\mathbf{u}\times\mathbf{v} &= (2\mathbf{i} - 3\mathbf{j} + \mathbf{k})\times(-\mathbf{i} + 2\mathbf{j} - 2\mathbf{k}) \\ &= -2\mathbf{i}\times\mathbf{i} + 4\mathbf{i}\times\mathbf{j} - 4\mathbf{i}\times\mathbf{k} + 3\mathbf{j}\times\mathbf{i} - 6\mathbf{j}\times\mathbf{j} + \\ &\quad 6\mathbf{j}\times\mathbf{k} - \mathbf{k}\times\mathbf{i} + 2\mathbf{k}\times\mathbf{j} - 2\mathbf{k}\times\mathbf{k} \\ &= \mathbf{0} + 4\mathbf{k} + 4\mathbf{j} - 3\mathbf{k} - \mathbf{0} + 6\mathbf{i} - \mathbf{j} - 2\mathbf{i} - \mathbf{0} \\ &= 4\mathbf{i} + 3\mathbf{j} + \mathbf{k}\ (=(4,3,1)).\end{aligned}$$

Of course, we could also use the determinant expression to compute $\mathbf{u}\times\mathbf{v}$ in this case. ❑

Now, by Theorem 3-15, $\mathbf{u}\times\mathbf{v}$ is a vector orthogonal to both $\mathbf{u}$ and $\mathbf{v}$, and its length is determined by property (6). In fact, if θ is the angle between $\mathbf{u}$ and $\mathbf{v}$, we know from properties of the dot product that $\mathbf{u}\cdot\mathbf{v} = |\mathbf{u}||\mathbf{v}|\cos\theta$. Thus, we may rewrite property (6) as

$$\begin{aligned}|\mathbf{u}\times\mathbf{v}|^2 &= |\mathbf{u}|^2\,|\mathbf{v}|^2 - |\mathbf{u}|^2\,|\mathbf{v}|^2\cos^2\theta \\ &= |\mathbf{u}|^2\,|\mathbf{v}|^2(1 - \cos^2\theta) \\ &= |\mathbf{u}|^2\,|\mathbf{v}|^2\sin^2\theta.\end{aligned}$$

COROLLARY

$|\mathbf{u}\times\mathbf{v}| = |\mathbf{u}||\mathbf{v}|\sin\theta =$ the area of the parallelogram whose sides are $\mathbf{u}$ and $\mathbf{v}$.

PROOF With reference to Figure 3-1, $|\mathbf{v}|\sin\theta$ is the altitude of the parallelogram with $\mathbf{u}$ considered as the base. So the area is $|\mathbf{u}||\mathbf{v}|\sin\theta$, as stated. ■

FIGURE 3-1

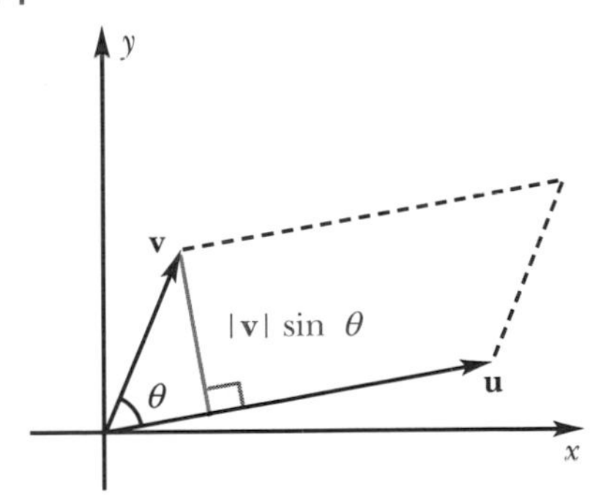

EXAMPLE 3-25

Let $P(-2,1,1)$, $Q(2,2,2)$, and $R(2,-2,1)$ be given points in R^3. We use the cross product to find the area of the triangle with these vertices. Let $\mathbf{u} = \{\mathbf{PQ}\} = (4,1,1)$ and $\mathbf{v} = \{\mathbf{PR}\} = (4,-3,0)$. Then,

$$\mathbf{u} \times \mathbf{v} = det \begin{bmatrix} \mathbf{i} & \mathbf{j} & \mathbf{k} \\ 4 & 1 & 1 \\ 4 & -3 & 0 \end{bmatrix} = 3\mathbf{i} + 4\mathbf{j} - 16\mathbf{k} = (3,4,-16).$$

Thus, the area of the parallelogram with sides $\mathbf{u}$ and $\mathbf{v}$ is $|\mathbf{u} \times \mathbf{v}| = \sqrt{9+16+256} \approx 16.76$ (square units), and the area of the triangle is $16.76/2 = 8.38$ (square units). ❑

EXAMPLE 3-26

If $\mathbf{u}$ and $\mathbf{v}$ are nonzero vectors in R^3, then $\mathbf{u} \times \mathbf{v} = \mathbf{0}$ iff $\mathbf{u}$ is parallel to $\mathbf{v}$. For $|\mathbf{u} \times \mathbf{v}| = |\mathbf{u}||\mathbf{v}| \sin \theta$, so $\mathbf{u} \times \mathbf{v} = \mathbf{0}$ iff $\theta = 0°$ or $180°$, and this is true iff $\mathbf{u}$ is parallel to $\mathbf{v}$. ❑

Next, let $\mathbf{u}$, $\mathbf{v}$, and $\mathbf{w}$ be any nonzero vectors in R^3. They are **coplanar** iff there is a plane $ax + by + cz = 0$ through the origin containing them, where at least one of a, b, and c is nonzero. This means that the components of $\mathbf{u}$, $\mathbf{v}$, and $\mathbf{w}$ are, respectively, solutions of the equations

$$\begin{aligned} au_1 + bu_2 + cu_3 &= 0 \\ av_1 + bv_2 + cv_3 &= 0 \\ aw_1 + bw_2 + cw_3 &= 0. \end{aligned}$$

This is a homogeneous system of three linear equations in a, b, and c with coefficient matrix

$$\begin{bmatrix} u_1 & u_2 & u_3 \\ v_1 & v_2 & v_3 \\ w_1 & w_2 & w_3 \end{bmatrix}.$$

As at least one of a, b, and c must be nonzero, the system has a nontrivial solution. Therefore, the coefficient matrix is not invertible (since, if it were, by Theorem 2-7, the system would have only the zero solution). By the properties of determinants, this is true iff

$$det \begin{bmatrix} u_1 & u_2 & u_3 \\ v_1 & v_2 & v_3 \\ w_1 & w_2 & w_3 \end{bmatrix} = 0.$$

Now the determinant above is, by Theorem 3-14, $(\mathbf{u}\times\mathbf{v})\cdot\mathbf{w} = \mathbf{w}\cdot(\mathbf{u}\times\mathbf{v})$. Using properties of determinants (interchanging rows), we see that

$$\mathbf{w}\cdot(\mathbf{u}\times\mathbf{v}) = \mathbf{u}\cdot(\mathbf{v}\times\mathbf{w}) = \mathbf{v}\cdot(\mathbf{w}\times\mathbf{u}).$$

Therefore, we see that if $\mathbf{u}$, $\mathbf{v}$, and $\mathbf{w}$ are coplanar, then $\mathbf{u}\cdot(\mathbf{v}\times\mathbf{w}) = 0$.

Conversely, if $\mathbf{u}\cdot(\mathbf{v}\times\mathbf{w}) = 0$, reversing the preceding argument shows that $\mathbf{u}$, $\mathbf{v}$, and $\mathbf{w}$ are coplanar. The scalar $\mathbf{u}\cdot(\mathbf{v}\times\mathbf{w})$ is often called a **triple product** of $\mathbf{u}$, $\mathbf{v}$, and $\mathbf{w}$, and may be denoted unambiguously (without brackets) as $\mathbf{u}\cdot\mathbf{v}\times\mathbf{w}$ (why?).

FIGURE 3-2

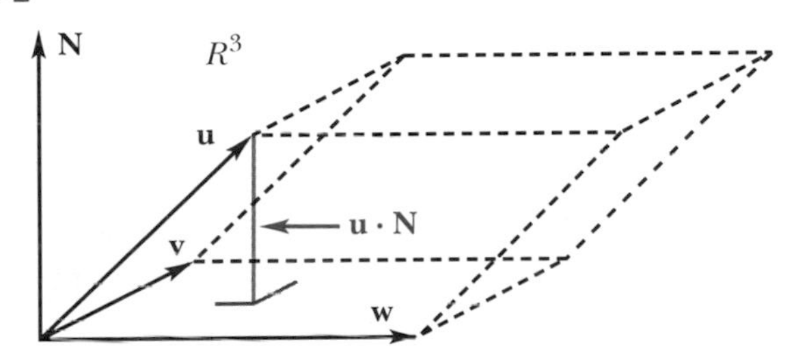

On the other hand, if $\mathbf{u}$, $\mathbf{v}$, and $\mathbf{w}$ are noncoplanar, as in Figure 3-2, then $|\mathbf{u}\cdot\mathbf{v}\times\mathbf{w}|$ is the volume V of the parallelepiped formed by $\mathbf{u}$, $\mathbf{v}$, and $\mathbf{w}$. This is so because V is the altitude times the area of the parallelogram formed by $\mathbf{v}$ and $\mathbf{w}$ (in the figure). Recall from Chapter 1 that the altitude is the absolute value of the scalar projection of $\mathbf{u}$ on a unit vector $\mathbf{N}$ perpendicular to the plane of $\mathbf{v}$ and $\mathbf{w}$. Therefore, it is $|\mathbf{u}\cdot\mathbf{N}|$. But, as $\mathbf{v}\times\mathbf{w}$ is orthogonal to both $\mathbf{v}$ and $\mathbf{w}$, we may take

$$\mathbf{N} = (1/|\mathbf{v}\times\mathbf{w}|)\ \mathbf{v}\times\mathbf{w},$$

so that

$$|\mathbf{u}\cdot\mathbf{N}| = |(1/|\mathbf{v}\times\mathbf{w}|)\ \mathbf{u}\cdot\mathbf{v}\times\mathbf{w}|.$$

By the corollary of Theorem 3-15, the area of the parallelogram with sides $\mathbf{v}$ and $\mathbf{w}$ is $|\mathbf{v}\times\mathbf{w}|$, so

$$V = |(1/|\mathbf{v}\times\mathbf{w}|)\ \mathbf{u}\cdot\mathbf{v}\times\mathbf{w}|\ |\mathbf{v}\times\mathbf{w}| = |\mathbf{u}\cdot\mathbf{v}\times\mathbf{w}|,$$

as stated. Therefore, we have proved

THEOREM 3-16

Vectors $\mathbf{u}$, $\mathbf{v}$, and $\mathbf{w}$ in R^3 are coplanar iff the triple product $\mathbf{u}\cdot\mathbf{v}\times\mathbf{w} = 0$. If they are noncoplanar, $|\mathbf{u}\cdot\mathbf{v}\times\mathbf{w}|$ is the volume of the parallelepiped formed by them. ■

EXAMPLE 3-27

(a) The volume of the parallelepiped formed by the vectors $\mathbf{u} = (1,1,0)$, $\mathbf{v} = (-1,3,1)$, and $\mathbf{w} = (1,1,1)$ is

$$|\mathbf{u}\cdot\mathbf{v}\times\mathbf{w}| = \left|det\begin{bmatrix} 1 & 1 & 0 \\ -1 & 3 & 1 \\ 1 & 1 & 1 \end{bmatrix}\right| = 4.$$

(b) The vectors $\mathbf{u} = (-3,5,1)$, $\mathbf{v} = (-1,3,1)$, and $\mathbf{w} = (1,1,1)$ are coplanar because

$$\mathbf{u}\cdot\mathbf{v}\times\mathbf{w} = det\begin{bmatrix} -3 & 5 & 1 \\ -1 & 3 & 1 \\ 1 & 1 & 1 \end{bmatrix} = 0. \ \square$$

Because $\mathbf{u}\times\mathbf{v}$ is orthogonal to both $\mathbf{u}$ and $\mathbf{v}$, it has numerous applications in vector geometry problems. The next two examples extend the ideas from Chapter 1.

EXAMPLE 3-28

(a) Let $\mathbf{u} = (1,1,-1)$ and $\mathbf{v} = (-1,0,1)$. A unit vector $\mathbf{N}$ orthogonal to both $\mathbf{u}$ and $\mathbf{v}$ is $\mathbf{N} = (1/|\mathbf{u}\times\mathbf{v}|)\ \mathbf{u}\times\mathbf{v}$. Since

$$\mathbf{u}\times\mathbf{v} = det\begin{bmatrix} \mathbf{i} & \mathbf{j} & \mathbf{k} \\ 1 & 1 & -1 \\ -1 & 0 & 1 \end{bmatrix} = (1,0,1),$$

it follows that $\mathbf{N} = (1/\sqrt{2},0,1/\sqrt{2})$.

(b) In Example 1-16, we constructed a unit vector orthogonal to $\mathbf{u} = (1,2,3)$. Of course, there are infinitely many such vectors, all lying in a plane through the origin. We may use the cross product to obtain one by selecting any other vector $\mathbf{v}$ that is not collinear with $\mathbf{u}$. For example, using $\mathbf{v} = (1,0,0) = \mathbf{i}$, we see that

$$\mathbf{u}\times\mathbf{v} = (\mathbf{i} + 2\mathbf{j} + 3\mathbf{k})\times\mathbf{i} = -2\mathbf{k} + 3\mathbf{j} = (0,3,-2)$$

is orthogonal to both $\mathbf{u}$ and $\mathbf{v}$. As $|\mathbf{u}\times\mathbf{v}| = \sqrt{13}$, a unit vector orthogonal to $\mathbf{u}$ is $(0,3/\sqrt{13},-2/\sqrt{13})$.

(c) Let $\mathbf{u} = (1,2,3)$ and $\mathbf{v} = (-2,-1,1)$. Since these vectors are not collinear, they determine a plane through the origin. As $\mathbf{u}\times\mathbf{v}$ is orthogonal to both

u and **v**, it is a normal vector to the plane. Since

$$\mathbf{u}\times\mathbf{v} = det\begin{bmatrix} \mathbf{i} & \mathbf{j} & \mathbf{k} \\ 1 & 2 & 3 \\ -2 & -1 & 1 \end{bmatrix} = (5,-7,3),$$

by the corollary of Theorem 1-8, the equation of the plane is

$$5x - 7y + 3z = 0. \square$$

EXAMPLE 3-29

Let $P = (1,2,2)$, $Q = (-1,1,1)$, and $R = (2,2,-1)$. We find the equation of the plane containing these points by two methods.

(a) $\mathbf{u} = \{\mathbf{PQ}\} = (-2,-1,-1)$ and $\mathbf{v} = \{\mathbf{PR}\} = (1,0,-3)$ are parallel to the plane, so $\mathbf{u}\times\mathbf{v}$ is orthogonal to it. Since

$$\mathbf{u}\times\mathbf{v} = det\begin{bmatrix} \mathbf{i} & \mathbf{j} & \mathbf{k} \\ -2 & -1 & -1 \\ 1 & 0 & -3 \end{bmatrix} = (3,-7,1),$$

and $P = (1,2,2)$ is on the plane, by the corollary of Theorem 1-8, the equation of the plane is

$$3(x - 1) - 7(y - 2) + 1(z - 2) = 0,$$

or,

$$3x - 7y + z = -9.$$

(b) Let $\mathbf{r} = (x,y,z)$ be any point on the plane (recall that $\mathbf{r}$ is often called the *position vector* of an arbitrary point on the plane). Then, with $\mathbf{r}_1 = \{\mathbf{OP}\} = (1,2,2)$, $\mathbf{r}_2 = \{\mathbf{OQ}\} = (-1,1,1)$, and $\mathbf{r}_3 = \{\mathbf{OR}\} = (2,2,-1)$, the vectors

$$\mathbf{r} - \mathbf{r}_1,\ \mathbf{r}_2 - \mathbf{r}_1\ (=\mathbf{u}),\ \text{and}\ \mathbf{r}_3 - \mathbf{r}_1\ (=\mathbf{v})$$

must be coplanar. By Theorem 3-16,

$$(\mathbf{r} - \mathbf{r}_1)\cdot(\mathbf{r}_2 - \mathbf{r}_1)\times(\mathbf{r}_3 - \mathbf{r}_1) = (\mathbf{r} - \mathbf{r}_1)\cdot\mathbf{u}\times\mathbf{v} = 0.$$

Using the result for $\mathbf{u}\times\mathbf{v}$ from (a) and $\mathbf{r} - \mathbf{r}_1 = (x - 1, y - 2, z - 2)$, this becomes

$$(x-1,y-2,z-2)\cdot(3,-7,1) = 0,$$

or,

$$3x - 7y + z = -9,$$

which is the same equation for the plane as obtained in (a). $\square$

EXERCISES 3-4

1. Find the cross product $\mathbf{u} \times \mathbf{v}$ for the following pairs of vectors in R^3.
   (a) $\mathbf{u} = (-1,2,1)$, $\mathbf{v} = (2,2,3)$ (b) $\mathbf{u} = (-1,2,4)$, $\mathbf{v} = (2,-4,-8)$
   (c) $\mathbf{u} = (2,3,-4)$, $\mathbf{v} = (-2,4,6)$
2. Evaluate the following expressions.
   (a) $(2\mathbf{i} + 3\mathbf{j} - 4\mathbf{k}) \times [2(\mathbf{i} - 2\mathbf{j} + 3\mathbf{k}) - 3(2\mathbf{i} - 4\mathbf{j} + \mathbf{k})]$
   (b) $[(\mathbf{i} + \mathbf{j}) \times (2\mathbf{i} - 3\mathbf{j} + 2\mathbf{k})] \cdot (\mathbf{i} - \mathbf{j} + 2\mathbf{k})$
3. Find the area of the triangle with vertices $(-2,-2,-3)$, $(2,3,4)$, and $(-3,1,0)$.
4. Find a unit vector orthogonal to each pair of vectors.
   (a) $\mathbf{u} = (-2,3,1)$ and $\mathbf{v} = (2,2,1)$
   (b) $\mathbf{u} = (0,5,-2)$ and $\mathbf{v} = (-2,2,0)$
   (c) $\mathbf{u} = (-1,1,1)$ and $\mathbf{v} = (2,-2,-2)$
5. Given $\mathbf{u} = (-2,1,1)$, $\mathbf{v} = (-1,1,2)$, and $\mathbf{w} = (-1,-1,1)$, compute $(\mathbf{u} \times \mathbf{v}) \times \mathbf{w}$ and $\mathbf{u} \times (\mathbf{v} \times \mathbf{w})$ to verify that they are not equal.
6. Find the volume of each parallelepiped with the following edges.
   (a) the vectors $\mathbf{u}$, $\mathbf{v}$, $\mathbf{w}$ in question 5
   (b) the vectors $\mathbf{u} = (-1,3,1)$, $\mathbf{v} = (-1,1,3)$, and $\mathbf{w} = (0,3,4)$
7. Prove properties (2), (3), (4), and (5) of Theorem 3-15.
8. Show that $\mathbf{w} \cdot (\mathbf{u} \times \mathbf{v}) = \mathbf{u} \cdot (\mathbf{v} \times \mathbf{w}) = \mathbf{v} \cdot (\mathbf{w} \times \mathbf{u})$.
9. Show that the vectors $\mathbf{u} = (-2,1,1)$, $\mathbf{v} = (-1,1,2)$, and $\mathbf{w} = (-4,3,5)$ are coplanar.
10. Prove that $(\mathbf{u} \times \mathbf{v}) \times \mathbf{w} = (\mathbf{u} \cdot \mathbf{w})\mathbf{v} - (\mathbf{v} \cdot \mathbf{w})\mathbf{u}$, while $\mathbf{u} \times (\mathbf{v} \times \mathbf{w}) = (\mathbf{u} \cdot \mathbf{w})\mathbf{v} - (\mathbf{u} \cdot \mathbf{v})\mathbf{w}$. Verify these formulae using the results of question 5.
11. Prove the identity $\mathbf{u} \times (\mathbf{v} \times \mathbf{w}) + \mathbf{v} \times (\mathbf{w} \times \mathbf{u}) + \mathbf{w} \times (\mathbf{u} \times \mathbf{v}) = \mathbf{0}$. (Hint: Use the results of question 10.)
12. (a) Find the equation of the plane through the origin that contains the vectors $\mathbf{u} = (3,0,2)$ and $\mathbf{v} = (0,2,-5)$.
    (b) Find the equation of the plane through $(-2,3,5)$ and parallel to the vectors $\mathbf{u}$ and $\mathbf{v}$ from (a).
13. Find the equation of the planes through the following points.
    (a) $P(1,2,3)$, $Q(-3,2,4)$, and $R(1,1,1)$
    (b) $P(-1,1,1)$, $Q(1,-1,1)$, and $R(1,1,-1)$
    (c) $P(1,0,2)$, $Q(0,2,1)$, and $R(1,2,0)$
14. Prove that $(\mathbf{u} \times \mathbf{v}) \cdot (\mathbf{v} \times \mathbf{w}) \times (\mathbf{w} \times \mathbf{u}) = (\mathbf{u} \cdot \mathbf{v} \times \mathbf{w})^2$.

4 VECTOR SPACES

4-1 Two Basic Examples
4-2 Vector Spaces and Subspaces
4-3 Generating Sets, Linear Independence, and Bases
4-4 Computational Techniques — Rank of a Matrix
4-5 Electrical Networks (Optional)
4-6 Complex Vector Spaces (Optional)

4-1 Two Basic Examples

In this chapter, we introduce the definition and basic properties of vector spaces. The concept of a vector space is an example of "abstract mathematics", a phrase many people probably find intimidating. Abstraction is perhaps the greatest power of mathematical thinking. Our interest here will be in certain algebraic properties possessed by vectors in R^2 and R^3, such as addition and scalar multiplication. These properties are shared by a great many entirely different sets of objects. When a set has these properties, it is called a **vector space** (or a linear space) and its elements are called **vectors**. For example, in addition to the sets of vectors in R^2 and R^3, we noted in Chapter 2 that the set of $m \times n$ matrices has these properties. It is because there are so many *different* sets sharing these common algebraic properties that the properties themselves become worthy of study. Thus, when we prove something about these properties, it applies to *all* specific examples, including vectors in the plane, matrices, and many other objects we shall encounter. Of course, the properties themselves (such as addition and scalar multiplication) have been suggested by, or taken or *abstracted* from, the many individual examples which possess them. It is in this sense (and only in this sense) that the study of *properties* (as opposed to the consideration of individual examples) is called abstract mathematics.

Before proceeding to the definition of a vector space, we consider two further examples of sets having the algebraic properties we shall need. The first of these is just a simple extension of vectors in R^3.

In Chapter 1, we saw that a vector **v** may be identified with a triple (x_1,x_2,x_3), the endpoint of its unique representative having initial point at the origin. Of course, x_1, x_2, and x_3 are the components of **v**. Since vectors in R^3 can be added and multiplied by scalars, the same naturally applies to the corresponding triples.

Similarly, in Chapter 2, we identified an n-tuple $(x_1,x_2,\ldots,x_n)$ (which was a solution for a system of equations) with the corresponding column vector. Because we can add column vectors and multiply them by scalars, it follows that we can do the same with n-tuples. We formalize these ideas with

DEFINITION 4-1

Let n be a positive integer. An n-tuple $\mathbf{x} = (x_1,\ldots,x_n)$ is an ordered set of n real numbers $x_1,x_2,\ldots,x_n$ called the **coordinates** of **x**. We call the set of all n-tuples **real n-space** and denote it by R^n.

For reasons that we shall see later in the chapter, we say that the **dimension** of R^n is n.

When $n = 1$, we obtain $R^1 = R$, the set of real numbers, or equivalently, the **real line**. When $n = 2$ and 3 we get back R^2 and R^3 respectively, which we considered in Chapter 1. When $n > 3$, we can no longer make the usual pictorial representations on paper, but this does not mean that R^n is any less real or useful in these cases. In fact, R^4 is the underlying set for the theory of relativity, and higher dimensional real n-spaces are important in many applications.

Now, as said earlier, we are identifying n-tuples with their corresponding column vectors, so that we are considering R^n to be "essentially the same" as the set $M(n \times 1)$ of all $n \times 1$ column vectors with real entries. By "translating" the addition and scalar multiplication of matrices into the language of n-tuples, we may *define* addition and scalar multiplication of n-tuples by

$$(x_1,\ldots,x_n) + (y_1,\ldots,y_n) = (x_1 + y_1,\ldots,x_n + y_n)$$
$$k(x_1,\ldots,x_n) = (kx_1,\ldots,kx_n), k \in R.$$

Of course, we say that two n-tuples $(x_1,\ldots,x_n)$ and $(y_1,\ldots,y_n)$ are *equal* iff $x_1 = y_1$, $x_2 = y_2$, ..., $x_n = y_n$.

For $n = 2$ and 3, these definitions amount to the definitions in Chapter 1 for vectors as "arrows", in terms of their components.

For later comparison, we list below the essential properties of addition and scalar multiplication for n-tuples. These should not be new to the reader, as they are just the properties (1) through (8) of Theorem 2-4 for matrices, but written here for n-tuples.

(1) $(x_1,\ldots,x_n) + (y_1,\ldots,y_n) = (y_1,\ldots,y_n) + (x_1,\ldots,x_n)$.

(2) $[(x_1,\ldots,x_n) + (y_1,\ldots,y_n)] + (z_1,\ldots,z_n) = (x_1,\ldots,x_n) + [(y_1,\ldots,y_n) + (z_1,\ldots,z_n)]$.

(3) There is a **zero n-tuple** $(0,0,\ldots,0)$ denoted $\mathbf{0}$ which is an **additive identity** for n-tuples: $(x_1,\ldots,x_n) + (0,\ldots,0) = (x_1 + 0,\ldots,x_n + 0) = (x_1,\ldots,x_n)$.

(4) Given any n-tuple $\mathbf{x}$, it has an unique **additive inverse** $-\mathbf{x} = (-x_1,\ldots,-x_n)$ such that $(x_1,\ldots,x_n) + (-x_1,\ldots,-x_n) = \mathbf{0} = (0,\ldots,0)$.

(5) If $a \in R$, $a[(x_1,\ldots,x_n) + (y_1,\ldots,y_n)] = a(x_1,\ldots,x_n) + a(y_1,\ldots,y_n)$.

(6) $(a + b)(x_1,\ldots,x_n) = a(x_1,\ldots,x_n) + b(x_1,\ldots,x_n)$, a, $b \in R$.

(7) $(ab)(x_1,\ldots,x_n) = a[b(x_1,\ldots,x_n)] = (abx_1,\ldots,abx_n)$, a, $b \in R$.

(8) $1(x_1,\ldots,x_n) = (x_1,\ldots,x_n)$, where $1 \in R$ is the number one.

Looking ahead, we remark that, as in the case of vectors in R^2 and R^3 and the case of $m \times n$ matrices, these eight properties imply that *R^n is a vector space with respect to the operations of addition and scalar multiplication defined above*.

Also, we shall use these properties in doing "algebra" with n-tuples.

EXAMPLE 4-1

In R^4, let $\mathbf{x} = (-1,2,0,3)$, $\mathbf{y} = (2,5,-1,0)$, $\mathbf{z} = (0,2,-4,1)$, $a = -2$, and $b = 3$. Then,

$$\begin{aligned} &a[b(\mathbf{x} + 4\mathbf{z}) - 5\mathbf{y}] \\ &= ab\mathbf{x} + 4ab\mathbf{z} - 5a\mathbf{y} \\ &= (-2)(3)(-1,2,0,3) + 4(-2)(3)(0,2,-4,1) - 5(-2)(2,5,-1,0) \\ &= (6,-12,0,-18) + (0,-48,96,-24) + (20,50,-10,0) \\ &= (26,-10,86,-42). \ \square \end{aligned}$$

The second example we are going to discuss is of a very general nature. In fact, it includes as particular cases all the examples of vector spaces we have considered so far (see questions 4, 5, and 6 in Exercises 4-1).

DEFINITION 4-2

Let S be any set. We denote by $F(S)$ the set of all functions $f:S \rightarrow R$.

For definiteness, the reader may wish to think of the specific case where S is an interval of numbers such as $[0,1]$, or even R itself; these are certainly important cases in applications. However, for each choice of S, we will get a different set of functions $F(S)$, and, as we shall see, a different vector space.

Given f and g in $F(S)$ we define their **sum** $f + g:S \rightarrow R$ to be the new function in $F(S)$ with $(f + g)(x) = f(x) + g(x)$ for $x \in S$. And, when $k \in R$, the **scalar multiple** of f by k is defined to be $kf:S \rightarrow R$ with $(kf)(x) = kf(x)$. It is important to observe that these definitions are possible because the values $f(x)$ of the function f in $F(S)$ are *numbers*, and so can be added to other numbers such as $g(x)$, or multiplied by a number such as k.

As a first specific case, suppose we take $S = [0,1]$; then $F(S)$ is the set of real-valued functions defined on the interval $\{x \mid 0 \leq x \leq 1\}$. Some examples of functions in $F(S)$ are $f(x) = \sqrt{x}$, $g(x) = e^x$, $h(x) = \log(x + 1)$, $k(x) = x^5$ $(x \in [0,1])$ to name but a few. Then, we have, for example,

(i) $(f + g)(x) = \sqrt{x} + e^x$,

so

$$(f + g)(1) = 1 + e,$$
$$(f + g)(1/2) = \sqrt{1/2} + e^{1/2};$$
$$5f(9) = (5)(3) = 15;$$

(ii) $(g + k)(x) = e^x + x^5$,

so

$$(g + k)(0) = e^0 + 0 = 1 + 0 = 1;$$

(iii) $(3h - k)(x) = 3\log(x+1) - x^5$,

so

$$(3h - k)(9) = 3\log 10 - 9^5 = 3 - 59049 = -59046.$$

The graphs of some of these functions are sketched in Figure 4-1.

In particular, we note that the operations of addition and scalar multiplication produce *new* functions in $F(S)$ from given functions in $F(S)$. The reader should check that the eight familiar properties we have seen also hold for $F(S)$ when $S = [0,1]$.

FIGURE 4-1

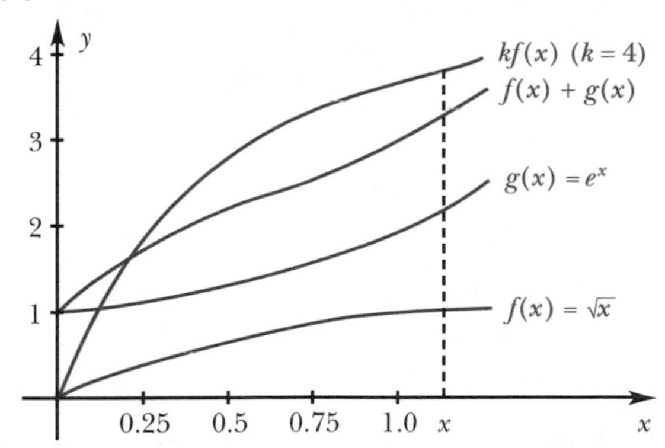

More generally, let S now be *any* set, f, g, and h be any elements of $F(S)$, and a, b be any numbers. Then, we have the following properties:

(1) $f + g = g + f$. Recall that this means $(f + g)(x) = (g + f)(x)$ for all $x \in S$, since two functions are said to be equal iff they have the same values at all points x of their common domain.

(2) $(f + g) + h = f + (g + h)$ (the same remark made for (1) applies here).

(3) There is a **zero function** 0 defined by $0(x) = 0 \in R$ for all $x \in S$. It has the property that $f + 0 = f$ for all $f \in F(S)$, because $(f + 0)(x) = f(x) + 0 = f(x)$ for all x.

(4) Given f in $F(S)$, there is a function $-f$ defined by $(-f)(x) = -f(x)$ and with the property $f + (-f) = 0$, the zero function.

(5) $a(f + g) = af + ag$.

(6) $(a + b)f = af + bf$.

(7) $(ab)f = a(bf)$.

(8) $1f = f$, where 1 is the number one.

The reader should note that all the properties above are true because of properties of numbers, and because the operations defined are in terms of *values* of functions, which are numbers.

EXAMPLE 4-2

To illustrate the generality of $F(S)$, we take as a second specific case the set $S = \{1,2\}$ consisting of only the two integers 1 and 2. Now, any function f in $F(S)$ has exactly two values, namely $f(1)$ and $f(2)$, both of which are numbers. In other words, any such function f is completely described by the ordered pair of numbers $(f(1),f(2))$. For example, if f is the function with $f(1) = 2$ and $f(2) = 7$, then f *corresponds* to the ordered pair (2,7). (For a discussion of 1-1 correspondence between sets, see Chapter 1.)

Therefore, it appears that with $S = \{1,2\}$, $F(S)$ is another way of describing R^2. This is the case because sums and scalar multiples in $F(S)$ and R^2 also correspond. For example, let g in $F(S)$ be the function with $g(1) = -5$ and $g(2) = 3$. Then, with f from above,

$$(f + g)(1) = f(1) + g(1) = 2 + (-5) = -3,$$

while,

$$(f + g)(2) = f(2) + g(2) = 7 + 3 = 10.$$

So, $f + g$ corresponds to the ordered pair $(-3,10)$, which is exactly the sum of the ordered pairs $(2,7)$ and $(-5,3)$ in R^2:

$$(-3,10) = (2,7) + (-5,3).$$

The reader should check that the same correspondence holds for scalar multiplication. Therefore, with $S = \{1,2\}$, $F(S)$ is "essentially" R^2. This is a situation we encountered earlier when we agreed to identify n-tuples with $n \times 1$ matrices. The property of being "essentially the same" is called **isomorphism** and is discussed in Chapter 5. ❑

EXERCISES 4-1

1. In R^3, let $\mathbf{x} = (1,2,3)$, $\mathbf{y} = (2,3,1)$, and $\mathbf{z} = (3,1,2)$. Verify that the eight properties making R^3 into a vector space are true for these triples when $a = 2$ and $b = -3$.

2. Verify that $R^1 = R$ is a vector space by checking each of the eight properties discussed for R^n, with $n = 1$.

3. Evaluate the following functions in $F(R)$ at the points $x = -2, 0, 1/2, 1, 2$.

(a) $(2f - 3g)(x)$, where $f(x) = x^2$ and $g(x) = x^3$

(b) $[(-1/2)h + 3k - 2l](x)$, where $h(x) = 2x + 1$, $k(x) = -3x^2 + 1$, and $l(x) = e^x$

(c) $(-4m + n)(x)$, with $m(x) = |x|$ and $n(x) = 1/(x^2 + 1)$

4. (a) Let $S = \{1,2,3\}$ be the set having the integers 1, 2, and 3 as its only members. Show that $F(S)$ is "essentially" R^3 and that the operations of addition and scalar multiplication in these two sets also correspond.

(b) Generalize the idea in (a) above to obtain a set S such that R^n is essentially $F(S)$.

5. In R^5, let $\mathbf{x} = (-1,3,-2,4,7)$, $\mathbf{y} = (0,5,-8,0,4)$, and $\mathbf{z} = (0,9,0,-3,11)$. Compute the following elements of R^5.

(a) $3\mathbf{x} - 7(4\mathbf{z} - 5\mathbf{y})$

(b) $(-2)[7\mathbf{y} - \mathbf{x} + 3(\mathbf{z} + 4\mathbf{y})]$

(c) $(-3)\{(-2)[(-1)(\mathbf{x} - \mathbf{y} + \mathbf{z}) - \mathbf{y}] + 2\mathbf{x}\}$

6. Show that $F(S)$, where $S = \{1\}$, is essentially R^1.

7. Show that the set of 2×3 matrices is essentially R^6.

4-2 Vector Spaces and Subspaces

We have seen that under very different circumstances, it is possible to define operations called addition and scalar multiplication on certain sets, such that the eight familiar properties are true. When this is possible, we say that the set in question is a **vector space** with respect to those operations.

DEFINITION 4-3

Let V be a set such that we can define operations called **addition** and **scalar multiplication** on V. This means that given any $\mathbf{u}$ and $\mathbf{v}$ in V and $a \in R$, we can define new elements $\mathbf{u} + \mathbf{v}$ (the **sum** of $\mathbf{u}$ and $\mathbf{v}$) and $a\mathbf{u}$ (the **scalar multiple** of $\mathbf{u}$ by a) in the set V. Let $\mathbf{u}, \mathbf{v}, \mathbf{w}, \ldots$ be any elements of V and $a, b, \ldots$ be any numbers. Then V (together with these operations) is called a **real vector space** provided the operations satisfy the following eight properties:

(1) $\mathbf{u} + \mathbf{v} = \mathbf{v} + \mathbf{u}$.

(2) $(\mathbf{u} + \mathbf{v}) + \mathbf{w} = \mathbf{u} + (\mathbf{v} + \mathbf{w})$.

(3) There is an unique element $\mathbf{0}$ in V called the **zero vector** such that $\mathbf{u} + \mathbf{0} = \mathbf{u}$ for any $\mathbf{u}$ in V.

(4) For any $\mathbf{u}$ in V there is an unique element denoted $-\mathbf{u}$ in V such that $\mathbf{u} + (-\mathbf{u}) = \mathbf{0}$. As usual, $-\mathbf{u}$ is called the **additive inverse** of $\mathbf{u}$. We will abbreviate expressions such as $\mathbf{v} + (-\mathbf{u})$ to $\mathbf{v} - \mathbf{u}$, which, in effect, defines **subtraction** of vectors.

(5) $a(\mathbf{u} + \mathbf{v}) = a\mathbf{u} + a\mathbf{v}$.

(6) $(a + b)\mathbf{u} = a\mathbf{u} + b\mathbf{u}$.

(7) $(ab)\mathbf{u} = a(b\mathbf{u})$.

(8) $1\mathbf{u} = \mathbf{u}$, where $1 \in R$.

When a set is a vector space, its elements are called **vectors**, and, where possible, these will conventionally be denoted by (boldface) letters such as $\mathbf{u}, \mathbf{v}, \mathbf{w}$. The numbers used in scalar multiplication are referred to as **scalars**. The eight properties of addition and scalar multiplication are called **axioms** for a vector space.

In more advanced treatments of linear algebra, scalars can be elements from certain sets other than the real numbers, the most important case being the set of complex numbers. The resulting structure is then called a **complex vector space**. In fact, almost all the results discussed in this book are valid for this type of vector space. The basic ideas about complex numbers are covered in Appendix A, and a brief introduction to complex vector spaces is given in the optional sections 4-6 and 7-6. However, unless otherwise noted, a vector space will mean a *real* vector space, as defined above.

Given a set V *together with* operations for addition and scalar multiplication (some sets may have several possible operations — see Exercises 4-2), in order to verify that V is indeed a vector space, we must check that the eight axioms hold for these particular operations. We have already done this type of verification for the following examples of vector spaces:

(1) Vectors ("arrows") in the plane and 3-space.

(2) The set of $m \times n$ matrices; we denote this vector space by $M(m \times n)$.

(3) R^1, R^2, R^3, and, in general, R^n.

(4) The vector spaces $F(S)$ where S is any set.

We consider two further examples which both lead very naturally to the important notion of a **subspace**.

EXAMPLE 4-3

Any plane through the origin in R^3 is a vector space. Recall that such a plane is a set of points in R^3 satisfying a linear equation

$$ax + by + cz = 0.$$

For definiteness, let us consider the plane through the origin,

$$x + 2y + 3z = 0.$$

With reference to the definition of a vector space, the set V in this case consists of all points in R^3 satisfying this equation. For example, $(2,-1,0)$, $(1,0,-1/3)$, and $(5,-1,-1)$ are some elements of V. Now, we already know how to add triples of numbers and multiply them by scalars. The important thing is that when we do this for elements of the plane V, *the resulting sums and scalar multiples are also in* V. For example, with $\mathbf{u} = (2,-1,0)$ and $\mathbf{v} = (1,0,-1/3)$, we find

$$\mathbf{u} + \mathbf{v} = (2,-1,0) + (1,0,-1/3) = (3,-1,-1/3),$$

which is in V because

$$3 + 2(-1) + 3(-1/3) = 0.$$

More generally, if $\mathbf{u} = (x',y',z')$ and $\mathbf{v} = (x'',y'',z'')$ are *any* elements of V, then $\mathbf{u} + \mathbf{v}$ and $a\mathbf{u}$ when $a \in R$ are also elements of V. For, since $\mathbf{u}$ is in V,

$$x' + 2y' + 3z' = 0,$$

and, since $\mathbf{v}$ is in V,

$$x'' + 2y'' + 3z'' = 0.$$

Adding the two equations yields

$$(x' + x'') + 2(y' + y'') + 3(z' + z'') = 0,$$

which shows that $\mathbf{u} + \mathbf{v} = (x' + x'', y' + y'', z' + z'')$ is in V. Similarly, we can show that $a\mathbf{u}$ is in V.

Thus, V may be a vector space with respect to these operations provided that the eight axioms hold. In fact, we already know from earlier examples that axioms (1),(2), and (5) through (8) are true for *all* triples, and therefore true for all elements of V. It remains to be verified that axioms (3) and (4) hold.

Axiom (3) We must check that there is a zero vector in V. The obvious candidate is $\mathbf{0} = (0,0,0)$, the zero vector for R^3. Since

$$0 + 2(0) + 3(0) = 0,$$

$\mathbf{0}$ is in V and clearly has the required property that $\mathbf{u} + \mathbf{0} = \mathbf{u}$ when $\mathbf{u} \in V$.

Axiom (4) Similarly, when $\mathbf{u} = (x',y',z')$ is in V, its additive inverse $-\mathbf{u} = (-x',-y',-z')$ is also in V and of course satisfies $\mathbf{u} + (-\mathbf{u}) = \mathbf{0}$. ❑

In the above example, we have verified that V is a vector space. As such, it is a special vector space, because it is a **subset** of a larger one, namely R^3, and the operations (addition, scalar multiplication) in V are just those of R^3 *restricted* to elements of V. For these reasons, V is called a **subspace** of R^3.

Let us now consider a similar example which illustrates one of the many ways a set may *fail* to be a vector space. Consider the plane V' described by the equation

$$x + 2y + 3z = 1,$$

which does not pass through the origin. Then V' is not a vector space with respect to the operations inherited as a subset of R^3. To see this, consider the points $\mathbf{u} = (0,-1,1)$ and $\mathbf{v} = (1,0,0)$ in V'. Then,

$$\mathbf{u} + \mathbf{v} = (0,-1,1) + (1,0,0) = (1,-1,1)$$

is *not* in V' because

$$1 + 2(-1) + 3(1) = 2,$$

which is not equal to the right-hand side of the equation for V'. Similarly, for example, $2\mathbf{v} = (2,0,0)$ is not in V'. We say that V' is **not closed** under these operations, and fails to be a vector space for these reasons.

EXAMPLE 4-4

By Definition 4-2, $F(R)$ denotes the vector space of all functions $f{:}R \rightarrow R$, real-valued functions of a real variable. We consider certain subsets of

$F(R)$ which are vector spaces themselves. These are the sets of **polynomial functions**, which are those f in $F(R)$ with the property that their values at any number $x \in R$ can be expressed as a specific finite sum of multiples of powers of x. Some examples are

$$\begin{aligned} f(x) &= -1 + 2x, \\ g(x) &= 1/2 - 5x + 2x^2, \\ h(x) &= -x^2 + 8x^3, x \in R. \end{aligned}$$

This means, for example, that

$$\begin{aligned} h(2) &= -2^2 + 8(2)^3 = 60, \\ h(0) &= 0, \\ h(1) &= -1 + 8 = 7, \end{aligned}$$

and so on.

In general, a function f in $F(R)$ is called a **polynomial function of degree n** (where n is a nonnegative integer) if there are numbers $a_0, a_1, \ldots, a_n$ with $a_n \neq 0$ such that

$$f(x) = a_0 + a_1x + a_2x^2 + \ldots + a_nx^n = \sum_{i=0}^{n} a_ix^i, x \in R.$$

The numbers $a_0, a_1, \ldots, a_n$ are called the **coefficients** of f and we write $deg\, f = n$ to denote the fact that the *highest* power of x occurring in the expression for $f(x)$ is the nth. For example, the polynomial functions f, g, and h listed above have degrees 1, 2, and 3 respectively. A polynomial function of degree 0 would have highest power of x equal to $x^0 = 1$, so it must have the form $f(x) = a_0$ for all x in R, i.e., it is a **constant function**.

Now, we do not give a name to the set of polynomial functions of a fixed degree n, because it turns out that this set is *not* a vector space with respect to the operations of addition and scalar multiplication which already exist in the larger set $F(R)$. To see this, let us consider the polynomial functions p and q both of degree 2 defined by

$$p(x) = -1 + 2x + 3x^2, q(x) = 2 + 6x - 3x^2, x \in R.$$

Then, by definition, $p + q$ is the function defined by

$$\begin{aligned} (p + q)(x) &= p(x) + q(x) \\ &= -1 + 2x + 3x^2 + 2 + 6x - 3x^2 \\ &= 1 + 8x, \end{aligned}$$

which is a polynomial function of degree 1, not 2. Therefore, it is not in the set from which p and q came, and this set is therefore *not closed* under addition. In other words, the set of polynomial functions of a fixed degree is not a vector space with respect to the natural operations inherited from $F(R)$.

The difficulty is that the degree of a sum may be *less* than the degrees of the terms in it. Thus, the set of interest is $P(n)$, *the set of all polynomial functions of degree at most* n. This is a subset of $F(R)$, and the operations of addition and scalar multiplication take on the following form in general. Let

$$f(x) = \sum_{i=0}^{n} a_i x^i, g(x) = \sum_{i=0}^{n} b_i x^i.$$

Then,

$$(f + g)(x) = \sum_{i=0}^{n} (a_i + b_i)x^i, (kf)(x) = \sum_{i=0}^{n} ka_i x^i \ (k \in R).$$

It should be clear from these formulae that $f + g$ and kf are also polynomial functions of degree at most n, and therefore are new elements of $P(n)$. It remains to be verified that the eight axioms hold for these operations.

As in Example 4-3, all but axioms (3) and (4) necessarily hold in $P(n)$ since they hold in the larger set $F(R)$. For (3), we need a "zero vector", which in this case must be a polynomial function of degree at most n. The natural choice is the "zero function" which we denoted by $0{:}R \rightarrow R$ and is defined by $0(x) = 0 \in R$ for all $x \in R$. Unfortunately, as a polynomial function, 0 has no degree, since no power of x occurs in its expression with nonzero coefficient. Nevertheless, we agree to regard it as a member of $P(n)$ for all $n \geq 0$. Since $f + 0 = f$ for any f in $P(n)$, it is the required zero vector.

Similarly, for axiom (4), we take the additive inverse inherited from $F(R)$. If f is in $P(n)$, $(-f)(x) = -f(x)$ for all $x \in R$. In terms of the expression for f, this becomes

$$(-f)(x) = -\sum_{i=0}^{n} a_i x^i = \sum_{i=0}^{n} (-a_i)x^i \text{ when } f(x) = \sum_{i=0}^{n} a_i x^i, x \in R.$$

Clearly, $f + (-f) = 0$, the zero polynomial function. ❑

Now we have constructed a vector space $P(n)$ for each integer $n \neq 0$; and each gets "bigger" as n increases:

$$P(0) \subseteq P(1) \subseteq P(2) \subseteq \ldots \subseteq P(n) \subseteq P(n+1) \subseteq \ldots \subseteq F(R).$$

In fact, the union

$$P(R) = \cup \{P(n) \mid n = 0, 1, 2, \ldots\}$$

of all these vector spaces is also a vector space; it consists of all polynomial functions of *all* degrees, and is also a subset of $F(R)$. We leave the verification of these remarks to the reader.

The vector spaces $P(n)$ are also all subspaces of $F(R)$ in the following sense:

DEFINITION 4-4

A subset W of a vector space V is called a **subspace** if W is itself a vector space when the operations of addition and scalar multiplication of V are restricted to W.

This means (among other things) that when $\mathbf{v}$ and $\mathbf{w}$ are any vectors in W and k is a scalar, then $\mathbf{v} + \mathbf{w}$ and $k\mathbf{v}$ are also in W. These conditions are often referred to as **closure**: we say W is **closed** under the operations "inherited" by being a subset of V.

As we can see from Examples 4-3 and 4-4, verifying that a subset W is a subspace of a vector space V does not require checking all eight axioms; in fact,

THEOREM 4-1

A subset W of a vector space V is a subspace iff it is closed under the inherited addition, subtraction (i.e., additive inverses) and scalar multiplication, and the zero vector $\mathbf{0}$ of V is in W.

PROOF The idea of the proof has been covered several times in preceding examples and is left for the reader (question 11, Exercises 4-2). ■

EXAMPLE 4-5

$P(n)$ is a subspace of $P(n+1)$ for $n = 0, 1, 2, \ldots$ and they are all subspaces of $P(R)$ and $F(R)$. Also, $P(R)$ is a subspace of $F(R)$. ❑

EXAMPLE 4-6

Let W be the subset of $M(2\times 2)$ defined by

$$W = \{\begin{bmatrix} a & b \\ 0 & 0 \end{bmatrix} \mid a, b \in R\}.$$

W consists of those 2×2 matrices whose 2nd row is zero. We check that W is a subspace of $M(2 \times 2)$ using Theorem 4-1. First, if

$$\begin{bmatrix} a & b \\ 0 & 0 \end{bmatrix}, \begin{bmatrix} a' & b' \\ 0 & 0 \end{bmatrix}$$

are two vectors (i.e., matrices in this particular example) in W, their sum (by definition) is the 2×2 matrix

$$\begin{bmatrix} a+a' & b+b' \\ 0 & 0 \end{bmatrix},$$

which is also in W. And, if k is a scalar (including, in particular, $k = -1$),

$$k \begin{bmatrix} a & b \\ 0 & 0 \end{bmatrix} = \begin{bmatrix} ka & kb \\ 0 & 0 \end{bmatrix}$$

is also in W. Finally, the zero matrix O is in W, so W is a subspace of $M(2 \times 2)$. ❑

The reader should construct some variants of this example: for instance, the set of 2×2 matrices whose first column is zero, and so on.

EXAMPLE 4-7

This example is of a general nature. Let V be *any* vector space. Then V always has at least two subspaces called **trivial subspaces**. These are

(a) V itself, since V is a subset of itself (in a "trivial" way) and is also a vector space;

(b) the **zero subspace** $\{\mathbf{0}\}$ consisting of only the zero vector of V. It is a subspace because $\mathbf{0} + \mathbf{0} = \mathbf{0}$ and $k\mathbf{0} = \mathbf{0}$ for any scalar k. As a vector space by itself, $\{\mathbf{0}\}$ is the **zero vector space**. ❑

The alert reader may have noticed that the fact $k\mathbf{0} = \mathbf{0}$ does not appear among the axioms for a vector space, so that despite its "obvious" truth (e.g., for numbers), it does require a proof in general. This is one of the principles we must observe when doing "abstract" mathematics: *we cannot allow our intuition to force us to conclude something is true in general simply because it's true in specific cases*. For example, certainly $k\mathbf{0} = \mathbf{0}$ when $\mathbf{0}$ is the zero vector in R^3 or when $\mathbf{0} = O$ is the $m \times n$ zero matrix. However, we must prove it is true for *any* vector space, using only the axioms, or other theorems proved from them. The proof of the next theorem illustrates this type of proof.

THEOREM 4-2 (Elementary Properties of Vector Spaces)
Let V be any vector space.
(1) $0\mathbf{v} = \mathbf{0}$ for any $\mathbf{v} \in V$.
(2) $k\mathbf{0} = \mathbf{0}$ for any scalar k.
(3) $k\mathbf{v} = \mathbf{0}$ iff $k = 0$ or $\mathbf{v} = \mathbf{0} \in V$.
(4) $(-k)\mathbf{v} = -(k\mathbf{v})$ for $\mathbf{v} \in V$ and $k \in R$.
(5) If $\mathbf{u} + \mathbf{w} = \mathbf{v} + \mathbf{w}$, then $\mathbf{u} = \mathbf{v}$, for $\mathbf{u}, \mathbf{v}, \mathbf{w}$ in V.

The last property means we can "cancel" identical terms appearing on the two sides of an equation. (Caution: $\mathbf{0}$ is the zero *vector* in V; 0 is the *number* zero!)

PROOF In the proofs of the first three properties, we justify every step by stating the axioms used; we leave it to the reader to justify the steps in the proofs of the other two.

Property (1)

$$\begin{aligned} 0\mathbf{v} &= (0 + 0)\mathbf{v} && (0 = 0 + 0 \text{ is true for } R) \\ &= 0\mathbf{v} + 0\mathbf{v} && (\text{axiom } (6)) \end{aligned} \tag{4-1}$$

Therefore,

$$\begin{aligned} \mathbf{0} &= 0\mathbf{v} - 0\mathbf{v} && (\text{axiom } (4)) \\ &= (0\mathbf{v} + 0\mathbf{v}) - 0\mathbf{v} && (\text{from equation 4-1}) \\ &= 0\mathbf{v} + (0\mathbf{v} - 0\mathbf{v}) && (\text{axiom } (2)) \\ &= 0\mathbf{v} + \mathbf{0} && (\text{axiom } (4)) \\ &= 0\mathbf{v} && (\text{axiom } (3)) \end{aligned}$$

which completes the proof.

Property (2)

$$\begin{aligned} k\mathbf{0} &= k(\mathbf{0} + \mathbf{0}) && (\text{axiom } (3)) \\ &= k\mathbf{0} + k\mathbf{0} && (\text{axiom } (5)) \end{aligned} \tag{4-2}$$

Therefore,

$$\begin{aligned} \mathbf{0} &= k\mathbf{0} - k\mathbf{0} && (\text{axiom } (4)) \\ &= (k\mathbf{0} + k\mathbf{0}) - k\mathbf{0} && (\text{by substitution from equation 4-2}) \\ &= k\mathbf{0} + (k\mathbf{0} - k\mathbf{0}) && (\text{axiom } (2)) \\ &= k\mathbf{0} + \mathbf{0} && (\text{axiom } (4)) \\ &= k\mathbf{0} && (\text{axiom } (3)) \end{aligned}$$

as claimed.

Property (3) We have proved that $k\mathbf{v} = \mathbf{0}$ if $k = 0$ or $\mathbf{v} = \mathbf{0}$ in (1) and (2) above. Conversely, suppose that $k\mathbf{v} = \mathbf{0}$ for some $\mathbf{v}$ in V and k in R; we show $\mathbf{v} = \mathbf{0}$ or $k = 0$.

If $k \neq 0$, then $1/k$ exists and we can write

$$\begin{aligned} \mathbf{v} &= 1\mathbf{v} && \text{(by axiom (8))} \\ &= (1/k)(k)\mathbf{v} \\ &= (1/k)(k\mathbf{v}) && \text{(by axiom (7))} \\ &= (1/k)\mathbf{0} && \text{(by hypothesis)} \\ &= \mathbf{0} && \text{(by part (2) of this theorem).} \end{aligned}$$

On the other hand, if $\mathbf{v} \neq \mathbf{0}$, we show that k must be 0. For, if $k \neq 0$, then by what we have just proved, we could conclude $\mathbf{v} = \mathbf{0}$, contrary to our supposition $\mathbf{v} \neq \mathbf{0}$. Thus $k\mathbf{v} = \mathbf{0}$ implies $k = 0$ or $\mathbf{v} = \mathbf{0}$ (or both), completing the proof.

Property (4) $k\mathbf{v} + (-k)\mathbf{v} = [k + (-k)]\mathbf{v} = 0\mathbf{v} = \mathbf{0}$,

by part (1). Thus $(-k)\mathbf{v}$ has the characteristic property of the additive inverse $-(k\mathbf{v})$ of $k\mathbf{v}$. But, by axiom (4), the additive inverse is unique, so that we must have

$$(-k)\mathbf{v} = -(k\mathbf{v})$$

as stated.

Property (5)

$$\begin{aligned} \mathbf{u} &= \mathbf{u} + \mathbf{0} \\ &= \mathbf{u} + (\mathbf{w} - \mathbf{w}) \\ &= (\mathbf{u} + \mathbf{w}) - \mathbf{w} \\ &= (\mathbf{v} + \mathbf{w}) - \mathbf{w} \\ &= \mathbf{v} + (\mathbf{w} - \mathbf{w}) \\ &= \mathbf{v} + \mathbf{0} \\ &= \mathbf{v} \end{aligned}$$

as required. Notice that this cancellation law is the converse of the substitution principle used earlier: if $\mathbf{u} = \mathbf{v}$, then $\mathbf{u} + \mathbf{w} = \mathbf{v} + \mathbf{w}$. The latter follows from the fact that addition produces an unique sum for each ordered pair of vectors. ■

We apply Theorem 4-2 to obtain an even shorter criterion for subspaces.

THEOREM 4-3

A **nonempty** subset W of a vector space V is a subspace iff $k\mathbf{u} + \mathbf{v}$ is in W whenever $\mathbf{u}$ and $\mathbf{v}$ are in W and $k \in R$.

PROOF It should be clear that if W *is* a subspace, then the stated condition holds. Conversely, with $k\mathbf{u} + \mathbf{v} \in W$, we use Theorem 4-1 to prove that W is a subspace.

Putting $k = 1$, we get

$$k\mathbf{u} + \mathbf{v} = 1\mathbf{u} + \mathbf{v} \in W.$$

Next, since W is not empty, we can choose some vector $\mathbf{w}$ in it. Let $k = -1$, $\mathbf{u} = \mathbf{w}$, and $\mathbf{v} = \mathbf{w}$ in the condition. Then,

$$k\mathbf{u} + \mathbf{v} = (-1)\,\mathbf{w} + \mathbf{w} = \mathbf{0} \in W.$$

And, putting $k = -1$ and $\mathbf{v} = \mathbf{0}$, we find that

$$k\mathbf{u} + \mathbf{v} = (-1)\mathbf{u} + \mathbf{0} = -\mathbf{u} \in W.$$

Finally, putting $\mathbf{v} = \mathbf{0}$ in the condition, we find that

$$k\mathbf{u} + \mathbf{v} = k\mathbf{u} \in W$$

whenever $\mathbf{u}$ is in W and k is a scalar. So, by Theorem 4-1, W is a subspace. ■

EXAMPLE 4-8

Recall that an $n \times n$ matrix A is called symmetric if A equals its own transpose (see Example 2-25 and Definition 2-11). A is called skew-symmetric if $A = -A^t$. For example,

$$\begin{bmatrix} 0 & -2 \\ 2 & 0 \end{bmatrix} \text{and} \begin{bmatrix} 0 & -1 & -2 \\ 1 & 0 & -3 \\ 2 & 3 & 0 \end{bmatrix}$$

are skew-symmetric. Note that the diagonal elements of a skew-symmetric matrix are necessarily 0, since $A_{ii} = -A_{ii}$ implies $A_{ii} = 0$. Symmetric and skew-symmetric matrices are important in applications, especially in physics and chemistry. In fact, *the sets of $n \times n$ symmetric and skew-symmetric matrices are subspaces of $M(n \times n)$*. We illustrate this for the 2×2 case using Theorem 4-3. Now, a 2×2 symmetric matrix necessarily has the form

$$\begin{bmatrix} a & b \\ b & c \end{bmatrix}$$

for some a, b, and $c \in R$. So if k is a scalar and

$$\begin{bmatrix} a & b \\ b & c \end{bmatrix}, \begin{bmatrix} a' & b' \\ b' & c' \end{bmatrix}$$

are two symmetric matrices, then

$$k\begin{bmatrix} a & b \\ b & c \end{bmatrix} + \begin{bmatrix} a' & b' \\ b' & c' \end{bmatrix} = \begin{bmatrix} ka + a' & kb + b' \\ kb + b' & kc + c' \end{bmatrix}$$

which is also symmetric. Thus, by Theorem 4-3, the 2×2 symmetric matrices are a subspace of $M(2 \times 2)$. We leave the similar verification for 2×2 skew-symmetric matrices as an exercise (question 3, Exercises 4-2).

The general proof for $n \times n$ matrices follows from properties of the transpose listed in Theorem 2-5:

(1) If A and B are $n \times n$ symmetric matrices and k is a scalar,

$$(kA + B)^t = kA^t + B^t = kA + B,$$

so $kA + B$ is symmetric.

(2) If A and B are skew-symmetric,

$$(kA + B)^t = kA^t + B^t = -kA - B = -(kA + B),$$

so $kA + B$ is skew-symmetric. ❑

EXAMPLE 4-9

This example illustrates certain subspaces of $F(R)$ which are important for calculus (no knowledge of calculus is assumed for the discussion). Roughly speaking, a function $f{:}R \rightarrow R$ is called **continuous** if its graph can be drawn without lifting the pen from the paper. It is called **differentiable** if its graph has a tangent line at every point. Some examples are shown in Figure 4-2.

FIGURE 4-2

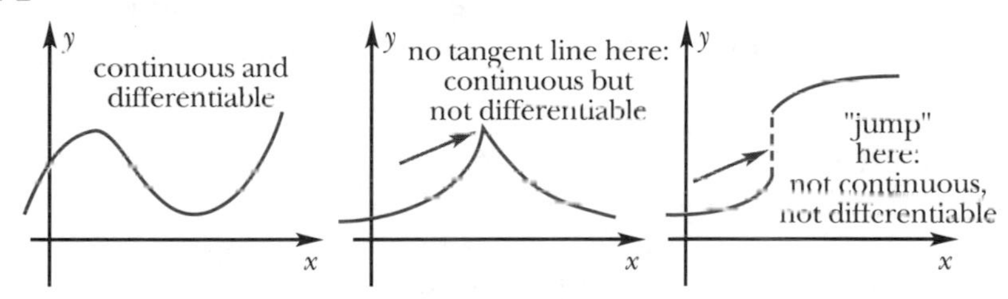

It is proved in calculus that if f and g are both continuous (or differentiable) and k is a number, then $kf + g$ is also continuous (respectively, differentiable). It follows from Theorem 4-3 that these sets of functions are subspaces of $F(R)$. ❑

The next example is so important that we state it as a theorem (see also Example 4-3).

THEOREM 4-4

Let A be an $m \times n$ matrix. Then the set of solutions of the homogeneous system of m linear equations in n unknowns $AX = O$ is a subspace of $M(n \times 1)$. It is called the **solution space** of the system.

PROOF If X' and X'' are solutions and k is a scalar, then

$$A(kX' + X'') = kAX' + AX'' = kO + O = O,$$

showing that $kX' + X''$ is also a solution. Thus, the set of solutions is a subspace of $M(n\times 1)$, by Theorem 4-3. ■

EXAMPLE 4-10

Consider the system

$$\begin{aligned} x_1 + 2x_2 + x_3 + 2x_4 &= 0 \\ 2x_1 + 5x_2 - x_3 + 3x_4 &= 0. \end{aligned}$$

We can check by substitution that $X' = (-7,3,1,0)$ and $X'' = (-4,1,0,1)$ are solutions. Then, for example, choosing $k = -2$,

$$kX' + X'' = -2X' + X'' = (10,-5,-2,1),$$

which is also a solution. In fact, using Gauss-Jordan reduction, it is easy to see that the solution space is the subset (note that it is a subspace) of $M(4\times 1) = R^4$ given by

$$\{(x_1,x_2,x_3,x_4) \mid x_1 = -7x_3 - 4x_4,\ x_2 = 3x_3 + x_4\}.$$

The solutions X' and X'' noted above were obtained by setting $x_3 = 1$, $x_4 = 0$ and $x_3 = 0$, $x_4 = 1$ respectively. ❑

EXAMPLE 4-11

Any line through the origin in R^3 is a subspace of R^3 and also of any plane through the origin containing that line. To see this, suppose a line L is contained in the plane P with equation $ax + by + cz = 0$. We know from Chapter 1 that a line is determined as an intersection of two planes, so suppose that L is also contained in the plane with equation $a'x + b'y + c'z = 0$. Then we also know from Chapter 1 that L is the solution set of the system consisting of these two equations. By Theorem 4-4, L is therefore a subspace of R^3. Furthermore, it is a vector space that is a subset of the plane P, so is a subspace of P. The reader may prove as an exercise that these remarks also follow using parametric equations for L. ❑

The importance of Theorem 4-4 stems from the fact that many subspaces we shall encounter in our later work will be expressed as solution spaces of certain homogeneous systems of equations. The associated computations of the problem will thus be done using the theory of systems of equations.

EXERCISES 4-2

In each of questions 1 through 8, either show that the given set V with the operations indicated is a vector space, or show that it is not by listing those axioms that fail to hold.

1. The set V of triples $(x,y,0)$ in R^3 with the usual operations of addition and scalar multiplication for R^3.

2. The set V of ordered pairs (x,y) in R^2 such that $x \geq 0$, with the usual operations for R^2.

3. The set of matrices of the form

$$\begin{bmatrix} a & b \\ c & -a \end{bmatrix}$$

with the usual operations for $M(2 \times 2)$.

4. The set $V = R^2$ with the usual addition but with scalar multiplication defined by

$$k(x,y) = \begin{cases} (0,0), & \text{if } k = 0, \\ (kx,y/k), & \text{if } k \neq 0. \end{cases}$$

5. The set $V = R^3$ with the usual addition but with scalar multiplication defined by

$$k(x_1,x_2,x_3) = (x_1,kx_2,x_3),\ k \in R.$$

6. The set $V = R^2$ with addition defined by

$$(x_1,x_2) + (y_1,y_2) = ((x_1^5 + y_1^5)^{1/5},(x_2^5 + y_2^5)^{1/5})$$

and scalar multiplication defined by

$$k(x_1,x_2) = (k^{1/5}x_1,k^{1/5}x_2),\ k \in R.$$

7. The set $V = R^2$ with addition defined by

$$(x_1,x_2) + (y_1,y_2) = ((x_1^2 + y_1^2)^{1/2},(x_2^2 + y_2^2)^{1/2})$$

and scalar multiplication defined by

$$k(x_1,x_2) = |k|^{1/2}(x_1,x_2),\ k \in R.$$

8. The set $V = R^2$ with the usual addition but with scalar multiplication defined by $k(x,y) = (0,0)$, $k \in R$.

9. Prove in detail that $P(1)$ is a vector space by verifying that all the axioms hold with the operations as defined in the text.

10. Prove in detail that $P(R)$ is a vector space.

11. Prove Theorem 4-1.

12. Let W be the subset of $P(2)$ consisting of polynomial functions of the form $f(x) = a_1x + a_2x^2$, $x \in R$. Determine whether or not W is a subspace of $P(2)$.

13. Which of the following sets of functions are subspaces of $F(R)$?

(a) $\{f \mid f(-1) = 0\}$ **(b)** $\{f \mid f(x^2) = (f(x))^2\}$

(c) $\{f \mid f(2) = 1 + f(-3)\}$ **(d)** $\{f \mid f(x) = -f(-x)\}$

(e) $\{f \mid f(0) = f(1)\}$ **(f)** $\{f \mid f(x) = f(-x)\}$

(g) $\{f \mid f(x) \leq 0\}$ **(h)** $\{f \mid f(x^2) = 2f(x)\}$

14. Show that the set of points on the line $y = mx + b$ is a subspace of R^2 iff $b = 0$.

15. Show that the plane in R^3 with equation $ax + by + cz = d$ is a subspace iff $d = 0$.

16. Which of the following subsets of $M(n \times n)$ are subspaces?

(a) the diagonal matrices **(b)** the upper (lower) triangular matrices

(c) the set of matrices whose trace is 0 (if A is an $n \times n$ matrix, its *trace*, $tr\,A$, is defined by $tr\,A = a_{11} + a_{22} + \ldots + a_{nn}$ = sum of the diagonal entries of A)

(d) the invertible $n \times n$ matrices

(e) the $n \times n$ matrices such that all entries are positive

17. Which of the following are subspaces of R^3?

(a) the set of triples (x_1,x_2,x_3) satisfying the system

$$\begin{aligned} x_1 \quad\quad - x_3 &= 1 \\ x_2 + x_3 &= 2 \end{aligned}$$

(b) the set of triples (x_1,x_2,x_3) such that $x_1x_2 = 0$

(c) the set of triples such that $x_1 = x_2 = 0$

18. **(a)** Let W_1 and W_2 be subspaces of a vector space V. Use Theorem 4-3 to show that $W_1 \cap W_2$ is also a subspace.

(b) Generalize (a) by showing that if W_i $(i = 1, 2, 3, \ldots)$ is any family of subspaces of V, then so is their intersection $\cap \{W_i \mid i = 1, 2, 3, \ldots\}$.

19. **(a)** Show that the x axis and the y axis are subspaces of R^2, but that their union is not.

(b) Generalize (a) to show that if W_1 and W_2 are subspaces of V, their union $W_1 \cup W_2$ is a subspace iff $W_1 \subseteq W_2$ or $W_2 \subseteq W_1$ (or both).

(c) The failure of a union of subspaces to be a subspace (in general) motivates the following definition: The *sum* $W_1 + W_2$ is

$$\{\mathbf{u} + \mathbf{v} \mid \mathbf{u} \in W_1 \text{ and } \mathbf{v} \in W_2\}.$$

Show that this is a subspace of V and is the smallest subspace of V containing W_1 and W_2 in the sense that if W were another subspace of V such that $W_1, W_2 \subseteq W$, then $W_1 + W_2 \subseteq W$.

20. A function $f:R \rightarrow R$ is called *bounded* if there exists a constant M such that $|f(x)| < M$ for all $x \in R$. Show that the set of bounded functions in $F(R)$ is a subspace of $F(R)$.

4-3 Generating Sets, Linear Independence, and Bases

In this section, we discuss three of the most important concepts in linear algebra. Consequently, the section is fairly theoretical. Furthermore, in order to elaborate the ideas and to enable the reader to understand them as completely as possible, we have included quite a few theorems and their proofs. Some readers may wish to skip the proofs on first reading and proceed to the next section dealing with computational aspects and illustrations of these concepts.

In the preceding sections, we have accumulated many examples of vector spaces. All these spaces, except the zero vector space, have infinitely many elements. For, if $\mathbf{v} \neq \mathbf{0}$ is a vector in V, then by "closure" $k\mathbf{v}$ is also in V for all numbers k, and there are infinitely many choices for k. It is therefore natural to try to find some "simple" *finite description* of the elements of a vector space V.

The idea we use is based on the observation that in the plane R^2, we need only two axes to assign coordinates to points; and in R^3, we need three. Stated somewhat differently, we can write any vector $\mathbf{u} = (x,y)$ in R^2 as

$$\mathbf{u} = (x,y) = x(1,0) + y(0,1) = x\mathbf{e}_1 + y\mathbf{e}_2 \tag{4-3}$$

where $\mathbf{e}_1 = \mathbf{i}$ and $\mathbf{e}_2 = \mathbf{j}$ are the standard basis vectors for R^2 introduced in Example 1-20. For example, if $\mathbf{u} = (-1,1/2)$, we can write (Figure 4-3)

$$\mathbf{u} = (-1)(1,0) + (1/2)(0,1) = -\mathbf{e}_1 + (1/2)\mathbf{e}_2.$$

FIGURE 4-3

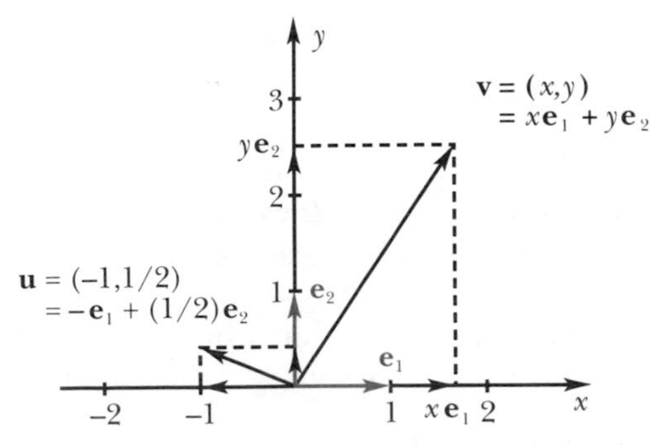

Similarly, any vector $\mathbf{v} = (x,y,z)$ in R^3 can be written as a sum of multiples of the standard basis vectors for R^3:

$$\mathbf{v} = (x,y,z) = x\mathbf{e}_1 + y\mathbf{e}_2 + z\mathbf{e}_3 \tag{4-4}$$

where $\mathbf{e}_1 = \mathbf{i} = (1,0,0)$, $\mathbf{e}_2 = \mathbf{j} = (0,1,0)$, and $\mathbf{e}_3 = \mathbf{k} = (0,0,1)$. This leads to

DEFINITION 4-5

Let V be a vector space and $\mathbf{v}_1, \mathbf{v}_2, \ldots, \mathbf{v}_n$ be n vectors in V. A vector $\mathbf{u}$ in V is called a **linear combination** of these vectors if there are scalars $k_1, k_2, \ldots, k_n$ such that

$$\mathbf{u} = k_1\mathbf{v}_1 + k_2\mathbf{v}_2 + \ldots + k_n\mathbf{v}_n. \tag{4-5}$$

For example, equations 4-3 and 4-4 above state that every vector in R^2 and R^3 is a linear combination of the standard basis vectors for those spaces. Notice that when $n > 2$, equation 4-5 is unambiguous without brackets because of the associative law of addition (see also Theorem 1-3 and its remarks). When $n = 1$, for $\mathbf{u}$ to be a linear combination of a single vector $\mathbf{v}_1$ means that it is a scalar multiple $\mathbf{u} = k_1\mathbf{v}_1$.

EXAMPLE 4-12

We generalize our observations about the standard basis vectors in R^2 and R^3 to R^n. Let $\mathbf{e}_1 = (1,0,0,\ldots,0)$, $\mathbf{e}_2 = (0,1,0,\ldots,0)$, ... and so on, with $\mathbf{e}_n = (0,0,\ldots,0,1)$. Here $\mathbf{e}_i$ has 1 in its ith spot (entry) and 0 elsewhere. Notice that in our identification of R^n with $M(n \times 1)$, $\mathbf{e}_i$ corresponds to the column vector which is the ith column of the $n \times n$ identity matrix I.

If $\mathbf{u} = (x_1,x_2,\ldots,x_n)$ is any vector in R^n, we can write

$$\begin{aligned}\mathbf{u} &= x_1(1,0,\ldots,0) + x_2(0,1,0,\ldots,0) + \ldots + x_n(0,\ldots,0,1)\\ &= x_1\mathbf{e}_1 + x_2\mathbf{e}_2 + \ldots + x_n\mathbf{e}_n = \sum_{i=1}^{n} x_i\mathbf{e}_i. \ \square\end{aligned}$$

The preceding example shows that any vector in R^n is a linear combination of $\mathbf{e}_1, \mathbf{e}_2, \ldots, \mathbf{e}_n$. They are the **standard basis vectors** for R^n. As noted earlier, it will always be clear from the context which space (i.e., which n) these basis vectors belong to.

As we shall see on numerous occasions, determining if a given vector $\mathbf{u}$ is a linear combination of a given set of vectors $\mathbf{v}_1, \mathbf{v}_2, \ldots, \mathbf{v}_n$ involves solving a system of linear equations.

EXAMPLE 4-13

In R^3, let $\mathbf{u} = (-1,7,0)$, $\mathbf{v}_1 = (1,-1,2)$, and $\mathbf{v}_2 = (0,3,1)$. We show $\mathbf{u}$ is a linear combination of $\mathbf{v}_1$ and $\mathbf{v}_2$, while (for example) $\mathbf{w} = (0,0,1) = \mathbf{e}_3$ is not. Starting with $\mathbf{u}$, we must show there are scalars k_1 and k_2 such that $\mathbf{u} = k_1\mathbf{v}_1 + k_2\mathbf{v}_2$, or, substituting the given expressions, that

$$(-1,7,0) = k_1(1,-1,2) + k_2(0,3,1) = (k_1, -k_1 + 3k_2, 2k_1 + k_2).$$

Therefore, we must show there are numbers k_1 and k_2 such that the above triples are equal. Using the definition of equality, this means that we must show the system of linear equations

$$\begin{aligned} k_1 \qquad &= -1 \\ -k_1 + 3k_2 &= 7 \\ 2k_1 + \ k_2 &= 0 \end{aligned}$$

has a solution. It is easy to see that the system does have an unique solution: $k_1 = -1$, $k_2 = 2$, so that $\mathbf{u} = -1\mathbf{v}_1 + 2\mathbf{v}_2$. Thus, $\mathbf{u}$ is a linear combination of $\mathbf{v}_1$ and $\mathbf{v}_2$ as stated.

In the case of $\mathbf{w} = (0,0,1)$, we have to solve the same system of equations but with $(-1,7,0)$ $(=\mathbf{u})$ replaced by $(0,0,1)$:

$$\begin{aligned} k_1 \qquad &= 0 \\ -k_1 + 3k_2 &= 0 \\ 2k_1 + \ k_2 &= 1. \end{aligned}$$

It should be clear that this system has no solution: putting $k_1 = 0$ into the second and third equations, we get a contradiction. Thus $\mathbf{w}$ is not a linear combination of $\mathbf{v}_1$ and $\mathbf{v}_2$ as k_1 and k_2 don't exist. ❑

The geometrical interpretation of Example 4-13 is that $\mathbf{u}$ lies in the plane determined by $\mathbf{v}_1$ and $\mathbf{v}_2$, while $\mathbf{w}$ does not. The plane here is, of course, the one through the origin of R^3 and the points of R^3 corresponding to $\mathbf{v}_1$ and $\mathbf{v}_2$ (see Figure 4-4). This plane has an equation of the form

$$ax + by + cz = 0,$$

if we use coordinates x, y, and z in R^3. And a, b, c are determined from the fact that $(1,-1,2)$ and $(0,3,1)$ are on the plane (the origin already satisfies the equation). Substituting these values into the equation we get

$$\begin{aligned} a - \ b + 2c &= 0 \\ 3b + \ c &= 0, \end{aligned}$$

which is a system of two linear equations with solutions $a = (-7/3)c$, $b = (-1/3)c$, and c arbitrary. Let us choose $c = -3$ for convenience, so $a = 7$, $b = 1$. Then the equation of the plane is

$$7x + y - 3z = 0.$$

Note that any other choice for c would have simply yielded a multiple of this equation and therefore the same plane. Now we see that $\mathbf{u} = (-1,7,0)$ satisfies this equation, while $\mathbf{w} = (0,0,1)$ does not. Therefore, $\mathbf{u}$ is in the plane, but $\mathbf{w}$ is not.

FIGURE 4-4

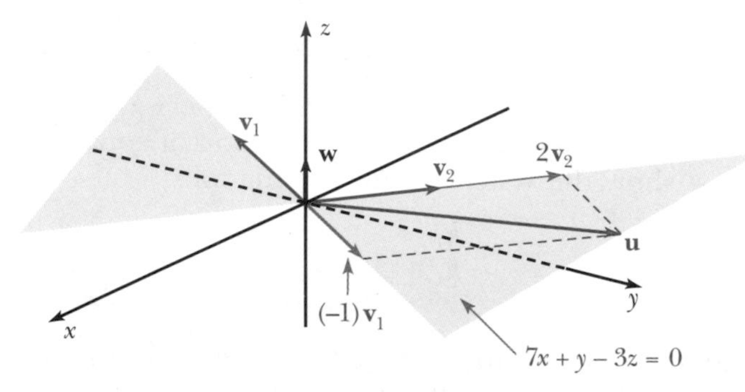

The previous discussion shows that sets of vectors such as the standard basis for R^n are very special, since *every* vector in R^n can be written as a linear combination of those vectors. This motivates the next definition.

DEFINITION 4-6

A set S of vectors in a vector space V is said to **generate** or **span** V if *every* vector in V can be expressed as a linear combination of a *finite* number of vectors from S. We then say that S is a **set of generators** for V.

Notice that so far we have not required S to be a finite set. However, our aim will be to find the smallest possible sets of generators S, preferably finite. For example, we have seen that $\{\mathbf{e}_1, \mathbf{e}_2, \ldots, \mathbf{e}_n\}$ is a set of n vectors, a finite set, which spans R^n.

EXAMPLE 4-14

Consider the vector space of 2×2 diagonal matrices A (the reader should check that this is indeed a vector space, and a subspace of $M(2 \times 2)$). Then,

$$A = \begin{bmatrix} a & 0 \\ 0 & b \end{bmatrix} = a \begin{bmatrix} 1 & 0 \\ 0 & 0 \end{bmatrix} + b \begin{bmatrix} 0 & 0 \\ 0 & 1 \end{bmatrix},$$

for some numbers a and b. The equation shows that the space of 2×2 diagonal matrices is generated by the two (diagonal) matrices

$$\begin{bmatrix} 1 & 0 \\ 0 & 0 \end{bmatrix} \text{ and } \begin{bmatrix} 0 & 0 \\ 0 & 1 \end{bmatrix}. \square$$

Now, when a set S generates a vector space V, then *V is precisely the set of all (finite) linear combinations of vectors of S*. For, every such linear combination is certainly in V by "closure" (since V is a vector space). And, conversely, every vector in V is such a linear combination, since this is what it means for S to generate V. This motivates

DEFINITION 4-7

If S is a set of vectors in V, we denote by *span* S the set of all finite linear combinations of vectors of S. If $S = \{\mathbf{u}, \mathbf{v}, \ldots\}$, we will also write *span* $\{\mathbf{u}, \mathbf{v}, \ldots\}$ for *span* S.

THEOREM 4-5

If S is a nonempty subset of a vector space V, then *span* S is a subspace of V.

PROOF Let $\mathbf{u}$ and $\mathbf{v}$ be any vectors in *span* S and let k be a number. Then, $\mathbf{u}$ and $\mathbf{v}$ are finite linear combinations of vectors from S:

$$\mathbf{u} = \sum_{i=1}^{n} a_i \mathbf{v}_i \text{ and } \mathbf{v} = \sum_{j=1}^{m} b_j \mathbf{w}_j$$

where $\mathbf{v}_i$, $\mathbf{w}_j$ are in S and a_i, b_j are some scalars, $1 \leq i \leq n$, $1 \leq j \leq m$; and m, n are positive integers. Then,

$$k\mathbf{u} + \mathbf{v} = k \sum_{i=1}^{n} a_i \mathbf{v}_i + \sum_{j=1}^{m} b_j \mathbf{w}_j = \sum_{i=1}^{n} k a_i \mathbf{v}_i + \sum_{j=1}^{m} b_j \mathbf{w}_j,$$

also a finite linear combination of vectors of S ($n + m$ of them, unless some of the $\mathbf{v}_i$ and $\mathbf{w}_j$ coincide). So, by Theorem 4-3, *span* S is a subspace. ■

In order that we can speak of *span* S for *any* set S, it is convenient to define *span* $\emptyset = \{\mathbf{0}\}$, the zero vector space, where $\emptyset$ is the empty set.

COROLLARY

A subset S of a vector space V generates V iff *span* $S = V$. ■

We henceforth denote the standard basis, $\{\mathbf{e}_1, \mathbf{e}_2, \ldots, \mathbf{e}_n\}$, for R^n by $B(n)$. Then we have seen that *span* $B(n) = R^n$.

There are, in fact, infinitely many subsets of R^n which generate it. For example, R^n generates itself since, if $\mathbf{u}$ is any vector in R^n, we may write

$$\mathbf{u} = 1\mathbf{u}$$

to express $\mathbf{u}$ as a linear combination of vectors of R^n. The same is true of any vector space. As a set of generators, the whole vector space is clearly far too large a set for our purposes. To be useful in describing the vectors in the simple way we suggested earlier, a set S of vectors of V must have two properties:

(1) *span* $S = V$.

(2) S is as small as possible, preferably finite.

This is indeed the case for the standard basis $B(n)$ for R^n:

(1) *span* $B(n) = R^n$.

(2) there is no set of vectors S in R^n with *fewer* than n elements which still generates R^n, so that $B(n)$ is the smallest set possible.

We shall prove this second fact later, but we can illustrate it now geometrically in the case of R^3. We emphasize that it is equivalent to saying that we need three (and no fewer) axes in R^3. Suppose S were a set of fewer than three vectors in R^3; then it contains at most two vectors, so that *span* S is at most a plane, as we saw in the specific case of Example 4-14. But not all vectors of R^3 are contained in a plane, so *span* S is not R^3.

It follows, in particular, that if S generates a vector space V and is as small as possible, then no **proper** (i.e., strictly smaller) subset S' of S generates V. This is clearly true for the standard basis $B(n)$ for R^n. For, if we leave out some $\mathbf{e}_i$, then the remaining set $B(n) - \{\mathbf{e}_i\}$ doesn't generate R^n because, for example, $\mathbf{e}_i$ is not in *span* $(B(n) - \{\mathbf{e}_i\})$. This observation leads to

DEFINITION 4-8(a)

A set of vectors B is called a **basis** for a vector space if *span* $B = V$, while no proper subset of B generates V.

The second condition in this definition is of sufficient importance that it is given a name to itself:

DEFINITION 4-9

A set S of vectors in V is called **linearly independent** if it has no proper subset S' such that *span* S' = *span* S (i.e., no smaller subset S' of S *spans* as much as S does). Otherwise, S is called **linearly dependent**.

Thus, S is linearly dependent if there *is* some proper subset S' of S such that *span* S' = *span* S. In particular, the set $\{\mathbf{v}\}$ is linearly independent if $\mathbf{v} \neq \mathbf{0}$, for any $\mathbf{v} \in V$! Other ways of stating the linear dependence of a set S are:

(1) S is larger than needed to generate the space *span* S.

(2) We can leave out at least one of the vectors of S so that the remaining set S' has the property that its linear combinations still produce all of *span* S.

With this terminology, Definition 4-8(a) can be reworded as follows.

DEFINITION 4-8(b)

A set B in a vector space V is a basis iff *span* $B = V$ and B is linearly independent.

EXAMPLE 4-15

Consider the subset $S = \{(1,2),(2,1),(3,4)\}$ of R^2. First, we will show that S generates R^2. If $\mathbf{u} = (x,y)$ is any vector in R^2, we must show that we can find scalars a, b, and c such that

$$\mathbf{u} = (x,y) = a(1,2) + b(2,1) + c(3,4),$$

or, such that

$$\begin{aligned} a + 2b + 3c &= x \\ 2a + b + 4c &= y. \end{aligned}$$

This is a system of two linear equations in a, b, and c with the augmented matrix

$$\begin{bmatrix} 1 & 2 & 3 & x \\ 2 & 1 & 4 & y \end{bmatrix}$$

which we transform by Gauss-Jordan reduction to

$$\begin{bmatrix} 1 & 0 & 5/3 & (-1/3)x + (2/3)y \\ 0 & 1 & 2/3 & (2/3)x - (1/3)y \end{bmatrix}.$$

This is a row-reduced echelon matrix corresponding to the system of equations

$$\begin{aligned} a + (5/3)c &= (-1/3)x + (2/3)y \\ b + (2/3)c &= (2/3)x - (1/3)y, \end{aligned} \tag{4-6}$$

which clearly has solutions for a, b, c for *any* x and y, i.e., for any $\mathbf{u} = (x,y)$. Thus, *span* $S = R^2$. ❑

We can conclude more from equation 4-6: we can always choose $c = 0$, regardless of the values of x and y. For example, if $x = 3$ and $y = 0$ (so that $\mathbf{u} = (3,0)$), we find (with $c = 0$) $a = -1$ and $b = 2$, which means that we can express $\mathbf{u}$ as

$$\mathbf{u} = -1(1,2) + 2(2,1) + 0(3,4) = -1(1,2) + 2(2,1),$$

a linear combination of *only* (1,2) and (2,1). And this holds for any $\mathbf{u}$ in R^2, so the proper subset $S' = \{(1,2),(2,1)\}$ of S generates R^2. Thus, S is linearly dependent.

In fact, it is easy to see that we could choose any one of a, b, or c to be 0 by solving for the other two variables in equation 4-6. This means that the proper subsets $\{(1,2),(3,4)\}$ and $\{(2,1),(3,4)\}$ also generate R^2. But, clearly *no single* vector can generate R^2, so that the three proper subsets of S obtained above are all linearly independent sets of vectors and hence are three different bases for R^2.

There is a second important observation. Since, for example, $B = \{(1,2),(2,1)\}$ generates R^2, the third vector (3,4) of the set S is expressible as a linear combination of the vectors of B: with $c = 0$, put $x = 3$ and $y = 4$ in 4-6 to get

$$a = (-1/3)(3) + (2/3)(4) = 5/3,$$
$$b = (2/3)(3) - (1/3)(4) = 2/3.$$

Therefore, we may write

$$(3,4) = (5/3)(1,2) + (2/3)(2,1).$$

This equation expresses (3,4) as a linear combination of the vectors of B. More generally,-we have

THEOREM 4-6

A set of vectors S is linearly dependent iff some vector $\mathbf{v}$ in it is expressible as a linear combination of the vectors $S - \{\mathbf{v}\}$.

PROOF First suppose $\mathbf{v}$ is expressible as a linear combination of the rest of S:

$$\mathbf{v} = c_1\mathbf{v}_1 + c_2\mathbf{v}_2 + \dots + c_k\mathbf{v}_k \tag{4-7}$$

with $\mathbf{v}$, $\mathbf{v}_i$ in S, $i = 1, 2, \dots, k$, k a positive integer and $c_1, c_2, \dots, c_k$ some scalars; $\mathbf{v} \neq \mathbf{v}_i$ for all $i = 1, \dots, k$. Note that $\mathbf{v}$ does not appear on the right side of this equation. Then, in fact, the proper subset $S - \{\mathbf{v}\}$ of S generates *span* S because, if $\mathbf{v}$ did appear in some element $\mathbf{w}$ in *span* S, we could eliminate it from $\mathbf{w}$ by substituting equation 4-7 for $\mathbf{v}$, i.e., $\mathbf{v}$ is never needed. Thus, S is linearly dependent.

Conversely, if S is linearly dependent, *span* S is generated by a proper subset S' of S. Choose a vector $\mathbf{v}$ in S but not in S' (S' is proper). Since *span* S' = *span* S and $\mathbf{v}$ is in *span* S (as $S \subseteq$ *span* S), we may express $\mathbf{v}$ as a linear combination of the vectors of S'. In other words, $\mathbf{v}$ is expressible as a linear combination of the vectors of $S - \{\mathbf{v}\}$. ■

An immediate consequence of this result is the fact that *two vectors are linearly dependent iff they are multiples of each other*. In addition,

COROLLARY 1

Any subset S of V containing the zero vector $\mathbf{0}$ is linearly dependent.

PROOF This follows from the theorem because $\mathbf{0}$ can be written as a linear combination of any vectors $\mathbf{v}_1, \mathbf{v}_2, \ldots, \mathbf{v}_k$:

$$\mathbf{0} = 0\mathbf{v}_1 + 0\mathbf{v}_2 + \ldots + 0\mathbf{v}_k. \blacksquare$$

Alternatively, we have

COROLLARY 2

No linearly independent set of vectors contains the zero vector. ■

The next corollary is often used as the definition of linear independence.

COROLLARY 3

A set of vectors $\{\mathbf{v}_1, \mathbf{v}_2, \ldots, \mathbf{v}_k\}$ is linearly independent iff whenever

$$c_1\mathbf{v}_1 + c_2\mathbf{v}_2 + \ldots + c_k\mathbf{v}_k = \mathbf{0}$$

for some scalars $c_1, c_2, \ldots, c_k$, then necessarily $c_1 = c_2 = \ldots = c_k = 0$.

This may be stated in other words as: the set $\{\mathbf{v}_1, \mathbf{v}_2, \ldots, \mathbf{v}_k\}$ is linearly independent iff the only linear combination of them equal to the zero vector $\mathbf{0}$ is the **trivial** one, i.e., with all coefficients = 0.

PROOF Suppose $\{\mathbf{v}_1, \mathbf{v}_2, \ldots, \mathbf{v}_k\}$ is linearly independent and

$$c_1\mathbf{v}_1 + c_2\mathbf{v}_2 + \ldots + c_k\mathbf{v}_k = \mathbf{0},$$

for some scalars $c_j \in R$. If one of them is not zero, say $c_j \neq 0$, $1 \leq j \leq k$, then we solve the above equation for $\mathbf{v}_j$ as

$$\mathbf{v}_j = \sum_{\substack{i=1 \\ i \neq j}}^{k} (-c_i/c_j)\mathbf{v}_i$$

where $\mathbf{v}_j$ does not appear in the right-hand sum. This expresses $\mathbf{v}_j$ as a linear combination of the other vectors in the set. It follows from Theorem 4-6 that the set is linearly dependent, which is contrary to our assumption, unless all $c_j = 0$, as required.

Conversely, suppose $\{\mathbf{v}_1, \mathbf{v}_2, \ldots, \mathbf{v}_k\}$ is not linearly independent, i.e., it is linearly dependent. Then, some $\mathbf{v}$ is a linear combination of the others,

$$\mathbf{v}_j = \sum_{\substack{i=1 \\ i \neq j}}^{k} c_i \mathbf{v}_i$$

where $c_1, c_2, \ldots, c_{j-1}, c_{j+1}, \ldots, c_k$ are scalars (again, $\mathbf{v}_j$ does not appear in the sum). This equation can be rewritten as

$$c_1\mathbf{v}_1 + c_2\mathbf{v}_2 + \ldots + c_{j-1}\mathbf{v}_{j-1} - \mathbf{v}_j + c_{j+1}\mathbf{v}_{j+1} + \ldots + c_k\mathbf{v}_k = \mathbf{0},$$

which is a nontrivial linear combination of $\mathbf{v}_1, \ldots, \mathbf{v}_k$ equal to $\mathbf{0}$, as the coefficient of $\mathbf{v}_j$ is -1. This completes the proof. ■

COROLLARY 4

Any subset of a finite linearly independent set is itself linearly independent. ■

EXAMPLE 4-16

$B(n)$ is linearly independent in R^n as we already know, since no proper subset of $B(n)$ generates R^n. We give a second proof of this fact to illustrate Corollary 3 of Theorem 4-6. Suppose

$$c_1\mathbf{e}_1 + c_2\mathbf{e}_2 + \ldots + c_n\mathbf{e}_n = \mathbf{0},$$

for some scalars c_i. Then

$$c_1(1,0,\ldots,0) + c_2(0,1,0,\ldots,0) + \ldots + c_n(0,\ldots,0,1) = (0,0,\ldots,0).$$

By the definitions of addition and scalar multiplication of n-tuples, this becomes

$$(c_1,c_2,\ldots,c_n) = (0,0,\ldots,0),$$

from which we conclude $c_1 = c_2 = \ldots = c_n = 0$. Thus, $B(n)$ is linearly independent. ❑

EXAMPLE 4-17

We construct a basis for $M(2\times 2)$. Let

$$E_1 = \begin{bmatrix} 1 & 0 \\ 0 & 0 \end{bmatrix}, E_2 = \begin{bmatrix} 0 & 1 \\ 0 & 0 \end{bmatrix}, E_3 = \begin{bmatrix} 0 & 0 \\ 1 & 0 \end{bmatrix}, E_4 = \begin{bmatrix} 0 & 0 \\ 0 & 1 \end{bmatrix}$$

and put $B = \{E_1, E_2, E_3, E_4\}$. Then, if $A = (a_{ij})$ is 2×2,

$$A = a_{11}E_1 + a_{12}E_2 + a_{21}E_3 + a_{22}E_4.$$

This expresses A as a linear combination of vectors of B, so B generates $M(2\times 2)$. Also, B is linearly independent. For, if

$$c_1E_1 + c_2E_2 + c_3E_3 + c_4E_4 = O,$$

the 2×2 zero matrix, for some scalars $c_1, \ldots, c_4$, then, after simplification,

$$\begin{bmatrix} c_1 & c_2 \\ c_3 & c_4 \end{bmatrix} = \begin{bmatrix} 0 & 0 \\ 0 & 0 \end{bmatrix},$$

so that all the c_i equal 0. So, B is a basis. ❑

The reader may have noticed the similarity to the standard basis for R^n: each E_i has exactly one nonzero entry, which is a 1. It is this feature that makes the verification of linear independence a simple matter. Also, the generalization to $M(m\times n)$ should be clear: this space has a basis of mn matrices each of the same type as those constructed for $M(2\times 2)$.

The next result states precisely the way in which bases for vector spaces provide a *simple description* of all vectors in a vector space.

THEOREM 4-7

If $B = \{\mathbf{v}_1, \mathbf{v}_2, \ldots, \mathbf{v}_n\}$ is a basis for a vector space V, then any vector can be expressed as a linear combination of vectors of B in *exactly* one way. In other words, the expression is unique.

PROOF Let $\mathbf{u}$ be any vector in V. Then, since B generates V, $\mathbf{u}$ is certainly a linear combination of vectors of B. We have to show there is only one way of expressing $\mathbf{u}$ as such. Suppose there were two such expressions for $\mathbf{u}$:

$$\mathbf{u} = \sum_{i=1}^{n} c_i\mathbf{v}_i,\ \mathbf{u} = \sum_{i=1}^{n} d_i\mathbf{v}_i,$$

where the c_i and d_i are scalars and the $\mathbf{v}_i$ are vectors in B. Then we would get

$$\sum_{i=1}^{n} c_i \mathbf{v}_i = \sum_{i=1}^{n} d_i \mathbf{v}_i$$

or,

$$\sum_{i=1}^{n} (c_i - d_i)\mathbf{v}_i = \mathbf{0},$$

the zero vector. But B is linearly independent, so all the coefficients in this linear combination must be 0:

$$c_i - d_i = 0,$$

or,

$$c_i = d_i,\ i = 1, \ldots, n.$$

So, the two hypothesized expressions for **u** are identical, thus completing the proof. ■

Remarks

(1) The result is true even when B is an infinite set.

(2) The converse of the theorem is also easily seen to be true: if every **v** in V has an unique expression as a linear combination of vectors from a set S, then S is a basis (question 18, Exercises 4-3).

Now, a number of questions arise which we have yet to address:

(1) Does every vector space possess a basis? We have seen that R^n and $M(m \times n)$ do.

(2) Are there vector spaces which do not have *finite* bases?

(3) We have seen that a vector space can have many bases (for example, R^2, see Example 4-15). Do all bases for a vector space have the same number of elements (as they did in that example)?

We shall see that the answer to all three questions is *yes*. We begin by introducing a definition related to question (2).

DEFINITION 4-10

A vector space is called **finite dimensional** if it is generated by some finite set of vectors. Otherwise, it is called **infinite dimensional**.

We have seen that R^n and $M(m \times n)$ are finite dimensional. The next example illustrates other finite dimensional spaces and an infinite dimensional one.

EXAMPLE 4-18

We obtain a finite basis $G(n)$ for $P(n)$ and an infinite basis S for $P(R)$. Let us start with $P(2)$ for definiteness. This is the set of all polynomial functions $f:R \to R$ of degree 2 or less, so a typical f has the form

$$f(x) = a_0 + a_1x + a_2x^2,$$

where the a_i are constants and $x \in R$. We put

$$p_0(x) = 1 = x^0,\ p_1(x) = x = x^1, \text{ and } p_2(x) = x^2,$$

for any number x. Then, clearly the p's are in $P(2)$, and we can write

$$f(x) = a_0p_0(x) + a_1p_1(x) + a_2p_2(x),$$

an equation which is true for all $x \in R$. This means that as a *function*,

$$f = a_0p_0 + a_1p_1 + a_2p_2,$$

which is a linear combination of the polynomial functions p_0, p_1, and p_2. For example, if

$$f(x) = -1 + 5x - 6x^2,$$

then

$$f = -p_0 + 5p_1 - 6p_2.$$

We see that $G(2) = \{p_0,p_1,p_2\}$ generates $P(2)$. Usually, p_0, p_1, and p_2 are denoted simply as 1, x, and x^2 respectively, so that $G(2)$ is written $\{1,x,x^2\}$.

In fact, $G(2)$ is linearly independent as well, and so is a basis for $P(2)$. To see this, suppose

$$a1 + bx + cx^2 = 0$$

(the zero function), for some scalars a, b, c (note that here the zero vector $\mathbf{0} = 0$). This means that

$$a + bx + cx^2 = 0 \in R, \text{ for all } x \in R.$$

Thus we can put

(i) $x = 0$ to get $a = 0$,
(ii) $x = 1$ to get $b + c = 0$, and
(iii) $x = -1$ to get $-b + c = 0$.

From the last two equations, it follows that $b = c = 0$. Therefore, $G(2)$ is linearly independent, and is a basis for $P(2)$.

Similarly, we could show that $G(n)$ is a basis for $P(n)$, where $G(n) = \{1,x,x^2,\ldots,x^{n-1},x^n\}$. Notice that it has $n + 1$ elements, so $P(n)$ is finite dimensional.

Next, recall that $P(R)$ is the union of all the spaces $P(n)$, so it consists of polynomial functions of all degrees. To get a contradiction, suppose that $P(R)$ were generated by a *finite* set $S' = \{f_1,f_2,...,f_n\}$ of polynomial functions. Now, each f_i has a degree d_i, the largest exponent of x appearing in $f_i(x)$. We put

$$d = 1 + \text{the } \textit{largest} \text{ of the } d_i,\ i = 1, 2, ..., n.$$

Then d is larger than the highest exponent of x in all the polynomial functions of S'. It follows that the polynomial function g with $g(x) = x^d$ for x in R cannot be a linear combination of the vectors of S'. This is so because any such linear combination has highest power of x at least one less than d, by the definition of d. Thus, $P(R)$ cannot be generated by a finite set.

To illustrate, suppose an $S' = \{f_1,f_2\}$ could be found with these two functions:

$$f_1(x) = 2 + 35x^{56} + 19x^{1001}$$

and

$$f_2(x) = 56x^{93} + x^{107}.$$

Then,

$$d = 1 + \text{largest of } \{1001,107\} = 1002,$$

so $g(x) = x^{1002}$, which cannot possibly be a linear combination of f_1 and f_2.

Thus, $P(R)$ is infinite dimensional. In fact, it can be shown that the (infinite) set given by $S = \{1,x,x^2,x^3,...,x^k,x^{k+1},...\}$ is an *infinite* basis for $P(R)$. ❑

We are mostly interested in finite dimensional spaces V. For such cases, the method of Example 4-15 provides an answer to question (1) above.

THEOREM 4-8

Let V be a vector space spanned by a finite set $S = \{\mathbf{v}_1,\mathbf{v}_2,...,\mathbf{v}_m\}$. Suppose $K = \{\mathbf{v}_1,...,\mathbf{v}_k\} \subseteq S$ is linearly independent, where $1 \leq k \leq m$. Then there is a basis B for V with $K \subseteq B \subseteq S$.

PROOF (Outline) The situation is illustrated schematically in Figure 4-5.

We briefly describe two possible approaches for the proof. As illustrated in Example 4-15, the first approach is to consider (decreasing) subsets S_d ($d = 1, 2, ...$) of S obtained by successively *removing* vectors from S, but in such a way that

(i) S_d still generates V, and

(ii) S_d still contains K (see the figure).

FIGURE 4-5

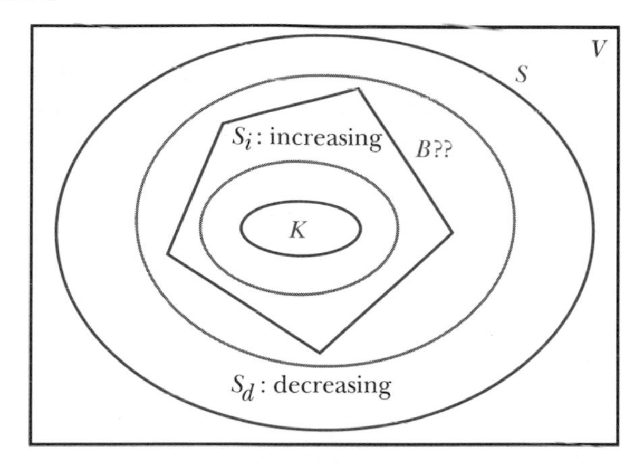

Then, since S is finite and contains K, after a finite number of such removals from S, the resulting set S_d must become linearly independent (certainly when S_d becomes K), so it will be the required basis B.

The second approach works from the inside out: we take (increasing) subsets S_i ($i = 1, 2, \ldots$) of S obtained by successively *adding* vectors to K in such a way that

(i) S_i is linearly independent, and

(ii) S_i is contained in S.

Again, since S is finite, eventually S_i must generate V (which certainly will happen when S_i gets as far as S). Then, that S_i will be the required basis B. ■

COROLLARY

Every finite dimensional vector space V ($\neq \{\mathbf{0}\}$) has a finite basis ($\varnothing$ is a basis for $\{\mathbf{0}\}$).

PROOF As $V \neq \{\mathbf{0}\}$, if S generates V (and S is finite), then there is a vector $\mathbf{v} \neq \mathbf{0}$ in S. Therefore, the set $K = \{\mathbf{v}\}$ is linearly independent and $K \subseteq S$. So there is a basis B for V with $K \subseteq B \subseteq S$. In particular, V has a finite basis. ■

Example 4-19 below illustrates the second approach discussed in the proof of Theorem 4-8. As noted before, the first approach is illustrated in Example 4-15. In the next section, we will develop a mechanical procedure for constructing bases by both approaches using matrix methods.

EXAMPLE 4-19

We construct a basis for $P(2)$ containing the vectors (functions) g and h, where $g(x) = x + 1$ and $h(x) = x - 1$, $x \in R$. First, to follow the point of view of Theorem 4-8, we note that g and h are linearly independent. For, if

$$ag + bh = \mathbf{0} = 0 \in P(2)$$

for some scalars a and b, then for all x in R (as in Example 4-18),

$$a(x + 1) + b(x - 1) = 0,$$

or,

$$(a + b)x + (a - b) = 0.$$

We leave it for the reader to show that by choosing $x = 0$ and $x = 1$, we can conclude $a = b = 0$, so that g and h are linearly independent. Thus $\{g,h\}$ plays the role of K in Theorem 4-8.

Next, we need a finite set S of generators for $P(2)$ with $K \subseteq S$. Now, we know from Example 4-18 that $G(2) = \{1,x,x^2\}$ is a set of three generators for $P(2)$, but K is not a subset of $G(2)$. We remedy this by putting

$$S = K \cup G(2),$$

which clearly contains K, is finite, and also generates $P(2)$ (why?). Then, following the approach discussed, we add vectors to K to construct linearly independent sets S_i. In order that the first such set S_1 ($i = 1$) be linearly independent, the vector added must come from $S - span\ K$ (why?), which rules out 1 and x since

$$1 = (1/2)(x + 1) - (1/2)(x - 1) = ((1/2)g - (1/2)h)(x),$$

and

$$x = (1/2)(x + 1) + (1/2)(x - 1) = ((1/2)g + (1/2)h)(x),$$

i.e., $\{1,g,h\}$ and $\{x,g,h\}$ are both linearly dependent sets. Thus, we are led to trying to add, for example, x^2 to K to get $S_1 = \{g,h,x^2\}$. Other choices are possible, but we notice that $g(x)$ and $h(x)$ do not contain x^2. It is clear that S_1 is linearly independent since no vector in it is expressible as a linear combination of the remaining two. Now, S_1 contains K, is linearly independent, and generates $P(2)$, as the reader may verify. Thus, it is the required basis B. The fact that it generates $P(2)$ also follows from the next results. ❑

THEOREM 4-9

Let $B = \{\mathbf{v}_1,\mathbf{v}_2,\ldots,\mathbf{v}_n\}$ be a basis for a vector space V. Then any linearly independent subset of V has at most n elements.

PROOF (Outline — Optional) We illustrate one method of proof for the particular case of $n = 2$; the proof for any arbitrary n is similar. (Another proof is indicated in question 17, Exercises 4-3). We let $B = \{\mathbf{v}_1, \mathbf{v}_2\}$ be a basis for V and suppose that $S = \{\mathbf{u}_1, \mathbf{u}_2, \mathbf{u}_3\}$ were a linearly independent subset of V with more than two elements ($n = 2$). We will obtain a contradiction to show this is impossible. Consider the set

$$S_1 = B \cup \{\mathbf{u}_1\}.$$

We may assume that $\mathbf{u}_1$ is not in B, since vectors of S can't all be in B. But as B is a basis, $\mathbf{u}_1$ is a linear combination of the vectors of B (so S_1 is linearly dependent):

$$\mathbf{u}_1 = c_1\mathbf{v}_1 + c_2\mathbf{v}_2, \tag{4-8}$$

for some $c_i \in R$. As $\mathbf{u}_1 \neq \mathbf{0}$, some $c_i \neq 0$; for convenience, let it be c_1. Then we may solve 4-8 for $\mathbf{v}_1$ as

$$\mathbf{v}_1 = (1/c_1)\mathbf{u}_1 - (c_2/c_1)\mathbf{v}_2, \tag{4-9}$$

expressing $\mathbf{v}_1$ as a linear combination of $\mathbf{u}_1$ and $\mathbf{v}_2$. So, any linear combination of vectors of B can be rewritten as a linear combination of $B_1 = \{\mathbf{u}_1, \mathbf{v}_2\}$ by using 4-9, i.e., B_1 also generates V. Next, we consider the set

$$S_2 = B_1 \cup \{\mathbf{u}_2\},$$

which is also linearly dependent, as B_1 generates V. Therefore, we can write

$$\mathbf{u}_2 = d_1\mathbf{u}_1 + d_2\mathbf{v}_2. \tag{4-10}$$

Now if $d_2 = 0$, then 4-10 would imply S is linearly dependent (why?), which is a contradiction to our supposition. So, $d_2 \neq 0$, and we may solve for $\mathbf{v}_2$:

$$\mathbf{v}_2 = (1/d_2)\mathbf{u}_2 - (d_1/d_2)\mathbf{u}_1.$$

Therefore, $B_2 = \{\mathbf{u}_1, \mathbf{u}_2\}$ also generates V (note it has $n = 2$ elements). Finally, we let

$$S_3 = B_2 \cup \{\mathbf{u}_3\} = S.$$

This must be linearly dependent, since B_2 generates V, thereby contradicting the original hypothesis and completing the proof. ■

Remark The above proof is called the **Steinitz replacement method**, since it replaces successively the elements of B by those of S.

COROLLARY 1

Any two bases for a finite dimensional vector space V are finite and have the same number of elements.

PROOF Since V is finite dimensional, it is generated by a finite set S. By the corollary of Theorem 4-8, V therefore has a basis $B \subseteq S$ which is also finite. Suppose it has n elements. We show that any other basis B' also has n elements. Suppose B' had n' elements. By the theorem $n' \leq n$ since B' is linearly independent and B is a basis. At the same time, $n \leq n'$ since B is linearly independent and B' is a basis. Consequently, $n' = n$ as claimed. ■

DEFINITION 4-11

If V is a finite dimensional vector space, we say that V has **dimension** n if it has a basis of n elements. In this case, we write $\dim V = n$.

By Corollary 1 of Theorem 4-9, every basis then has n elements, so this definition produces exactly one positive integer n for the dimension of V. By convention, we set $\dim \{\mathbf{0}\} = 0$. In earlier examples, we have seen the following:

(1) $\dim R^n = n$.

(2) $\dim M(m \times n) = mn$.

(3) $\dim P(n) = n+1$.

EXAMPLE 4-20

Recall that any 2×2 symmetric matrix has the form

$$\begin{bmatrix} a & b \\ b & c \end{bmatrix}.$$

This can be rewritten as

$$a\begin{bmatrix} 1 & 0 \\ 0 & 0 \end{bmatrix} + b\begin{bmatrix} 0 & 1 \\ 1 & 0 \end{bmatrix} + c\begin{bmatrix} 0 & 0 \\ 0 & 1 \end{bmatrix}$$

for some $a, b, c \in R$. It is easy to see that the three matrices in this linear combination form a linearly independent set. It follows that they constitute a basis and the dimension of this subspace of $M(2 \times 2)$ is 3.

Similarly, any 2×2 skew-symmetric matrix has the form

$$\begin{bmatrix} 0 & a \\ -a & 0 \end{bmatrix} = a\begin{bmatrix} 0 & 1 \\ -1 & 0 \end{bmatrix},$$

from which we see that the dimension of this subspace is 1. ❑

These examples illustrate

COROLLARY 2

If W is a subspace of a finite dimensional vector space V, then W is also finite dimensional and $dim\ W \leq dim\ V$.

PROOF The result is clearly true if $W = \{\mathbf{0}\}$. But suppose W is nontrivial. Let $\mathbf{w}$ be a nonzero vector in W. Now, any finite set of generators S for V necessarily generates W, since W is part of V. Then, by Theorem 4-8, there is a basis B' for W with

$$\{\mathbf{w}\} \subseteq B' \subseteq S \cup \{\mathbf{w}\}.$$

Note that $\{\mathbf{w}\}$ is linearly independent and $S \cup \{\mathbf{w}\}$ generates W. Furthermore, since B' is linearly independent and contained in $S \cup \{\mathbf{w}\}$, another application of the same theorem tells us there is a basis B for V with

$$B' \subseteq B \subseteq S \cup \{\mathbf{w}\}.$$

It follows that $dim\ W$ (= number of elements in B') $\leq dim\ V$ (= number of elements in B), proving the result. ■

COROLLARY 3

If $dim\ V = n$, the following conditions regarding a set of n vectors, B, in V are equivalent:

(1) B is a basis.

(2) B is linearly independent.

(3) B generates V.

PROOF We leave the proof as an exercise for the reader. ■

The corollary implies that in a space of dimension n, in order to show that n vectors constitute a basis, it is sufficient to show that *one* of these is true: (a) they generate V; or (b) they are linearly independent.

EXAMPLE 4-21

(a) *Without any calculation*, we know that the three functions x^3, $1 + x$, $-2x^2 + x^3$ cannot be a basis for $P(3)$ because $dim\ P(3) = 4$.

(b) Similarly, (1,2,3), (−1,0,3), (8,2,0), and (−5,0,4) cannot be a basis for R^3 as *dim* $R^3 = 3$.

(c) Consider the vectors $\mathbf{u} = (1,0,2)$, $\mathbf{v} = (-1,1,3)$, and $\mathbf{w} = (0,1,1)$ in R^3. To determine whether or not they constitute a basis we need only check for linear independence *or* that they generate R^3 (because there are three of them and this is the dimension of R^3). In this case, it is probably easier to do the former (we leave the other for the reader). Suppose

$$a\mathbf{u} + b\mathbf{v} + c\mathbf{w} = \mathbf{0} = (0,0,0)$$

for some scalars $a, b, c \in R$. This leads to the system of equations

$$\begin{aligned} a - b &= 0 \\ b + c &= 0 \\ 2a + 3b + c &= 0 \end{aligned}$$

whose only solution is easily seen to be $a = b = c = 0$. Thus $\{\mathbf{u},\mathbf{v},\mathbf{w}\}$ is a basis for R^3. ❑

EXERCISES 4-3

1. In each of the following cases, express the given vector $\mathbf{u}$ as a linear combination of $\mathbf{v} = (-1,2,0)$ and $\mathbf{w} = (1,1,2)$ or show that it is not a linear combination of $\mathbf{v}$ and $\mathbf{w}$.

   (a) $\mathbf{u} = (5,-1,6)$ **(b)** $\mathbf{u} = (-2,-1/2,-3)$

   (c) $\mathbf{u} = (-1,-1,2)$ **(d)** $\mathbf{u} = (1,1,1)$

2. Determine whether or not the following polynomial functions in $P(2)$ are linear combinations of f and g where $f(x) = 1 + 2x + x^2$ and $g(x) = 1 - 2x + x^2$, $x \in R$.

   (a) $h(x) = 2$ **(b)** $h(x) = -1 + 4x - x^2$

   (c) $h(x) = -x + x^2$ **(d)** $h(x) = (1/2)x + (1/2)x^2$

3. Express, if possible, each of the matrices D, E, F, and G as a linear combination of A, B, and C.

$$A = \begin{bmatrix} -1 & 1 \\ 1 & 0 \end{bmatrix}, B = \begin{bmatrix} 3 & 0 \\ 2 & 1 \end{bmatrix}, C = \begin{bmatrix} -2 & 1 \\ 1 & 2 \end{bmatrix}$$

$$D = \begin{bmatrix} 7 & -3 \\ -1 & -1 \end{bmatrix}, E = \begin{bmatrix} 4 & -3 \\ -3 & -2 \end{bmatrix}, F = \begin{bmatrix} 2 & -2 \\ -2 & 1 \end{bmatrix}, G = \begin{bmatrix} 0 & 0 \\ 0 & 1 \end{bmatrix}$$

4. **(a)** Show that the vectors $\mathbf{u} = (-1,1)$, $\mathbf{v} = (2,3)$, and $\mathbf{w} = (-2,-2)$ generate R^2 but are linearly dependent.

   (b) Find all proper subsets of $\{\mathbf{u},\mathbf{v},\mathbf{w}\}$ that span R^2.

5. **(a)** Show that the polynomial functions $f_1(x) = 2 + x$, $f_2(x) = 1 - x + 2x^2$, and $f_3(x) = 3x - 4x^2$ are linearly dependent.

   (b) Find conditions for a_0, a_1, and a_2 so that $f(x) = a_0 + a_1x + a_2x^2$ will be in the subspace generated by f_1, f_2, and f_3.

6. **(a)** Prove that any subset of a linearly independent set of vectors is linearly independent.

   (b) Show that any set of vectors containing a linearly dependent subset is itself linearly dependent.

   (c) Show that a set of generators for a finite dimensional space V generates any subspace of V.

7. **(a)** Show that the matrices

$$A = \begin{bmatrix} 1 & 0 \\ 0 & 2 \end{bmatrix}, B = \begin{bmatrix} 2 & 0 \\ 0 & 1 \end{bmatrix}, C = \begin{bmatrix} 0 & 2 \\ 2 & 0 \end{bmatrix}$$

   form a basis for the space of symmetric 2×2 matrices.

   (b) Find a fourth matrix D so that $\{A,B,C,D\}$ is a basis for $M(2 \times 2)$.

8. Determine which of the following sets of vectors in R^3 are linearly independent.

   (a) $\{(1,-1,3),(5,-2,3),(-3,3,-9)\}$

   (b) $\{(1,1,1),(-1,2,0),(4,-1,2)\}$

   (c) $\{(2,1,-1),(5,4,-2),(-3,-3,1)\}$

   (d) $\{(2,-3,1),(1,-1,0),(2,-4,2)\}$

9. Find a basis for the space of diagonal $n \times n$ matrices. What is the dimension of this subspace of $M(n \times n)$?

10. Find bases for the spaces of upper and lower triangular matrices in $M(n \times n)$. What are the dimensions of these subspaces?

11. **(a)** Construct a basis for R^3 containing the vectors $\mathbf{v}$ and $\mathbf{w}$ given in question 1.

    (b) Construct a basis for $P(2)$ containing the polynomial functions f and g defined in question 2.

    (c) Construct a basis for $M(2 \times 2)$ containing the matrices A, B, and C given in question 3.

    (d) Construct a basis for $M(2 \times 2)$ from the matrices A, B, C, D, E, F, and G given in question 3.

12. Let W be the subset of $P(3)$ consisting of polynomial functions f of the form $f(x) = a_0 + a_1x + a_2x^2 + a_3x^3$, with $a_0 = a_1 = a_2 = a_3$.

    (a) Show that W is a subspace of $P(3)$.

    (b) Construct a basis for W.

    (c) Find *dim* W.

13. Let $S = \{a,b,c\}$. Find a basis for $F(S)$.

14. Let Z be the subspace of $M(2\times 2)$ consisting of matrices having trace 0 (see question 16(c), Exercises 4-2). Construct a basis for Z and find *dim* Z.

15. Generalize Example 4-20 to show the following.

(a) The space of symmetric $n\times n$ matrices has dimension $(n/2)(n+1)$.

(b) The space of $n\times n$ skew-symmetric matrices has dimension $(n/2)(n-1)$.

16. Prove Theorem 4-9 for *dim* $V = n$ using the Steinitz replacement method.

17. Prove Theorem 4-9 using the result of question 8, Exercises 2-2 relating to the existence of nontrivial solutions for homogeneous systems of n equations in $m > n$ unknowns as follows: Let $S = \{\mathbf{u}_1, \mathbf{u}_2, \ldots, \mathbf{u}_m\}$ be a set of $m > n$ vectors in a space V of dimension n. Then S is linearly dependent iff there are $x_i \in R$ (not all 0) such that

$$\sum_{i=1}^{m} x_i \mathbf{u}_i = \mathbf{0}.$$

By expressing the $\mathbf{u}_i$ in terms of the n basis vectors, this yields a system of n equations in the m unknowns x_i, which must have a nontrivial solution, by the result quoted above. (See also the corollary of Theorem 4-14 in the next section.)

18. Let V be a finite dimensional vector space and $S \subseteq V$. Suppose that every $\mathbf{v} \in V$ has an *unique* expression as a linear combination of vectors of S. Show that S is a basis for V.

4-4 Computational Techniques — Rank of a Matrix

In this section, we establish mechanical methods for dealing with problems relating to generating sets, linear independence, and bases. The ideas are based on Gauss-Jordan reduction of appropriate matrices: in essence, we arrange sets of vectors in a matrix. But first, we need the following definition.

DEFINITION 4-12

Let A be an $m\times n$ matrix. The **row space** of A, denoted $row(A)$, is the subspace of R^n generated by the rows $A_{[1]}, A_{[2]}, \ldots, A_{[m]}$ of A, each row $A_{[i]}$ being considered as an n-tuple (recall Definition 2-9). Similarly, we can define the **column space** $col(A)$ of A as the subspace of R^m generated by the columns $A^{[1]}, A^{[2]}, \ldots, A^{[n]}$ of A, each $A^{[j]}$ being regarded as an m-tuple.

EXAMPLE 4-22

Let A be the 2×3 matrix given by

$$A = \begin{bmatrix} 1 & -1 & 0 \\ 2 & 3 & 1 \end{bmatrix}.$$

Then $row(A)$ is the subspace of R^3 generated by the vectors $A_{[1]} = (1, -1, 0)$ and $A_{[2]} = (2, 3, 1)$. The column space $col(A)$ is the subspace of R^2 generated by $A^{[1]} = (1, 2)$, $A^{[2]} = (-1, 3)$, and $A^{[3]} = (0, 1)$. ❑

In fact, we shall usually be working with the row space because $col(A) = row(A^t)$: *the column space of A is the row space of* A^t.

The main idea of the methods to be discussed rests in the next two theorems.

THEOREM 4-10

Row equivalent $m \times n$ matrices have the same row spaces, so that these spaces have the same bases and dimensions.

PROOF (Outline) Let A and A' be row-equivalent $m \times n$ matrices. As noted in Chapter 2, it is sufficient to consider the case where A' has been obtained from A by a single elementary row operation. We consider the three possibilities for this operation:

Type 1 Then A and A' have the same rows except for the order in which they appear, so the result is true.

Type 2 If A' has been obtained from A by a type 2 operation, then some row of A' is a nonzero scalar multiple of a row of A, and vice-versa, so again $row(A') = row(A)$, since vector spaces are closed under scalar multiplication.

Type 3 Now some row of A' is a linear combination of two rows of A, and vice-versa, so the result again follows (we leave the details of the proof for the reader). ■

EXAMPLE 4-23

Consider the matrix

$$A = \begin{bmatrix} 2 & 1 & -4 & 3 \\ 0 & 2 & 6 & -4 \\ 2 & 3 & 2 & -1 \end{bmatrix}.$$

The reader should check that it is row-equivalent to

$$A' = \begin{bmatrix} 1 & 0 & -7/2 & 5/2 \\ 0 & 1 & 3 & -2 \\ 0 & 0 & 0 & 0 \end{bmatrix}.$$

Thus, these matrices have the same row space. Here A' is a row-reduced echelon matrix, so its row space is generated by the first two rows, as the last row is $\mathbf{0}$.

Furthermore, $A'_{[1]} = (1,0,-7/2,5/2)$ and $A'_{[2]} = (0,1,3,-2)$ are linearly independent because of the convenient form of row-reduced echelon matrices (all entries in a column containing a leading row entry of 1 are zero, etc.). To see this, suppose

$$c_1A'_{[1]} + c_2A'_{[2]} = \mathbf{0} = (0,0,0,0)$$

for some scalars c_i. Then

$$(c_1,c_2,(-7/2)c_1 + 3c_2,(5/2)c_1 - 2c_2) = (0,0,0,0),$$

from which we see at once that $c_1 = c_2 = 0$. ❑

More generally,

THEOREM 4-11

If A is a row-reduced echelon matrix, its nonzero rows form a basis for the row space of A. If A has r nonzero rows, $\mathit{dim}\ \mathit{row}(A) = r$.

PROOF (Optional) If A is an $m \times n$ row-reduced echelon matrix with r nonzero rows ($1 \le r \le m$), then A must have the form

$$\begin{array}{rr} & \text{col:} \\ \text{row:} & 1 \\ & 2 \\ & \\ & r \\ & r+1 \\ & \\ & m \end{array} \begin{array}{c} \\ \left[\begin{array}{cccccccc} 1 & & j_i & & j_2 & & j_r & \\ \hline 0 & \dots & 1 & *** & 0 & *** & 0 & *** \\ 0 & \dots & 0 & \dots & 1 & *** & 0 & *** \\ \dots & \dots & \dots & \dots & \dots & \dots & \dots & \dots \\ 0 & \dots & 0 & \dots & 0 & \dots & 1 & *** \\ 0 & \dots & \dots & \dots & \dots & \dots & \dots & 0 \\ \dots & \dots & \dots & \dots & \dots & \dots & \dots & \dots \\ 0 & \dots & \dots & \dots & \dots & \dots & \dots & 0 \end{array}\right] \end{array}, \quad (4\text{-}11)$$

where $1 \le j_1 < j_2 < \dots < j_r \le n$, and $***$ indicate numbers, which are not necessarily 0. Consequently, if $c_1, c_2, \dots, c_r$ are scalars,

$$\begin{aligned}
c_1 A_{[1]} &+ c_2 A_{[2]} + \ldots + c_r A_{[r]} \\
&= c_1(0,\ldots,0,1,***,0,***,0,***) \\
&\quad + c_2(0,\ldots\ldots\ldots,0,1,***,0,***) \\
&\quad + \ldots\ldots\ldots\ldots\ldots\ldots \\
&\quad + c_r(0,\ldots\ldots\ldots\ldots\ldots\ldots,0,1,***) \\
&= (0,\ldots,0,c_1,***,c_2,***,c_r,***). \qquad (4\text{-}12)
\end{aligned}$$

Therefore, if

$$c_1 A_{[1]} + c_2 A_{[2]} + \ldots + c_r A_{[r]} = \mathbf{0} = (0,0,\ldots,0),$$

it necessarily follows that $c_1 = c_2 = \ldots = c_r = 0$, so $A_{[1]}, A_{[2]}, \ldots, A_{[r]}$ are linearly independent, and form a basis for *row*(A). ■

Remark A similar proof shows that the result is also true when A is only a row echelon matrix (question 20, Exercises 4-4). However, we will always work with the row-reduced echelon form to obtain bases for the row space of A.

EXAMPLE 4-24

From Example 4-23, $(1,0,-7/2,5/2)$ and $(0,1,3,-2)$ form a basis for the row space of A'. They also form a basis for *row*(A) because A and A' have the same row space. ❑

Example 4-24 illustrates the next statement.

COROLLARY

If A' is a row-reduced echelon matrix row-equivalent to a matrix A, the nonzero rows of A' form a basis for *row*(A). ■

DEFINITION 4-13

The **rank** of an $m \times n$ matrix A, denoted *rank* A, is the dimension of its row space.

In fact, the rank of A is also the dimension of its column space, because, as we shall see in Chapter 5, $dim\ col(A) = dim\ row(A)$ for every matrix A. As the dimension cannot exceed the number of generators, it follows that $rank\ A \leq$ the *lesser* of m and n. In addition, it follows that $rank\ A^t = rank\ A$.

EXAMPLE 4-25

Let A be the 3×5 matrix given by

$$A = \begin{bmatrix} 1 & 2 & -4 & 3 & -1 \\ 1 & 2 & -2 & 2 & 1 \\ 2 & 4 & -2 & 3 & 4 \end{bmatrix}.$$

The reader should verify that A is row-equivalent to

$$A' = \begin{bmatrix} 1 & 2 & 0 & 1 & 3 \\ 0 & 0 & 1 & -1/2 & 1 \\ 0 & 0 & 0 & 0 & 0 \end{bmatrix}.$$

It follows that rank $A = 2$. It would be worthwhile for the reader to check that the column space of A also has a basis of two vectors (which are triples). This is done by row-reducing the transpose of A, a 5×3 matrix. The result must then be a 5×3 matrix with only two nonzero rows. ❑

We now apply these ideas to problems relating to linear independence of sets of vectors in a vector space V. First, we consider the case $V = R^n$. Suppose we are given an ordered set $\{\mathbf{u}_1, \mathbf{u}_2, \ldots, \mathbf{u}_m\}$ of m vectors in R^n. We can arrange them in an $m \times n$ matrix U with rows $\mathbf{u}_1$, $\mathbf{u}_2$, ..., $\mathbf{u}_m$ (in that order) and denoted

$$U = \begin{bmatrix} \mathbf{u}_1 \\ \mathbf{u}_2 \\ \cdots \\ \mathbf{u}_m \end{bmatrix}.$$

Then, *span* $\{\mathbf{u}_1, \mathbf{u}_2, \ldots, \mathbf{u}_m\} = row(U)$, and we have (by the preceding corollary) a mechanical procedure for finding a basis for $row(U)$, namely, Gauss-Jordan reduction.

EXAMPLE 4-26

Given $\mathbf{u}_1 = (1,2,-1,4)$, $\mathbf{u}_2 = (2,-1,4,2)$, $\mathbf{u}_3 = (0,1,0,-1)$ in R^4, we get the 3×4 matrix U:

$$U = \begin{bmatrix} 1 & 2 & -1 & 4 \\ 2 & -1 & 4 & 2 \\ 0 & 1 & 0 & -1 \end{bmatrix}.$$

Using Gauss-Jordan reduction, this is transformed to

$$U' = \begin{bmatrix} 1 & 0 & 0 & 25/6 \\ 0 & 1 & 0 & -1 \\ 0 & 0 & 1 & -11/6 \end{bmatrix}.$$

It follows that $\mathbf{v}_1 = (1,0,0,25/6)$, $\mathbf{v}_2 = (0,1,0,-1)$, and $\mathbf{v}_3 = (0,0,1,-11/6)$ constitute a basis for $row(U)$ and therefore also for the subspace of R^4 generated by $\mathbf{u}_1$, $\mathbf{u}_2$, and $\mathbf{u}_3$. ❑

Another important type of problem we have considered earlier is that of determining whether or not a given vector belongs to the subspace spanned by certain vectors.

EXAMPLE 4-27

Continuing Example 4-26, we determine whether or not the vector $\mathbf{u} = (1,3,-1,3)$ lies in the subspace of R^4 generated by $\{\mathbf{u}_1,\mathbf{u}_2,\mathbf{u}_3\}$. Now, we already know how to do this by the basic principles: we would check whether or not scalars c_1, c_2, c_3 can be found so that

$$\mathbf{u} = c_1\mathbf{u}_1 + c_2\mathbf{u}_2 + c_3\mathbf{u}_3.$$

This amounts to determining whether a system of equations has a solution. (See also questions 15 and 16, Exercises 4-4, in this regard.)

Alternatively, we can use the fact that $\mathbf{u}$ is in $span\ \{\mathbf{u}_1,\mathbf{u}_2,\mathbf{u}_3\}$ iff $span\ \{\mathbf{u}_1,\mathbf{u}_2,\mathbf{u}_3,\mathbf{u}\} = span\ \{\mathbf{u}_1,\mathbf{u}_2,\mathbf{u}_3\}$. Therefore, we construct the matrix U'' with rows $\mathbf{u}_1$, $\mathbf{u}_2$, $\mathbf{u}_3$, and $\mathbf{u}$:

$$U'' = \begin{bmatrix} 1 & 2 & -1 & 4 \\ 2 & -1 & 4 & 2 \\ 0 & 1 & 0 & -1 \\ 1 & 3 & -1 & 3 \end{bmatrix}$$

which is easily seen to be row-equivalent to

$$\begin{bmatrix} 1 & 0 & 0 & 25/6 \\ 0 & 1 & 0 & -1 \\ 0 & 0 & 1 & -11/6 \\ 0 & 0 & 0 & 0 \end{bmatrix}.$$

This matrix has the same row space as U' in Example 4-26. Hence $\mathbf{u}$ is in the subspace generated by $\{\mathbf{u}_1,\mathbf{u}_2,\mathbf{u}_3\}$. ❑

So far we have applied these matrix methods only to R^n. We next extend them to any arbitrary V using the important property of bases (Theorem 4-7) that any vector $\mathbf{v}$ in V has an unique expression as a linear combination of the vectors of a basis B. When $dim\ V = n$, this yields an unique n-tuple corresponding to the vector $\mathbf{v}$ (the coefficients in the expression for $\mathbf{v}$), provided we specify a fixed order for the elements in the basis B.

DEFINITION 4-14(a)

Let V be a finite dimensional space. An **ordered basis** B for V is a basis $\{\mathbf{v}_1,\mathbf{v}_2,\ldots,\mathbf{v}_n\}$ for V with a specific order prescribed for its vectors. In other words, it is a sequence of n vectors, which, as a set, is a basis for V.

Thus, for example, $\{\mathbf{v}_1,\mathbf{v}_2,\ldots,\mathbf{v}_n\}$ and $\{\mathbf{v}_2,\mathbf{v}_1,\ldots,\mathbf{v}_n\}$ are *different* ordered bases, but are the same unordered basis. Henceforth, bases will always be ordered; *we will always use the term basis to mean ordered basis*. Now let $B = \{\mathbf{v}_1,\mathbf{v}_2,\ldots,\mathbf{v}_n\}$ be an (ordered) basis for V. Then, for $\mathbf{v}$ in V, there are n scalars c_1, c_2, ..., c_n such that

$$\mathbf{v} = c_1\mathbf{v}_1 + c_2\mathbf{v}_2 + \ldots + c_n\mathbf{v}_n$$

and this expression is uniquely determined by $\mathbf{v}$ (given B). These scalars may be arranged as an $n \times 1$ matrix

$$\begin{bmatrix} c_1 \\ c_2 \\ \cdots \\ c_n \end{bmatrix}.$$

DEFINITION 4-14(b)

The $n \times 1$ matrix above is called the (coordinate, or, component) **matrix of v with respect to the (ordered) basis *B***. It is denoted $[\mathbf{v}]_B$. The scalars c_1, ..., c_n are the **components** of $\mathbf{v}$ with respect to B.

EXAMPLE 4-28

(a) The standard (ordered) basis for R^n is $B(n) = \{\mathbf{e}_1,\ldots,\mathbf{e}_n\}$, so if $\mathbf{v} = (c_1,c_2,\ldots,c_n)$ is any vector in R^n,

$$[\mathbf{v}]_{B(n)} = \begin{bmatrix} c_1 \\ c_2 \\ \cdots \\ c_n \end{bmatrix}.$$

For example, if $\mathbf{v} = (1,2,3)$ in R^3, then

$$[\mathbf{v}]_{B(3)} = \begin{bmatrix} 1 \\ 2 \\ 3 \end{bmatrix}.$$

(b) The standard (ordered) basis for $M(2\times 2)$ is $\{E_1,E_2,E_3,E_4\}$, where

$$E_1 = \begin{bmatrix} 1 & 0 \\ 0 & 0 \end{bmatrix}, E_2 = \begin{bmatrix} 0 & 1 \\ 0 & 0 \end{bmatrix}, E_3 = \begin{bmatrix} 0 & 0 \\ 1 & 0 \end{bmatrix}, \text{ and } E_4 = \begin{bmatrix} 0 & 0 \\ 0 & 1 \end{bmatrix}.$$

Let us denote this basis by B. If A is given by

$$A = \begin{bmatrix} -1 & 2 \\ 3 & 0 \end{bmatrix},$$

then,

$$[A]_B = \begin{bmatrix} -1 \\ 2 \\ 3 \\ 0 \end{bmatrix}.$$

(c) The standard (ordered) basis for $P(n)$ is $\{1,x,x^2,\ldots,x^n\} = B$, say. Then (with $n = 3$), in $P(3)$, the vector f with $f(x) = 2x^3 + 3x^2 - 2x + 4$ $(x \in R)$ has matrix

$$[f]_B = \begin{bmatrix} 4 \\ -2 \\ 3 \\ 2 \end{bmatrix}.$$

(d) Another ordered basis for $P(3)$ is $\{x^3,x^2,x,1\} = B'$. Then, for the polynomial function f from (c) above,

$$[f]_{B'} = \begin{bmatrix} 2 \\ 3 \\ -2 \\ 4 \end{bmatrix}.$$

(e) Still another ordered basis for $P(3)$ is $B'' = \{x + 1, x - 1, x^2, x^3\}$. Since

$$x = (1/2)(x + 1) + (1/2)(x - 1)$$

and

$$1 = (1/2)(x + 1) - (1/2)(x - 1),$$

we can rewrite f (defined in (c) above) as

$$\begin{aligned} f(x) &= 2x^3 + 3x^2 - 2[(1/2)(x + 1) + (1/2)(x - 1)] + 4[(1/2)(x + 1) - (1/2)(x - 1)] \\ &= 2x^3 + 3x^2 + (x + 1) - 3(x - 1). \end{aligned}$$

So (watching the order of the elements in B''),

$$[f]_{B''} = \begin{bmatrix} 1 \\ -3 \\ 3 \\ 2 \end{bmatrix}. \square$$

THEOREM 4-12

Let V be an n dimensional space with basis B. Then a set of k vectors $\{\mathbf{v}_1,\mathbf{v}_2,\ldots,\mathbf{v}_k\}$ ($k \leq n$) is linearly independent in V iff $\{[\mathbf{v}_1]_B,[\mathbf{v}_2]_B,\ldots,[\mathbf{v}_k]_B\}$ is a linearly independent set in $M(n \times 1)$.

PROOF The proof is a particular case of a more general result that we will discuss in Chapter 5, so we will leave it until then. However, it is worth noting now that the operation of taking the coordinate matrix of a vector $\mathbf{v}$ in V establishes a one-to-one, onto **correspondence** (see Chapter 1) $V \leftrightarrow M(n \times 1)$ with $\mathbf{v} \leftrightarrow [\mathbf{v}]_B$, such that

(i) linear combinations of vectors in V correspond to the *same* linear combinations of the matrices of those vectors in $M(n \times 1)$;

and

(ii) linearly independent sets of vectors in V correspond to linearly independent sets of matrices in $M(n \times 1)$, namely, the matrices of the vectors with respect to the basis B.

This is another instance of **isomorphism** (see the remarks at the end of Example 4-2). ∎

EXAMPLE 4-29

Find a basis for the subspace W of $P(2)$ generated by p, q, and r where $p(x) = x - 3x^2$, $q(x) = -2 + x - 3x^2$, and $r(x) = 4 + x - 3x^2$, $x \in R$.

Solution Let B be the standard (ordered) basis for $P(2)$, i.e., $B = \{1,x,x^2\}$. Then,

$$[p]_B = \begin{bmatrix} 0 \\ 1 \\ -3 \end{bmatrix}, [q]_B = \begin{bmatrix} -2 \\ 1 \\ -3 \end{bmatrix}, [r]_B = \begin{bmatrix} 4 \\ 1 \\ -3 \end{bmatrix}.$$

We construct the 3×3 matrix U with these triples as *rows* (because we are identifying the matrices of the vectors p, q, and r with the corresponding triples in R^3):

$$U = \begin{bmatrix} 0 & 1 & -3 \\ -2 & 1 & -3 \\ 4 & 1 & -3 \end{bmatrix}.$$

By Gauss-Jordan reduction, we see that this matrix is row-equivalent to the row-reduced echelon matrix U':

$$U' = \begin{bmatrix} 1 & 0 & 0 \\ 0 & 1 & -3 \\ 0 & 0 & 0 \end{bmatrix}.$$

It follows that the polynomial functions with matrices equal to the (transposes of the) *nonzero* rows are linearly independent and constitute a basis for the subspace W. In other words, if we put

$$g(x) = 1,\ h(x) = x - 3x^2,$$

then $\{g,h\}$ is a basis for $W = span\ \{p,q,r\}$. In particular, $dim\ W = 2$, so, in fact, any two of p, q, and r would also constitute a basis. ❑

We remark that we could, alternatively, arrange the matrices of the vectors as *columns* of a matrix U and apply **elementary column operations** to change U to a matrix in **column reduced echelon form**. We leave the formulation of the appropriate definitions and procedure to the interested reader. In this text, we will use row operations exclusively.

EXAMPLE 4-30

We construct a basis for $M(2 \times 2)$ which contains the matrices

$$A_1 = \begin{bmatrix} -1 & -1 \\ 2 & 1 \end{bmatrix}, A_2 = \begin{bmatrix} 2 & 3 \\ 1 & 1 \end{bmatrix}.$$

As $dim\ M(2 \times 2) = 4$, we need two additional vectors (matrices) A_3, A_4 such that $\{A_1,A_2,A_3,A_4\}$ is linearly independent. Let $B = \{E_1,E_2,E_3,E_4\}$ be the standard basis for $M(2 \times 2)$. Then,

$$[A_1]_B = \begin{bmatrix} -1 \\ -1 \\ 2 \\ 1 \end{bmatrix}, [A_2]_B = \begin{bmatrix} 2 \\ 3 \\ 1 \\ 1 \end{bmatrix} \text{ and } [A_3]_B, [A_4]_B = \begin{bmatrix} * \\ * \\ * \\ * \end{bmatrix}$$

where the *'s indicate numbers to be found so that the A's are linearly independent. We construct the usual matrix with these column vectors (i.e., 4-tuples) as rows:

$$U = \begin{bmatrix} -1 & -1 & 2 & 1 \\ 2 & 3 & 1 & 1 \\ * & * & * & * \\ * & * & * & * \end{bmatrix}.$$

Now we want to choose the *'s so that U has rank $= 4 =$ the dimension of $M(2\times 2)$. Applying Gauss-Jordan reduction to the first two rows of this matrix yields

$$U' = \begin{bmatrix} 1 & 0 & -7 & -4 \\ 0 & 1 & 5 & 3 \\ * & * & * & * \\ * & * & * & * \end{bmatrix}.$$

We see that U' will have rank 4 if we choose the *'s so that (for example)

$$U' = \begin{bmatrix} 1 & 0 & -7 & -4 \\ 0 & 1 & 5 & 3 \\ 0 & 0 & 1 & 0 \\ 0 & 0 & 0 & 1 \end{bmatrix}.$$

(Other choices for the last two rows are possible.) In other words, one possible choice for $[A_3]_B$ and $[A_4]_B$ is

$$[A_3]_B = \begin{bmatrix} 0 \\ 0 \\ 1 \\ 0 \end{bmatrix}, [A_4]_B = \begin{bmatrix} 0 \\ 0 \\ 0 \\ 1 \end{bmatrix}.$$

Then,

$$A_3 = \begin{bmatrix} 0 & 0 \\ 1 & 0 \end{bmatrix}, A_4 = \begin{bmatrix} 0 & 0 \\ 0 & 1 \end{bmatrix}.$$

The reader should examine other possible choices. ❑

We conclude this section with some important applications of the concept of rank of a matrix.

THEOREM 4-13

An $n\times n$ matrix is nonsingular iff its rank is n.

PROOF We saw in Theorem 2-11 that an $n \times n$ matrix A is nonsingular iff it is row-equivalent to the $n \times n$ identity matrix I. Consequently, if A is nonsingular, $rank\ A = rank\ I = n$, since the rows of I are the standard basis vectors for R^n.

Conversely, if $rank\ A = n$, A is row-equivalent to an $n \times n$ row-reduced echelon matrix A' with n linearly independent rows. A little reflection should convince the reader that A' can only be I. Thus, A is row-equivalent to I, and is therefore nonsingular. ■

COROLLARY

If A is an $n \times n$ matrix, the following conditions are equivalent:

(1) A is nonsingular.

(2) $rank\ A = n$.

(3) The system of homogeneous equations $AX = O$ has only the trivial solution.

(4) The system of equations $AX = B$ has an unique solution for any $n \times 1$ matrix B.

PROOF This is simply a list of results proved earlier. ■

THEOREM 4-14

If A is an $m \times n$ matrix of rank r, then the solution space of the system of homogeneous equations $AX = O$ has dimension $n - r$.

PROOF Suppose A is row-equivalent to the row-reduced echelon matrix A'. Then we know that the systems of equations $AX = O$ and $A'X = O$ have the same solutions and therefore their solution spaces are the same and have the same dimensions. Now A', being in row-reduced echelon form, has the special form set out in equation 4-11; in particular, there are exactly r nonzero rows, and thus r leading entries of "1", which, in equation 4-11, we placed in columns $j_1, j_2, ..., j_r$, where $1 \leq j_1 < j_2 < ... < j_r \leq n$. This means that in the system $A'X = O$, only r of the variables, namely, $x_{j_1}, x_{j_2}, ..., x_{j_r}$ are determined and the remaining $n - r$ can be assigned arbitrary values. For each assignment of values to those $n - r$ variables, we compute the values of the r variables $x_{j_1}, x_{j_2}, ..., x_{j_r}$ to get the corresponding solution, an n-tuple. The important point is that we can assign values to the $n - r$ undetermined variables systematically so as to get $n - r$ linearly independent solutions for the system $A'X = O$ (and thus also for $AX = O$). ■

EXAMPLE 4-31

Consider the system

$$\begin{aligned} x_1 \qquad + x_3 \qquad + x_5 &= 0 \\ x_2 \qquad - x_4 \qquad &= 0 \\ x_1 + x_2 + x_3 - x_4 + x_5 &= 0 \end{aligned} \tag{4-13}$$

of $m = 3$ equations in $n = 5$ variables. The coefficient matrix A can be reduced to A' in row-reduced echelon form as follows and has rank 2:

$$A = \begin{bmatrix} 1 & 0 & 1 & 0 & 1 \\ 0 & 1 & 0 & -1 & 0 \\ 1 & 1 & 1 & -1 & 1 \end{bmatrix} \rightarrow \begin{bmatrix} 1 & 0 & 1 & 0 & 1 \\ 0 & 1 & 0 & -1 & 0 \\ 0 & 0 & 0 & 0 & 0 \end{bmatrix} = A'.$$

The leading entries occur in columns $j_1 = 1$ and $j_2 = 2$. Thus the system $A'X = O$ determines x_1 and x_2:

$$\begin{aligned} x_1 &= -x_3 - x_5 \\ x_2 &= x_4 \end{aligned} \tag{4-14}$$

and the remaining $n - r = 5 - 2 = 3$ variables, x_3, x_4, and x_5, can be assigned arbitrary values, which are then substituted into equation 4-14 to compute x_1 and x_2. By our earlier remarks, we choose $n - r = 5 - 2 = 3$ sets of values for x_3, x_4, and x_5 to get three solutions (5-tuples) that are linearly independent. This is conveniently done in a table as shown below:

Solution	*Values Assigned*			*Values Computed*	
	x_3	x_4	x_5	x_1	x_2
$\mathbf{u}_1$	1	0	0	−1	0
$\mathbf{u}_2$	0	1	0	0	1
$\mathbf{u}_3$	0	0	1	−1	0

Notice that the three sets of values assigned were chosen to guarantee that the resulting solutions $\mathbf{u}_1$, $\mathbf{u}_2$, and $\mathbf{u}_3$ would be linearly independent because of the relative positions of the 1's and 0's. Thus, $\mathbf{u}_1 = (-1,0,1,0,0)$, $\mathbf{u}_2 = (0,1,0,1,0)$, and $\mathbf{u}_3 = (-1,0,0,0,1)$ are linearly independent and constitute a basis for the solution space, which has dimension 3. ❑

EXAMPLE 4-32

Consider the system

$$\begin{aligned} x + 2y + 5z &= 0 \\ 3x - \ y + \ z &= 0 \\ 2x + \ y + 4z &= 0. \end{aligned}$$

The coefficient matrix is

$$A = \begin{bmatrix} 1 & 2 & 5 \\ 3 & -1 & 1 \\ 2 & 1 & 4 \end{bmatrix}.$$

By Gauss-Jordan reduction, this changes to the row-reduced echelon matrix

$$A' = \begin{bmatrix} 1 & 0 & 1 \\ 0 & 1 & 2 \\ 0 & 0 & 0 \end{bmatrix}.$$

Here $m = n = 3$ and $rank\ A' = r = 2$. Thus, we expect the dimension of the solution space to be $3 - 2 = 1$. The corresponding system $A'X = O$ is

$$\begin{aligned} x &= -z \\ y &= -2z, \end{aligned}$$

so the undetermined variable is z. For example, putting $z = 1$, we compute $x = -1, y = -2$. Thus, $\{(-1,-2,1)\}$ is a basis for the solution space. Of course, any nonzero value for z would do: $z = -1$ yields the basis $\{(1,2,-1)\}$, and so on. ❑

EXAMPLE 4-33

Find the dimension of the subspace W of R^5 defined by

$$W = \{(a,b,c,d,e) \mid 2a - e = b - 4c + e = b + d = 0\}.$$

Solution We first recognize that W is the solution space of the homogeneous system

$$\begin{aligned} 2a \qquad\qquad\qquad - e &= 0 \\ b - 4c \qquad + e &= 0 \\ b \qquad + d \qquad &= 0. \end{aligned}$$

The coefficient matrix is

$$\begin{bmatrix} 2 & 0 & 0 & 0 & -1 \\ 0 & 1 & -4 & 0 & 1 \\ 0 & 1 & 0 & 1 & 0 \end{bmatrix}$$

which is transformed to the row-reduced echelon matrix

$$\begin{bmatrix} 1 & 0 & 0 & 0 & -1/2 \\ 0 & 1 & 0 & 1 & 0 \\ 0 & 0 & 1 & 1/4 & -1/4 \end{bmatrix}$$

having rank 3. Since here $n = 5$, $dim\ W = 5 - 3 = 2$. ❑

COROLLARY

If A is an $m \times n$ matrix and $m < n$, then the homogeneous system $AX = O$ has a nontrivial solution.

PROOF By Theorem 4-14, the dimension of the solution space is $n - r$, where $r = rank\ A$. Now, as we saw earlier, r cannot be greater than the lesser of m and n, i.e. $r < m < n$. Thus, $n - r > 0$. Therefore the solution space of the system has positive dimension, and consequently contains a nonzero vector (a solution), proving the result. ■

The final theorem below shows how we may use the preceding results for homogeneous systems to write explicitly *all* solutions of a **nonhomogeneous** system $AX = B$.

THEOREM 4-15

Let $AX = B$ be a system of m linear equations in n unknowns having a solution X'. Suppose that $rank\ A = r$. Then, *every* solution X may be written as

$$X = X' + \sum_{i=1}^{n-r} c_i H_i,$$

where c_i are constants and $\{H_1, H_2, \ldots, H_{n-r}\}$ is a basis for the solution space of $AX = O$.

PROOF The proof is outlined in question 16, Exercises 4-4. The postulated solution X' which must first be found is called a **particular solution** of the system. The homogeneous system with the (same) coefficient matrix A is called the **associated homogeneous system** of $AX = B$. ■

EXAMPLE 4-34

We consider the system $AX = B$ having the coefficient matrix of the system $AX = O$ solved in Example 4-31, and B given by

$$B = \begin{bmatrix} 1 \\ 2 \\ 3 \end{bmatrix}.$$

The reader may check that an example of a particular solution is given by $X' = (1,2,0,0,0)$. A basis for the associated homogeneous system was found (in Example 4-31) to be $\{\mathbf{u}_1,\mathbf{u}_2,\mathbf{u}_3\}$. Therefore, by Theorem 4-15, *every* solution $X = (x_1,x_2,x_3,x_4,x_5)$ of the system $AX = B$ may be expressed as (using $H_i = \mathbf{u}_i$)

$$\begin{aligned} X &= X' + c_1H_1 + c_2H_2 + c_3H_3 \\ &= (1,2,0,0,0) + c_1(-1,0,1,0,0) + c_2(0,1,0,1,0) + c_3(-1,0,0,0,1) \\ &= (1 - c_1 - c_3, 2 + c_2, c_1, c_2, c_3),\ c_i \in R, i = 1, 2, 3. \ \square \end{aligned}$$

EXERCISES 4-4

1. **(a)** Show that the matrices A and B below have the same row space.

$$A = \begin{bmatrix} 1 & 1 & 0 & 1 \\ 2 & 1 & 0 & 0 \\ 1 & 0 & 1 & 0 \end{bmatrix}, B = \begin{bmatrix} -2 & 1 & 1 & 5 \\ 3 & 2 & 3 & 4 \\ 1 & 2 & -1 & 2 \end{bmatrix}$$

(b) Find bases for the row spaces of A and B above.

2. **(a)** Find bases for the column spaces of the following matrices.

$$\textbf{(i)}\ A = \begin{bmatrix} 2 & -1 \\ 3 & 2 \end{bmatrix} \quad \textbf{(ii)}\ B = \begin{bmatrix} 1 & -1 & 2 & 4 \\ 2 & 0 & 1 & 3 \\ -1 & 2 & 0 & 1 \end{bmatrix}$$

(b) What are the ranks of A and B above?

3. Find the ranks of the following matrices using Gauss-Jordan reduction to row-reduced echelon form.

$$\textbf{(a)} \begin{bmatrix} -1 & 2 \\ 3 & -1 \\ 1 & 3 \end{bmatrix} \qquad \textbf{(b)} \begin{bmatrix} 2 & -1 & 0 \\ -1 & 0 & 1 \\ 0 & 1 & 1 \end{bmatrix}$$

$$\textbf{(c)} \begin{bmatrix} 2 & 0 & -1 & 3 \\ 4 & -1 & 0 & 2 \\ 3 & 3 & -1 & 2 \\ 3 & -4 & 0 & 3 \end{bmatrix} \qquad \textbf{(d)} \begin{bmatrix} 4 & -1 & 2 & 0 \\ 0 & -1 & -2 & 0 \\ 0 & 2 & 3 & -1 \\ 4 & 5 & -2 & 1 \\ 0 & -4 & 7 & 2 \end{bmatrix}$$

$$\textbf{(e)} \begin{bmatrix} 1 & -2 & 0 & 3 & 5 & -2 \\ 2 & -2 & 1 & -4 & 5 & 7 \\ 1 & 0 & 0 & 2 & 0 & 3 \end{bmatrix}$$

4. Prove that if a matrix is not square, then either its rows or its columns must be linearly dependent.

In questions 5 to 7 below, use Gauss-Jordan reduction of an appropriate matrix to find a basis for the vector spaces V indicated.

5. V is the subspace of R^3 generated by $\mathbf{u} = (-1,2,0)$, $\mathbf{v} = (1,1,2)$, $\mathbf{w} = (5,-1,6)$, and $\mathbf{z} = (5,-4,4)$.

6. V is the subspace of $P(2)$ generated by the functions $f(x) = 1 + 2x + x^2$, $g(x) = 1 - 2x + x^2$, and $h(x) = x^2$, $x \in R$.

7. V is the subspace of $M(2\times 2)$ spanned by the matrices

$$A = \begin{bmatrix} -1 & 1 \\ 1 & 0 \end{bmatrix}, B = \begin{bmatrix} 3 & 0 \\ 2 & 1 \end{bmatrix}, C = \begin{bmatrix} -2 & 1 \\ 1 & 2 \end{bmatrix},$$

$$D = \begin{bmatrix} 7 & -3 \\ -1 & -1 \end{bmatrix}, E = \begin{bmatrix} 7 & -1 \\ 3 & 2 \end{bmatrix}.$$

8. Use Gauss-Jordan reduction of an appropriate matrix to find whether or not each of the following vectors $\mathbf{u}$ is in the subspace of R^3 generated by $\mathbf{v} = (-1,2,0)$ and $\mathbf{w} = (1,1,2)$.

(a) $\mathbf{u} = (5,-4,4)$

(b) $\mathbf{u} = (1,1,1)$

(c) $\mathbf{u} = (1,23,-13)$

9. By Gauss-Jordan reduction of an appropriate matrix, determine whether or not each of the matrices D, E, and F below is in the subspace of $M(2\times 2)$ generated by A, B, and C.

$$A = \begin{bmatrix} -1 & 1 \\ 1 & 0 \end{bmatrix}, B = \begin{bmatrix} 3 & 0 \\ 2 & 1 \end{bmatrix}, C = \begin{bmatrix} -2 & 1 \\ 1 & 2 \end{bmatrix}$$

$$D = \begin{bmatrix} 0 & -2 \\ 0 & -1 \end{bmatrix}, E = \begin{bmatrix} -2 & 1 \\ 3 & 1 \end{bmatrix}, F = \begin{bmatrix} 2 & -2 \\ -1 & 1 \end{bmatrix}$$

10. By considering an appropriate row-reduced echelon matrix, find a basis for R^3 containing the vector $(-1,1,2)$.

11. Find a basis for $M(2\times 3)$ containing the matrices

$$A = \begin{bmatrix} -1 & 2 & 0 \\ 0 & 1 & 1 \end{bmatrix}, B = \begin{bmatrix} 0 & 1 & 0 \\ -1 & 2 & 1 \end{bmatrix}, \text{ and } C = \begin{bmatrix} 0 & 0 & 1 \\ -1 & 1 & 2 \end{bmatrix}.$$

12. Find bases for the solution spaces of the systems of linear equations below. What are their dimensions?

(a)
$$\begin{aligned} x_1 + 2x_2 - x_3 &= 0 \\ 2x_1 \qquad - 2x_3 &= 0 \end{aligned}$$

(b)
$$\begin{aligned} x + 3y + 2z &= 0 \\ x + 5y + z &= 0 \\ 3x + 5y + 8z &= 0 \end{aligned}$$

(c)
$$\begin{aligned} 2x_1 + x_2 &= 0 \\ 5x_1 + 3x_2 &= 0 \end{aligned}$$

(d)
$$\begin{aligned} a + 2b - c + 3d - 4e &= 0 \\ 2a + 4b - 2c - d + 5e &= 0 \\ 2a + 4b - 2c + 4d - 2e &= 0 \end{aligned}$$

(e)
$$\begin{aligned} x + y - z \qquad &= 0 \\ x - y + z \qquad &= 0 \\ 2x \qquad - z - w &= 0 \\ x - y \qquad - w &= 0 \end{aligned}$$

13. Find bases for the subspaces of R^4 below. Also compute their dimensions.

(a) $W = \{(a,b,c,d) \mid a = b \text{ and } c + d = 0\}$

(b) $W' = \{(a,b,c,d) \mid a + b + c + d = 0\}$

14. Show that the (only) subspaces of R^3 are **(a)** the origin; **(b)** lines through the origin; **(c)** planes through the origin; and **(d)** R^3 itself.

15. Show that the system $AX = B$ of m equations in n unknowns has a solution iff *rank* A = *rank* $(A{:}B)$. (Hint: Show there is an X such that $AX = B$ iff B is in the column space of A.)

16. This question outlines a proof of Theorem 4-15. Let $AX = B$ be a system of m equations in n unknowns which has a solution X'.

(a) Show that $X' + H$ is also a solution for any solution H of the associated homogeneous system $AX = O$.

(b) Show that *every* solution of $AX = B$ is of the form $X' + H$. (Hint: If X is a solution of $AX = B$, show $X - X'$ is a solution of $AX = O$.)

(c) Suppose *rank* $A = r$. Show that every solution X of $AX = B$ can be written as

$$X = X' + \sum_{i=1}^{n-r} c_i H_i \, ,$$

where c_i are constants and $\{H_1, H_2, \ldots, H_{n-r}\}$ is a basis for the solution space of $AX = O$.

17. For each system $AX = B$ given in (a) and (b) below, use Theorem 4-15 to find all solutions of the system as follows:

(i) First find a particular solution X' of $AX = B$.

(ii) Next, find a basis for the solution space of $AX = O$.

(a)
$$\begin{aligned} x_1 + 2x_2 + x_3 &= 1 \\ 2x_1 + x_2 + 8x_3 &= 3 \end{aligned}$$

(b)
$$\begin{aligned} x_1 + 2x_2 + x_3 - 3x_4 - x_5 &= 6 \\ 5x_2 + x_3 - 2x_4 + 5x_5 &= 3 \\ x_1 + 2x_2 + x_3 - 2x_4 &= 5 \end{aligned}$$

18. **(a)** Show that the matrix A below has rank 1 by transforming it to a row-reduced echelon matrix A'.

$$A = \begin{bmatrix} 2 & -1 \\ -6 & 3 \\ 4 & -2 \end{bmatrix}$$

(b) Using the form of A', show that A' and also A can be expressed as a product of a column vector and a row vector.

(c) Generalize the observation of (b) to any $m \times n$ matrix of rank 1.

19. If W is a subspace of V (finite dimensional), show $W = V$ iff $dim\ W = dim\ V$.

20. Prove that if A is a row echelon matrix, the nonzero rows are a basis for $row(A)$.

4-5 Electrical Networks (Optional)

We examine an application of the ideas developed in the previous section. This material is presented solely for the stimulation and motivation of the interested reader, and is not required reading for other parts of the text.

An electrical network is a physical system consisting of various kinds of "network elements" such as voltage sources (e.g., batteries or generators), resistors and capacitors, to name but a few. These are generally connected together by means of wires or other conductors of electricity. Of course, radios, television sets, and computers are examples of complex electrical networks. The behaviour of an electrical network is characterized in terms of the **electrical current** i through each network element and the **voltage** v across each element. The current i is a measure of the actual number of electrons flowing through the element, while the voltage v across the element is, roughly speaking, a measure of the force driving the electrons through the element. Current and voltage are measured in units called amperes and volts respectively, and by instruments called ammeters and voltmeters respectively.

Network elements are characterized by the relationship between the voltage across them and the current through them. For example, the simplest element is a "resistor", for which the relationship takes the form $v = Ri$, where R is a constant called the **resistance** of the resistor. This relationship is known as **Ohm's Law** after its discoverer, Georg Simon Ohm (1787–1845), a German high-school teacher and experimenter. For other types of elements, the relationship generally involves calculus, so we restrict our attention to networks involving only resistors and batteries. Two such networks are shown in Figure 4-6. In the figure, the R_i indicate resistors, while each circle labelled "v" is a battery of voltage v volts. They are connected by wires represented

FIGURE 4-6

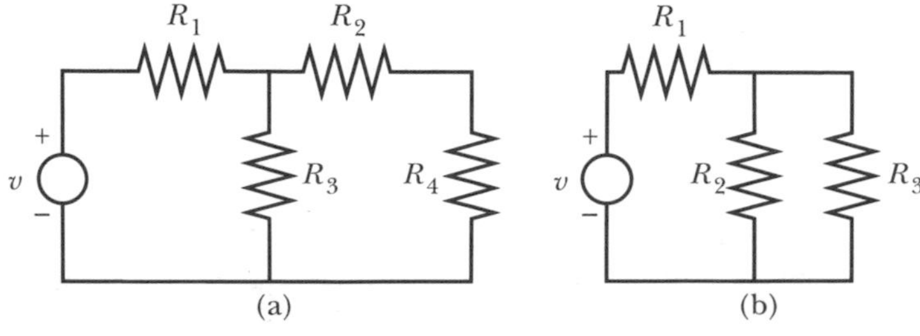

v = voltage source, R = resistor

by line segments in the figure. Of course, the battery drives electrons through the circuit. Such circuits, though simple, are often found as part of larger networks.

Electrical networks such as these may be analysed by regarding them as graphs (section 2-3). When this is done, the conventional representations are simplified by omitting the straight line segments ("wires") connecting the elements to produce equivalent graphical representations. Thus, for example, the network of Figure 4-6(a) yields a graph with 4 vertices and 5 edges as shown in Figure 4-7(a).

FIGURE 4-7

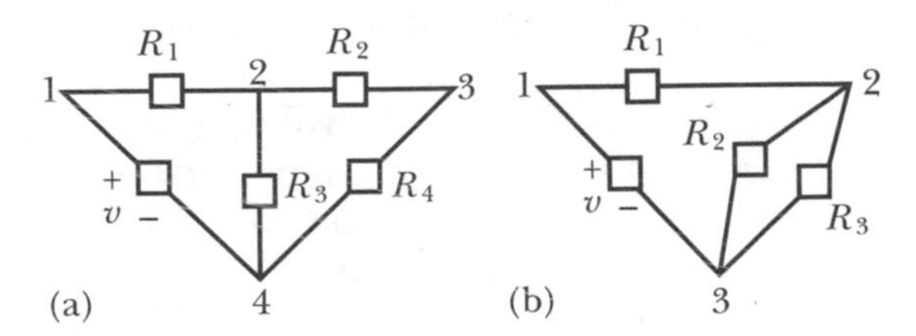

In addition to Ohm's Law, the equations that describe the behaviour of an electrical network such as the ones in the figures are determined by intuitively plausible physical principles known as **Kirchhoff's Laws**, after Gustav Robert Kirchhoff (1824–1887), an eminent German physicist. We illustrate these principles using the network of Figure 4-6(a). The graphical representation in Figure 4-7(a) has been redrawn in Figure 4-8 with a simplified notation for the analysis in terms of graphs.

FIGURE 4-8

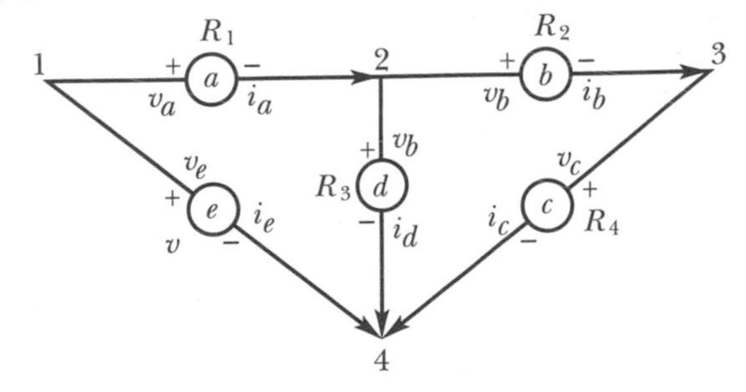

First, Kirchhoff's **current law** states that the sum of the currents flowing into any vertex must equal the sum of the currents flowing out. To implement this, we arbitrarily assign directions (arrows) of current flow through each element, thereby making the network into a **directed graph**. We arbitrarily adopt the convention that currents flowing into a vertex are given negative signs (respectively, positive signs for those flowing out). This does not affect the system of equations as a whole.

Let i_z denote the current flow through element z. The **current equations** for this network are

$$\begin{array}{llcl} \text{vertex 1:} & i_a & + i_e & = 0 \\ \text{vertex 2:} & -i_a + i_b \quad + i_d & & = 0 \\ \text{vertex 3:} & \quad - i_b + i_c & & = 0 \\ \text{vertex 4:} & \qquad - i_c - i_d - i_e & & = 0. \end{array}$$

We leave as an exercise for the reader the verification that the matrix form for these equations is $AJ = O$, where A is the 4×5 incidence matrix of the directed graph and J is the 5×1 matrix of currents in the order labelled. Notice that the last equation is the sum of the first three; we say the equations are *not independent*. In fact, the rank of the incidence matrix A is 3, which is one less than the number of vertices. This observation is pursued in the exercises.

The second physical principle is Kirchhoff's **voltage law**, which states that the algebraic sum of the voltages around any **loop** must be zero. A loop (or circuit) is a path (i.e., a sequence of contiguous edges) that begins and ends at the same vertex, and doesn't pass through any vertex more than once. For example, in Figure 4-8, there are three loops, namely, *ade*, *bcd*, and *abce*.

To implement the voltage law, we adopt the convention that current flows from (+) to (−) across each element. If this coincides with the direction of the loop, the voltage across the given element is taken to be positive, and negative if otherwise. Let v_z denote the voltage across element z. Then the **voltage equations** are (with $v_e = v =$ the voltage supplied by the battery)

$$\begin{array}{ll} \text{loop } ade: & v_a \qquad\qquad + v_d - v_e = 0 \\ \text{loop } bcd: & \qquad v_b + v_c - v_d \qquad = 0 \\ \text{loop } abce: & v_a + v_b + v_c \qquad - v_e = 0. \end{array}$$

The coefficient matrix B of this system of equations is called a **circuit matrix** of the directed graph. In general, Kirchhoff's voltage equations for a network can be written $BV = O$, where B is a circuit matrix for the graph of the network and V is the column vector of voltages (it has entries v_a, v_b, v_c, v_d, v_e in the example above). The definition of B and some of its properties are discussed in the exercises. In particular, its rank is seen to be equal to the number of edges less the number of vertices, plus one.

To conclude, we apply these ideas to an analysis of the circuit in Figure 4-8, where the elements have values $v = v_e = 10$ volts, and $R_1 = 200$, $R_2 = 200$, $R_3 = 400$, $R_4 = 600$ ohms. Ohm's law for the resistors yields the equations:

$$\begin{aligned} R_1 &: v_a = 200i_a \\ R_2 &: v_b = 200i_b \\ R_3 &: v_d = 400i_d \\ R_4 &: v_c = 600i_c \end{aligned}$$

where voltages are in volts and currents in amperes. We substitute into the first two voltage equations to get

$$\begin{aligned} 200i_a \qquad\qquad\qquad + 400i_d - 10 &= 0 \\ 200i_b + 600i_c - 400i_d \qquad &= 0. \end{aligned}$$

Together with the first three current equations, namely,

$$\begin{aligned} i_a \qquad\qquad\quad + i_e &= 0 \\ -i_a + i_b \qquad + i_d \qquad &= 0 \\ - i_b + i_c \qquad\qquad &= 0, \end{aligned}$$

we now have five linear equations in i_a, i_b, i_c, i_d, i_e with augmented matrix

$$\begin{bmatrix} 200 & 0 & 0 & 400 & 0 & 10 \\ 0 & 200 & 600 & -400 & 0 & 0 \\ 1 & 0 & 0 & 0 & 1 & 0 \\ -1 & 1 & 0 & 1 & 0 & 0 \\ 0 & -1 & 1 & 0 & 0 & 0 \end{bmatrix}.$$

The reader should verify that this is row-equivalent to

$$\begin{bmatrix} 1 & 0 & 0 & 0 & 0 & 3/140 \\ 0 & 1 & 0 & 0 & 0 & 1/140 \\ 0 & 0 & 1 & 0 & 0 & 1/140 \\ 0 & 0 & 0 & 1 & 0 & 1/70 \\ 0 & 0 & 0 & 0 & 1 & -3/140 \end{bmatrix}.$$

Thus, the unique solution is $i_a = 3/140$, $i_b = 1/140$, $i_c = 1/140$, $i_d = 1/70$, and $i_e = -3/140$ (amperes). In addition, for example, the voltage across resistor R_4 is $v_c = 600i_c = 600/140 \approx 4.29$ volts.

EXERCISES 4-5

1. A graph is said to be *connected* if any two vertices i and i' can be joined by a path, which is a sequence of contiguous edges joining i and i'. For example, in Figure 2-5(b), *abd*, *ae*, *f*, and *cdeaf* are all paths from vertex 1 to vertex 4; but *acd* is not. It can be shown that the incidence matrix of a connected digraph with v vertices has rank $v - 1$. (Note: Here v denotes the *number* of vertices, not a voltage.)

   (a) Determine which of the following incidence matrices can correspond to connected digraphs.

   (i) $\begin{bmatrix} -1 & 0 & -1 \\ 1 & 1 & 0 \\ 0 & -1 & 1 \end{bmatrix}$ **(ii)** $\begin{bmatrix} -1 & 0 \\ 0 & 1 \\ 1 & 0 \\ 0 & -1 \end{bmatrix}$ **(iii)** $\begin{bmatrix} -1 & 0 & 0 & 1 \\ 1 & -1 & 0 & 0 \\ 0 & 1 & -1 & 0 \\ 0 & 0 & 1 & -1 \end{bmatrix}$

(iv) $$\begin{bmatrix} -1 & -1 & 0 & 0 & 0 & 0 \\ 0 & 1 & -1 & 0 & 0 & 0 \\ 0 & 0 & 0 & -1 & 1 & 0 \\ 1 & 0 & 1 & 0 & 0 & 0 \\ 0 & 0 & 0 & 1 & 0 & 1 \\ 0 & 0 & 0 & 0 & -1 & -1 \end{bmatrix}$$

(b) Draw the digraphs having the above incidence matrices and verify your answers to (a).

2. A *circuit matrix B* for a connected digraph is an $l \times e$ matrix, where l is the number of loops in the graph and e is the number of edges. It is defined by

$$B_{ij} = \begin{cases} 1, & \text{if edge } j \text{ is in loop } i \text{ and has the direction of the loop;} \\ -1, & \text{if edge } j \text{ is in loop } i \text{ and has the opposite direction;} \\ 0, & \text{if edge } j \text{ is not in loop } i. \end{cases}$$

For example, as noted in the preceding section, a circuit matrix for the digraph in Figure 4-8 is

$$B = \begin{array}{c} \\ \textit{loop} \\ ade \\ bcd \\ abce \end{array} \begin{array}{c} \textit{edge} \\ \begin{array}{ccccc} a & b & c & d & e \end{array} \\ \begin{bmatrix} 1 & 0 & 0 & 1 & -1 \\ 0 & 1 & 1 & -1 & 0 \\ 1 & 1 & 1 & 0 & -1 \end{bmatrix} \end{array}.$$

Write the circuit matrices for the graphs in Figure 4-9.

FIGURE 4-9

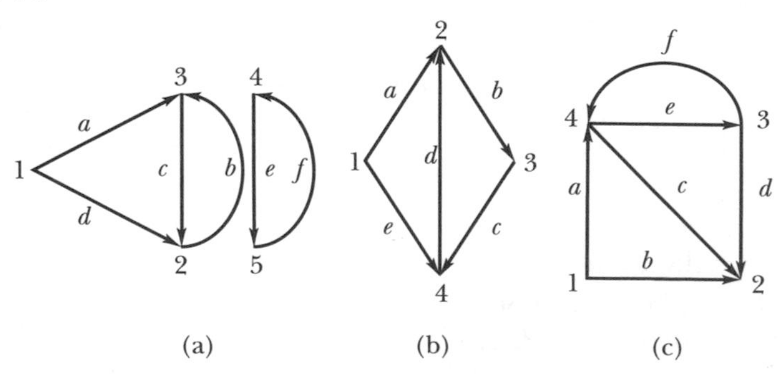

3. It can be shown that the rank of a circuit matrix for a connected digraph is $e - v + 1$, where e is the number of edges and v is the number of vertices (see the note in question 1). Verify this for the graphs in Figure 4-9(b) and (c). What is the rank of the circuit matrix for Figure 4-9(a)?

4. It can be shown that if A is an incidence matrix and B is a circuit matrix, then $AB^t = O$ and $BA^t = O$, provided the edges are enumerated in the same order for both matrices. Verify this fact for the graphs in Figure 4-9.

5. **(a)** Explain why, in an electrical network having a connected digraph with e edges and v vertices, exactly $e - v + 1$ of Kirchhoff's voltage equations are *independent*: no smaller number will yield all of them by forming linear combinations.

(b) Verify the property stated in (a) above for the network in Figure 4-8.

6. The circuit in Figure 4-10(a) consists of a voltage source and two resistors connected "in series".

FIGURE 4-10

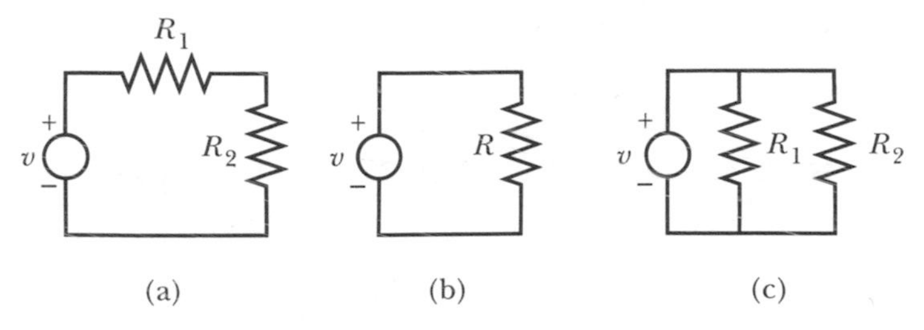

(a) Represent this circuit by a digraph with appropriate notation, as was done in the text.

(b) Write Kirchhoff's current and voltage equations from the digraph obtained in (a) above.

(c) Use Ohm's Law for the resistors to obtain an equation for the current through the voltage source.

(d) Do the same for the network in Figure 4-10(b) and show that the current through the voltage source will be the same as that in Figure 4-10(a) if $R = R_1 + R_2$. (When this is the case, networks (a) and (b) are said to be "equivalent".)

(e) Compute the current through the voltage source when $v = 10$ volts, $R_1 = 200$ ohms, and $R_2 = 300$ ohms.

7. The circuit in Figure 4-10(c) consists of a voltage source and two resistors connected "in parallel". Repeat the instructions of question 6 (a),(b),(c), and (e) for this network. Also show that the current through the voltage source will be the same as that for the network in Figure 4-10(b) when $1/R = 1/R_1 + 1/R_2$.

8. The circuit in Figure 4-11 is called a *Wheatstone's bridge*, after Sir Charles Wheatstone (1802–1875), an English physicist and inventor. This device is used for the accurate measurement of resistance.

FIGURE 4-11

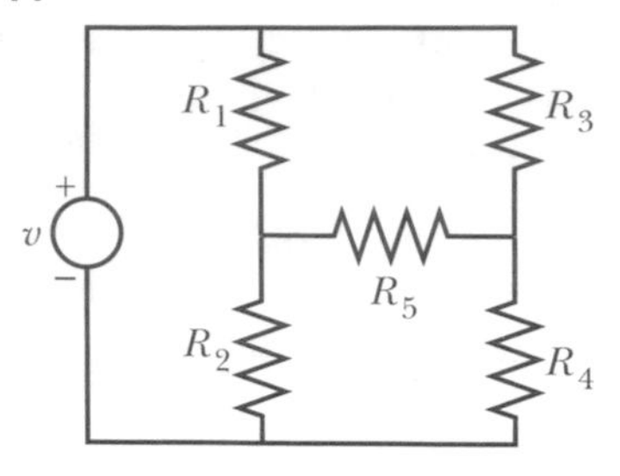

(a) Draw a graph for this circuit.

(b) The bridge is said to be "balanced" when the voltage across R_5 is zero. Use Ohm's Law and Kirchhoff's equations to show that this happens when $R_1R_4 = R_2R_3$. Note that this allows us to compute R_1, given the values of R_2, R_3, and R_4.

4-6 Complex Vector Spaces (Optional)

As noted earlier in the chapter pursuant to the definition of a real vector space (Definition 4-3), it is both possible and useful to generalize the theory of vector spaces to allow multiplication by scalars other than real numbers. In mathematics and its applications, notably physics and engineering, complex numbers arise naturally and are indispensable in the solution and interpretation of many problems. For example, complex matrices are used routinely in calculations concerning electrical networks, while the study of quantum mechanics and atomic physics would be impossible without the ideas of complex vector spaces.

Since the axioms for a complex vector space are identical to those for a real vector space, except that scalar multiplication is by elements of C, the set of complex numbers (see Appendix A for the definition and basic properties), we may summarize the definition of a complex vector space as

DEFINITION 4-15

A **complex vector space** is a set V together with two operations called **addition** and **scalar multiplication** such that

(i) for any $\mathbf{u}, \mathbf{v} \in V$ and $z \in C$, there are new vectors $\mathbf{u} + \mathbf{v} \in V$ (the **sum** of $\mathbf{u}$ and $\mathbf{v}$) and $z\mathbf{u} \in V$ (the **scalar multiple** of $\mathbf{u}$ by z), and

(ii) the eight axioms of Definition 4-3 are true.

Note that $1\mathbf{u} = \mathbf{u}$ (axiom 8) continues to make sense since R is a subset of C, so $1 \in C$.

By simply substituting complex numbers for real ones, we obtain the following examples of complex vector spaces analogous to the earlier real examples.

EXAMPLE 4-35

The most important example of a complex vector space is C^n, the space of n-tuples of complex numbers:

$$C^n = \{(z_1, z_2, \ldots, z_n) \mid z_i \in C\}.$$

C^n is also called **complex *n*-space**. Frequently it will be convenient to denote such n-tuples by $\mathbf{z} = (z_1, z_2, \ldots, z_n)$, as in the case of R^n, but now it will be understood that the components z_i are complex. The definitions of addition and scalar multiplication are analogous to those for R^n.

For example, in C^3, let

$$\mathbf{u} = (2 - i, 1 + 3i, -4 + 2i),\ \mathbf{v} = (-3 - 2i, 2, -5i),\ z = 2 + 2i.$$

Then (recalling that $i^2 = -1$),

$$\begin{aligned} z\mathbf{u} + \mathbf{v} &= (2 + 2i)(2 - i, 1 + 3i, -4 + 2i) + (-3 - 2i, 2, -5i) \\ &= (6 + 2i, -4 + 8i, -12 - 4i) + (-3 - 2i, 2, -5i) \\ &= (3, -2 + 8i, -12 - 9i). \end{aligned}$$

Note that $i = \sqrt{-1}$ should not be confused with the (boldface) standard basis vector $\mathbf{i} = (1,0,0)$ in R^3. ❑

EXAMPLE 4-36

The space of **complex $m \times n$ matrices** is usually denoted $M_{m\times n}(C)$, while the space of complex (square) $n \times n$ matrices is $M_n(C)$. When both real and complex matrices are under consideration, the spaces of real $m \times n$ and $n \times n$ matrices are usually denoted by $M_{m\times n}(R)$ and $M_n(R)$ respectively. The definitions of addition and scalar multiplication in $M_{m\times n}(C)$ are analogous to those for $M(m \times n)$.

For example, in $M_2(C)$, let

$$A = \begin{bmatrix} -i & 2 + i \\ -4 - 3i & 6 \end{bmatrix}, B = \begin{bmatrix} 7 - 3i & 4 + 4i \\ -2 - 2i & 3i \end{bmatrix}, z = -1 + 2i.$$

Then,

$$\begin{aligned} zA + B &= \begin{bmatrix} 2 + i & -4 + 3i \\ 10 - 5i & -6 + 12i \end{bmatrix} + \begin{bmatrix} 7 - 3i & 4 + 4i \\ -2 - 2i & 3i \end{bmatrix} \\ &= \begin{bmatrix} 9 - 2i & 7i \\ 8 - 7i & -6 + 15i \end{bmatrix}. \end{aligned}$$

Of course, we can also apply multiplication to complex matrices. The product of complex matrices A and B is formed by the same rules as for the real case, but AB is complex. For example, for A and B above, we get

$$AB = \begin{bmatrix} -i(7-3i) + (2+i)(-2-2i) & -i(4+4i) + (2+i)(3i) \\ (-4-3i)(7-3i) + 6(-2-2i) & (-4-3i)(4+4i) + 6(3i) \end{bmatrix}$$
$$= \begin{bmatrix} -5-13i & 1+2i \\ -49-21i & -4-10i \end{bmatrix}.$$

Inverses and determinants of square matrices are found as before, but using the rules of arithmetic for complex numbers. For example, readers may check that for the matrix A above, *det* $A = 5 + 4i$ and

$$A^{-1} = \begin{bmatrix} \frac{30}{41} - \frac{24}{41}i & -\frac{14}{41} + \frac{3}{41}i \\ \frac{32}{41} - \frac{1}{41}i & -\frac{4}{41} - \frac{5}{41}i \end{bmatrix}.$$

The same criteria for invertibility hold (e.g., *det* $A \neq 0$ and Theorem 4-13, Corollary). ❑

EXAMPLE 4-37

We may also generalize the space $F(S)$ of all functions $f:S \rightarrow R$, where S is any set. This was introduced at the beginning of the chapter. Let us denote by $F(S,C)$ the set of all functions $f:S \rightarrow C$. These are the **complex-valued functions** on the set S. This is a complex vector space for reasons analogous to those that make $F(S)$ a real vector space. For f and $g \in F(S,C)$, $z \in C$, and $x \in S$, the operations are defined by

$$(f+g)(x) = f(x) + g(x),$$
$$(zf)(x) = zf(x),$$

where it is important to note that $f(x)$, $g(x)$, and z are in C, so the arithmetic operations on the right-hand side of these equations are those in C.

Since every $z \in C$ can be expressed as the sum $z = a + bi$ (or, $a + ib$), $a, b \in R$, the values $f(x)$ of any function $f:S \rightarrow C$ can be decomposed as

$$f(x) = f_1(x) + if_2(x), x \in S, \tag{4-15}$$

where $f_1(x)$ and $f_2(x)$ are *real* numbers, the **real** and **imaginary parts** of $f(x)$. Notice that f_1 and f_2 are therefore functions in $F(S)$. Conversely, given two functions f_1 and $f_2 \in F(S)$, we can define a function $f \in F(S,C)$ by equation 4-15.

The spaces $F(S,C)$ yield many important vector spaces for various choices of the set S. For example, as in Example 4-2, we see that if $S = \{1,2\}$, $F(S,C)$ is essentially C^2, since each $f:S \to C$ corresponds to the unique ordered pair $(f(1),f(2)) \in C^2$, where $f(1)$ and $f(2) \in C$ are the values of f.

Two cases of $F(S,C)$ which are important for advanced mathematics and its applications are obtained by choosing the set S as follows:

(a) $S = R$: $F(R,C)$ is the space of *complex-valued* functions $f:R \to C$ of a real variable. As noted in equation 4-15, each $f \in F(R,C)$ can be written as $f(x) = f_1(x) + if_2(x)$, where now f_1 and f_2 are real functions of a real variable.

For example, if we let $f_1(x) = \cos x$ and $f_2(x) = \sin x$, then we obtain the important function (studied in calculus)

$$f(x) = \cos x + i \sin x = e^{ix},$$

where $e \approx 2.71828$ is the base of the natural logarithm.

(b) $S = C$: $F(C,C) = F(C)$ is the space of complex-valued functions $f:C \to C$ of a complex variable. ❑

Next, we note that the definitions and results for subspaces of complex vector spaces are identical to those for the real case, keeping in mind that scalars are now complex numbers.

EXAMPLE 4-38

(a) A particularly important subspace of $F(C)$ is the space $P_n(C)$ of polynomial functions $f:C \to C$ of degree of at most n, $n \geq 0$. Again, the definition is analogous to the real case: $f \in P_n(C)$ iff there are scalars $a_i \in C$ such that the (complex) values $f(z)$ of f can be written as

$$f(z) = a_0 + a_1z + a_2z^2 + \dots + a_nz^n = \sum_{i=0}^{n} a_iz^i, z \in C.$$

The fact that $P_n(C)$ is a subspace of $F(C)$ can be proved as in Example 4-4, or, using Theorem 4-3, which remains valid for complex vector spaces.

(b) As in the real case, an $n \times n$ complex matrix is called *symmetric* if $A^t = A$. These matrices form a subspace of $M_n(C)$ by Theorem 4-3. For, if A and $B \in M_n(C)$ and $z \in C$,

$$(zA + B)^t = zA^t + B^t = zA + B.$$

(c) Because of the complex conjugate operation in C, we can define a corresponding operation in $M_{m \times n}(C)$ by taking complex conjugates of *all* entries of a matrix. So, if A is $m \times n$, $\bar{A}$ is the $m \times n$ matrix whose entries are the complex conjugates of those of A. We may therefore ask if $\{A \mid \overline{A} = A\}$ is a subspace of $M_{m\times n}(C)$. It is easy to see that these matrices must be real (since $z = \bar{z}$ iff $z \in R$) and therefore, *not* a subspace of $M_{m\times n}(C)$ (as zA is not real if z is not). ❑

The methods for calculating generators, and determining linear independence, bases, and dimension are identical to the real case. But, we must always remember that the scalars are now complex numbers. Therefore, for example, Definition 4-5 for linear combination now reads as

DEFINITION 4-16

Let V be a complex vector space and $\mathbf{v}_1, \mathbf{v}_2, \ldots, \mathbf{v}_n$ be n vectors in V. A vector $\mathbf{u}$ in V is called a **linear combination** of these vectors if there are scalars $k_1, k_2, \ldots, k_n \in C$ such that

$$\mathbf{u} = k_1\mathbf{v}_1 + k_2\mathbf{v}_2 + \ldots + k_n\mathbf{v}_n.$$

EXAMPLE 4-39

Let $\mathbf{v}_1 = (-1 + i, 2 + 3i, -2 - i)$, $\mathbf{v}_2 = (-i, 3 - 3i, -1 - 2i)$, and $\mathbf{u} = (-5 - 2i, -4 - 10i, -9 - 4i)$. These are vectors in C^3. To determine whether $\mathbf{u}$ is a linear combination of $\mathbf{v}_1$ and $\mathbf{v}_2$, we try (as usual) to find k_1 and $k_2 \in C$ such that

$$\mathbf{u} = k_1\mathbf{v}_1 + k_2\mathbf{v}_2.$$

Substituting values, the equation becomes

$$\begin{aligned}&(-5 - 2i, -4 - 10i, -9 - 4i)\\ &= k_1(-1 + i, 2 + 3i, -2 - i) + k_2(-i, 3 - 3i, -1 - 2i).\end{aligned}$$

After equating the three components, this leads to the following system of three equations in two (complex) unknowns:

$$\begin{aligned}(-1 + i)k_1 + \quad (-i)k_2 &= -5 - 2i\\ (2 + 3i)k_1 + \quad (3 - 3i)k_2 &= -4 - 10i\\ (-2 - i)k_1 + (-1 - 2i)k_2 &= -9 - 4i.\end{aligned}$$

The augmented matrix is

$$\begin{bmatrix} -1 + i & -i & -5 - 2i \\ 2 + 3i & 3 - 3i & -4 - 10i \\ -2 - i & -1 - 2i & -9 - 4i \end{bmatrix},$$

which is reduced by Gauss-Jordan reduction to

$$\begin{bmatrix} 1 & 0 & 1+i \\ 0 & 1 & 2-3i \\ 0 & 0 & 0 \end{bmatrix}.$$

The solution is therefore $k_1 = 1 + i$ and $k_2 = 2 - 3i$, showing that **u** is a linear combination of $\mathbf{v}_1$ and $\mathbf{v}_2$.

By solving the same system, but with right-hand sides set to 0, we find that $k_1 = 0 = k_2$. This means that

$$k_1\mathbf{v}_1 + k_2\mathbf{v}_2 = \mathbf{0} = (0,0,0) = (0 + 0i, 0 + 0i, 0 + 0i)$$

implies $k_1 = k_2 = 0$. Therefore, we see that $\mathbf{v}_1$ and $\mathbf{v}_2$ are linearly independent and constitute a basis for the subspace *span* $\{\mathbf{v}_1, \mathbf{v}_2\}$ of C^3. So, the dimension of this subspace is 2. ❑

EXAMPLE 4-40

We find all solutions of the system of equations

$$\begin{aligned} z_1 + (1+i)z_2 - \quad iz_3 &= 2 + i \\ (1-i)z_1 - \quad 2z_2 + (3-2i)z_3 &= 2 - i. \end{aligned}$$

The augmented matrix is

$$\begin{bmatrix} 1 & 1+i & -i & 2+i \\ 1-i & -2 & 3-2i & 2-i \end{bmatrix},$$

which is reduced to

$$\begin{bmatrix} 1 & 0 & 5/4 - (1/4)i & 7/4 + (3/4)i \\ 0 & 1 & -1 + (1/4)i & 1/4 \end{bmatrix}.$$

The corresponding system of equations is

$$\begin{aligned} z_1 \quad + [5/4 - (1/4)i]z_3 &= 7/4 + (3/4)i \\ z_2 + [-1 + (1/4)i]z_3 &= 1/4, \end{aligned}$$

so that all solutions are given by

$$\begin{aligned} z_1 &= 7/4 + (3/4)i + [-5/4 + (1/4)i]z_3 \\ z_2 &= 1/4 + \quad [1 - (1/4)i]z_3, \end{aligned}$$

where z_3 may be any complex number. A particular solution is

$$z_1 = 7/4 + (3/4)i,\ z_2 = 1/4,\ z_3 = 0.$$

We also see that the dimension of the solution space of the associated homogeneous system is 1. ❑

As is painfully clear from working the details of the preceding examples, calculations for complex vector spaces are much more complicated than for real ones. However, the underlying concepts are the same. In fact, our main interest will be in the spaces C^n and $M_n(C)$, which arise most frequently in applications.

THEOREM 4-16

(1) $dim\ C^n = n$.

(2) $dim\ M_{m\times n}(C) = mn$.

(3) $dim\ P_n(C) = n + 1$.

PROOF (1) As in the case of R^n, we define the standard basis $B(n) = \{\mathbf{e}_1,\ldots,\mathbf{e}_n\}$ for C^n by

$$\mathbf{e}_1 = (1,0,\ldots,0),\ \mathbf{e}_2 = (0,1,0,\ldots,0),\ \mathbf{e}_n = (0,0,\ldots,0,1).$$

The $\mathbf{e}_i$ are in C^n because 0 and 1 are in C (as well as in R). The proof that $B(n)$ is linearly independent is identical to the real case and is left for the reader. Similarly, it generates C^n because any $\mathbf{z} = (z_1,\ldots,z_n) \in C^n$ can be written as

$$\mathbf{z} = z_1(1,0,\ldots,0) + \ldots + z_n(0,0,\ldots,0,1) = z_1\mathbf{e}_1 + \ldots + z_n\mathbf{e}_n.$$

Thus, $B(n)$ is a basis and $dim\ C^n = n$.

(2) For $M_{m\times n}(C)$, we define, for each $1 \le k \le m$ and $1 \le j \le n$, the matrices E_{kj} to have all entries 0 except in row k and column j, where the entry is 1. Then, these mn matrices constitute the standard basis for $M_{m\times n}(C)$. The details of the proof will be similar to Example 4-17 and are left for the reader.

(3) The proof is left for the reader. ■

Remark Since R^n is a proper subset of C^n (any real n-tuple $(x_1,\ldots,x_n)$ is automatically in C^n), C^n is in a sense "larger" than R^n. This being the case, it may seem puzzling that

(i) $dim\ C^n = dim\ R^n = n$, and

(ii) the same set of vectors $B(n)$ is a basis for two different vector spaces, R^n and C^n.

The explanation lies in the fact that linear combinations in C^n require *complex coefficients*, while those in R^n require *real coefficients*. Thus, the "larger size" of C^n may be considered to arise from the fact that complex coefficients are "larger" than real ones. However, the *dimensions* of R^n and C^n are the same, since the dimension of a vector space is simply the *number of elements* in a basis.

In this connection, it is interesting to observe that C^n can also be regarded as a *real* vector space, since we may certainly define a scalar multiplication in C^n by $k \in R$ as

$$k(z_1,\ldots,z_n) = (kz_1,\ldots,kz_n).$$

We leave for the reader the verification that this makes C^n, together with the usual addition, into a real vector space (Definition 4-3) and that its dimension as a real vector space is $2n$ (question 15 below). This reveals that we are really using the notation "*dim V*" in two senses, one for real vector spaces and one for complex vector spaces. In more advanced books, these are often distinguished by writing $dim_R\ V$ and $dim_C\ V$, when a set V is both a real and a complex vector space.

EXERCISES 4-6

1. Use Definition 4-15 to verify that **(a)** C^2 and **(b)** $M_2(C)$ are complex vector spaces with respect to the operations defined in this section.

2. Consider C^2 with the usual scalar multiplication, but with addition defined by

$$(z_1,z_2) + (z'_1,z'_2) = (z_1 + \bar{z}'_1, z_2 + \bar{z}'_2).$$

Determine whether or not C^2 is a vector space with respect to these operations.

3. Let V be the set of 2×2 complex matrices of the form

$$\begin{bmatrix} a & \bar{a} \\ 0 & 0 \end{bmatrix},$$

with the usual addition but with scalar multiplication defined for $z \in C$ by

$$z\begin{bmatrix} a & \bar{a} \\ 0 & 0 \end{bmatrix} = \begin{bmatrix} za & \overline{za} \\ 0 & 0 \end{bmatrix}.$$

Determine whether or not V is a complex vector space.

4. Suppose V is a real vector space and we define scalar multiplication for $z \in C$ by **(a)** $z\mathbf{v} = |z|\mathbf{v}$ when $\mathbf{v} \in V$. Does this make V a complex vector space? What if we define **(b)** $z\mathbf{v} = Re(z)\mathbf{v}$, or **(c)** $z\mathbf{v} = Im(z)\mathbf{v}$?

5. Which of the following sets are subspaces of $F(R,C)$? Explain your answer.

(a) $\{f \mid |f(x)| = 0, x \in R\}$

(b) $\{f \mid f(x) \in R, x \in R\}$

(c) $\{f \mid f(x) = x + i, x \in R\}$

(d) $\{f \mid f(x) = \overline{f(x)}, x \in R\}$

(e) $\{f \mid f(x) = -f(x), x \in R\}$

6. Which of the following sets of matrices are subspaces of $M_2(C)$? Explain your answer.

(a) the matrices of the form

$$\begin{bmatrix} z & 0 \\ 0 & \bar{z} \end{bmatrix}$$

(b) the matrices of the form

$$\begin{bmatrix} z_1 & z_2 \\ z_3 & z_4 \end{bmatrix}$$

where $z_1 + z_4 = 0$

(c) the matrices of the form

$$\begin{bmatrix} z_1 & z_2 \\ z_3 & z_4 \end{bmatrix}$$

where z_1 and z_4 are real, and $z_2 = \bar{z}_3$

7. Let

$$\mathbf{u} = (-1 + i, 3i, 2 + 4i),\ \mathbf{v} = (-3, 5 - 2i, 2 - i),$$
$$\mathbf{w} = (-3 - i, 2i, 4 + i).$$

Compute the following expressions in C^3.

(a) $\mathbf{u} - i\mathbf{v}$

(b) $(3 + 2i)\mathbf{u} - 5\mathbf{v} + (1 + i)\mathbf{w}$

(c) $(2 + i)\{i\mathbf{v} - (2 - i)[3\mathbf{u} + (1 - i)\mathbf{w}]\}$

8. Which of the following sets of vectors are linearly independent in C^2?

(a) $(1,-1), (-i,i)$

(b) $(1 - i, 2 + i), (3 - 4i, 2 - 5i)$

(c) $(2 - 3i, 4 + 3i), (13, -1 + 18i)$

9. Show that the vectors $\mathbf{u} = (1,i,0)$, $\mathbf{v} = (1 + i, 1, 1)$, and $\mathbf{w} = (i, 1, 1 - i)$ are linearly independent and a basis for C^3.

10. Express each of the following vectors $\mathbf{z}$ as a linear combination of $B = \{\mathbf{u},\mathbf{v},\mathbf{w}\}$, where $\mathbf{u}$, $\mathbf{v}$, and $\mathbf{w}$ are defined in question 9. Also write $[\mathbf{z}]_B$.

(a) $\mathbf{z} = (i,i,i)$

(b) $\mathbf{z} = (1,2,3)$

(c) $\mathbf{z} = (1 + i, 1 - i, 2)$

11. Which of the following sets of vectors are bases for C^3?

(a) $(i,1,1), (i,i,1), (i,i,i)$

(b) $(2 - i, 3 - 4i, 2i), (2 + i, 3 + 4i, -2i), (2i, 3i, 0)$

(c) $(1 - i, 1 + i, 0), (1 - i, 0, 1 + i), (0, 1 - i, 1 + i)$

12. Find a basis for the solution space of each of the following systems of equations. Also state the dimension of the solution space.

(a)
$$\begin{aligned} iz_1 + (2-i)z_2 &= 0 \\ (2+i)z_1 - \quad 5iz_2 &= 0 \end{aligned}$$

(b)
$$\begin{aligned} (-2-i)z_1 - (3-4i)z_2 &= 0 \\ (2+i)z_1 - \quad (5+i)z_2 &= 0 \end{aligned}$$

(c)
$$\begin{aligned} z_1 + \qquad\quad iz_2 - z_3 &= 0 \\ (1-i)z_1 + \qquad 2z_2 + iz_3 &= 0 \\ (2+i)z_1 + (1+2i)z_2 - z_3 &= 0 \end{aligned}$$

13. Find all solutions of the systems of equations $AZ = B$ whose left-hand sides AZ are those in question 12 above and whose right-hand sides B are given as follows.

(a) $B = \begin{bmatrix} 2+3i \\ 8-i \end{bmatrix}$ **(b)** $B = \begin{bmatrix} 5-3i \\ -2-7i \end{bmatrix}$ **(c)** $B = \begin{bmatrix} -2i \\ 1+i \\ -2-i \end{bmatrix}$

14. Find all solutions of the system of equations

$$\begin{aligned} (1-i)z_1 + \qquad iz_2 + (2-5i)z_3 &= 2-i \\ z_1 + (1+i)z_2 - \qquad 2iz_3 &= -3+i. \end{aligned}$$

15. (a) Verify that C^n is a real vector space with (real) dimension $2n$.

(b) Verify that any complex vector space of dimension n is a real vector space with (real) dimension $2n$.

5 LINEAR TRANSFORMATIONS

5-1 Introduction
5-2 Properties of Linear Transformations
5-3 Isomorphism (Optional)
5-4 Basis Change and Similarity
5-5 Linear Functionals and Applications (Optional)

5-1 Introduction

Frequently in mathematics and its applications, we need to describe how one or more variables depend on other variables. The appropriate mathematical notion for this is a function $y = f(x)$. Readers will be familiar with the case in which $y = f(x)$ is a real-valued function of a single real variable, $f:R \rightarrow R$. We can use vectors to formulate the idea of functions that depend on several variables. The following example illustrates this idea in an important application.

EXAMPLE 5-1

A company manufactures two products A and B using three ingredients a, b, and c. The numbers of units of these ingredients required to manufacture one unit of A and B are given in the table below:

Ingredient	*Product*	
	A	B
a	4	2
b	2	6
c	3	3

The quantities of these ingredients available daily are limited to 80, 120, and 75 units respectively. The company would like to decide how many units of each of A and B to produce each day so as to maximize its daily profit, given that A and B yield unit profits of 15 and 12 dollars respectively. This is a **linear programming problem** (LP for short; see also section 2-3). In fact, this type of problem is applicable to a rather wide variety of situations. For example, we may have a situation in which the "ingredients" are not materials, but are the times required in various stages of the manufacture of A and B.

To formulate the problem described above, we let x and y be the numbers of units of A and B respectively, which the company will manufacture each day. These are the **decision variables** for the problem. Then the daily profit is

$$P = 15x + 12y,$$

which is a real-valued function of the two variables, x and y. If we now let

$$X = \begin{bmatrix} x \\ y \end{bmatrix}, C = \begin{bmatrix} 15 & 12 \end{bmatrix},$$

then, we can write

$$P = P(X) = \begin{bmatrix} 15 & 12 \end{bmatrix} \begin{bmatrix} x \\ y \end{bmatrix} = CX.$$

This shows that P may be thought of as a real-valued function of the **vector variable** X (the "production vector"), whose entries are the daily quantities of A and B produced. So, P is a function between vector spaces: $P{:}M(2 \times 1) \rightarrow R$. Notice that the "values" $P(X)$ of P are obtained by matrix multiplication.

Continuing the formulation of the LP problem, it is clear that if there were no limitations on x and y, the company could make its profit as large as it pleased by simply increasing x and y without bound. However, the profit is constrained by the limited quantities of the ingredients available each day to make A and B. For example, in regard to ingredient a, since each unit of A requires 4 units of a and each unit of B requires 2 units of a, the total daily consumption of ingredient a in the manufacture of x units of A and y units of B is $4x + 2y$. This cannot exceed the daily availability of 80 units of ingredient a. Therefore, we obtain the **constraint**

$$4x + 2y \leq 80. \qquad \text{(i)}$$

Similarly, the constraints imposed by the availabilities of b and c are

$$2x + 6y \leq 120, \qquad \text{(ii)}$$

$$3x + 3y \leq 75. \qquad \text{(iii)}$$

The three inequalities, (i), (ii), and (iii), together form the **system of constraints** for the LP problem. In addition, we require that x and y be nonnegative. We let

$$M = \begin{bmatrix} 4 & 2 \\ 2 & 6 \\ 3 & 3 \end{bmatrix}, N = \begin{bmatrix} 80 \\ 120 \\ 75 \end{bmatrix},$$

and, as before,

$$X = \begin{bmatrix} x \\ y \end{bmatrix}.$$

Then, we may write the system of constraints as $MX \leq N$, with the obvious definition of "$\leq$" applying to the 3×1 matrices on the two sides. For each X, the vector (i.e., matrix) MX has as its three entries the daily consumptions of ingredients a, b, and c for that particular X. It therefore defines a vector-valued function D (for "daily consumption") with $D(X) = MX$. This D is a function that maps from the vector space $M(2 \times 1)$ to the vector space $M(3 \times 1)$. The complete LP problem may thus be written as

maximize $P(X) = CX$,
subject to $D(X) = MX \leq N$ and $X \geq O$.

It is interesting to interpret the formulation geometrically (Figure 5-1). Identifying $n \times 1$ matrices with n-tuples as usual, we are looking for a vector X in the first quadrant of the vector space R^2 such that (i) its "value" $D(X)$ under D lies in the box in R^3 shown, and (ii) the profit $P(X)$ is as large as possible.

FIGURE 5-1

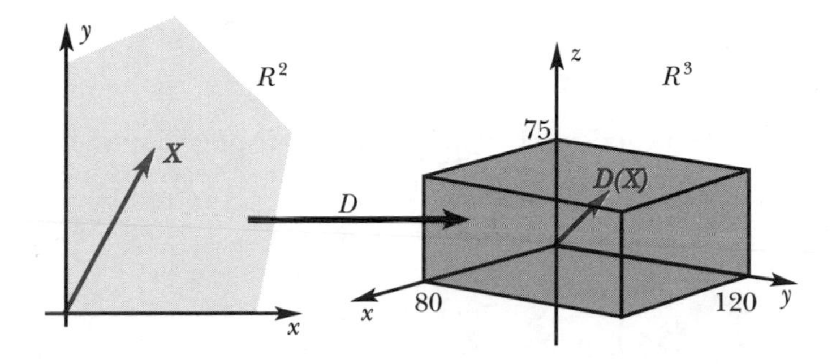

A discussion and solution of this and other LP problems are provided in Appendix C of this book. ❑

In the preceding example, we encountered two functions mapping vector spaces into vector spaces. Both were defined by matrix multiplication. We generalize this type of function as follows.

> **DEFINITION 5-1**
>
> Let A be an $m \times n$ matrix. Then A determines a function $T_A{:}M(n \times 1) \rightarrow M(m \times 1)$ by $T_A(X) = AX$ when $X \in M(n \times 1)$. Identifying $n \times 1$ matrices with n-tuples, we also write this as $T_A{:}R^n \rightarrow R^m$.

Because of the algebraic properties of matrices (section 2-5), the map T_A has two important properties. Let c be a scalar and $X_1, X_2, X \in M(n \times 1)$. Then,

(1) $T_A(X_1 + X_2) = A(X_1 + X_2) = T_A(X_1) + T_A(X_2)$,

(2) $T_A(cX) = A(cX) = cAX = cT_A(X)$.

EXAMPLE 5-2

Let

$$A = \begin{bmatrix} 1 & 2 & 3 \\ 4 & 5 & 6 \end{bmatrix}.$$

Then, $T_A: R^3 \to R^2$ with

$$\begin{aligned} T_A(x_1,x_2,x_3) &= AX \\ &= \begin{bmatrix} 1 & 2 & 3 \\ 4 & 5 & 6 \end{bmatrix} \begin{bmatrix} x_1 \\ x_2 \\ x_3 \end{bmatrix} \\ &= \begin{bmatrix} x_1 + 2x_2 + 3x_3 \\ 4x_1 + 5x_2 + 6x_3 \end{bmatrix} \\ &= (x_1 + 2x_2 + 3x_3, 4x_1 + 5x_2 + 6x_3), \end{aligned}$$

having identified $n \times 1$ matrices with n-tuples. Thus, for example,

$$T_A(1,-1,3) = \begin{bmatrix} 1 & 2 & 3 \\ 4 & 5 & 6 \end{bmatrix} \begin{bmatrix} 1 \\ -1 \\ 3 \end{bmatrix} = (8,17),$$

$$T_A(2,0,-2) = (-4,-4).$$

And,

$$T_A((1,-1,3) + (2,0,-2)) = T_A(3,-1,1) = (4,13),$$

which is equal to

$$T_A(1,-1,3) + T_A(2,0,-2) = (8,17) + (-4,-4).$$

Similarly, with $c = -2$, and $X = (1,-1,3)$,

$$\begin{aligned} T_A(cX) &= T_A(-2(1,-1,3)) = (-16,-34) \\ &= -2T_A(1,-1,3) = cT_A(X). \square \end{aligned}$$

We "abstract" these two properties to obtain the following definition.

DEFINITION 5-2

Given vector spaces V and W, a function $T: V \to W$ is called a **linear transformation** if

(1) $T(\mathbf{v} + \mathbf{w}) = T(\mathbf{v}) + T(\mathbf{w})$ and

(2) $T(c\mathbf{v}) = cT(\mathbf{v})$,

for all scalars c and vectors $\mathbf{v}, \mathbf{w} \in V$.

Another phrase for "linear transformation" is "linear map", and we will often simply say "T is linear". $T(\mathbf{v})$, the value of T at $\mathbf{v}$ in V, is usually called the **image** of $\mathbf{v}$ under T. Sometimes (as in Examples 5-8 and 5-9) it is convenient to write $T(\mathbf{v})$ as simply $T\mathbf{v}$ in order to avoid a confusing number of parentheses. When $W = V$, so that $T: V \to V$, we will often call T a **linear operator** (on V).

EXAMPLE 5-3

As a result of matrix properties, the functions T_A defined by matrices are linear transformations. We shall often call them **matrix transformations**. In fact, we shall see later on in this section that every linear transformation can be obtained as a matrix transformation. ❑

Now, if $T:V \to W$ is linear, it maps the zero vector of V to the zero vector of W. For,

$$T(\mathbf{0}) = T(\mathbf{0} + \mathbf{0}) = T(\mathbf{0}) + T(\mathbf{0}) = 2T(\mathbf{0}),$$

so that $T(\mathbf{0}) = \mathbf{0}$, the zero vector of W. Notice that we don't use different symbols for the (different) zero vectors of V and W, because it will always be clear from the context which space $\mathbf{0}$ belongs to.

EXAMPLE 5-4

We determine all linear transformations $T:R \to R$ (i.e., all linear operators on R). Recall that R is a vector space with respect to ordinary addition and multiplication of numbers. Now, by our observation just before this example, $T(0) = 0$, and therefore the graph of $y = T(x)$ must pass through the origin. In fact, the graph is seen to be a straight line as follows.

By property (2) of a linear transformation given in Definition 5-2,

$$y = T(x) = T(x1) = xT(1)$$

Thus, if we put $m = T(1)$ (a constant determined by T), we may write $y = mx$, which is the equation of a straight line through the origin.

As a function, T simply multiplies any argument x by the number m. So, T may be thought of as multiplication by a 1×1 matrix $[m]$. We leave it for the reader to verify that the functions $y = mx$ are linear according to the linear algebra definition, and that the function $y = mx + b$, which is called "linear" in analytic geometry, is *not* a linear transformation unless $b = 0$. ❑

Given a function $T:V \to W$ between arbitrary vector spaces, to verify that it is linear, we must check that it satisfies the two properties in Definition 5-2. In practice, it is faster to use the following equivalent criterion.

LEMMA 5-1

$T:V \to W$ is linear iff

$$T(c\mathbf{v} + \mathbf{w}) = cT(\mathbf{v}) + T(\mathbf{w}) \qquad (5\text{-}1)$$

for all scalars c and vectors $\mathbf{v}$ and $\mathbf{w}$ in V.

PROOF It should be clear that if T is linear, then certainly equation 5-1 is true. Conversely, if equation 5-1 is true, then, putting $c = 1$ yields property (1) of linearity, while putting $\mathbf{w} = \mathbf{0}$ gives

$$T(c\mathbf{v}) = T(c\mathbf{v} + \mathbf{0}) = cT(\mathbf{v}) + T(\mathbf{0}) = cT(\mathbf{v}),$$

which is property (2) in Definition 5-2. ■

EXAMPLE 5-5

(a) Let $T:R^2 \to R^3$ be defined by $T(x,y) = (x + 2y, -y, 2x - y)$. We use the lemma to show that T is linear. With $\mathbf{v} = (x,y)$, $\mathbf{w} = (x',y')$, and $c \in R$,

$$\begin{aligned}
&T(c\mathbf{v} + \mathbf{w}) \\
&= T(c(x,y) + (x',y')) \\
&= T(cx + x', cy + y') \\
&= (cx + x' + 2(cy + y'), -(cy + y'), 2(cx + x') - (cy + y')) \\
&= (c(x + 2y) + (x' + 2y'), -cy - y', c(2x - y) + (2x' - y')) \\
&= c(x + 2y, -y, 2x - y) + (x' + 2y', -y', 2x' - y') \\
&= cT(\mathbf{v}) + T(\mathbf{w}),
\end{aligned}$$

as required.

(b) Let $F:R^2 \to R^2$ be given by $F(x,y) = (-3x + y, x + 4y + 1)$. Then F is not linear since $F(c(x,y) + (x',y')) \neq cF(x,y) + F(x',y')$ as shown below:

$$\begin{aligned}
&F(c(x,y) + (x',y')) \\
&= F(cx + x', cy + y') \\
&= (-3(cx + x') + cy + y', cx + x' + 4(cy + y') + 1).
\end{aligned}$$

But,

$$\begin{aligned}
&cF(x,y) + F(x',y') \\
&= c(-3x + y, x + 4y + 1) + (-3x' + y', x' + 4y' + 1) \\
&= (c(-3x + y) - 3x' + y', c(x + 4y + 1) + x' + 4y' + 1) \\
&= (-3(cx + x') + cy + y', cx + x' + 4(cy + y') + c + 1),
\end{aligned}$$

which differs from the earlier expression. Thus F is not linear.

Alternatively, the reader may note that $F(0,0) \neq (0,0)$, so that F cannot be linear. ❑

EXAMPLE 5-6

(a) Given any vector spaces V and W, there is always a "trivial" linear transformation between them. This is the zero map $0:V \to W$ defined by $0(\mathbf{v}) = \mathbf{0}$, the zero vector of W, for every $\mathbf{v}$ in V. The reader should verify that it is linear.

(b) The identity function from a vector space V to itself is a linear operator on V. We write $1_V:V\to V$. Then, $1_V(c\mathbf{v} + \mathbf{w}) = c\mathbf{v} + \mathbf{w} = c1_V(\mathbf{v}) + 1_V(\mathbf{w})$, so 1_V is linear by Lemma 5-1. ❑

The next three examples concern important linear transformations that occur in various applications.

EXAMPLE 5-7

We consider the function $T:R^2\to R^2$ which operates by rotating a nonzero vector $\mathbf{v}$ in the plane counterclockwise through an angle θ with $0 \le \theta \le \pi$, say. This is illustrated in Figure 5-2. We also define $T(\mathbf{0}) = \mathbf{0}$. Then T is seen to be linear by the following geometrical arguments.

FIGURE 5-2

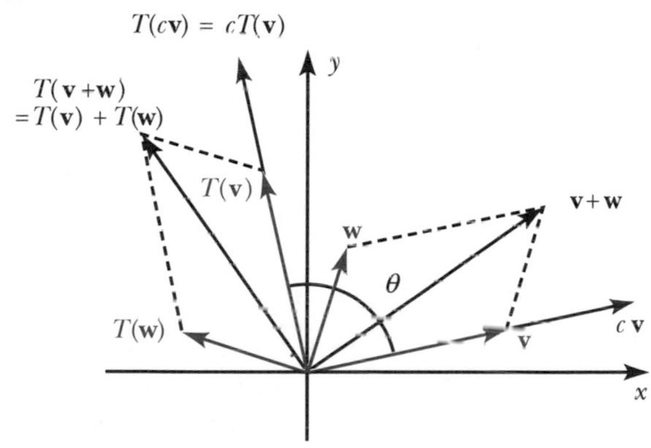

(1) Since T rotates the entire parallelogram formed by $\mathbf{v}$ and $\mathbf{w}$ to the parallelogram formed by $T(\mathbf{v})$ and $T(\mathbf{w})$, it maps the diagonal $\mathbf{v} + \mathbf{w}$ of the former to the diagonal $T(\mathbf{v}+\mathbf{w})$ of the latter. Thus, $T(\mathbf{v}+\mathbf{w}) = T(\mathbf{v}) + T(\mathbf{w})$.

(2) If c is a nonzero scalar, then $c\mathbf{v}$ is collinear with $\mathbf{v}$, so $T(c\mathbf{v})$ is collinear with $T(\mathbf{v})$. By the properties of rotation, $T(c\mathbf{v})$ has the same length and direction as $cT(\mathbf{v})$. So, $T(c\mathbf{v}) = cT(\mathbf{v})$. Finally, if $c = 0$, then $c\mathbf{v} = \mathbf{0}$, so $T(c\mathbf{v}) = \mathbf{0} = cT(\mathbf{v})$.

Therefore, T is a linear transformation. We can obtain expressions for $T(x,y)$ using analytic geometry as shown in Figure 5-3. Let (x',y') be the coordinates of $T(x,y) = T(\mathbf{v})$. Then we see from the figure that

$$x' = |\mathbf{v}|\cos(\theta + \alpha) = |\mathbf{v}|\cos\theta\cos\alpha - |\mathbf{v}|\sin\theta\sin\alpha = x\cos\theta - y\sin\theta.$$
$$y' = |\mathbf{v}|\sin(\theta + \alpha) = |\mathbf{v}|\sin\theta\cos\alpha + |\mathbf{v}|\cos\theta\sin\alpha = x\sin\theta + y\cos\theta.$$

Thus, $T(x,y) = (x\cos\theta - y\sin\theta, x\sin\theta + y\cos\theta)$. ❑

FIGURE 5-3

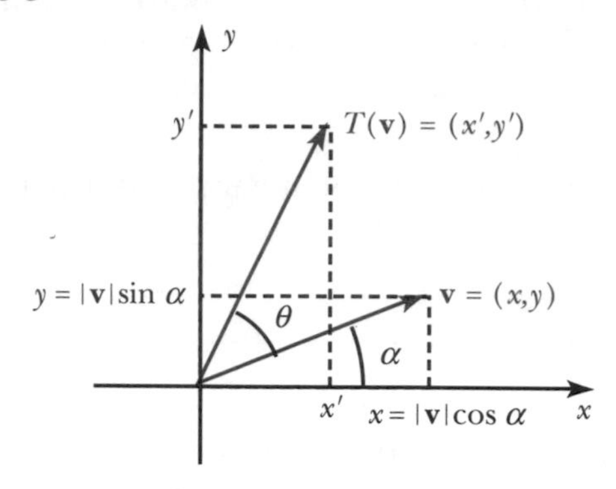

We will obtain this result in another way in the next section. Notice that if $\theta = 0$, we get the identity transformation $I_{R^2}:R^2 \to R^2$, $I_{R^2}(x,y) = (x,y)$, since $\cos 0 = 1$ and $\sin 0 = 0$. And, if $\theta = \pi$, we obtain the **reflection** in the origin: $T(x,y) = (-x,-y)$.

EXAMPLE 5-8

Let $V = F[0,1]$, the vector space of all real-valued functions defined on the interval $[0,1]$, and let $W = F(0,1]$, the space of all real-valued functions defined on $(0,1] = \{x \mid 0 < x \leq 1\}$. We define $\Delta:V \to W$ by

$$\Delta f(x) = (\Delta f)(x) = (1/x)[f(x) - f(0)], x \in (0,1].$$

We call $\Delta f(x)$ "the average rate of change of the function f over the interval $[0,x]$". Geometrically, $\Delta f(x)$ is the slope of the so-called "secant line" L shown in Figure 5-4. In fact, Δ is a linear transformation as shown below. If f and g are in V and c is a scalar,

$$\begin{aligned}\Delta(cf + g)(x) &= (1/x)[(cf + g)(x) - (cf + g)(0)] \\ &= c(1/x)[f(x) - f(0)] + (1/x)[g(x) - g(0)] \\ &= (c\Delta f + \Delta g)(x), \text{ for any } x \in (0,1].\end{aligned}$$

FIGURE 5-4

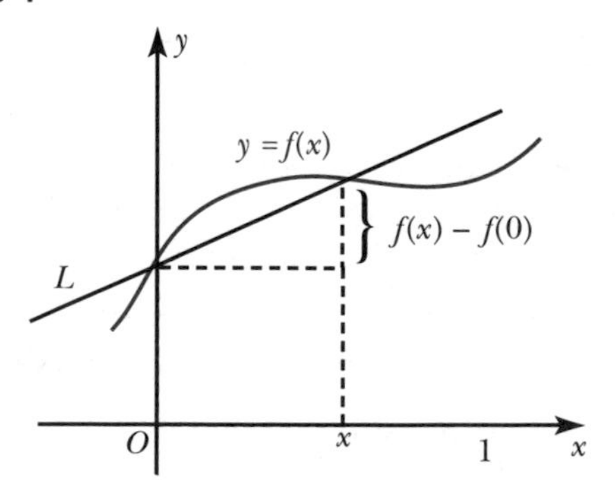

So,

$$\Delta(cf + g) = c\Delta f + \Delta g$$

as required for linearity. (Recall that in order to show that two functions are equal, we must show they have the same values at every x in their common domain.) Readers familiar with calculus will recognize that this example has important applications, since Δf is used to define the **derivative** of f. ❑

EXAMPLE 5-9

We consider $V = P(1)$, the space of polynomial functions of degree at most 1, and let $W = F[0,1]$. Given any $f \in P(1)$ with $f(x) = ax + b$ for $x \in R$, we define $A{:}P(1) \rightarrow F[0,1]$ by $Af(x) =$ the "signed" area bounded by the line $y = ax + b$, the x axis, $x = 0$, and the vertical line through x, as shown in Figure 5-5 ($0 \leq x \leq 1$). In other words, Af is the function, with domain $[0,1]$, whose value at any x is the *area* of the shaded region shown. We agree to regard the area of a region below the x axis as negative, as illustrated below. Of course, for any f in $P(1)$, $Af(0) = 0$.

FIGURE 5-5

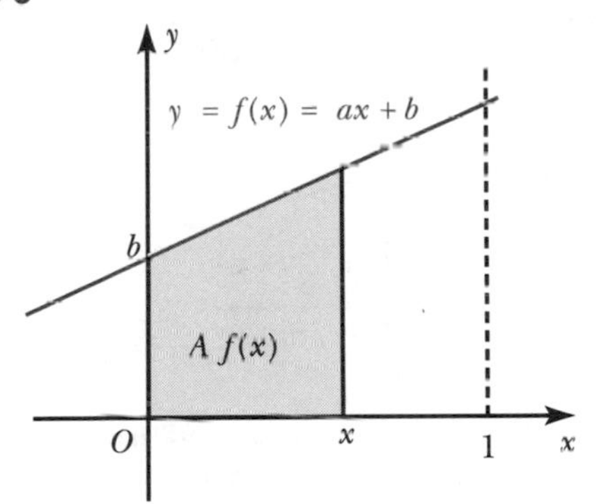

Figure 5-6 shows the graph of $f(x) = 2x - 1$. Using the formula for the area of a triangle, we obtain

$$Af(1/2) = (-1)(1/2)(1/2) = -1/4.$$

From the figure, we see that

$$Af(1) = -1/4 + 1/4 = 0,$$

which is the "algebraic sum" of the areas above and below the x axis.

In fact, the reader may show (question 6, Exercises 5-1) that, if $f(x) = ax + b$, then

$$Af(x) = ax^2/2 + bx, \; x \in [0,1].$$

FIGURE 5-6

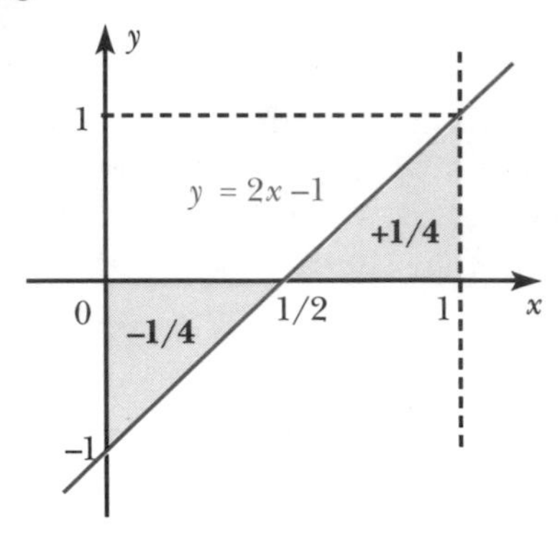

The diagrams in Figure 5-7 illustrate that A as defined above is a linear transformation. This linear transformation is also important in calculus: Af is called an **antiderivative** of f. ❑

FIGURE 5-7

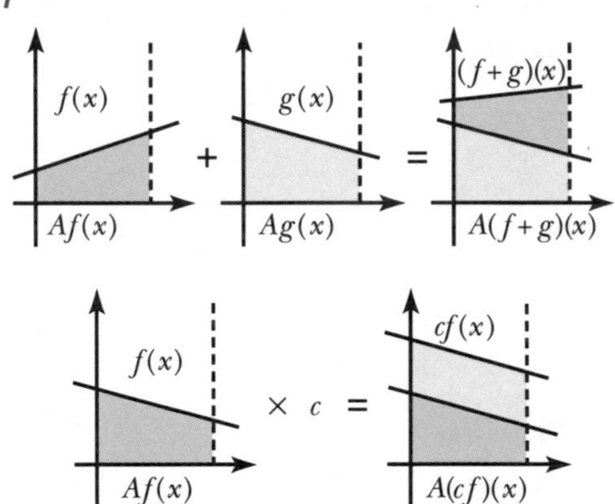

The following theorem summarizes the elementary properties of linear transformations. We have already considered property (1). The proofs of properties (2) to (4) are left as exercises for the reader (question 7, Exercises 5-1).

THEOREM 5-1

Let V, W be any vector spaces and $T: V \to W$ a linear transformation. Then,

(1) $T(\mathbf{0}) = \mathbf{0}$,

(2) $T(-\mathbf{v}) = -T(\mathbf{v})$ for any $\mathbf{v} \in V$,

(3) $T(\mathbf{v} - \mathbf{w}) = T(\mathbf{v}) - T(\mathbf{w})$ for any $\mathbf{v}, \mathbf{w} \in V$,

(4) $T(c_1\mathbf{v}_1 + \ldots + c_m\mathbf{v}_m) = c_1T(\mathbf{v}_1) + \ldots + c_mT(\mathbf{v}_m)$ for any $\mathbf{v}_i \in V$, $c_i \in R$, and any integer $m > 0$. ■

Remarks

(1) Property (3) states that "the image of a difference of vectors in V is the difference of their images in W".

(2) Property (4) states that "the image of a linear combination of vectors in V is the same linear combination of their images in W".

EXERCISES 5-1

1. **(a)** For each of the following matrices A, find m and n so that $T_A:R^n \to R^m$ and write the expression for $T_A(X)$, where $X = (x_1,\ldots,x_n)$.

(i) $\begin{bmatrix} -1 & 2 & 3 \end{bmatrix}$ **(ii)** $\begin{bmatrix} 1 & 2 \\ 3 & 4 \end{bmatrix}$ **(iii)** $\begin{bmatrix} 2 \\ 5 \\ 0 \\ 3 \end{bmatrix}$ **(iv)** $\begin{bmatrix} 2 & 3 & 0 \\ 4 & 1 & -2 \end{bmatrix}$

(b) For each of the matrices A above, and with n as determined in (a), compute $T_A(\mathbf{e}_i)$, $i = 1, 2, \ldots, n$, where $\{\mathbf{e}_1,\ldots,\mathbf{e}_n\}$ is the standard basis for R^n.

2. Let S and $T:R^2 \to R^2$ be defined by $S(x,y) = (-x,y)$ and $T(x,y) = (x,-y)$ respectively.

(a) Show that S and T are linear.

(b) Sketch the images of $(1,0)$, $(0,1)$, and $(1,1)$ under the mappings S and T. (S and T are called *reflections* in the y and x axes respectively.)

3. Determine whether or not the following functions are linear.

(a) $T:R^2 \to R^2$ with $T(x,y) = (2x + y, y - 2x)$

(b) $T:R^3 \to R$ with $T(x,y,z) = x + y + z + 1$

(c) $T:R^2 \to R^3$ with $T(x,y) = (xy, x + y, y)$

(d) $T:R^3 \to R^3$ with $T(x,y,z) = (2x - y + z, x - 2z, y)$

(e) $T:R \to R^3$ with $T(x) = (x, -2x, 0.5x)$

4. If V is any vector space and c is a scalar, let $T:V \to V$ be the function defined by $T(\mathbf{v}) = c\mathbf{v}$.

(a) Show that T is a linear operator (it is called the *scalar transformation* by c).

(b) For $V = R^2$, sketch $T(1,0)$ and $T(0,1)$ in the following cases:
(i) $c = 2$; **(ii)** $c = 1/2$; **(iii)** $c = -1$.

(When $c > 1$, T is called a *dilatation* since it stretches any vector by the factor c; when $0 < c < 1$, T is called a *contraction*; and when $c = -1$, T is called *reflection in the origin* — compare with Example 5-7.)

5. Let A and B be $p \times m$ and $n \times q$ matrices respectively, and define $T:M(m \times n) \to M(p \times q)$ by $T(M) = AMB$, for M in $M(m \times n)$. Use the properties of matrix multiplication to show that T is a linear transformation.

6. **(a)** With reference to Example 5-9, if f is in $P(1)$ with $f(x) = ax + b$, verify that $Af(x) = ax^2/2 + bx$, for $x \in [0,1]$. (Hint: Consider the three possible locations of the graph of $y = ax + b$ with respect to the x axis.)

   (b) Prove that the function $A:P(1) \to P(2)$ defined by the rule that $Af(x) = ax^2/2 + bx$ for any f in $P(1)$, with $f(x) = ax + b$, is a linear transformation.

   (c) More generally, define $A:P(n) \to P(n+1)$ by

$$Af(x) = \sum_{i=0}^{n} \frac{a_i x^{i+1}}{i+1}, \text{ when } f(x) = \sum_{i=0}^{n} a_i x^i, \text{ for } x \in R.$$

   Show that A is a linear transformation.

7. Prove parts (2), (3), and (4) of Theorem 5-1.

8. Let V be a vector space of dimension n and having basis B. Define $M_B:V \to M(n \times 1)$ by $M_B(\mathbf{v}) = [\mathbf{v}]_B$. Show that M_B is a linear transformation.

5-2 Properties of Linear Transformations

In the preceding section, we encountered many specific examples of linear transformations. The next theorem implies that there are, in fact, *infinitely* many linear transformations between two vector spaces V and W.

> **THEOREM 5-2**
>
> If $B = \{\mathbf{v}_1, \ldots, \mathbf{v}_n\}$ is a basis for a vector space V and $\mathbf{w}_1, \ldots, \mathbf{w}_n$ are *any* vectors in W (not necessarily even distinct), then there is *exactly one* linear transformation $T:V \to W$ such that $T(\mathbf{v}_j) = \mathbf{w}_j$, for $j = 1, 2, \ldots, n$.

PROOF To define any function T with domain V, we must say what it does to any $\mathbf{v}$ in V. Since B is a basis, any $\mathbf{v}$ in V may be expressed as

$$\mathbf{v} = \sum_{i=1}^{n} c_i \mathbf{v}_i,$$

for some $c_i \in R$ which are uniquely determined by $\mathbf{v}$ (and B). We *define* $T(\mathbf{v})$ by

$$T(\mathbf{v}) = \sum_{i=1}^{n} c_i \mathbf{w}_i. \tag{5-2}$$

Then, certainly $T(\mathbf{v}_j) = \mathbf{w}_j$ for $j = 1, 2, ..., n$, since

$$\mathbf{v}_j = 0\mathbf{v}_1 + ... + 0\mathbf{v}_{j-1} + 1\mathbf{v}_j + 0\mathbf{v}_{j+1} + ... + 0\mathbf{v}_n$$

means that for $\mathbf{v}_j$, the scalars $c_i = 0$ when $i \neq j$, and $c_j = 1$.

In fact, T is then linear, for if

$$\mathbf{v}' = \sum_{i=1}^{n} c'_i \mathbf{v}_i$$

is another vector in V and c is a scalar, we obtain

$$T(c\mathbf{v} + \mathbf{v}') = T(c\sum_{i=1}^{n} c_i\mathbf{v}_i + \sum_{i=1}^{n} c'_i\mathbf{v}_i) = T(\sum_{i=1}^{n} (cc_i + c'_i)\mathbf{v}_i)$$

$$\overset{*}{=} \sum_{i=1}^{n} (cc_i + c'_i)\mathbf{w}_i = c\sum_{i=1}^{n} c_i\mathbf{w}_i + \sum_{i=1}^{n} c'_i\mathbf{w}_i = cT(\mathbf{v}) + T(\mathbf{v}'),$$

where we have used the definition of T to obtain the equality (*). Therefore, T is linear by Lemma 5-1.

If T' were another linear transformation with $T'(\mathbf{v}_i) = \mathbf{w}_i$ for all $i = 1, 2, ..., n$, then for any $\mathbf{v} \in V$ as above,

$$T'(\mathbf{v}) = T'(\sum_{i=1}^{n} c_i\mathbf{v}_i) = \sum_{i=1}^{n} c_iT'(\mathbf{v}_i) = \sum_{i=1}^{n} c_i\mathbf{w}_i = T(\mathbf{v}).$$

So $T' = T$, showing that T is unique. ■

Furthermore, equation 5-2 implies that

COROLLARY

A linear transformation $T:V \rightarrow W$ is completely determined by its values

$$T(\mathbf{v}_1) = \mathbf{w}_1, ..., T(\mathbf{v}_n) = \mathbf{w}_n$$

at a basis $B = \{\mathbf{v}_1, ..., \mathbf{v}_n\}$ for V. ■

In other words, we know all about T (i.e., we can determine all its values $T(\mathbf{v})$) when we know only its values at a basis. This is a very strong property indeed. For example, we cannot calculate all the values of $y = x^2$ in this way when we know only its value $y = 1$ at the basis $\{1\}$ for R. Of course, this is *not* a linear transformation.

EXAMPLE 5-10

We find the unique linear transformation $T:R^3 \to R^3$ such that

$$T(\mathbf{e}_1) = (1,1,1),\ T(\mathbf{e}_2) = (-1,2,0), \text{ and } T(\mathbf{e}_3) = (1,1,1).$$

Now, if $\mathbf{v} = (x,y,z)$ is any vector in R^3, then it can be written as

$$\mathbf{v} = x\mathbf{e}_1 + y\mathbf{e}_2 + z\mathbf{e}_3,$$

so that the scalars c_1, c_2, and c_3 referred to in the proof of Theorem 5-2 are x, y, and z respectively. So, by definition,

$$\begin{aligned} T(\mathbf{v}) &= T(x,y,z) = x(1,1,1) + y(-1,2,0) + z(1,1,1) \\ &= (x - y + z, x + 2y + z, x + z). \end{aligned}$$

This completely determines T, meaning that we can calculate $T(\mathbf{v})$ for *any* $\mathbf{v}$ in R^3 from this formula. For example,

$$T(1,2,3) = (1 - 2 + 3, 1 + 2(2) + 3, 1 + 3) = (2,8,4). \ \square$$

Linear transformations are important theoretical tools for studying properties of matrices and systems of linear equations. The concepts introduced next have immediate application in these areas.

DEFINITION 5-3

If $T:V \to W$ is linear,

(1) the **kernel** (or null space) of T, denoted $Ker\ T$, is the subset of V defined by

$$Ker\ T = \{\mathbf{v} \in V \mid T(\mathbf{v}) = \mathbf{0} \in W\};$$

(2) the **image** (or range) of T, denoted $Im\ T$, is the subset of W defined by

$$Im\ T = \{T(\mathbf{v}) \mid \mathbf{v} \in V\}.$$

Thus, *Ker T consists of the vectors in V whose images under T are the zero vector of W*, while *Im T consists of those vectors in W that are images under T of vectors in V*. Notice that neither set is empty, since $T(\mathbf{0}) = \mathbf{0}$ implies that the zero vector of V is in $Ker\ T$ and the zero vector of W is in $Im\ T$.

EXAMPLE 5-11

Let $T:R^2 \to R^2$ be the linear map defined by $T(x,y) = (x,0)$. T is called the **projection** of R^2 onto the x axis, for reasons that are apparent from

Figure 5-8. Then,

$$\begin{aligned} Ker\ T &= \{(x,y) \mid x = 0\} = \text{the } y \text{ axis}, \\ Im\ T &= \{T(x,y) \mid (x,y) \in R^2\} = \{(x,0) \mid (x,y) \in R^2\} = \text{the } x \text{ axis}. \end{aligned}$$

In fact, T maps all points (x,y) on the vertical line through $(x,0)$ onto the vector $(x,0)$. Notice that *Ker T* and *Im T* are subspaces. Of course, we can similarly define the projection of R^2 onto the y axis by $(x,y) \rightarrow (0,y)$. ❑

FIGURE 5-8

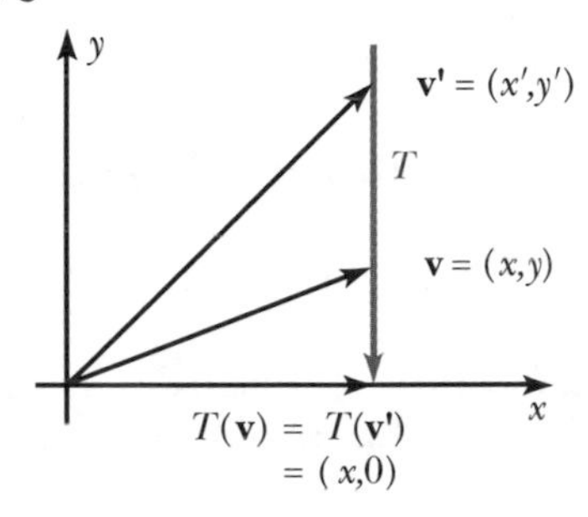

EXAMPLE 5-12

Let $T_A: R^3 \rightarrow R^2$ be defined by the matrix

$$A = \begin{bmatrix} 1 & 2 & 0 \\ -1 & 1 & 3 \end{bmatrix}.$$

Then $Ker\ T_A = \{X | AX = O\}$, which is just the solution space of the homogeneous system of linear equations with coefficient matrix A:

$$\begin{aligned} x_1 + 2x_2 &= 0 \\ -x_1 + x_2 + 3x_3 &= 0. \end{aligned}$$

So, we know "all about" $Ker\ T_A$. For example, we know it is a subspace of R^3. Furthermore, since A is row-equivalent to

$$A' = \begin{bmatrix} 1 & 0 & -2 \\ 0 & 1 & 1 \end{bmatrix},$$

which is a matrix in row-reduced echelon form, $Ker\ T_A$ is also the solution space of

$$\begin{aligned} x_1 - 2x_3 &= 0 \\ x_2 + x_3 &= 0. \end{aligned}$$

A basis for this solution space is, for example, $\{(2,-1,1)\}$. Therefore, $Ker\ T_A$ is the subspace of R^3 with basis $\{(2,-1,1)\}$.

Next, $Im\ T_A$ consists of all vectors in R^2 of the form

$$T(X) = AX = \begin{bmatrix} 1 & 2 & 0 \\ -1 & 1 & 3 \end{bmatrix} \begin{bmatrix} x_1 \\ x_2 \\ x_3 \end{bmatrix},$$

where $X = (x_1,x_2,x_3)$ is in R^3. For example, with $X = \mathbf{e}_1 = (1,0,0)$, we obtain the vector

$$\begin{bmatrix} 1 \\ -1 \end{bmatrix} = (1,-1) \in Im\ T_A.$$

Similarly, with $X = \mathbf{e}_2$ and $\mathbf{e}_3$, we see that the other two columns of A are also in $Im\ T_A$. ❑

In fact, it is not hard to see (Lemma 5-3 below) that $Im\ T_A$ is precisely $col(A)$, the column space of A. Therefore, we see that it is a subspace of R^2.

THEOREM 5-3

If $T{:}V \to W$ is linear, then $Ker\ T$ is a subspace of V and $Im\ T$ is a subspace of W.

PROOF Let $\mathbf{v}$ and $\mathbf{v}'$ be vectors in $Ker\ T$ and c be a scalar. Then,

$$T(c\mathbf{v} + \mathbf{v}') = cT(\mathbf{v}) + T(\mathbf{v}') = c\mathbf{0} + \mathbf{0} = \mathbf{0},$$

showing that $c\mathbf{v} + \mathbf{v}'$ is in $Ker\ T$. Thus $Ker\ T$ is a subspace of V.

Next, let $\mathbf{w}$ and $\mathbf{w}'$ be vectors in $Im\ T$. Then,

$$\mathbf{w} = T(\mathbf{v}) \text{ and } \mathbf{w}' = T(\mathbf{v}'),$$

for some $\mathbf{v}$ and $\mathbf{v}'$ in V. Now if c is a scalar,

$$T(c\mathbf{v} + \mathbf{v}') = cT(\mathbf{v}) + T(\mathbf{v}') = c\mathbf{w} + \mathbf{w}',$$

showing that $c\mathbf{w} + \mathbf{w}'$ is also in $Im\ T$. Therefore, $Im\ T$ is a subspace of W. ■

DEFINITION 5-4

If $T{:}V \to W$ is linear, $dim\ Ker\ T$ is called the **nullity** of T and $dim\ Im\ T$ is called the **rank** of T.

EXAMPLE 5-13

(a) Let $T:R^2 \to R^2$ be the projection onto the x axis, defined in Example 5-11. Then the nullity of T is 1, the dimension of the y axis; and the rank of T is also 1, the dimension of the x axis.

(b) Let $T_A:R^3 \to R^2$ be the map of Example 5-12. We saw that the kernel of this linear transformation has a basis $\{(2,-1,1)\}$. Thus the nullity of T_A is 1. Also, we noted that $Im\ T_A$ is the column space of A, which is the row space of the transpose of A. Now,

$$A^t = \begin{bmatrix} 1 & -1 \\ 2 & 1 \\ 0 & 3 \end{bmatrix}$$

is row-equivalent to

$$\begin{bmatrix} 1 & 0 \\ 0 & 1 \\ 0 & 0 \end{bmatrix},$$

so the rank of $T_A = 2$. ❑

Arithmetically minded readers may have noticed that in both examples, the sum of rank and nullity is the dimension of the domain of the linear transformation: in (a), they add to 2, the dimension of R^2; and in (b) they add to 3, the dimension of R^3. This is no coincidence, as the next theorem shows.

THEOREM 5-4

If V and W are finite dimensional and $T:V \to W$ is linear, then

$$dim\ V = \text{rank of } T + \text{nullity of } T.$$

The proof below can be omitted without loss of continuity. However, the underlying idea is quite simple and is illustrated schematically in Figure 5-9. We first prove a preparatory result which is useful in its own right.

LEMMA 5-2

If $\mathbf{v}_1, \ldots, \mathbf{v}_m$ generate V, then their images $T(\mathbf{v}_1), \ldots, T(\mathbf{v}_m)$ generate $Im\ T$.

PROOF If $T(\mathbf{v})$ is an arbitrary vector in $Im\ T$, then as $\mathbf{v}_1, \ldots, \mathbf{v}_m$

generate V,

$$\mathbf{v} = \sum_{i=1}^{m} c_i \mathbf{v}_i$$

for some scalars c_i. So,

$$T(\mathbf{v}) = T(\sum_{i=1}^{m} c_i \mathbf{v}_i) = \sum_{i=1}^{m} c_i T(\mathbf{v}_i),$$

showing that $T(\mathbf{v})$ is a linear combination of $T(\mathbf{v}_1)$, ..., $T(\mathbf{v}_m)$, thus proving the lemma. ■

PROOF of Theorem 5-4 Let $B' = \{\mathbf{v}_1,...,\mathbf{v}_k\}$ be a basis for *Ker T*, a subspace of V, and let $dim\ V = n$. We know by Theorems 4-8 and 4-9 (Corollary 2) that B' is part of a basis $B = \{\mathbf{v}_1,...,\mathbf{v}_k,\mathbf{v}_{k+1},...,\mathbf{v}_n\}$ for all of V. We show that $B'' = \{T(\mathbf{v}_{k+1}),...,T(\mathbf{v}_n)\}$ is a basis for *Im T*. The idea is illustrated in Figure 5-9.

FIGURE 5-9

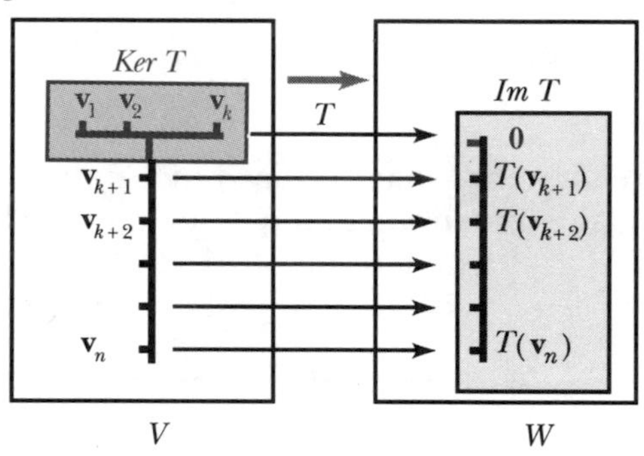

First, it follows from the lemma that

$$\begin{aligned} Im\ T &= span\ \{T(\mathbf{v}_1),...,T(\mathbf{v}_k),T(\mathbf{v}_{k+1}),...,T(\mathbf{v}_n)\} \\ &= span\ \{T(\mathbf{v}_{k+1}),...,T(\mathbf{v}_n)\}, \end{aligned}$$

since $T(\mathbf{v}_i) = \mathbf{0}$, for $1 \leq i \leq k$.

Second, this set is linearly independent as shown below. If

$$\sum_{i=k+1}^{n} c_i T(\mathbf{v}_i) = \mathbf{0},$$

then

$$T(\sum_{i=k+1}^{n} c_i \mathbf{v}_i) = \mathbf{0},$$

so

$$\sum_{i=k+1}^{n} c_i \mathbf{v}_i$$

is in *Ker T*. As B' is a basis for *Ker T*, this vector is therefore a linear combination of the vectors of B':

$$\sum_{i=k+1}^{n} c_i\mathbf{v}_i = \sum_{j=1}^{k} d_j\mathbf{v}_j,$$

for some $d_j \in R$. We rewrite this as

$$\sum_{j=1}^{k} d_j\mathbf{v}_j - \sum_{i=k+1}^{n} c_i\mathbf{v}_i = \mathbf{0},$$

which is a linear combination of vectors of B equal to the zero vector. Since B is linearly independent, all $d_j, c_i = 0$. Therefore, B'' is a basis for *Im T*.

Let us denote the number of elements in a set S by $\#(S)$. Then,

$$\begin{aligned} \text{rank of } T &= \#(B'') = n - k, \\ \text{nullity of } T &= \#(B') = k, \end{aligned}$$

and their sum is n, the dimension of V, as stated. ■

EXAMPLE 5-14

(a) Consider the linear transformation $T{:}R^3 \rightarrow R^3$ defined by

$$T(x,y,z) = (2x - y - z, x + y + z, x - y - z).$$

Ker T is the solution space of the homogeneous system

$$\begin{aligned} 2x - y - z &= 0 \\ x + y + z &= 0 \\ x - y - z &= 0. \end{aligned}$$

The reader may check that this has, for example, the solution $(0,1,-1)$, so *Ker T* is the one-dimensional space with basis $\{(0,1,-1)\}$.

To find *Im T*, we use Lemma 5-2.

$$\begin{aligned} Im\ T &= span\ \{T(\mathbf{e}_1), T(\mathbf{e}_2), T(\mathbf{e}_3)\} \\ &= span\ \{(2,1,1),(-1,1,-1),(-1,1,-1)\} \\ &= span\ \{(2,1,1),(-1,1,-1)\}. \end{aligned}$$

Since this last set of two vectors is clearly linearly independent, it follows that *Im T* is the two-dimensional subspace of R^3 with $\{(2,1,1),(-1,1,-1)\}$ as a basis. We note that

$$\text{nullity of } T + \text{rank of } T = 1 + 2 = 3 = dim\ R^3,$$

as predicted by Theorem 5-4.

(b) This example uses the ideas and notation of Example 5-9. We define a function called $Area{:}P(1) \rightarrow R$ by $Area(f) = Af(1)$. Recall that this is the algebraic sum of the areas bounded by the graph of $y = f(x)$, the x axis,

$x = 0$, and $x = 1$ (see Figure 5-10). Just as A is a linear transformation, so also is *Area*.

Now, *Ker Area* consists of those f in $P(1)$ with $Area(f) = 0$. If $f(x) = ax + b, x \in R$, the geometry in Figure 5-10 shows that $b = -a/2$ for f in the kernel. So

$$Ker\ Area = \{f \in P(1) \mid f(x) = ax - a/2, a \in R\}.$$

This is a 1-dimensional subspace of $P(1)$ with basis vector f given by (e.g., setting $a = 1$) $f(x) = x - 1/2$. Thus the nullity of *Area* is 1. By Theorem 5-4, we conclude that the rank of $Area = 2 - 1 = 1$, since $dim\ P(1) = 2$. This implies that $Im\ Area = R$, as it is a subspace of R having the same dimension as R. Geometrically, this simply means that there is a polynomial function $y = ax + b$ having *Area* as defined equal to any given number. We say that the linear transformation *Area* is **onto** the vector space R. ❑

FIGURE 5-10

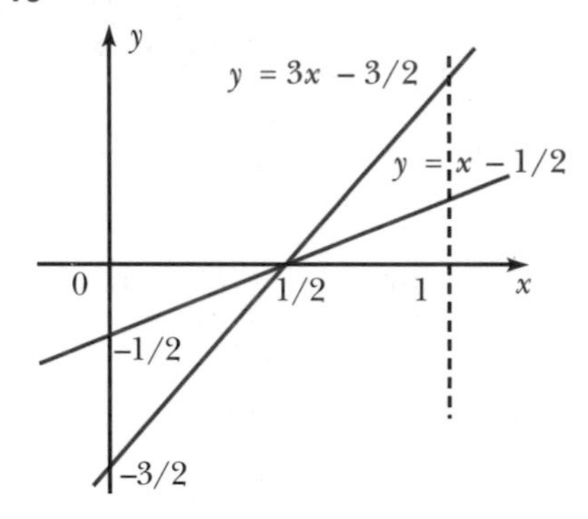

> **THEOREM 5-5**
>
> If A is an $m \times n$ matrix, then the row rank of A is equal to its column rank. Equivalently, a matrix and its transpose have the same rank.

This is the result first mentioned when we defined the rank of a matrix as its row rank (Definition 4-13). Its proof is a nice application of Theorem 5-4 and illustrates the usefulness of linear transformations. However, it may be omitted on first reading. The following fact used in the proof is a formalization of an earlier observation.

> **LEMMA 5-3**
>
> The column space of A is $Im\ T_A$.

PROOF By Lemma 5-2, $Im\ T_A$ is generated by $T_A(\mathbf{e}_1)$, $T_A(\mathbf{e}_2)$, ..., $T_A(\mathbf{e}_n)$ and these are precisely columns 1 through n respectively of A. ■

PROOF of Theorem 5-5 (Optional) As noted in Example 5-12, $Ker\ T_A$ is the solution space of the system $AX = O$. This space has dimension $n - r$, where r is the (row) rank of A, by Theorem 4-14. Therefore,

$$\text{nullity of } T_A = n - r. \tag{5-3}$$

On the other hand, $Im\ T_A$ is the column space of A, so

$$\text{rank of } T_A = \text{column rank of } A. \tag{5-4}$$

Finally, by Theorem 5-4,

$$\text{nullity of } T_A + \text{rank of } T_A = dim\ R^n = n.$$

Substituting equations 5-3 and 5-4 into this last equation yields

$$(n - r) + \text{column rank of } A = n,$$

i.e., column rank of $A = n - (n - r) = r =$ row rank of A. ■

Notice that in the course of the proof we also showed that rank of $T_A =$ column rank of $A =$ rank of A.

The next corollary is a useful application.

COROLLARY

Let A be an $m \times n$ matrix and B be $m \times 1$. Then, the system of equations $AX = B$ is consistent iff $rank\ (A{:}B) = rank\ A$.

(Note that this is question 15, Exercises 4-4.)

PROOF (Optional) $AX = B$ is consistent iff there is an X in $M(n \times 1)$ such that $T_A(X) = B$; i.e., iff B is in $Im\ T_A$. By Lemma 5-3,

$$Im\ T_A = col(A) = span\ \{A^{[1]},\ldots,A^{[n]}\}.$$

Therefore, $B \in Im\ T_A$ iff $B \in span\ \{A^{[1]},\ldots,A^{[n]}\}$, i.e., iff

$$span\ \{A^{[1]},\ldots,A^{[n]}\} = span\ \{A^{[1]},\ldots,A^{[n]},B\}.$$

Now, the left-hand side of the last equation is the column space of A, while the right-hand side is the column space of $(A{:}B)$. Since the former is a subspace of the latter, they are equal iff they have the same dimension. In other words, iff

$$\text{column rank } A = \text{column rank } (A{:}B).$$

Since we now know that column and row ranks are the same, it follows that $AX = B$ is consistent iff *rank* A = *rank* $(A:B)$, as required. (See also questions 14 and 15, Exercises 5-2.) ■

We next describe the method whereby the action of *any linear transformation* $T:V \to W$ *can be described as a multiplication by some matrix* A. One of the reasons for doing this is that we can then find $T(\mathbf{v})$ by means of the matrix A. We begin with the special case $V = M(n \times 1) = R^n$, $W = M(m \times 1) = R^m$, and $T:R^n \to R^m$.

We already know that in the case of a matrix transformation $T_A:R^n \to R^m$, the columns of A are just the vectors $T_A(\mathbf{e}_1)$, $T_A(\mathbf{e}_2)$, ..., $T_A(\mathbf{e}_n)$. This suggests that, given an arbitrary linear map $T:R^n \to R^m$, the column vectors

$$[T(\mathbf{e}_1)]_{B(m)},\ [T(\mathbf{e}_2)]_{B(m)},\ \ldots,\ [T(\mathbf{e}_n)]_{B(m)}$$

might serve as the *columns* of the matrix A that we want for T. This is, in fact, the case.

THEOREM 5-6

If $T:R^n \to R^m$ is linear, then $T = T_A$, where A is the $m \times n$ matrix with columns $[T(\mathbf{e}_1)]_{B(m)}$, $[T(\mathbf{e}_2)]_{B(m)}$, ..., $[T(\mathbf{e}_n)]_{B(m)}$.

PROOF By Theorem 5-2, it will be sufficient to show that T and T_A agree on the elements of the basis $B(n)$. But this is the case by the definition of A, since $T_A(\mathbf{e}_i)$ is the ith column of A. This is $T(\mathbf{e}_i)$, written as a matrix, i.e., $[T(\mathbf{e}_i)]_{B(m)}$. ■

DEFINITION 5-5

We call A the matrix of T with respect to the standard bases for R^n and R^m. It is denoted by $[T]$.

EXAMPLE 5-15

We compute the matrices of some linear transformations from previous examples.

(a) Example 5-5(a)
Here $T:R^2 \to R^3$ was defined by $T(x,y) = (x + 2y, -y, 2x - y)$. Therefore,

$$T(\mathbf{e}_1) = T(1,0) = (1,-1,2),$$
$$T(\mathbf{e}_2) = T(0,1) = (2,-1,-1),$$

and

$$[T] = \begin{bmatrix} 1 & 2 \\ -1 & -1 \\ 2 & -1 \end{bmatrix}.$$

Knowing this, we see, for example, that *Ker T* is the solution space of $AX = O$, where $A = [T]$, and *Im T* is the column space of A.

(b) Example 5-7
Here $T:R^2 \to R^2$ was defined by a rotation through an angle θ in a counterclockwise direction. We obtain $[T]$ in two ways:

(i) First, we saw that

$$T(x,y) = (x\cos\theta - y\sin\theta, x\sin\theta + y\cos\theta).$$

It follows that

$$T(\mathbf{e}_1) = (\cos\theta, \sin\theta),$$
$$T(\mathbf{e}_2) = (-\sin\theta, \cos\theta).$$

Therefore,

$$[T] = \begin{bmatrix} \cos\theta & -\sin\theta \\ \sin\theta & \cos\theta \end{bmatrix}.$$

(ii) Second, consider Figure 5-11. We see that the coordinates of $T(\mathbf{e}_1)$ are $(\cos\theta, \sin\theta)$, because $|\mathbf{e}_1|$ (= the length of $\mathbf{e}_1$) = 1. Similarly, the coordinates of $T(\mathbf{e}_2)$ are $(-\sin\theta, \cos\theta)$. Thus $[T]$ is as stated above.

FIGURE 5-11

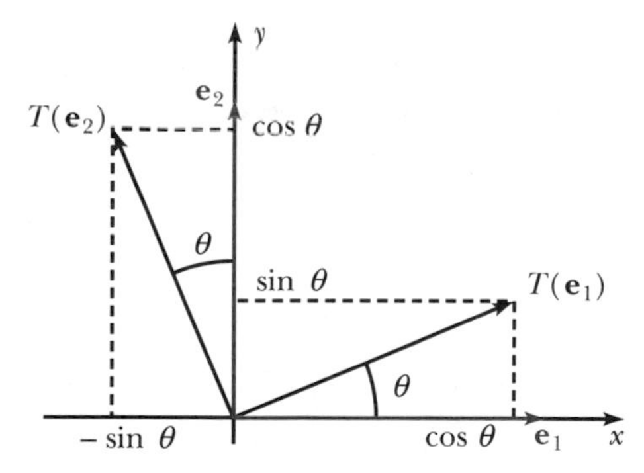

(c) Example 5-11
Here $T:R^2 \to R^2$ was defined by $T(x,y) = (x,0)$, the projection onto the x axis. Since

$$T(\mathbf{e}_1) = (1,0) \text{ and } T(\mathbf{e}_2) = (0,0),$$

it follows that

$$[T] = \begin{bmatrix} 1 & 0 \\ 0 & 0 \end{bmatrix},$$

a 2×2 matrix. ❑

Now we extend this idea to linear transformations $T:V \to W$ between *arbitrary* vector spaces of finite dimension. Let $B = \{\mathbf{v}_1,\ldots,\mathbf{v}_n\}$ be an (ordered) basis for V and $B' = \{\mathbf{w}_1,\ldots,\mathbf{w}_m\}$ be an (ordered) basis for W. We imitate the definition for linear maps from R^n to R^m to get

DEFINITION 5-6

The matrix of T with respect to the bases B and B', denoted $[T;B/B']$, is the $m \times n$ matrix with columns $[T(\mathbf{v}_1)]_{B'}, \ldots, [T(\mathbf{v}_n)]_{B'}$.

When $V = W$ (so T is a linear operator), we will frequently be interested in $[T;B/B]$ (i.e., the special case $B' = B$). We abbreviate this to $[T;B]$. In some books, $[T;B/B']$ is denoted $[T]_B{}^{B'}$.

EXAMPLE 5-16

Let $D:P(2) \to P(1)$ be the function such that if $f \in P(2)$, $f(x) = ax^2 + bx + c$, then Df in $P(1)$ is defined by

$$Df(x) = 2ax + b.$$

We leave it for the reader to verify that D is linear (readers familiar with calculus will recognize D as the derivative operator on $P(2)$).

Let $B = \{1,x,x^2\}$ and $B' = \{1,x\}$ be the usual ordered bases for $P(2)$ and $P(1)$ respectively. We determine the matrix of D with respect to B and B' as follows.

$D(1) = D(1 + 0x + 0x^2) = 2(0)x + 0 = (0)1 + (0)x$, so

$$[D(1)]_{B'} = \begin{bmatrix} 0 \\ 0 \end{bmatrix}.$$

$D(x) = D(0 + 1x + 0x^2) = 2(0)x + 1 = (1)1 + (0)x$, so

$$[D(x)]_{B'} = \begin{bmatrix} 1 \\ 0 \end{bmatrix}.$$

$D(x^2) = D(0 + 0x + 1x^2) = (0)1 + 2(1)x$, so

$$[D(x^2)]_{B'} = \begin{bmatrix} 0 \\ 2 \end{bmatrix}.$$

Therefore,

$$[D;B/B'] = \begin{bmatrix} 0 & 1 & 0 \\ 0 & 0 & 2 \end{bmatrix}.$$

Notice that it is a 2×3 matrix since $P(2)$ and $P(1)$ have dimensions 3 and 2 respectively. ❑

EXAMPLE 5-17

Let $T:M(2\times 2) \rightarrow M(2\times 2)$ be the linear map defined by $T(M) = AM$, where A is the matrix:

$$A = \begin{bmatrix} 1 & 2 \\ -2 & 1 \end{bmatrix}.$$

We leave it for the reader to verify that T is linear (see also question 5, Exercises 5-1).

Let $B = \{E_1,...,E_4\}$ be the standard basis for $M(2\times 2)$. We determine $[T;B]$ as follows.

$$T(E_1) = \begin{bmatrix} 1 & 2 \\ -2 & 1 \end{bmatrix}\begin{bmatrix} 1 & 0 \\ 0 & 0 \end{bmatrix} = \begin{bmatrix} 1 & 0 \\ -2 & 0 \end{bmatrix} = 1E_1 - 2E_3.$$

Similarly,

$$T(E_2) = \begin{bmatrix} 0 & 1 \\ 0 & -2 \end{bmatrix} = E_2 - 2E_4,$$

$$T(E_3) = \begin{bmatrix} 2 & 0 \\ 1 & 0 \end{bmatrix} = 2E_1 + E_3,$$

$$T(E_4) = \begin{bmatrix} 0 & 2 \\ 0 & 1 \end{bmatrix} = 2E_2 + E_4.$$

Therefore,

$$[T;B] = \begin{bmatrix} 1 & 0 & 2 & 0 \\ 0 & 1 & 0 & 2 \\ -2 & 0 & 1 & 0 \\ 0 & -2 & 0 & 1 \end{bmatrix}. \ ❑$$

EXAMPLE 5-18

(a) Let V and W be any finite dimensional vector spaces, with bases B and B' respectively. If $0{:}V \to W$ is the zero map (Example 5-6), then $[0;B/B'] = O$ is the zero matrix of order $(dim\ W) \times (dim\ V)$.

(b) Let $1_V{:}V \to V$ be the identity linear operator on a vector space V of dimension n. Then $[1_V;B] = I$ is the $n \times n$ identity matrix for any basis B for V. This is the case since $1_V(\mathbf{v}_i) = \mathbf{v}_i$ for any $\mathbf{v}_i$ in B implies that $[1_V(\mathbf{v}_i)]_B$ is the ith column of I. ❑

THEOREM 5-7

If $T{:}V \to W$ is linear and V and W are finite dimensional with bases B and B' respectively, then $[T(\mathbf{v})]_{B'} = A[\mathbf{v}]_B = T_A[\mathbf{v}]_B$, where $A = [T;B/B']$.

Note that the theorem shows that the coordinate matrix of $T(\mathbf{v})$ is the value of the matrix transformation T_A at the coordinate matrix of $\mathbf{v}$. Therefore, this result represents T as a matrix multiplication. If T is a linear operator on V, the formula reduces to

$$[T(\mathbf{v})]_B = [T;B][\mathbf{v}]_B,$$

where B is a basis for V.

PROOF (Optional) Let $B = \{\mathbf{v}_1,\ldots,\mathbf{v}_n\}$. We first prove the result for the case $\mathbf{v} \in B$, i.e., $\mathbf{v} = \mathbf{v}_i$, $i = 1, 2, \ldots, n$. By definition, $[T(\mathbf{v}_i)]_{B'}$ is the ith column of $[T;B/B']$. Since $[\mathbf{v}_i]_B$ has only a 1 in the ith spot and 0's elsewhere, it follows that

$$[T;B/B'][\mathbf{v}_i]_B = i^{\text{th}} \text{ column of } [T;B/B'],$$

proving the result for this case.

An arbitrary $\mathbf{v} \in V$ is a linear combination of the vectors of B,

$$\mathbf{v} = c_1\mathbf{v}_1 + \ldots + c_n\mathbf{v}_n,$$

for some $c_i \in R$. By the result of question 8, Exercises 5-1 (see also the next section), the operation of taking the matrix of a vector is linear, so

$$[T(\mathbf{v})]_{B'} = [T(\sum_{i=1}^{n} c_i\mathbf{v}_i)]_{B'} = [\sum_{i=1}^{n} c_iT(\mathbf{v}_i)]_{B'} = \sum_{i=1}^{n} c_i[T(\mathbf{v}_i)]_{B'}$$

$$= \sum_{i=1}^{n} c_i[T;B/B'][\mathbf{v}_i]_B = [T;B/B'][\sum_{i=1}^{n} c_i\mathbf{v}_i]_B = [T;B/B'][\mathbf{v}]_B,$$

proving the result in general. ■

COROLLARY 1

With the hypotheses stated in Theorem 5-7 above,

(1) *Ker T* consists of those vectors **v** in V such that $[\mathbf{v}]_B$ is in $Ker\ T_A$;

(2) *Im T* consists of those **w** in W such that $[\mathbf{w}]_{B'}$ is in $Im\ T_A$.

PROOF This follows from the fact that the operation of taking matrices of vectors is one-to-one and onto, a fact which we also used in our computations in Chapter 4 (Theorem 4-12) and which we will prove in the next section. ■

COROLLARY 2

The rank of $T = rank\ [T;B/B']$. ■

EXAMPLE 5-19

Consider again the map $D:P(2)\rightarrow P(1)$ defined in Example 5-16. We saw that

$$[D;B/B'] = \begin{bmatrix} 0 & 1 & 0 \\ 0 & 0 & 2 \end{bmatrix}$$

$= A$, say. If f in $P(2)$ is defined by $f(x) = ax^2 + bx + c$, $x \in R$, then

$$[f]_B = \begin{bmatrix} c \\ b \\ a \end{bmatrix}$$

and $f \in Ker\ D$ iff

$$\begin{bmatrix} 0 & 1 & 0 \\ 0 & 0 & 2 \end{bmatrix} \begin{bmatrix} c \\ b \\ a \end{bmatrix} = O,$$

which holds true iff $b = a = 0$. Thus, *Ker D* consists of the f in $P(2)$ of the form $f(x) = c$, for some number c, so $Ker\ D = P(0) = R$. Of course, this is clear from the definition of D as well.

Next, the column space of A has dimension 2, which is the dimension of $P(1)$. Therefore, $Im\ D = P(1)$. In fact, if $g(x) = ax + b \in P(1)$, then defining f in $P(2)$ by

$$f(x) = ax^2/2 + bx + c,$$

where c is any scalar, we find that

$$Df(x) = ax + b = g(x).$$

This shows that there are infinitely many vectors f in $P(2)$ that are mapped to any given g in $P(1)$ by D. And, any two such vectors differ only by a scalar c, an element of *Ker D*. ❑

EXAMPLE 5-20

(a) If $V = R^n$, $W = R^m$, $T{:}V \to W$, then $[T;B(n)/B(m)] = [T]$, where $B(n)$, $B(m)$ are the standard bases, as before. This shows that the general notion of matrix of a linear transformation (Definition 5-6) agrees with the special case (Theorem 5-6).

(b) If B or B' is *not* a standard basis, and $T{:}R^n \to R^m$ is linear, then $[T;B/B']$ is *not* $[T]$. For example, let $T{:}R^2 \to R^2$ be defined by $T(x,y) = (2x + y, -x + 2y)$. Then,

$$[T] = \begin{bmatrix} 2 & 1 \\ -1 & 2 \end{bmatrix}.$$

Now let $B = \{\mathbf{v}_1, \mathbf{v}_2\}$, $B' = \{\mathbf{w}_1, \mathbf{w}_2\}$, where $\mathbf{v}_1 = (4,1)$, $\mathbf{v}_2 = (2,2)$, $\mathbf{w}_1 = (-1,1)$, $\mathbf{w}_2 = (1,1)$. Then,

$$T(\mathbf{v}_1) = T(4,1) = (9,-2),$$
$$T(\mathbf{v}_2) = T(2,2) = (6,2).$$

To find $[T;B/B']$, we must first find the coordinate matrices of these vectors with respect to the basis B'. That is, we must express $(9,-2)$ and $(6,2)$ as linear combinations of $\mathbf{w}_1$ and $\mathbf{w}_2$. In the case of $(9,-2)$, this means finding x and y such that

$$(9,-2) = x(-1,1) + y(1,1).$$

As usual, this leads to a system of equations for x and y, whose solution is easily seen to be $x = -11/2$, $y = 7/2$. Therefore,

$$[T(\mathbf{v}_1)]_{B'} = \begin{bmatrix} -11/2 \\ 7/2 \end{bmatrix}.$$

Similarly, we find that

$$[T(\mathbf{v}_2)]_{B'} = \begin{bmatrix} -2 \\ 4 \end{bmatrix},$$

so,

$$[T;B/B'] = \begin{bmatrix} -11/2 & -2 \\ 7/2 & 4 \end{bmatrix} \neq [T]. \ ❑$$

It is not unreasonable at this point for readers to wonder why, given $T:R^n \to R^m$, we would be interested in finding $[T;B/B']$ for bases B, B' different from the standard bases. This question is pursued in Chapter 6, and is analogous to the case of choosing new axes in the plane. Readers familiar with the analytic geometry of conics will recall that frequently the equation of, say, a parabola, may be simplified by an appropriate choice of new axes. Similarly, we shall see that in many cases, $[T;B/B']$ may be much simpler than $[T]$ (see question 19, Exercises 5-2).

EXERCISES 5-2

In questions 1 through 5 below, find formulae for the linear transformations specified and compute the values as indicated.

1. $T:R^2 \to R$ such that $T(\mathbf{e}_1) = 1$ and $T(\mathbf{e}_2) = -2$. Find $T(x,y)$, $T(-2,3)$, and $T(2,1)$.

2. $T:R^2 \to R^2$ such that $T(\mathbf{e}_1) = (1,3)$ and $T(\mathbf{e}_2) = (-4,2)$. Find $T(x,y)$, $T(-1,-1)$, and $T(5,-2)$.

3. $T:R^2 \to R^4$ such that $T(\mathbf{e}_1) = (1,-1,0,2)$ and $T(\mathbf{e}_2) = (-1,1,2,0)$. Find $T(x,y)$, $T(2,6)$, and $T(3,-2)$.

4. $T:R^4 \to R^2$ such that $T(\mathbf{e}_1) = T(\mathbf{e}_2) = (1,1)$ and $T(\mathbf{e}_3) = T(\mathbf{e}_4) = (-1,-1)$. Find $T(w,x,y,z)$ and $T(2,-2,3,4)$.

5. $T:R^2 \to R^2$ such that $T(\mathbf{e}_1) = \mathbf{e}_2$ and $T(\mathbf{e}_2) = 2\mathbf{e}_2$. Find $T(x,y)$, $T(0,4)$, and $T(2,2)$.

6. For each of the linear transformations defined in questions 1 through 5 above, find **(a)** $[T]$; **(b)** a basis for $Ker\ T$; **(c)** a basis for $Im\ T$; **(d)** the nullity of T; **(e)** the rank of T.

7. Let $T:V \to W$ be linear with V finite dimensional. If $T(\mathbf{v}) = \mathbf{0}$ for all $\mathbf{v}$ in a basis B for V, show that $T = 0$ (the zero map).

8. Let $T:R^2 \to R^2$ be the rotation through an angle θ in a counterclockwise direction, with $0 \le \theta \le \pi/2$ (see Examples 5-7 and 5-15(b)). Show that $Ker\ T = \{\mathbf{0}\}$ and $Im\ T = R^2$ in two ways: **(a)** by a geometrical argument; **(b)** by considering $[T]$.

9. Let T_A be the matrix transformation defined by

$$A = \begin{bmatrix} 1 & -1 & 2 & 3 & 0 \\ 2 & 0 & 1 & -1 & 1 \\ 1 & 1 & -1 & -4 & -1 \\ 0 & 2 & -3 & -7 & 1 \end{bmatrix}.$$

Find a basis for $Ker\ T_A$ and for $Im\ T_A$, and verify that Theorem 5-4 holds in this case.

10. Let $T:M(2\times 2)\to M(2\times 2)$ be the linear operator defined by $T(M) = AM$, where A is given by

$$A = \begin{bmatrix} 1 & 2 \\ -1 & -2 \end{bmatrix}.$$

(a) Find $[T;B]$.

(b) Find bases for *Ker* T and *Im* T.

(c) Verify that Theorem 5-4 holds.

11. Find a formula for the linear operator $S:R^2\to R^2$ with $S(1,1) = (2,1)$ and $S(-1,1) = (1,2)$. Explain why such an S exists and find $[S]$.

12. **(a)** Let $T:R^5\to R^3$ be a linear transformation with $[T] = A$ and nullity 2. Explain why the system $AX = B$ is consistent for any 3×1 matrix B.

(b) Let $T:R^3\to R^3$ be a linear operator having matrix A and nullity 0. Explain why the system $AX = O$ has an unique solution.

(c) Let $T:R^6\to R^4$ be a linear map with matrix A and rank 3.

(i) Explain why the system of equations $AX = B$ need not be consistent.

(ii) What is the nullity of T?

(iii) What is the dimension of the solution space of the system $AX = O$?

13. **(a)** Construct two *different* linear operators on R^3 having as kernels the plane with equation $x + 2y - z = 0$ and as images the line with parametric equations

$$\begin{aligned} x &= t \\ y &= 2t \\ z &= -t, t \in R. \end{aligned}$$

(b) Show that there is exactly one linear operator T on R^3 with the properties described in (a), and such that $T(1,1,1) = (-1/2,-1,1/2)$.

(c) Is there a linear operator on R^3 with kernel as described in (a) and having as image the line with parametric equations

$$\begin{aligned} x &= 1 + t \\ y &= 1 + 2t \\ z &= -t, t \in R? \end{aligned}$$

14. Let A be an $m\times n$ matrix. Show that the set of B in $M(m\times 1)$ such that $AX = B$ is consistent is a subspace.

15. Let A be an $m\times n$ matrix and B be $m\times 1$.

(a) Show that $AX = B$ has an unique solution iff *rank* $(A{:}B) =$ *rank* $A = n$.

(b) Show that $AX = B$ has infinitely many solutions iff *rank* $(A{:}B) =$ *rank* $A < n$.

16. Let $D:P(2) \to P(1)$ be the linear map described in Examples 5-16 and 5-19. Let B be the usual basis for $P(2)$ and $B' = \{x + 1, x - 1\}$.

(a) Verify that B' is a basis for $P(1)$.

(b) Find $[D;B/B']$.

17. Let $T:R^3 \to R^3$ be defined by $T(x,y,z) = (x - z, 2y + z, x + y + z)$ and let $B = \{(1,0,0),(1,1,0),(1,1,1)\}$ and $B' = \{(2,1,1),(1,2,1),(1,1,2)\}$.

(a) Find $[T]$.

(b) Find $[T;B/B']$.

(c) Use (b) to find the nullity of T and the rank of T.

18. Recall that if $\mathbf{N} = (l,m,n)$ is a vector in R^3, the plane through the origin and perpendicular to $\mathbf{N}$ has the equation $lx + my + nz = 0$.

(a) Find a formula for the projection operator $T:R^3 \to R^3$ that maps any $\mathbf{v}$ in R^3 to its orthogonal projection on this plane.

(b) What is *Ker* T?

19. Let $S:R^2 \to R^2$ be the linear operator given by $S(x,y) = (x + 2y, 2x + y)$.

(a) Find $[S]$.

(b) Let $B = \{(1,1),(1,-1)\}$ and find $[S,B]$. Is it "simpler" than $[S]$?

5-3 Isomorphism (Optional)

Isomorphism is a concept we have used informally on several occasions in earlier sections. For example, in Chapter 2, we agreed to "identify" R^n with $M(n \times 1)$ because, as vector spaces, they are *essentially the same*. We say that they are **isomorphic** vector spaces. Another informal way of thinking of this is that R^n and $M(n \times 1)$ are "different names for the same set of objects".

The notion of isomorphism is important because it helps us to *classify* the truly distinct examples of vector spaces we encounter. Thus, we shall see that

(1) any vector space of dimension n is essentially the same as R^n (i.e., is isomorphic to R^n);

(2) linear transformations are essentially the same as matrices.

To say that vector spaces V and W are isomorphic means that there is a one-to-one correspondence (see Chapter 1) between vectors of V and W such that "doing algebra in V *corresponds* to doing algebra in W". We can now make this idea precise using linear transformations.

Recall that a function $f:S \to S'$ (where S, S' are sets) is called

(1) 1-1 (**"one-to-one"**) if $x \neq y$ in S implies that $f(x) \neq f(y)$ in S', for all $x, y \in S$. In other words, no two elements of S are mapped to the same image in S'. Alternatively, no element of S' is the image of more than one element of S;

(2) **onto** if every y in S' is $f(x)$ for some x in S. In other words, every y in S' is the image of some (and possibly several) x in S.

So, if f is 1-1 *and* onto, every y in S' is the image of *exactly one* element x in S. Therefore, we say that a 1-1, onto function f establishes a *one-to-one correspondence* between S and S'. (This elaborates Definition 1-9.)

> **DEFINITION 5-7**
>
> A linear transformation $T{:}V \to W$ is called an **isomorphism** (from V to W) if T is 1-1 and onto. In this case (i.e., if such a T exists), V and W are said to be **isomorphic**.

EXAMPLE 5-21

(a) Let $T{:}R^2 \to R^2$ be the counterclockwise rotation through an angle θ (see Example 5-7). In this case, it is intuitively clear that T is both 1-1 and onto. As we can see, every vector $\mathbf{w} \neq \mathbf{0}$ in R^2 is the image of exactly one vector $\mathbf{v}$ in R^2: $\mathbf{v}$ is the unique vector which when rotated through an angle θ coincides with $\mathbf{w}$. And by definition, $T(\mathbf{0}) = \mathbf{0}$, so T is an isomorphism.

(b) Let $T{:}R^2 \to R$ be the projection onto the x axis, $T(x,y) = x$ (see also Example 5-11). Then although T is onto, it is not 1-1, since all vectors (x,y) lying vertically over $(x,0)$ are mapped to x by T. Therefore, T is not an isomorphism. ❑

EXAMPLE 5-22

Let $T{:}R^2 \to R^3$ be defined by $T(x,y) = (x + y, x - y, x + 2y)$. Then, T is 1-1 as shown below. If $T(x,y) = T(x',y')$, we get

$$(x + y, x - y, x + 2y) = (x' + y', x' - y', x' + 2y'),$$

which implies

$$\begin{aligned} x + y &= x' + y' \\ x - y &= x' - y'. \end{aligned}$$

Adding these two equations, we get $2x = 2x'$, hence $x = x'$ and also $y = y'$. Thus $(x,y) = (x',y')$, so T is 1-1.

However, T is not onto. One way of seeing this is to recall that $Im\ T$ is the column space of $[T]$, a 3×2 matrix, so

$$dim\ Im\ T = rank\ [T] = 2,$$

which is less than $dim\ R^3\ (=3)$. Thus, T is not onto. ❑

Notice that the linearity of T plays a large role in this argument: T being linear implies $Im\ T$ is a subspace, so we can speak of its dimension. More generally,

THEOREM 5-8

Let $T:V \to W$ be linear. Then

(1) T is 1-1 iff $Ker\ T = \{\mathbf{0}\}$;

(2) T is onto iff $Im\ T = W$.

PROOF (1) Suppose first that T is 1-1. Then the unique vector mapped to $\mathbf{0}$ must be $\mathbf{0}$, since it is always true that $T(\mathbf{0}) = \mathbf{0}$. So $Ker\ T = \{\mathbf{0}\}$. Conversely, if $Ker\ T$ is $\{\mathbf{0}\}$, suppose $T(\mathbf{v}) = T(\mathbf{v}')$ for some $\mathbf{v}$, $\mathbf{v}'$ in V. Then, by linearity, $T(\mathbf{v} - \mathbf{v}') = \mathbf{0}$, so $\mathbf{v} - \mathbf{v}'$ is in $Ker\ T$. Therefore, $\mathbf{v} - \mathbf{v}' = \mathbf{0}$, or $\mathbf{v} = \mathbf{v}'$. It follows that T is 1-1.

(2) The result is obvious from the definitions. ■

COROLLARY 1

$T:V \to W$ is 1-1 iff the nullity of T is 0 and onto iff the rank of T is $dim\ W$. ■

EXAMPLE 5-23

(a) Let us consider again the linear transformation T of Example 5-22. Its matrix is

$$\begin{bmatrix} 1 & 1 \\ 1 & -1 \\ 1 & 2 \end{bmatrix},$$

which has rank 2. So T is not onto ($dim\ R^3 = 3$). However, it is 1-1, as the nullity of T is $2 - 2 = 0$.

(b) Let T be the rotation through an angle θ, with matrix (see Example 5-15)

$$[T] = \begin{bmatrix} \cos\theta & -\sin\theta \\ \sin\theta & \cos\theta \end{bmatrix}.$$

This matrix always has rank 2, so T is 1-1 and onto.

(c) Let $T_A:R^3 \to R^2$ be the matrix transformation with

$$A = \begin{bmatrix} 1 & 2 & 0 \\ -1 & 1 & 3 \end{bmatrix}.$$

The nullity of T_A is 1 and the rank of T_A is 2 (see Examples 5-12 and 5-15), so T_A is onto but not 1-1.

(d) We saw in Example 5-19 that the linear mapping $D:P(2) \to P(1)$ with $D(ax^2 + bx + c) = 2ax + b$ is onto but not 1-1 (using basic principles). Also, its matrix was seen to be

$$[D;B/B'] = \begin{bmatrix} 0 & 1 & 0 \\ 0 & 0 & 2 \end{bmatrix}$$

which has rank 2. Thus, D has rank 2 and nullity 1, so we see again that D is onto but not 1-1. ❑

Now, we saw in Theorem 5-4 that the rank and nullity of a linear transformation are related — they must add to the dimension of V. This leads to the following result.

COROLLARY 2

If $dim\ V = dim\ W$, then a linear map $T:V \to W$ is 1-1 iff it is onto.

PROOF T is 1-1 iff the nullity of $T = 0$. By Theorem 5-4, this is true iff the rank of T is $dim\ V = dim\ W$. As $Im\ T$ is a subspace of W, rank of $T = dim\ W$ iff $Im\ T = W$, i.e., iff T is onto. ■

EXAMPLE 5-24

We use the preceding corollary to show that $V = M(m \times n)$ is isomorphic to $W = M(n \times m)$. Both have dimension mn, so we only need to find a linear map $T:M(m \times n) \to M(n \times m)$ which is either 1-1 or onto (in which case it is automatically both). The linear map we need is provided by the transpose operation on matrices: we put $T(A) = A^t$, an $n \times m$ matrix when A is $m \times n$. Furthermore, T is linear since, by Theorem 2-5,

$$T(cA + B) = (cA + B)^t = cA^t + B^t = cT(A) + T(B),$$

for any scalar c, and matrices A, B in $M(m \times n)$. Finally, T is onto, for if C is in $M(n \times m)$, then C^t is in $M(m \times n)$ and

$$T(C^t) = (C^t)^t = C,$$

again by Theorem 2-5. Thus T is an isomorphism, and $M(m \times n)$ and $M(n \times m)$ are isomorphic vector spaces. ❑

With respect to the intuitive meaning of isomorphism given earlier, the above example certainly makes sense, for, given mn numbers, we can arrange them in m rows with n columns, or in n rows with m columns. Also, algebraically these are essentially the same, in the sense we have now made precise (Definition 5-7). The following result summarizes the ways in which isomorphic vector spaces are essentially the same.

THEOREM 5-9

Let $T:V \rightarrow W$ be an isomorphism. Then T establishes a 1-1 correspondence $\mathbf{v} \leftrightarrow T(\mathbf{v})$ between V and W with these properties:

(1) $\mathbf{v} + \mathbf{w} \leftrightarrow T(\mathbf{v}) + T(\mathbf{w})$.

(2) $c\mathbf{v} \leftrightarrow cT(\mathbf{v}) \quad (c \in R)$.

(3) $\sum_{i=1}^{n} c_i\mathbf{v}_i \leftrightarrow \sum_{i=1}^{n} c_iT(\mathbf{v}_i) \quad (c_i \in R)$.

(4) A set of vectors $\{\mathbf{v}_1,...,\mathbf{v}_k\}$ in V is linearly independent in V (respectively, generates V; is a basis for V) iff $\{T(\mathbf{v}_1),...,T(\mathbf{v}_k)\}$ is linearly independent in W (respectively, generates W; is a basis for W).

PROOF We leave the proof for the reader (question 9, Exercises 5-3). ■

The next theorem is the fundamental result regarding isomorphism of vector spaces.

THEOREM 5-10

A vector space V has dimension n iff it is isomorphic to $M(n \times 1)$.

PROOF We already have a candidate for an isomorphism since, given a basis $B = \{\mathbf{v}_1,...,\mathbf{v}_n\}$ for V, we can associate the $n \times 1$ matrix $[\mathbf{v}]_B$ with any $\mathbf{v}$ in V. We denote the function defined in this way by $mat:V \rightarrow M(n \times 1)$:

$$mat(\mathbf{v}) = [\mathbf{v}]_B, \text{ for } \mathbf{v} \in V.$$

(Note that different choices of B will lead to *different* functions *mat*. We assume a specific B throughout.) Now, *mat* is certainly onto $M(n \times 1)$, since if $C = (c_i)$ is in $M(n \times 1)$, then $mat(\mathbf{v}) = C$ if we let $\mathbf{v} = c_1\mathbf{v}_1 + ... + c_n\mathbf{v}_n$. Therefore, *mat* will be an isomorphism if it is shown to be linear. We leave this as an exercise for the reader.

Conversely, if $T:V \rightarrow M(n \times 1)$ is an isomorphism, suppose that V had a basis of m elements. Then, their images under T would be a basis of m elements for $M(n \times 1)$, by Theorem 5-9. Since $dim\ M(n \times 1) = n$, it follows that $m = n$, completing the proof. ■

Note in particular that R^n is isomorphic to $M(n \times 1)$, so Theorem 5-10 provides the justification for our identification of R^n with $M(n \times 1)$ in earlier sections, as well as for our use of matrices of vectors in computations in section 4-5, and for the corollary of Theorem 5-7. Moreover, it follows from the theorem that

COROLLARY

Two finite dimensional vector spaces are isomorphic iff they have the same dimension.

PROOF This is left for the reader (question 20, Exercises 5-3). ■

We next apply these ideas to the set of all linear transformations $T:V \to W$, where V and W are finite dimensional vector spaces. Denote this set by $L(V,W)$. Our goal will be to show that it is a vector space which is isomorphic to $M(m \times n)$, when $n = \dim V$ and $m = \dim W$. We already have a function mapping linear transformations to matrices, for, if B and B' are bases for V and W respectively, we can associate with any T in $L(V,W)$ its matrix $[T;B/B']$. However, we must first show that $L(V,W)$ is indeed a vector space.

DEFINITION 5-8

(1) If T and T' are in $L(V,W)$, their **sum** is the function $T + T':V \to W$ defined by

$$(T + T')(\mathbf{v}) = T(\mathbf{v}) + T'(\mathbf{v}), \text{ for } \mathbf{v} \in V.$$

(2) If c is a scalar, the **scalar multiple** of T by c is the function $cT:V \to W$ defined by

$$(cT)(\mathbf{v}) = cT(\mathbf{v}), \ \mathbf{v} \in V.$$

Notice that the definitions are possible because $T(\mathbf{v})$ and $T'(\mathbf{v})$ are vectors in W, so can be added and multiplied by scalars. The reader should compare these definitions with the definitions in $F(S)$ (see the discussion following Definition 4-2). However, in this case we must verify that $T + T'$ and cT are also linear, so that they belong to $L(V,W)$.

Now, when T and T' are matrix transformations from $M(n \times 1)$ to $M(m \times 1)$, the linearity of $T + T'$ and cT' follows by matrix algebra, as the next example shows.

EXAMPLE 5-25

Let $T = T_A$ and $T' = T_{A'}$ where A and A' are $m \times n$ matrices. Then for X and $X' \in M(n \times 1)$ and $k \in R$,

$$\begin{aligned}(T + T')(kX + X') &= T_A(kX + X') + T_{A'}(kX + X') \\ &= A(kX + X') + A'(kX + X') \\ &= k(A + A')X + (A + A')X' \\ &= kT_{A+A'}(X) + T_{A+A'}(X') \\ &= k(T + T')(X) + (T + T')(X'),\end{aligned}$$

since $T + T' = T_{A+A'}$. Therefore, $T + T'$ is linear. The proof for the linearity of cT is similar. ❑

More generally, we have

> **THEOREM 5-11**
>
> $T + T'$ and cT are linear for any T and T' in $L(V,W)$ and scalars c. Furthermore, $L(V,W)$ is a vector space with respect to these operations.

PROOF Let **u** and **v** be vectors in V and k be a scalar. Then

$$\begin{aligned}&(T + T')(k\mathbf{u} + \mathbf{v}) \\ &= T(k\mathbf{u} + \mathbf{v}) + T'(k\mathbf{u} + \mathbf{v}) && \text{(by definition of addition of transformations)} \\ &= kT(\mathbf{u}) + T(\mathbf{v}) + kT'(\mathbf{u}) + T'(\mathbf{v}) && \text{(by linearity of } T \text{ and } T'\text{)} \\ &= k(T + T')(\mathbf{u}) + (T + T')(\mathbf{v}) && \text{(by definition of addition of transformations),}\end{aligned}$$

showing that $T + T'$ is linear. The proof for cT is similar.

We leave the details of the verification that $L(V,W)$ is a vector space for the reader. The most important points to note are

(1) the **zero vector** of $L(V,W)$ is the zero map $0{:}V \to W$, which we know to be linear;

(2) the **additive inverse** of T is $-T$, with $(-T)(\mathbf{v}) = -T(\mathbf{v})$. ■

EXAMPLE 5-26

Let $V = R^2$, $W = R^3$, $T, T'{:}R^2 \to R^3$ with

$$\begin{aligned}T(x,y) &= (x + y, -x + y, 2x + y), \\ T'(x,y) &= (2x - y, x, x - 2y).\end{aligned}$$

Then,

$$\begin{aligned}(T + T')(x,y) &= T(x,y) + T'(x,y)\\ &= (x + y, -x + y, 2x + y) + (2x - y, x, x - 2y)\\ &= (3x, y, 3x - y);\end{aligned}$$

and

$$\begin{aligned}cT(x,y) &= c(x + y, -x + y, 2x + y)\\ &= (cx + cy, -cx + cy, 2cx + cy).\end{aligned}$$

With $c = 2$ and $(x,y) = (-1,2)$,

$$2T(-1,2) = (-2 + 4, 2 + 4, -4 + 4) = (2,6,0).$$

In anticipation of the next theorem, let us consider the matrices of $T + T'$ and $2T$:

$$[T + T'] = \begin{bmatrix} 3 & 0 \\ 0 & 1 \\ 3 & -1 \end{bmatrix} = \begin{bmatrix} 1 & 1 \\ -1 & 1 \\ 2 & 1 \end{bmatrix} + \begin{bmatrix} 2 & -1 \\ 1 & 0 \\ 1 & -2 \end{bmatrix} = [T] + [T'];$$

$$[2T] = \begin{bmatrix} 2 & 2 \\ -2 & 2 \\ 4 & 2 \end{bmatrix} = 2\begin{bmatrix} 1 & 1 \\ -1 & 1 \\ 2 & 1 \end{bmatrix} = 2[T]. \square$$

In general, we have

THEOREM 5-12

Let $\dim V = n$, $\dim W = m$, and B and B' be bases of V and W respectively. Then the function $MAT: L(V,W) \rightarrow M(m \times n)$, defined by $MAT(T) = [T;B/B']$, is an isomorphism.

PROOF We leave it for the reader to prove that MAT is linear. It is 1-1 because, if $MAT(T) = O$, then $[T;B/B'] = O$, so $T = 0$, the zero map, by Theorems 5-2 and 5-9. Finally, it can be shown (question 12(b), Exercises 5-3) that $\dim L(V,W) = mn = \dim M(m \times n)$, so that MAT is onto, hence an isomorphism. ■

Note that different choices for B and/or B' will yield different functions MAT. Also, the fact that MAT is linear can be stated in other words as *the matrix of a linear combination of linear transformations is the same linear combination of their matrices*.

Now, we know that under certain circumstances, matrices can also be multiplied: if A' is $m \times n$ and A is $n \times p$, then $A'A$ is defined and has order $m \times p$. It is therefore natural to consider the following question: If T' and T are linear transformations which correspond (under MAT) to A' and A respectively, what linear transformation corresponds to $A'A$?

Let us look first at the case of matrix transformations:

$$T_A:M(p\times 1)\rightarrow M(n\times 1) \text{ and } T_{A'}:M(n\times 1)\rightarrow M(m\times 1).$$

We consider the matrix transformation $T_{A'A}$ corresponding to $A'A$. Since $A'A$ is an $m\times p$ matrix, $T_{A'A}:M(p\times 1)\rightarrow M(m\times 1)$ and

$$T_{A'A}(X) = (A'A)X = A'(AX) = T_{A'}(AX) = T_{A'}(T_A(X)).$$

Therefore, the value of $T_{A'A}$ at X in $M(p\times 1)$ is computed as the value of $T_{A'}$ at the argument $T_A(X)$. For this reason, $T_{A'A}$ is the **composition** $T_{A'}T_A$ of $T_{A'}$ and T_A. In general,

DEFINITION 5-9

If $T:V\rightarrow V'$ and $T':V'\rightarrow V''$ are linear transformations, their **composition** $T'T:V\rightarrow V''$ is the function defined by $T'T(\mathbf{v}) = T'(T(\mathbf{v}))$ for all $\mathbf{v}$ in V.

This is just the usual definition of composition of functions, except that here T' and T are *linear*.

EXAMPLE 5-27

Let T be the rotation in R^2 through α and T' be the rotation through β, as shown in Figure 5-12. Then, it is obvious from the figure that $T'T$ is a rotation through $\alpha + \beta$, so that $T'T$ is also a linear transformation. ❑

FIGURE 5-12

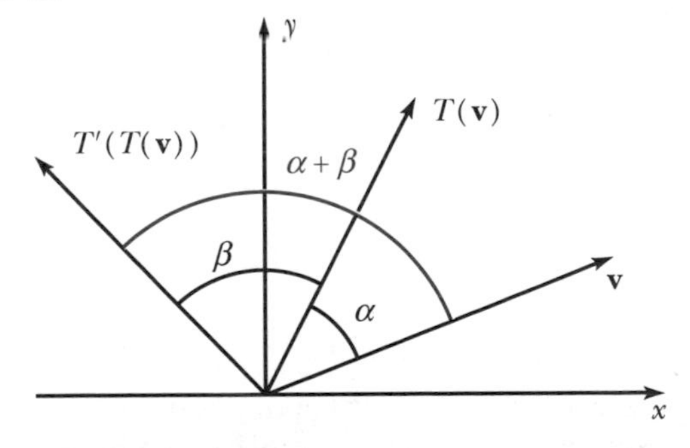

LEMMA 5-4

$T'T$ is linear when T and T' are linear.

PROOF This is a simple exercise using the linearity of T and T', and is therefore left for the reader. ∎

EXAMPLE 5-28

Let $T:R^2 \rightarrow R^2$ and $T':R^2 \rightarrow R^3$ be defined by

$$T(x,y) = (x + 2y, 2x + y),$$
$$T'(x,y) = (x + y, x - y, 2x - y).$$

Then,

$$\begin{aligned} T'T(x,y) &= T'(x + 2y, 2x + y) \\ &= ((x + 2y) + (2x + y), (x + 2y) - (2x + y), 2(x + 2y) - (2x + y)) \\ &= (3x + 3y, -x + y, 3y). \ \square \end{aligned}$$

In anticipation of the next theorem, it is useful to consider the matrices of these linear transformations with respect to the standard bases:

$$[T] = \begin{bmatrix} 1 & 2 \\ 2 & 1 \end{bmatrix}, [T'] = \begin{bmatrix} 1 & 1 \\ 1 & -1 \\ 2 & -1 \end{bmatrix}, \text{so } [T'][T] = \begin{bmatrix} 3 & 3 \\ -1 & 1 \\ 0 & 3 \end{bmatrix} = [T'T].$$

In fact, we have also seen that for matrix transformations, $[T_{A'}T_A] = A'A = [T_{A'}][T_A]$. More generally,

> **THEOREM 5-13**
>
> If V, V', and V'' are finite dimensional vector spaces with bases B, B', and B'' respectively, and $T:V \rightarrow V'$, $T':V' \rightarrow V''$ are linear, then $[T'T;B/B''] = [T';B'/B''][T;B/B']$.

Thus, *the matrix of a composition is the product of the corresponding matrices, in the same order*.

PROOF Let the ith vector in B be $\mathbf{v}_i$. Then, the ith column of $[T'T;B/B'']$ is, by definition,

$$\begin{aligned} [T'T(\mathbf{v}_i)]_{B''} &= [T'(T(\mathbf{v}_i))]_{B''} \\ &= [T';B'/B''][T(\mathbf{v}_i)]_{B'} \quad \text{(by Theorem 5-7)} \\ &= [T';B'/B''][T;B/B'][\mathbf{v}_i]_B \quad \text{(also by Theorem 5-7)} \\ &= \text{the } i^{\text{th}} \text{ column of } [T';B'/B''][T;B/B'], \end{aligned}$$

because $[\mathbf{v}_i]_B$ is a column vector with 1 in the ith row and 0 in all the other rows. ∎

EXAMPLE 5-29

Consider again the composition of rotations in the plane (Example 5-27). We use Theorem 5-13 to obtain the so-called "expansion formulae" for sine and cosine. We have seen that

$$[T] = \begin{bmatrix} \cos\alpha & -\sin\alpha \\ \sin\alpha & \cos\alpha \end{bmatrix}, [T'] = \begin{bmatrix} \cos\beta & -\sin\beta \\ \sin\beta & \cos\beta \end{bmatrix},$$

so

$$[T'T] = \begin{bmatrix} \cos\beta & -\sin\beta \\ \sin\beta & \cos\beta \end{bmatrix}\begin{bmatrix} \cos\alpha & -\sin\alpha \\ \sin\alpha & \cos\alpha \end{bmatrix}$$

$$= \begin{bmatrix} \cos\alpha\cos\beta - \sin\alpha\sin\beta & -\sin\alpha\cos\beta - \cos\alpha\sin\beta \\ \cos\alpha\sin\beta + \sin\alpha\cos\beta & -\sin\alpha\sin\beta + \cos\alpha\cos\beta \end{bmatrix}.$$

But since $T'T$ is the rotation through $\alpha + \beta$, we have

$$[T'T] = \begin{bmatrix} \cos(\alpha+\beta) & -\sin(\alpha+\beta) \\ \sin(\alpha+\beta) & \cos(\alpha+\beta) \end{bmatrix}.$$

Comparing the two matrix expressions for $[T'T]$, we see that

$$\cos(\alpha+\beta) = \cos\alpha\cos\beta - \sin\alpha\sin\beta,$$
$$\sin(\alpha+\beta) = \cos\alpha\sin\beta + \sin\alpha\cos\beta,$$

which are the expansion formulae for sine and cosine. ❑

EXAMPLE 5-30

Let V be the vector space $P(1)$. In Example 5-9, we constructed the linear transformation $A{:}V \to F[0,1]$ with $Af(t)$ = the algebraic area bounded by $y = f(x)$, the x axis, $x = 0$, and $x = t$, for $t \in [0,1]$ (Figure 5-13).

FIGURE 5-13

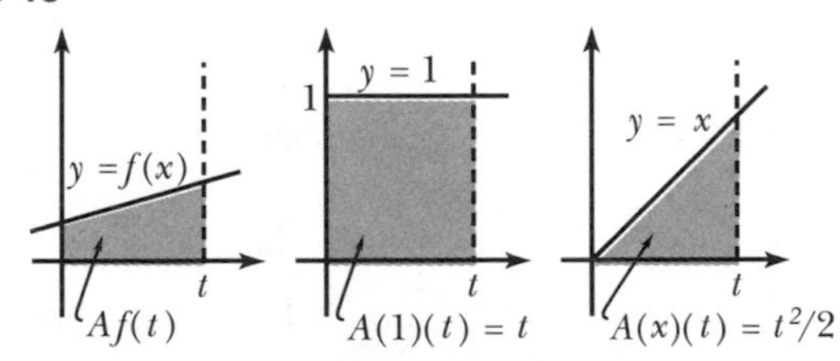

Let $B = \{1,x\}$ be the usual ordered basis for V. Then, as shown in the figure,

$$A(1)(t) = t \text{ for } t \in [0,1],$$

so $A(1) = x$, the function on $[0,1]$ with $x(t) = t$. Similarly,

$$A(x)(t) = t^2/2,$$

so $A(x) = x^2/2$. As A is linear, this shows that $A:V \to W$, where W is the space of polynomial functions of degree ≤ 2 with domain $[0,1]$ (see also question 6, Exercises 5-1). Therefore,

$$[A;B/B'] = \begin{bmatrix} 0 & 0 \\ 1 & 0 \\ 0 & 1/2 \end{bmatrix}$$

where B' is the usual basis for W.

Next, recall (Example 5-16) the linear transformation $D:W \to V$ with $D(ax^2 + bx + c) = 2ax + b$. Then,

$$[D;B'/B] = \begin{bmatrix} 0 & 1 & 0 \\ 0 & 0 & 2 \end{bmatrix}.$$

By Theorem 5-13,

$$\begin{aligned} [DA;B] &= [D;B'/B][A;B/B'] \\ &= \begin{bmatrix} 0 & 1 & 0 \\ 0 & 0 & 2 \end{bmatrix} \begin{bmatrix} 0 & 0 \\ 1 & 0 \\ 0 & 1/2 \end{bmatrix} = \begin{bmatrix} 1 & 0 \\ 0 & 1 \end{bmatrix} = I. \end{aligned}$$

It follows that $DA = 1_V$, the identity transformation of V. This means that for any f in V, $DA(f) = f$. So, D "undoes the work" of the area function A. These ideas are important in calculus. ❑

Now, the composition $T'T$ of two linear transformations is only defined when T maps into the domain of T' (so that $T'(T(\mathbf{v}))$ is defined). This parallels the situation for matrices: a product $A'A$ only exists when the number of columns of $A' =$ the number of rows of A.

The case in which compositions and products are *always* defined is when $V = W$, so that T and T' are linear operators and their matrices are square: $n \times n$, if $dim\ V = n$.

DEFINITION 5-10

We denote by $L(V)$ the set of linear operators on V. We also abbreviate $M(n \times n)$ to $M(n)$.

Since $L(V) = L(V,V)$, it is a vector space, by Theorem 5-11. And, by Theorem 5-12, $L(V)$ and $M(n)$ are isomorphic vector spaces. In fact, more is true, because these spaces have, in addition to the usual operations on vector spaces, a "product" operation:

(1) in $L(V)$, the "product" is the composition of linear operators;
(2) in $M(n)$, the "product" is the usual matrix product.

Furthermore, these product operations behave "nicely". For example, in the case of $M(n)$, we have (for $A, B, C \in M(n)$ and $k \in R$)

(1) $A(BC) = (AB)C$ (the associative law of multiplication);

(2) $A(B + C) = AB + AC$ (a distributive law);

(3) $k(AB) = (kA)B = A(kB)$ (also a distributive law).

Similar properties hold in $L(V)$ for the composition of linear operators (see question 21, Exercises 5-3), so we say that $L(V)$ and $M(n)$ are **linear algebras**. In fact, by Theorem 5-13, the "product" in $L(V)$ corresponds (under MAT) to the product in $M(n)$. Therefore, we say that they are **isomorphic linear algebras**. In summary,

THEOREM 5-14

If B is a basis for V, then $MAT{:}L(V) \to M(n)$ with $MAT(T) = [T;B]$ is an isomorphism of linear algebras:

$$[T'T;B] = [T';B][T;B]$$

for $T', T \in L(V)$. ∎

EXAMPLE 5-31

We show some computations in $L(R^2)$ to illustrate the linear algebra of linear operators on R^2. Let T be the counterclockwise rotation through 90° ($\pi/2$) and T' be the projection onto the x axis (Figure 5-14).

FIGURE 5-14

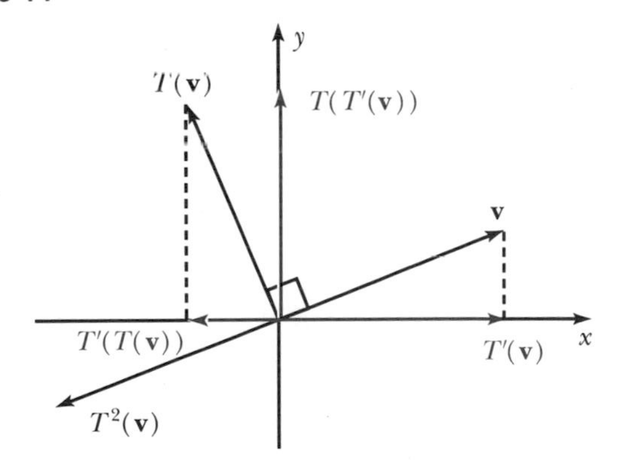

$$[T] = \begin{bmatrix} 0 & -1 \\ 1 & 0 \end{bmatrix}, [T'] = \begin{bmatrix} 1 & 0 \\ 0 & 0 \end{bmatrix},$$

$$[TT'] = \begin{bmatrix} 0 & -1 \\ 1 & 0 \end{bmatrix}\begin{bmatrix} 1 & 0 \\ 0 & 0 \end{bmatrix} = \begin{bmatrix} 0 & 0 \\ 1 & 0 \end{bmatrix},$$

$$[T'T] = \begin{bmatrix} 1 & 0 \\ 0 & 0 \end{bmatrix} \begin{bmatrix} 0 & -1 \\ 1 & 0 \end{bmatrix} = \begin{bmatrix} 0 & -1 \\ 0 & 0 \end{bmatrix}.$$

So, $TT' \neq T'T$. Of course, we can verify this from the figure by the fact that $TT'(\mathbf{v})$ and $T'T(\mathbf{v})$ are different.

Next, consider the linear operator $T^2 = TT$ (in general, for $T \in L(V)$, $T^n = TT^{n-1}$ as for matrices, and $T^0 = 1_V$). Geometrically, it is clear that T^2 is a rotation through $2 \times 90° = 180°$, or a reflection in the origin:

$$T^2(x,y) = (-x,-y).$$

Let us verify this using the fact that $[T^2] = [T]^2$ (Theorem 5-14):

$$[T^2] = \begin{bmatrix} 0 & -1 \\ 1 & 0 \end{bmatrix}^2 = \begin{bmatrix} -1 & 0 \\ 0 & -1 \end{bmatrix},$$

so, with B equal to the standard basis for R^2,

$$[T^2(x,y)]_B = \begin{bmatrix} -1 & 0 \\ 0 & -1 \end{bmatrix} \begin{bmatrix} x \\ y \end{bmatrix} = \begin{bmatrix} -x \\ -y \end{bmatrix}. \square$$

We conclude this section with an application to matrix theory. To describe it, we need one further concept, which arises from the following observation. Suppose $T:V \to W$ is a linear transformation whose matrix $[T;B/B'] = A$ is invertible. Then, A is square, so $dim\ V = dim\ W$. We also know there is an unique linear transformation $T':W \to V$ such that

$$[T';B'/B] = A^{-1} \text{ (since } MAT \text{ is 1-1 and onto).}$$

And $T'T = 1_V:V \to V$, because

$$[T'T;B] = [T';B'/B][T;B/B'] = A^{-1}A = I.$$

Similarly, $[TT';B'] = I$ implies $TT' = 1_W:W \to W$.

DEFINITION 5-11

A linear transformation $T:V \to W$ is said to be **invertible** (nonsingular) iff there is a linear transformation $T':W \to V$ such that $TT' = 1_W:W \to W$ and $T'T = 1_V:V \to V$. In this case, we denote T' by T^{-1}, and call it the **inverse** of T.

Intuitively, we see that the linear transformations T and T^{-1} are related by the fact that each "undoes the work" of the other, since

$$T^{-1}T(\mathbf{v}) = \mathbf{v} \text{ for } \mathbf{v} \in V, \text{ and } TT^{-1}(\mathbf{w}) = \mathbf{w} \text{ for } \mathbf{w} \in W.$$

In practice, we can always find T^{-1} using the fact that $[T^{-1};B'/B] = [T;B/B']^{-1}$.

In fact, we may legitimately speak of *the* inverse, since there is exactly one such linear transformation when T is invertible. To see this, suppose also $T''T = 1_V$ for some $T'':W \to V$. Then, for all $\mathbf{w} \in W$,

$$T''(\mathbf{w}) = T''(1_W(\mathbf{w})) = T''(TT'(\mathbf{w})) = (T''T)(T'(\mathbf{w})) = 1_V(T'(\mathbf{w})) = T'(\mathbf{w}).$$

It follows that $T'' = T'$.

EXAMPLE 5-32

(a) Intuitively, we see that the inverse of a counter-clockwise rotation T through θ is a clockwise rotation through θ, or, a counter-clockwise rotation through $-\theta$. We can verify this by the fact that a rotation through $-\theta$ would have matrix

$$\begin{bmatrix} \cos(-\theta) & -\sin(-\theta) \\ \sin(-\theta) & \cos(-\theta) \end{bmatrix} = \begin{bmatrix} \cos\theta & \sin\theta \\ -\sin\theta & \cos\theta \end{bmatrix}$$

$$= \begin{bmatrix} \cos\theta & -\sin\theta \\ \sin\theta & \cos\theta \end{bmatrix}^{-1} = [T]^{-1}.$$

(b) Let $T:R^2 \to R^2$ be defined by $T(x,y) = (2x + 5y, x + 3y)$. Therefore,

$$[T] = \begin{bmatrix} 2 & 5 \\ 1 & 3 \end{bmatrix}; [T]^{-1} = \begin{bmatrix} 3 & -5 \\ -1 & 2 \end{bmatrix}.$$

(See Example 2-27.) Thus,

$$T^{-1}(x,y) = \begin{bmatrix} 3 & -5 \\ -1 & 2 \end{bmatrix}\begin{bmatrix} x \\ y \end{bmatrix} = \begin{bmatrix} 3x - 5y \\ -x + 2y \end{bmatrix} = (3x - 5y, -x + 2y).$$

Let us verify that T^{-1} indeed "undoes the work" of T.

$$\begin{aligned} T^{-1}(T(x,y)) &= T^{-1}(2x + 5y, x + 3y) \\ &= (3(2x + 5y) - 5(x + 3y), -(2x + 5y) + 2(x + 3y)) \\ &= (x,y). \end{aligned}$$

(c) Let $T:R^2 \to R^2$ be defined by $T(x,y) = (x - 2y, -2x + 4y)$. Then T is not invertible because

$$[T] = \begin{bmatrix} 1 & -2 \\ -2 & 4 \end{bmatrix}$$

is not invertible (see Example 2-27). ❑

THEOREM 5-15

If $T:V \to W$ is a linear transformation with finite $dim\ V = dim\ W$, the following are equivalent:

(1) T is invertible.

(2) T is 1-1.

(3) T is onto.

(4) T is an isomorphism.

PROOF We already know that (2), (3), and (4) are equivalent by Theorem 5-8, Corollary 2. We shall prove (1) is equivalent to (3). Suppose T is invertible. Then, if $\mathbf{w}$ is a vector in W, $T(T^{-1}(\mathbf{w})) = 1_W(\mathbf{w}) = \mathbf{w}$, showing that T is onto. Conversely, if T is onto, we define $T':W \to V$ by $T'(\mathbf{w}) = \mathbf{v}$, where $T(\mathbf{v}) = \mathbf{w}$. We leave it for the reader to verify that T' is a well defined function, and is, in fact, the inverse of T, to complete the proof. ∎

We now use the notion of the inverse of a linear transformation to prove the fact that an $n \times n$ matrix A is invertible iff *one* of the following is true (see the discussion following Definition 2-12):

(1) There is an $n \times n$ matrix P such that $AP = I$.

(2) There is an $n \times n$ matrix Q such that $QA = I$.

In fact, if one of these is true, so is the other, and $P = Q = A^{-1}$. This result is sometimes called the "one-sided inverse theorem". It means that in checking whether or not a matrix B is the inverse of A, we need only show $AB = I$ *or* $BA = I$.

To prove (1), suppose $AP = I$. Then, $T_{AP} = T_I = 1_V$, the identity operator on V where $V = M(n \times 1)$. But then $T_{AP} = T_A T_P = 1_V$ implies $T_A(T_P(X)) = X$ for any X in $M(n \times 1)$, showing that T_A is onto and hence invertible. Thus $A = [T_A]$ is invertible because $[T_A{}^{-1}] = A^{-1}$. The proof of (2) is similar.

EXERCISES 5-3

1. For each of the following matrices A, determine whether or not T_A is **(a)** 1-1; **(b)** onto; **(c)** an isomorphism.

(i) $\begin{bmatrix} 1 & 2 \\ 2 & 1 \end{bmatrix}$ **(ii)** $\begin{bmatrix} 1 & -1/2 \\ -2 & 1 \end{bmatrix}$ **(iii)** $[2]$ **(iv)** $[1 \quad -1 \quad 2]$

(v) $\begin{bmatrix} 1 & 2 \\ -2 & 4 \\ 0 & 8 \end{bmatrix}$ **(vi)** $\begin{bmatrix} 1 & 3 & 1 \\ 1 & 0 & 0 \\ 0 & 1 & 1 \end{bmatrix}$ **(vii)** $\begin{bmatrix} 1 & 2 & 3 \\ 2 & 3 & 4 \\ 0 & 1 & 2 \end{bmatrix}$

2. **(a)** Let $T:R^2 \to R^3$ be defined by $T(x,y) = (2x + y, x + y, x - y)$. Show that T is 1-1 by **(i)** using the definition of a 1-1 function; **(ii)** using $[T]$.

(b) Let $T:R^3 \to R^2$ be defined by $T(x,y,z) = (x + y - 2z, x + y + z)$. Show that T is not 1-1 by **(i)** using the definition of a 1-1 function; **(ii)** using $[T]$.

3. Show that a linear map $T:R^n \to R^m$ **(a)** cannot be onto if $n < m$; **(b)** cannot be 1-1 if $n > m$. Generalize these facts to linear maps $T:V \to W$ where V and W have dimensions n and m respectively.

4. Determine whether or not the reflections in the x and y axes (question 2, Exercises 5-1) are isomorphisms of R^2.

5. For the linear transformations T in questions 1 to 5, Exercises 5-2, determine whether each is **(a)** 1-1; **(b)** onto; **(c)** an isomorphism.

6. Show that if A is $m \times n$, the system of equations $AX = B$ has an unique solution iff T_A is 1-1 and B is in $Im\ T_A$.

7. Let $P:M(2\times2) \to P(3)$ be the function defined by $P(A) = a + bx + cx^2 + dx^3$ when

$$A = \begin{bmatrix} a & b \\ c & d \end{bmatrix}.$$

Show that P is linear, onto, and hence an isomorphism.

8. Let $T:V \to W$ be linear, where V and W are finite dimensional.

(a) Show that T is 1-1 iff $T(S)$ is a linearly independent set in W whenever S is linearly independent in V.

(b) Show that T is onto iff $T(S)$ generates W whenever S generates V.

9. Use question 8 above to complete the proof of Theorem 5-9.

10. Let $T, T':R^2 \to R^3$ be linear with matrices given by

$$[T] = \begin{bmatrix} 1 & 2 \\ -1 & 1 \\ -2 & 4 \end{bmatrix}, [T'] = \begin{bmatrix} 0 & -2 \\ 3 & -1 \\ 1 & -2 \end{bmatrix}.$$

Find **(a)** $(T + T')(x,y)$; **(b)** $3T(x,y)$; **(c)** $(2T - 4T')(x,y)$.

11. **(a)** Define $T, T':M(n\times n) \to M(n\times n)$ by $T(A) = (1/2)(A + A^t)$, $T'(A) = (1/2)(A - A^t)$, $A \in M(n\times n)$.

(i) Show that T and T' are linear.

(ii) Show that $Ker\ T = Im\ T'$ is the space of skew-symmetric $n\times n$ matrices, and that $Ker\ T' = Im\ T$ is the space of symmetric matrices.

(iii) Show that $T + T' = 1_V$, where $V = M(n\times n)$.

(b) For $n = 2$, find the matrices A and A' of T and T' respectively, with respect to the usual basis B for $M(2\times2)$. Verify that $A + A' = I$, the 4×4 identity.

12. **(a)** Define four functions $T_{ij}:R^2 \to R^2$ by $T_{ij}(\mathbf{e}_k) = \mathbf{e}_i$ if $j = k$ and $\mathbf{0}$ if $j \neq k$, $i, j, k = 1, 2$.

(i) Show that the T_{ij}, $i, j = 1, 2$, constitute a basis for $L(R^2,R^2)$.

(ii) Show that $[T_{11}] = E_1$, $[T_{12}] = E_2$, $[T_{21}] = E_3$, and $[T_{22}] = E_4$, where the E's are the usual ordered basis for $M(2 \times 2)$.

(b) Generalize the above to construct a basis $\{T_{ij} \mid 1 \leq i \leq m, 1 \leq j \leq n\}$ for $L(R^n,R^m)$.

(c) Generalize (b) to construct a basis for $L(V,W)$, given bases B and B' for finite dimensional vector spaces V and W respectively.

13. Given $T:R^3 \to R^2$ with $T(x,y,z) = (x + 2y - z, 2x - y + 2z)$, and $T':R^2 \to R^3$ with $T'(x,y) = (2x + y, x + y, x - y)$, find the compositions $T'T(x,y,z)$ and $TT'(x,y)$ by **(a)** using the definition of composition; and **(b)** finding $[T'T]$ and $[TT']$.

14. Let $T:R^2 \to R^2$ be the counterclockwise rotation through 45° and $T':R^2 \to R^2$ be the projection onto the y axis.

(a) Find $T'T(x,y)$ and $TT'(x,y)$ and show they are not equal.

(b) Find $[T'T]$ and $[TT']$ to verify the conclusion in (a).

15. Let $T:R^2 \to R^2$ be the counterclockwise rotation through an angle θ. Use Theorem 5-13 to find formulae for $\cos 3\theta$ and $\sin 3\theta$ from the fact that T^3 is a rotation through 3θ.

16. Let $P'(2)$ be the subspace of $P(2)$ consisting of all polynomials of the form $ax^2 + bx$ (question 12, Exercises 4-3). Let $D:P'(2) \to P(1)$ be defined by $D(ax^2 + bx) = 2ax + b$ and $A:P(1) \to P'(2)$ be defined by $A(ax + b) = ax^2/2 + bx$ (Example 5-30). Show that A and D are inverse isomorphisms on these spaces.

17. Let $T_{mn}:M(m \times n) \to M(n \times m)$ be the transpose map: $T_{mn}(A) = A^t$, when A is $m \times n$. Show that T_{mn} is invertible with inverse T_{nm}.

18. For each of the following linear operators T with $[T]$ as given, either find a formula for T^{-1} as directed, or show that T is not invertible.

(a) $[T] = \begin{bmatrix} 2 & 1 \\ 1 & 2 \end{bmatrix}$, find $T^{-1}(x,y)$, if possible.

(b) $[T] = \begin{bmatrix} 1 & -1 & 0 \\ 2 & -1 & 2 \\ 4 & -3 & 2 \end{bmatrix}$, find $T^{-1}(x,y,z)$, if possible.

(c) $[T] = \begin{bmatrix} 0 & 1 & 0 & 1 \\ 1 & 1 & 0 & 1 \\ 0 & 0 & 1 & 1 \\ 1 & 1 & 1 & 1 \end{bmatrix}$, find $T^{-1}(w,x,y,z)$, if possible.

19. Suppose a linear operator satisfies $T^k = 0$ for some positive integer k. Show that T is singular.

20. **(a)** Show that if $T:V \to W$ is an isomorphism, then so is its inverse.

(b) Show that if $T':W \to U$ is another isomorphism, then so is the composition $T'T:V \to U$.

(c) Use (a) and (b) above to complete the proof of Theorem 5-10, Corollary.

21. Let V be a vector space, $T, T', T'' \in L(V)$, and $k \in R$. Show that

(a) $T(T'T'') = (TT')T''$;

(b) $T(T' + T'') = TT' + TT''$;

(c) $k(TT') = T(kT')$.

5-4 Basis Change and Similarity

So far, we have introduced several concepts such as vector spaces and linear transformations. The reasons for studying them are largely theoretical, since these notions provide a unified conceptual framework for studying many different phenomena. However, when it comes to computing with vectors and linear transformations, it is often convenient to represent them by matrices. Thus, we defined the idea of matrices of vectors and of linear transformations with respect to bases. Of course, the matrices we get depend on the bases we choose. In particular, as noted earlier, it will often be to our advantage to choose the bases so that the matrices are "simple" as possible for the purpose at hand. This point of view is pursued for linear operators in Chapter 6.

In this section, we study the relationships between the various matrix representations of vectors and linear operators (the case of general linear transformations is less important and is considered in the exercises). We begin with the relationship between the matrices of a vector $\mathbf{v}$ (in a vector space V) with respect to two different bases B and B' for V.

EXAMPLE 5-33

The standard basis $B = \{\mathbf{e}_1, \mathbf{e}_2\}$ is only one of infinitely many bases for R^2 (albeit a very useful one). Let us consider another basis $B' = \{\mathbf{v}'_1, \mathbf{v}'_2\}$, where

$$\mathbf{v}'_1 = \mathbf{e}_1 + \mathbf{e}_2, \mathbf{v}'_2 = -\mathbf{e}_1 + \mathbf{e}_2. \tag{5-5}$$

Thus,

$$[\mathbf{v}'_1]_B = \begin{bmatrix} 1 \\ 1 \end{bmatrix} \text{and} [\mathbf{v}'_2]_B = \begin{bmatrix} -1 \\ 1 \end{bmatrix}.$$

Now, any vector $\mathbf{v} = (x,y) \in R^2$ may be expressed as

$$\mathbf{v} = x'\mathbf{v}'_1 + y'\mathbf{v}'_2,$$

with $x', y' \in R$, so

$$[\mathbf{v}]_{B'} = \begin{bmatrix} x' \\ y' \end{bmatrix}. \tag{5-6}$$

Substituting the expressions for $\mathbf{v}'_1$ and $\mathbf{v}'_2$ given in 5-5, we get

$$\mathbf{v} = x'(\mathbf{e}_1 + \mathbf{e}_2) + y'(-\mathbf{e}_1 + \mathbf{e}_2) = (x' - y')\mathbf{e}_1 + (x' + y')\mathbf{e}_2.$$

Therefore,

$$[\mathbf{v}]_B = \begin{bmatrix} x' - y' \\ x' + y' \end{bmatrix}. \tag{5-7}$$

Combining equations 5-6 and 5-7 yields

$$[\mathbf{v}]_B = \begin{bmatrix} x' - y' \\ x' + y' \end{bmatrix} = \begin{bmatrix} 1 & -1 \\ 1 & 1 \end{bmatrix}\begin{bmatrix} x' \\ y' \end{bmatrix} = \begin{bmatrix} 1 & -1 \\ 1 & 1 \end{bmatrix}[\mathbf{v}]_{B'}, \tag{5-8}$$

which is an equation relating $[\mathbf{v}]_B$ to $[\mathbf{v}]_{B'}$. ❑

Notice that we obtained the expression 5-8 for $[\mathbf{v}]_B$ in terms of $[\mathbf{v}]_{B'}$ by expressing the vectors of B' as linear combinations of those of B (equation 5-5). This is exactly how we would obtain the matrix $[1_V;B'/B]$ of the identity operator $1_V:R^2 \to R^2$ ($V = R^2$) with respect to the bases B' and B for R^2. This is because the two columns of this matrix are, by definition, $[1_V(\mathbf{v}'_1)]_B = [\mathbf{v}'_1]_B$ and $[1_V(\mathbf{v}'_2)]_B = [\mathbf{v}'_2]_B$ respectively.

In general, we have

THEOREM 5-16

If V is a finite dimensional vector space with bases B and B', and $\mathbf{v}$ is a vector in V, then

$$[\mathbf{v}]_B = [1_V;B'/B][\mathbf{v}]_{B'}.$$

The matrix $[1_V;B'/B]$ is usually called the **transition matrix** from the basis B' to the basis B because it is obtained by expressing the vectors of B' as linear combinations of the vectors of B, i.e., its ith column is $[\mathbf{v}'_i]_B$ when $B' = \{\mathbf{v}'_1,\ldots,\mathbf{v}'_n\}$. In many books, this matrix is denoted by a letter such as P or Q. However, for mnemonic reasons, we will abbreviate it to $[B'/B]$. Thus, the theorem states that

$$[\mathbf{v}]_B = [B'/B][\mathbf{v}]_{B'}.$$

PROOF This is just a particular case of Theorem 5-7 with $V = W$ and $T = 1_V$. ■

In fact, $[B'/B]$ is an invertible matrix with inverse $[B/B'] = [1_V;B/B']$. This follows immediately from Theorem 5-13 and Definition 5-11, since $1_V:V \to V$ is an invertible linear operator with itself as the inverse. However, for readers who have not read section 5-3, we give the following independent argument. By two applications of Theorem 5-16,

$$[B/B'][B'/B][\mathbf{v}]_{B'} = [B/B'][\mathbf{v}]_B = [\mathbf{v}]_{B'}.$$

Since this is true for all $\mathbf{v}$ in V, it follows that

$$[B/B'][B'/B]X = X \text{ for } all\ n \times 1 \text{ matrices } X$$

(as $\mathbf{v} \leftrightarrow [\mathbf{v}]_{B'} = X$ is a 1-1 correspondence). By Theorem 2-6, we get

$$[B/B'][B'/B] = I,$$

so $[B/B'] = [B'/B]^{-1}$, as stated. Thus we have proved

COROLLARY

$[B/B']$ is the inverse of $[B'/B]$ and

$$[\mathbf{v}]_{B'} = [B/B'][\mathbf{v}]_B = [B'/B]^{-1}[\mathbf{v}]_B. \blacksquare$$

This last formula is the one most often used in practice.

EXAMPLE 5-34

We continue with Example 5-33 where B is the standard basis and $B' = \{\mathbf{v}'_1, \mathbf{v}'_2\}$. We have seen that

$$[B'/B] = \begin{bmatrix} 1 & -1 \\ 1 & 1 \end{bmatrix},$$

so

$$[\mathbf{v}]_{B'} = \begin{bmatrix} 1 & -1 \\ 1 & 1 \end{bmatrix}^{-1} [\mathbf{v}]_B = (1/2)\begin{bmatrix} 1 & 1 \\ -1 & 1 \end{bmatrix}[\mathbf{v}]_B$$

for any vector $\mathbf{v}$ in V. For example, if $\mathbf{v} = (2,3)$, then

$$[\mathbf{v}]_B = \begin{bmatrix} 2 \\ 3 \end{bmatrix}, [\mathbf{v}]_{B'} = (1/2)\begin{bmatrix} 1 & 1 \\ -1 & 1 \end{bmatrix}\begin{bmatrix} 2 \\ 3 \end{bmatrix} = \begin{bmatrix} 5/2 \\ 1/2 \end{bmatrix}.$$

This *means* that, with respect to the basis B',

$$\mathbf{v} = (2,3) = (5/2)\mathbf{v}'_1 + (1/2)\mathbf{v}'_2. \square$$

EXAMPLE 5-35

Consider the polynomials f_1, f_2, and f_3 in $P(2)$ given by

$$f_1(x) = \frac{(x-2)(x-3)}{(1-2)(1-3)} = 3 - (5/2)x + (1/2)x^2,$$

$$f_2(x) = \frac{(x-1)(x-3)}{(2-1)(2-3)} = -3 + 4x - x^2,$$

$$f_3(x) = \frac{(x-1)(x-2)}{(3-1)(3-2)} = 1 - (3/2)x + (1/2)x^2.$$

The reader should be able to show that they are a basis B' for $P(2)$. (They are "Lagrange interpolation polynomials", named after Joseph-Louis Lagrange, 1736–1813; see Example 5-47, section 5-5.) Let $B = \{1,x,x^2\}$ be the usual basis for $P(2)$. Then,

$$[B'/B] = \begin{bmatrix} 3 & -3 & 1 \\ -5/2 & 4 & -3/2 \\ 1/2 & -1 & 1/2 \end{bmatrix},$$

so

$$[B/B'] = [B'/B]^{-1} = \begin{bmatrix} 1 & 1 & 1 \\ 1 & 2 & 4 \\ 1 & 3 & 9 \end{bmatrix}.$$

Now, if $f(x) = ax^2 + bx + c$ is any polynomial function in $P(2)$,

$$[f]_B = \begin{bmatrix} c \\ b \\ a \end{bmatrix},$$

so

$$[f]_{B'} = [B'/B]^{-1}[f]_B = \begin{bmatrix} 1 & 1 & 1 \\ 1 & 2 & 4 \\ 1 & 3 & 9 \end{bmatrix} \begin{bmatrix} c \\ b \\ a \end{bmatrix} = \begin{bmatrix} c + b + a \\ c + 2b + 4a \\ c + 3b + 9a \end{bmatrix}.$$

For example, if $f(x) = -x^2 + 2x + 2$,

$$[f]_{B'} = \begin{bmatrix} 3 \\ 2 \\ -1 \end{bmatrix}.$$

This means, of course, that

$$f = 3f_1 + 2f_2 - f_3.$$

Notice that, from the given forms of the f_i, f has the property

$$f(1) = 3, f(2) = 2, \text{ and } f(3) = -1,$$

so its graph passes through the points (1,3), (2,2), and (3, −1). ❑

We next consider the relationship between the matrices of a linear operator $T:V \to V$ with respect to different bases B and B' for V (for general linear transformations, see question 11, Exercises 5-4).

EXAMPLE 5-36

Let $B' = \{\mathbf{v'}_1, \mathbf{v'}_2\}$ be the basis for R^2 with $\mathbf{v'}_1 = (3,1)$ and $\mathbf{v'}_2 = (-1,1)$. By Theorem 5-2, there is an unique linear operator $T:R^2 \to R^2$ such that

$$T(\mathbf{v'}_1) = 2\mathbf{v'}_1, T(\mathbf{v'}_2) = -\mathbf{v'}_2.$$

The action of T is very simple in terms of the basis B':T stretches vectors in the $\mathbf{v'}_1$ direction by a factor of 2 and reflects those in the $\mathbf{v'}_2$ direction in the origin. Consequently, the matrix of T in the basis B' is very "simple":

$$[T;B'] = \begin{bmatrix} 2 & 0 \\ 0 & -1 \end{bmatrix}.$$

Now let us find $[T;B] = [T]$, where B is the standard basis for R^2. In order to do this, we must express $T(\mathbf{e}_1)$ and $T(\mathbf{e}_2)$ as linear combinations of $\mathbf{e}_1$ and $\mathbf{e}_2$. Since we only have formulae for $T(\mathbf{v'}_1)$ and $T(\mathbf{v'}_2)$, we must first write $\mathbf{e}_1$ and $\mathbf{e}_2$ in terms of $\mathbf{v'}_1$ and $\mathbf{v'}_2$:

$$\mathbf{v'}_1 = (3,1) = 3\mathbf{e}_1 + \mathbf{e}_2; \mathbf{v'}_2 = (-1,1) = -\mathbf{e}_1 + \mathbf{e}_2, \tag{5-9}$$

so

$$[B'/B] = \begin{bmatrix} 3 & -1 \\ 1 & 1 \end{bmatrix}.$$

Since $[B/B'] = [B'/B]^{-1}$, we find

$$[B/B'] = \begin{bmatrix} 1/4 & 1/4 \\ -1/4 & 3/4 \end{bmatrix}.$$

This means that

$$\mathbf{e}_1 = (1/4)\mathbf{v'}_1 - (1/4)\mathbf{v'}_2; \mathbf{e}_2 = (1/4)\mathbf{v'}_1 + (3/4)\mathbf{v'}_2. \tag{5-10}$$

Therefore, using equations 5-9 and 5-10,

$$\begin{aligned} T(\mathbf{e}_1) &= T((1/4)\mathbf{v'}_1 - (1/4)\mathbf{v'}_2) \\ &= (1/4)2\mathbf{v'}_1 - (1/4)(-1)\mathbf{v'}_2 \\ &= (1/2)(3\mathbf{e}_1 + \mathbf{e}_2) + (1/4)(-\mathbf{e}_1 + \mathbf{e}_2) \\ &= (5/4)\mathbf{e}_1 + (3/4)\mathbf{e}_2. \end{aligned}$$

Similarly, $T(\mathbf{e}_2) = (9/4)\mathbf{e}_1 - (1/4)\mathbf{e}_2$. We therefore find that

$$[T;B] = \begin{bmatrix} 5/4 & 9/4 \\ 3/4 & -1/4 \end{bmatrix} (= [T]).$$

Clearly, $[T;B']$ is "simpler" than $[T]$ in this case. ❑

Let us retrace the steps that led from $[T;B']$ to $[T]$.

(1) We expressed the $\mathbf{e}_i$ in terms of the $\mathbf{v}'_j$ (i.e., we found $[B/B']$).

(2) We applied T to the result, using $[T;B']$.

(3) We expressed the result in terms of B using $[B'/B]$, which we already had.

The same steps allow us to derive a formula for $[T;B]$ in terms of $[T;B']$, where T is an arbitrary linear operator on a vector space V. Let $B = \{\mathbf{v}_1,\ldots,\mathbf{v}_n\}$. Then, the ith column of $[T;B]$ is

$$\begin{aligned}[T(\mathbf{v}_i)]_B &= [B'/B][T(\mathbf{v}_i)]_{B'} \\ &= [B'/B][T;B'][\mathbf{v}_i]_{B'} \\ &= [B'/B][T;B'][B/B'][\mathbf{v}_i]_B,\end{aligned}$$

which is the ith column of $[B'/B][T;B'][B/B']$ (since $[\mathbf{v}_i]_B$ has a 1 in row i and 0 elsewhere). This gives the required formula. Notice that the three "=" signs above correspond to the steps used in the example.

THEOREM 5-17

If B, B' are bases for a finite dimensional vector space V and T is a linear operator on V, then

$$[T;B] = [B'/B][T;B'][B/B']. \blacksquare$$

The formula above is called the **change of basis formula** for linear operators. It is more commonly written as

$$[T;B'] = [B'/B]^{-1}[T;B][B'/B],$$

which gives the "unknown" matrix of T in a "new" basis B' in terms of the "known" matrix of T in an "old" basis B and the transition matrix $[B'/B]$ from the new basis to the old basis. It is often convenient to write it as

$$A' = P^{-1}AP,$$

where A' and A are the matrices of T in the new and old bases respectively, and P is the transition matrix.

EXAMPLE 5-37

Let us return to Example 5-36 and verify the formula for $[T;B']$ above. Now, the right-hand side of the change of basis formula is

$$\begin{bmatrix} 1/4 & 1/4 \\ -1/4 & 3/4 \end{bmatrix} \begin{bmatrix} 5/4 & 9/4 \\ 3/4 & -1/4 \end{bmatrix} \begin{bmatrix} 3 & -1 \\ 1 & 1 \end{bmatrix}$$

$$= \begin{bmatrix} 1/2 & 1/2 \\ 1/4 & -3/4 \end{bmatrix} \begin{bmatrix} 3 & -1 \\ 1 & 1 \end{bmatrix} = \begin{bmatrix} 2 & 0 \\ 0 & -1 \end{bmatrix},$$

which is the left-hand side, $[T;B']$, as previously obtained. ❑

EXAMPLE 5-38

We apply the foregoing theory to solve the following geometrical problem. Let T be the linear operator on R^2 that reflects a vector $\mathbf{v}$ in the line L inclined at an angle θ to the x axis as shown in Figure 5-15. We want to determine a formula for $T(x,y)$.

FIGURE 5-15

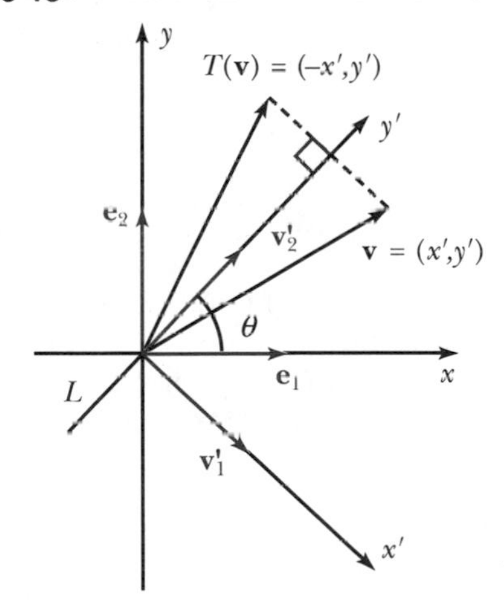

If we establish x' and y' axes as shown (with L in the direction of the y' axis), the action of T is simply described in these coordinates by

$$T(x',y') = (-x',y').$$

In other words,

$$[T;B'] = \begin{bmatrix} -1 & 0 \\ 0 & 1 \end{bmatrix},$$

where $B' = \{\mathbf{v}'_1, \mathbf{v}'_2\}$ and $\mathbf{v}'_1$, $\mathbf{v}'_2$ are unit vectors as shown, constituting a basis for R^2. We find $[T;B]$, where $B = B(2)$, using Theorem 5-17. Now,

$$\mathbf{v}'_1 = (\sin\theta)\mathbf{e}_1 - (\cos\theta)\mathbf{e}_2 \text{ and } \mathbf{v}'_2 = (\cos\theta)\mathbf{e}_1 + (\sin\theta)\mathbf{e}_2,$$

so

$$[B'/B] = \begin{bmatrix} \sin\theta & \cos\theta \\ -\cos\theta & \sin\theta \end{bmatrix}.$$

Therefore,

$$[B/B'] = [B'/B]^{-1} = \begin{bmatrix} \sin\theta & -\cos\theta \\ \cos\theta & \sin\theta \end{bmatrix}.$$

It follows that

$$\begin{aligned}
[T] = [T;B] &= [B'/B][T;B'][B/B'] \\
&= \begin{bmatrix} \sin\theta & \cos\theta \\ -\cos\theta & \sin\theta \end{bmatrix}\begin{bmatrix} -1 & 0 \\ 0 & 1 \end{bmatrix}\begin{bmatrix} \sin\theta & -\cos\theta \\ \cos\theta & \sin\theta \end{bmatrix} \\
&= \begin{bmatrix} -\sin\theta & \cos\theta \\ \cos\theta & \sin\theta \end{bmatrix}\begin{bmatrix} \sin\theta & -\cos\theta \\ \cos\theta & \sin\theta \end{bmatrix} \\
&= \begin{bmatrix} -\sin^2\theta + \cos^2\theta & 2\sin\theta\cos\theta \\ 2\sin\theta\cos\theta & -\cos^2\theta + \sin^2\theta \end{bmatrix} \\
&= \begin{bmatrix} \cos 2\theta & \sin 2\theta \\ \sin 2\theta & -\cos 2\theta \end{bmatrix}.
\end{aligned}$$

So,

$$T(x,y) = (x\cos 2\theta + y\sin 2\theta, x\sin 2\theta - y\cos 2\theta).$$

Notice that, while $[T]$ has much in common with the matrix of a rotation, it is *not*, in fact, a rotation (as some "reflection" should prove!). For example, if we let $\theta = \pi/4$ (45°), then $2\theta = \pi/2$, $\cos 2\theta = 0$, $\sin 2\theta = 1$, and therefore $T(x,y) = (y,x)$. ❑

The relationship between the matrices of a linear operator in different bases is an important one and is called **similarity**.

DEFINITION 5-12

An $n \times n$ matrix A' is said to be **similar** to an $n \times n$ matrix A iff there is an invertible matrix P such that $A' = P^{-1}AP$. (5-11)

Notice that when A' is similar to A, then A is also similar to A', because we can rewrite 5-11 as $A = (P^{-1})^{-1}A'P^{-1}$. So, we can simply say "A' and A are similar". In fact, similarity is an equivalence relation on $M(n \times n)$

(question 14, Exercises 5-4). Of course, any matrix is similar to itself because we can write $A = I^{-1}AI$.

EXAMPLE 5-39

From Examples 5-37 and 5-38, we find that

$$\begin{bmatrix} 2 & 0 \\ 0 & -1 \end{bmatrix} \text{ is similar to } \begin{bmatrix} 5/4 & 9/4 \\ 3/4 & -1/4 \end{bmatrix} \text{ (Example 5-37).}$$

$$\begin{bmatrix} -1 & 0 \\ 0 & 1 \end{bmatrix} \text{ is similar to } \begin{bmatrix} \cos 2\theta & \sin 2\theta \\ \sin 2\theta & -\cos 2\theta \end{bmatrix} \text{ (Example 5-38).} \square$$

From Theorem 5-17, it follows that *the matrices of a linear operator in different bases are all similar.* In fact,

THEOREM 5-18

Two $n \times n$ matrices A, A' are similar iff they are the matrices of the same linear operator T with respect to two bases for R^n.

PROOF Let A and A' be similar. We leave the proof of the following remark for the reader: There is a basis B' for R^n such that $[B'/B] = P$, where B is the standard basis for R^n, and P arises in $A' = P^{-1}AP$. (Hint: B' is obtained from the columns of P.) Then,

$$[T_A;B'] = P^{-1}AP = A',$$

so A and A' are matrices of $T = T_A$. ■

COROLLARY

Similar matrices have the same rank.

PROOF The rank is that of the operator they (both) represent. ■

EXAMPLE 5-40

The matrices below are not similar, since they have different ranks:

$$\begin{bmatrix} -1 & 2 & 1 \\ 2 & 0 & 1 \\ 1 & 1 & 1 \end{bmatrix}, \begin{bmatrix} -1 & 2 & 1 \\ 2 & 0 & 1 \\ 1 & 2 & 2 \end{bmatrix}. \square$$

We can say that the space of $n \times n$ matrices is partitioned into nonoverlapping sets of matrices, with all matrices in any such set being similar and representing *one* linear operator on R^n. In this sense, *a single linear operator represents a whole class of (similar) matrices.*

EXERCISES 5-4

In each of questions 1 through 5, use the change of basis formula to find the coordinate matrices of the vectors as indicated.

1. Let B be the standard basis for R^2 and let $B' = \{\mathbf{v}'_1, \mathbf{v}'_2\}$ with $\mathbf{v}'_1 = (1,0)$ and $\mathbf{v}'_2 = (1,1)$.

(a) Find $[B'/B]$.

(b) Find $[\mathbf{v}]_{B'}$, where **(i)** $\mathbf{v} = (x,y)$; **(ii)** $\mathbf{v} = (-1,3)$; **(iii)** $\mathbf{v} = (4,-8)$.

2. In R^2, let $B'' = \{\mathbf{u}, \mathbf{v}\}$ with $\mathbf{u} = (-2,3)$ and $\mathbf{v} = (4,5)$, and let B be the standard basis.

(a) Find $[B''/B]$.

(b) Find $[\mathbf{w}]_{B''}$, where **(i)** $\mathbf{w} = (x,y)$; **(ii)** $\mathbf{w} = (-4,3)$; **(iii)** $\mathbf{w} = (-3,7)$.

3. In R^3, let $B' = \{\mathbf{v}'_1, \mathbf{v}'_2, \mathbf{v}'_3\}$ with $\mathbf{v}'_1 = (-1,0,1)$, $\mathbf{v}'_2 = (2,1,1)$, and $\mathbf{v}'_3 = (-1,1,3)$.

(a) Find the transition matrix from B to B', where B is the standard basis.

(b) Find $[\mathbf{v}]_{B'}$, where **(i)** $\mathbf{v} = (x,y,z)$; **(ii)** $\mathbf{v} = (-2,1,3)$.

4. In $P(2)$, let $B' = \{f'_1, f'_2, f'_3\}$ where $f'_1(x) = -x + 2x^2$, $f'_2(x) = 2 - x^2$, and $f'_3(x) = 1 + x + x^2$. Let $B = \{1, x, x^2\}$.

(a) Find $[B'/B]$.

(b) Find $[f]_{B'}$, where **(i)** $f(x) = ax^2 + bx + c$; **(ii)** $f(x) = -3 + x - x^2$.

5. In $M(2 \times 2)$, let B be the usual basis and let B' consist of

$$M_1 = \begin{bmatrix} 1 & 0 \\ 0 & 1 \end{bmatrix}, M_2 = \begin{bmatrix} 1 & 1 \\ 0 & 0 \end{bmatrix}, M_3 = \begin{bmatrix} 1 & 1 \\ 1 & 0 \end{bmatrix}, M_4 = \begin{bmatrix} 1 & 1 \\ 1 & 1 \end{bmatrix}.$$

(a) Find $[B'/B]$ and $[B/B']$.

(b) Find $[M]_{B'}$ where

$$M = \begin{bmatrix} 2 & -1 \\ 3 & -4 \end{bmatrix}.$$

6. Let $B' = \{\mathbf{v}'_1, \mathbf{v}'_2\}$, and $B'' = \{\mathbf{v}''_1, \mathbf{v}''_2\}$ be the bases for R^2 given by $\mathbf{v}'_1 = (1,2)$, $\mathbf{v}'_2 = (2,1)$, $\mathbf{v}''_1 = (-1,1)$, and $\mathbf{v}''_2 = (1,1)$.

(a) Find the transition matrices $[B'/B]$ and $[B''/B]$ where B is the standard basis for R^2.

(b) Using (a) or otherwise, find the transition matrix $[B''/B']$.

(c) Find $[\mathbf{v}]_{B'}$ and $[\mathbf{v}]_{B''}$ where **(i)** $\mathbf{v} = (x,y)$; **(ii)** $\mathbf{v} = (6,2)$.

7. Recall that a circle with radius r and center at the origin in R^2 has equation $x^2 + y^2 = r^2$, with respect to the axes defined by the standard basis $B(2)$. Find the equation of this circle with respect to a coordinate system (x',y') defined by a basis B' for R^2 which is **(a)** obtained by a rotation of the standard basis; **(b)** arbitrary.

8. Let $T:R^2 \rightarrow R^2$ be the linear operator defined by

$$T(x,y) = (-x + 2y,\ 3x + y).$$

(a) Find $[T]$.

(b) Find $[T;B']$ where B' is the basis defined in question 1.

(c) Find $[T;B'']$ where B'' is the basis defined in question 2.

(d) Find $T(\mathbf{v})$ where $\mathbf{v}$ is the vector whose components are -1 and 3 with respect to the basis B''.

9. Let $D:P(2) \rightarrow P(2)$ be the linear operator defined by $D(ax^2 + bx + c) = 2ax + b$. Use Theorem 5-17 to find $[D;B']$, where B' is defined in Example 5-35.

10. Let $T:M(2\times 2) \rightarrow M(2\times 2)$ be defined by $T(A) = (1/2)(A + A^t)$, where A is a 2×2 matrix.

(a) Find $[T;B]$ where B is the usual basis for $M(2\times 2)$.

(b) Let $B' = \{E'_1,E'_2,E'_3,E'_4\}$ be defined by $E'_1 = E_1$, $E'_4 = E_4$, and

$$E'_2 = \begin{bmatrix} 0 & 1 \\ 1 & 0 \end{bmatrix}, E'_3 = \begin{bmatrix} 0 & 1 \\ -1 & 0 \end{bmatrix}.$$

Find $[B'/B]$.

(c) Find $[T;B']$. (Notice that it is diagonal.)

11. Let $T:V \rightarrow W$ be a linear transformation where V and W are finite dimensional vector spaces. Suppose B_1 and B'_1 are two bases for V and B_2, and B'_2 are two bases for W. Extend Theorem 5-17 to obtain the change of basis theorem for linear transformations: $[T;B'_1/B'_2] = [B'_2/B_2]^{-1}[T;B_1/B_2][B'_1/B_1]$.

12. Use the result of question 11 to find $[T;B'/B'']$ where T is the linear operator defined in question 8 and B' and B'' are the bases defined in questions 1 and 2 respectively.

13. Use the method of Example 5-38 to find a formula for $T(x,y)$, where T is the orthogonal projection on the line L defined in that example.

14. Complete the verification that similarity is an equivalence relation by showing that if A is similar to A' and A' is similar to A'', then A is similar to A''.

15. Show that A and B defined below are not similar:

$$A = \begin{bmatrix} 2 & -1 & 3 \\ 1 & 0 & 1 \\ -1 & 2 & -2 \end{bmatrix}, B = \begin{bmatrix} -1 & 1 & 2 \\ 2 & -1 & 0 \\ 2 & 0 & 4 \end{bmatrix}.$$

16. Prove that similar matrices have the same trace. (Hint: First show that $tr\ (AB) = tr\ (BA)$.)

17. Show that if A is similar to B, then

(a) A^2 is similar to B^2;

(b) A^k is similar to B^k, for any nonnegative integer k. (Hint: Use the invertibility of the matrix P in the definition of similarity.)

18. We have seen that similar matrices have the same rank. Show that the converse is not true by constructing two 2×2 matrices of the same rank that are not similar. (Hint: Use question 16.)

19. Let A and B be similar. Show that A is invertible iff B is, and that, in this case, the inverses are also similar.

5-5 Linear Functionals and Applications (Optional)

A **linear functional** is a linear transformation $f: V \to R$ from a vector space to the set of real numbers. Here, of course, $R = R^1$ is regarded as a vector space with respect to the usual addition and multiplication of numbers. Thus, f must satisfy

$$f(c\mathbf{v} + \mathbf{v}') = cf(\mathbf{v}) + f(\mathbf{v}')$$

for all vectors $\mathbf{v}, \mathbf{v}' \in V$ and scalars $c \in R$.

DEFINITION 5-13

The set $L(V,R)$ of all linear functionals is called the **dual space** of V. It is usually denoted by V^*.

By Theorem 5-11, V^* is a vector space, and, by Theorem 5-12, if $dim\ V = n$, it is isomorphic to $M(1 \times n)$, since $dim\ R = 1$. Thus, $dim\ V^* = dim\ V = n$. Recall that for each basis B of V, the correspondence $f \leftrightarrow [f;B/B(1)]$ (where $B(1) = \{1\}$ is the usual basis for R) establishes this isomorphism between V^* and $M(1 \times n)$. Therefore, *linear functionals may be thought of as $1 \times n$ matrices*.

EXAMPLE 5-41

By the preceding remarks, we may identify a linear functional f in R^{n*} with the $1 \times n$ matrix $[f]$. Therefore, if $A = (a_i) \in M(1 \times n)$, it defines a linear functional f_A on R^n by

$$f_A(x_1,...,x_n) = \begin{bmatrix} a_1 & \dots & a_n \end{bmatrix} \begin{bmatrix} x_1 \\ \dots \\ x_n \end{bmatrix} = \sum_{i=1}^{n} a_i x_i.$$

And every linear functional on R^n is obtained in this way. We may therefore identify R^{n*} with $M(1 \times n)$ just as we identified R^n itself with $M(n \times 1)$.

As a particular case, consider f_A on R^3 defined by the matrix

$$A = \begin{bmatrix} 2 & -3 & 1 \end{bmatrix}.$$

Then,

$$f_A(x_1,x_2,x_3) = 2x_1 - 3x_2 + x_3. \square$$

An interesting application is that any plane through the origin in R^3 is the kernel of some linear functional $f:R^3 \rightarrow R$. For example, the kernel of f_A defined above is the solution set of the equation

$$2x_1 - 3x_2 + x_3 = 0.$$

We know this is a plane in R^3 through the origin and perpendicular to the vector $(2,-3,1)$. Conversely, any plane through $\mathbf{0}$ is the solution set of an equation

$$a_1x_1 + a_2x_2 + a_3x_3 = 0,$$

which is $Ker\, f_A$, where $A = (a_i)$.

More generally, the kernel of a linear functional f_A on R^n with $A = (a_i)$ is an $(n - 1)$-dimensional subspace described by the equation

$$a_1x_1 + a_2x_2 + \dots + a_nx_n = 0.$$

Such a subspace is called a **hyperplane** in R^n.

EXAMPLE 5-42

Let S be any set and recall the vector space $F(S)$ of all functions $g:S \rightarrow R$. For any $x \in S$, we may define a linear functional $E_x:F(S) \rightarrow R$ by

$$E_x(g) = g(x), \; g \in F(S).$$

E_x is called **evaluation** at x. It is linear because, for any scalar c, and g and $g' \in F(S)$,

$$E_x(cg + g') = (cg + g')(x) = cg(x) + g'(x) = cE_x(g) + E_x(g').$$

For example, since $P(2)$ is a subspace of $F(R)$, we have $E_a{:}P(2) \rightarrow R$ for any number a. In particular, suppose $g(x) = -3x^2 + 5x - 1$ and $a = 6$. Then,

$$E_6(g) = -3(6)^2 + 5(6) - 1 = -79.$$

Similarly, $E_0(g) = -1$ and $E_{-1}(g) = -9$. ❑

EXAMPLE 5-43

If A is an $n \times n$ matrix, its **trace**, denoted $tr\ A$, is defined to be the sum of the diagonal elements of A:

$$tr\ A = a_{11} + \ldots + a_{nn}.$$

Using matrix algebra, it follows that $tr{:}M(n \times n) \rightarrow R$ is a linear functional. ❑

Linear functionals have important applications in many areas. While a discussion of most of these is beyond the scope of this book, we have already encountered some particular cases in earlier sections.

EXAMPLE 5-44

Typically, the objective function Z in a LP problem is a linear functional on R^n:

$$Z = c_1x_1 + \ldots + c_nx_n,$$

where $x_1,\ldots,x_n$ are the decision variables for the problem and the c's are constants. In Example 5-1, Z is the daily profit function from the manufacture of x_1 and x_2 units per day of two products. We saw that $Z = 15x_1 + 12x_2$ in that problem, so the linear functional is $Z{:}R^2 \rightarrow R$. ❑

EXAMPLE 5-45

In Example 5-14, we defined a linear transformation $Area{:}P(1) \rightarrow R$, with $Area(f)$ equal to the algebraic sum of the areas bounded by the graph of $y = f(x)$, the x axis, $x = 0$, and $x = 1$. Thus, $Area$ is a linear functional on $P(1)$. Clearly, we can define such a linear functional using intervals other than $[0,1]$.

As noted earlier, this linear functional is imporant in calculus, where the following more general situation is considered. Let $f:[a,b]\rightarrow R$ be a function, whose graph might have the form as shown in Figure 5-16. For each positive integer n, we divide $[a,b]$ into n subintervals $[x_i,x_{i+1}]$, $i = 0, 1, ..., n-1$, each of length $(b - a)/n$, and $x_0 = a$, $x_n = b$. Now, for each subinterval $[x_i,x_{i+1}]$, we may approximate the area of the region bounded by $y = f(x)$, the x axis, $x = x_i$, and $x = x_{i+1}$ by $Area(f_i)$, where f_i in $P(1)$ is the linear function whose graph passes through $(x_i,f(x_i))$ and $(x_{i+1},f(x_{i+1}))$ as shown (shaded region). Then, adding the areas for all n regions, we get an approximation $A_n(f)$ for the (algebraic) area of the region bounded by $y = f(x)$, the x axis, $x = a$, and $x = b$.

FIGURE 5-16

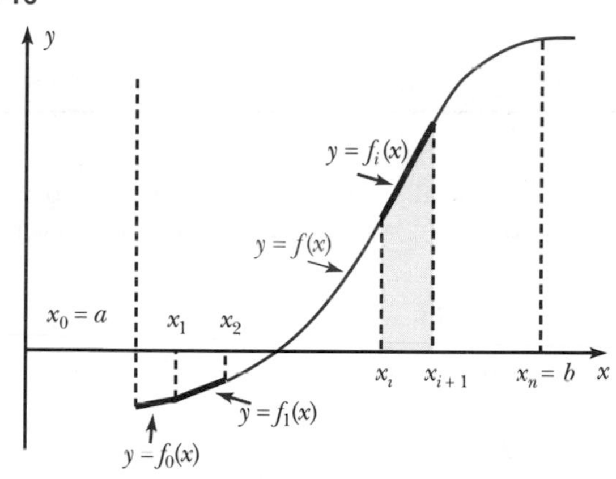

For each n, the function $A_n:F[a,b]\rightarrow R$ obtained in this way is a linear functional because *Area* is linear. Now, in calculus it is shown that for a certain subspace of $F[a,b]$ (say $I[a,b]$, called the **integrable** functions on $[a,b]$), there is a number $Area(f)$ for f in $I[a,b]$ such that the approximations $A_n(f)$ can be made as close as desired to $Area(f)$ by making n ever larger. From the figure, it would appear that as we increase n (= the number of subdivisions), the approximations $A_n(f)$ should get better and better. The number $Area(f)$ obtained in this way is therefore considered to be the actual area of the region bounded by $y = f(x)$, the x axis, $x = a$, and $x = b$. It is called the **definite integral** of f (from $x = a$ to $x = b$). And since each A_n is linear, it follows that $Area:I[a,b]\rightarrow R$ defined in this way is a linear functional. ❑

We next consider bases for dual spaces. We have seen that $f \leftrightarrow [f;B/B(1)]$ is an isomorphism from V^* to $M(1\times n)$, so we can obtain any basis for V^* from a basis B for $M(1\times n)$. For example, let us denote the standard (ordered) basis for $M(1\times n)$ by $B(n)^* = \{\mathbf{e}_1^*,\mathbf{e}_2^*,...,\mathbf{e}_n^*\}$ with

$$\mathbf{e}_1^* = [1 \quad 0 \quad \dots \quad 0], \dots, \mathbf{e}_n^* = [0 \quad 0 \quad \dots \quad 1].$$

Then, the linear functionals f_i:$R^n \to R$ with $[f_i] = \mathbf{e}_i^*$ constitute a basis for R^{n*} by Theorem 5-9, and

$$f_i(x_1,\ldots,x_n) = \begin{bmatrix} 0 & \ldots & 0 & \underset{(i)}{1} & 0 & \ldots & 0 \end{bmatrix} \begin{bmatrix} x_1 \\ \ldots \\ x_n \end{bmatrix} = x_i,$$

which is the ith coordinate of the n-tuple. For this reason, these linear functionals are often called the **coordinate functions** on R^n. Furthermore, they have the property that

$$f_i(\mathbf{e}_j) = \delta_{ij} = \begin{cases} 1, \text{if } i = j; \\ 0, \text{if } i \neq j. \end{cases}$$

More generally,

> **THEOREM 5-19**
>
> Let V be an n-dimensional vector space with basis $B = \{\mathbf{v}_1,\ldots,\mathbf{v}_n\}$. There is an unique basis $B^* = \{f_1,\ldots,f_n\}$ for V^* characterized by the fact that
>
> $$f_i(\mathbf{v}_j) = \delta_{ij},\ 1 \leq i, j \leq n. \tag{5-12}$$
>
> Furthermore, if f is in V^*,
>
> $$f = \sum_{i=1}^{n} f(\mathbf{v}_i) f_i, \tag{5-13}$$
>
> while if $\mathbf{v}$ is in V,
>
> $$\mathbf{v} = \sum_{i=1}^{n} f_i(\mathbf{v})\mathbf{v}_i. \tag{5-14}$$

PROOF Since $B(n)^* = \{\mathbf{e}_1^*,\ldots,\mathbf{e}_n^*\}$ is a basis for $M(1\times n)$, we define $f_1, \ldots, f_n$ to be the unique linear functionals on V such that $[f_i;B/B(1)] = \mathbf{e}_i^*$, $i = 1, 2, \ldots, n$. This means that

$$f_i(\mathbf{v}_j) = \begin{bmatrix} 0 & \ldots & 1 & \ldots & 0 \end{bmatrix} \begin{bmatrix} 0 \\ \ldots \\ 1 \\ \ldots \\ 0 \end{bmatrix} = \delta_{ij},$$

by Theorem 5-7 (the 1 appears in the ith spot in the row vector and in the jth spot in the column vector). Alternatively, the f_i:$V \to R$ are the unique linear transformations defined on the basis B and having values prescribed by equation 5-12 (Theorem 5-2).

Now, if f is in V^*, it is a linear combination:

$$f = \sum_{i=1}^{n} c_i f_i,\ c_i \in R.$$

Applying both sides (linear functionals) to $\mathbf{v}_j$, we get

$$f(\mathbf{v}_j) = \sum_{i=1}^{n} c_i f_i(\mathbf{v}_j) = c_j,$$

proving equation 5-13. A similar proof (left for the reader) will establish equation 5-14. ■

Remark Equation 5-13 shows that if f is in V^*,

$$[f]_{B^*} = \begin{bmatrix} f(\mathbf{v}_1) \\ \cdots \\ f(\mathbf{v}_n) \end{bmatrix}.$$

Note the distinction between this $n \times 1$ matrix of the vector f in the basis B^* and the $1 \times n$ matrix $[f;B/B(1)]$ of f as a linear transformation, $f:V \to R$. In fact,

LEMMA 5-5

$$[f]_{B^*} = [f;B/B(1)]^t.$$

PROOF The proof is left as an exercise (question 7, Exercises 5-5). ■

DEFINITION 5-14

The basis B^* constructed above is called the **dual basis** of B.

Thus, the dual basis functionals of the standard basis $B(n)$ for R^n may be thought of as the $1 \times n$ matrices $\mathbf{e}_1^*, \ldots, \mathbf{e}_n^*$.

EXAMPLE 5-46

Consider the basis $B = \{\mathbf{v}_1, \mathbf{v}_2\}$ for R^2 with $\mathbf{v}_1 = (2,1)$ and $\mathbf{v}_2 = (-1,2)$. We compute the dual basis $\{f_1, f_2\}$ of B. To do this, we must be able to specify $f_1(x,y)$ and $f_2(x,y)$ for any $\mathbf{v} = (x,y)$ in R^2. Now, f_1 and f_2 are completely characterized by equation 5-12 above:

$$f_1(\mathbf{v}_1) = 1, f_1(\mathbf{v}_2) = 0, \tag{5-15}$$
$$f_2(\mathbf{v}_1) = 0, f_2(\mathbf{v}_2) = 1. \tag{5-16}$$

Since any f in R^{2*} has the form $f(x,y) = ax + by$ for some scalars a and b, substitution of $\mathbf{v}_1$ and $\mathbf{v}_2$ into 5-15 yields

$$\begin{aligned} f_1(\mathbf{v}_1) &= 2a + b = 1 \\ f_1(\mathbf{v}_2) &= -a + 2b = 0. \end{aligned}$$

This is a system of equations that determines the scalars a and b for f_1. Similarly, substitution into 5-16 yields

$$\begin{aligned} f_2(\mathbf{v}_1) &= 2a' + b' = 0 \\ f_2(\mathbf{v}_2) &= -a' + 2b' = 1, \end{aligned}$$

which is a system of equations that determines $f_2(x,y) = a'x + b'y$. Notice that both systems have the same left-hand sides, so we can solve both of them at once by Gauss-Jordan reduction of the (combined) augmented matrix

$$\begin{bmatrix} 2 & 1 & 1 & 0 \\ -1 & 2 & 0 & 1 \end{bmatrix}.$$

This matrix is row-equivalent to

$$\begin{bmatrix} 1 & 0 & 2/5 & -1/5 \\ 0 & 1 & 1/5 & 2/5 \end{bmatrix},$$

so that for f_1, $a = 2/5$ and $b = 1/5$, while for f_2, $a' = -1/5$ and $b' = 2/5$. Therefore,

$$\begin{aligned} f_1(x,y) &= (2/5)x + (1/5)y \\ f_2(x,y) &= (-1/5)x + (2/5)y. \ \square \end{aligned}$$

EXAMPLE 5-47

In this example, we describe an application of dual bases known as **Lagrange interpolation**, which is concerned with the following problem. Given $n + 1$ points $(a_1,b_1), ..., (a_{n+1},b_{n+1})$ in the plane, with $a_1, ..., a_{n+1}$ *distinct*, find a polynomial function of degree at most n whose graph passes through these points. For simplicity, we consider the case $n = 2$.

Let $F_i:P(2) \to R$ be the "evaluation at a_i" functionals with $F_i(f) = f(a_i)$, $i = 1, 2, 3$. (See Example 5-42; we write F_i for E_{a_i}.) Now, we can show directly that $B^* = \{F_1,F_2,F_3\}$ is a basis for $P(2)^*$ (see question 10, Exercises 5-5). However, let us do this indirectly by constructing a basis $B = \{f_1,f_2,f_3\}$ for $P(2)$ such that B^* above is the dual of B; i.e., such that $F_i(f_j) = \delta_{ij}$, $i, j = 1, 2, 3$. Then f_1 is determined by the equations

$$\begin{aligned} 1 &= F_1(f_1) = f_1(a_1), \\ 0 &= F_2(f_1) = f_1(a_2), \\ 0 &= F_3(f_1) = f_1(a_3). \end{aligned}$$

So, f_1 is to be a polynomial of degree 2 or less whose values at a_1, a_2, and a_3 are 1, 0, and 0 respectively. Using the fact that $(x - a)/(b - a)$ (where $a \neq b$) is a polynomial of degree $= 1$ whose value at $x = b$ is 1 and whose value at $x = a$ is 0, some reflection leads us to put

$$f_1(x) = \frac{(x - a_2)(x - a_3)}{(a_1 - a_2)(a_1 - a_3)}, \tag{5-17}$$

which clearly has the required properties. Similarly, from

$$\begin{aligned} 0 &= F_1(f_2) = f_2(a_1), \\ 1 &= F_2(f_2) = f_2(a_2), \\ 0 &= F_3(f_2) = f_2(a_3), \end{aligned}$$

we find

$$f_2(x) - \frac{(x - a_1)(x - a_3)}{(a_2 - a_1)(a_2 - a_3)}. \tag{5-18}$$

Finally, we put

$$f_3(x) = \frac{(x - a_1)(x - a_2)}{(a_3 - a_1)(a_3 - a_2)}, \tag{5-19}$$

Now, f_1, f_2, f_3 are linearly independent because, if

$$c_1 f_1 + c_2 f_2 + c_3 f_3 = 0 \text{ (the zero polynomial function)}$$

for some scalars, then we may evaluate both sides at a_1, a_2, and a_3 to get

$$\begin{aligned} 0 &= c_1 f_1(a_1) + c_2 f_2(a_1) + c_3 f_3(a_1) = c_1, \\ 0 &= c_1 f_1(a_2) + c_2 f_2(a_2) + c_3 f_3(a_2) = c_2, \\ 0 &= c_1 f_1(a_3) + c_2 f_2(a_3) + c_3 f_3(a_3) = c_3, \end{aligned}$$

that is, all the c's are 0, as required. Therefore, $B = \{f_1, f_2, f_3\}$ is a basis for $P(2)$. The f_i are called **Lagrange interpolation polynomials**. By Theorem 5-12, B^* is a basis for $P(2)^*$ since the matrices $[F_i; B/B(1)]$ are $\mathbf{e}_1^*$, $\mathbf{e}_2^*$, and $\mathbf{e}_3^*$ respectively, for $i = 1, 2, 3$.

We now return to the interpolation problem. We want f in $P(2)$ such that $f(a_i) = b_i$, $i = 1, 2, 3$. By Theorem 5-19,

$$f = \sum_{i=1}^{3} F_i(f) f_i = \sum_{i=1}^{3} f(a_i) f_i,$$

since $F_i(f) = f(a_i)$. Substituting $f(a_i) = b_i$, we obtain

$$f = \sum_{i=1}^{3} b_i f_i, \tag{5-20}$$

which describes the polynomial we want since b_i and f_i are known. For example, suppose the three points are $(1,1)$, $(2,3)$, and $(3,-2)$ respectively. Then, from equation 5-20, the required polynomial function is

$$\begin{aligned}
f(x) &= 1f_1(x) + 3f_2(x) - 2f_3(x)\\
&= 1\frac{(x-2)(x-3)}{(1-2)(1-3)} + 3\frac{(x-1)(x-3)}{(2-1)(2-3)} + (-2)\frac{(x-1)(x-2)}{(3-1)(3-2)}\\
&= (1/2)(x^2 - 5x + 6) - 3(x^2 - 4x + 3) - (x^2 - 3x + 2)\\
&= (1/2)(-7x^2 + 25x - 16).
\end{aligned}$$

The reader should make a sketch of the graph (a parabola) to verify that it passes through the three given points. ❑

We conclude with one more idea that has useful applications. Let $T:V \rightarrow W$ be a linear transformation and suppose that V and W have dimensions n and m respectively. Then, we have seen that T corresponds to an $m \times n$ matrix A when we choose bases B and B' for V and W respectively. Also, we saw that the transpose operation, mapping A to A^t, is an isomorphism from $M(m \times n)$ to $M(n \times m)$. It is therefore natural to ask if there is a linear transformation related to T whose matrix is A^t. The answer is provided by the next definition.

DEFINITION 5-15

With the notation above, the **transpose** of T is the linear transformation $T^t:W^* \rightarrow V^*$ defined by $T^t(g) = gT$ for $g \in W^*$ (the composition of T and g).

Notice first that gT is indeed a linear functional on V (i.e., $T^t(g)$ is in V^*; see Lemma 5-4). And, T^t is a linear transformation, because if g and h are in W^* and c is a scalar,

$$T^t(cg + h) = (cg + h)T = cgT + hT = cT^t(g) + T^t(h).$$

EXAMPLE 5-48

Let $T:V \rightarrow V$ be the linear operator with $T(\mathbf{v}) = c\mathbf{v}$ for $\mathbf{v}$ in V, where c is a fixed scalar. T is a scalar transformation. Then, if g is in V^*,

$$T^t(g)(\mathbf{v}) = (gT)(\mathbf{v}) = g(T(\mathbf{v})) = g(c\mathbf{v}) = (cg)(\mathbf{v}),$$

so that $T^t(g) = cg$, showing that T^t is also a scalar transformation. ❑

EXAMPLE 5-49

Let $T:R^2 \rightarrow R^3$ be defined by $T(x,y) = (x + y, x - y, 2x - y)$. Then, $T^t:R^{3*} \rightarrow R^{2*}$. We completely describe T^t by saying what it does to the standard basis $B(3)^* = \{\mathbf{e}_1^*, \mathbf{e}_2^*, \mathbf{e}_3^*\}$ for R^{3*}. Now,

$$T^t(\mathbf{e}_1^*)(x,y) = \mathbf{e}_1^* T(x,y) = \mathbf{e}_1^*(x + y, x - y, 2x - y)$$

$$= [1 \quad 0 \quad 0]\begin{bmatrix} x + y \\ x - y \\ 2x - y \end{bmatrix} = x + y.$$

Similarly,

$$T^t(\mathbf{e}_2^*)(x,y) = x - y,\ T^t(\mathbf{e}_3^*)(x,y) = 2x - y.$$

So, we can write $T^t(\mathbf{e}_i^*)$ in terms of $B(2)^* = \{\mathbf{e}_1^*, \mathbf{e}_2^*\}$ $(1 \leq i \leq 3)$ as

$$T^t(\mathbf{e}_1^*) = \mathbf{e}_1^* + \mathbf{e}_2^*,\ T^t(\mathbf{e}_2^*) = \mathbf{e}_1^* - \mathbf{e}_2^*,\ T^t(\mathbf{e}_3^*) = 2\mathbf{e}_1^* - \mathbf{e}_2^*.$$

It follows that

$$[T^t;B(3)^*/B(2)^*] = \begin{bmatrix} 1 & 1 & 2 \\ 1 & -1 & -1 \end{bmatrix} = \begin{bmatrix} 1 & 1 \\ 1 & -1 \\ 2 & -1 \end{bmatrix}^t = [T;B(2)/B(3)]^t.$$

Alternatively, since $T = T_A$, where $A = [T;B(2)/B(3)]$, we may visualize $T^t = T_A{}^t$ more easily by identifying R^{3*} with $M(1 \times 3)$ and R^{2*} with $M(1 \times 2)$ as noted in Example 5-41. Then,

$$T_A{}^t([x_1 \quad x_2 \quad x_3]) = [x_1 \quad x_2 \quad x_3]\, T_A = [x_1 \quad x_2 \quad x_3]A$$

$$= [x_1 \quad x_2 \quad x_3]\begin{bmatrix} 1 & 1 \\ 1 & -1 \\ 2 & -1 \end{bmatrix} = [x_1 + x_2 + 2x_3 \quad x_1 - x_2 - x_3],$$

a 1×2 matrix, which we regard as a linear functional on R^2. ❑

In general, we have

THEOREM 5-20
If B and B' are bases for V and W respectively, and $T{:}V \rightarrow W$ is linear, then $[T^t;B'^*/B^*] = [T;B/B']^t$.

In other words, *the matrix of the transpose of T is the transpose of the matrix of T.*

PROOF The steps are given in question 14, Exercises 5-5. ■

As a consequence, we see that *rank* T^t = *rank* T, by Theorem 5-5.

EXERCISES 5-5

1. Which of the following are linear functionals?
   (a) $f{:}R^2 \to R$ with $f(x,y) = x - 2y + 1$
   (b) $f{:}P(2) \to R$ with $f(ax^2 + bx + c) = a + b + c$
   (c) $f{:}M(2\times 2) \to R$ with $f(A) = a_{12} + a_{21}$
   (d) $f{:}M(2\times 2) \to R$ with $f(A) = a_{11}a_{22} - a_{12}a_{21}$
   (e) $f{:}M(n\times 1) \to R$ with $f(X) = X^tX$
2. Express the linear functional f on R^2 with $f(x,y) = 2x - y$ as a linear combination of
   (a) the linear functionals defined by the matrices $[1 \quad 0]$ and $[0 \quad 1]$;
   (b) the linear functionals defined by $[1 \quad 0]$ and $[1 \quad 1]$.
3. Find the linear functional g on R^3 such that the plane through the origin and perpendicular to $\mathbf{N} = (2,1,3)$ is $Ker\ g$.
4. Show that any line through the origin in R^3 is the intersection of $Ker\ f$ and $Ker\ g$ for some linear functionals f and g on R^3.
5. For the interval $[a,b] = [0,2]$ (Example 5-45), find a formula for $A_2(f)$, where f in $P(2)$ is given by $f(x) = ax^2 + bx + c$. Show that $A_2{:}P(2) \to R$ so defined is linear using the formula obtained.
6. Let $B = \{\mathbf{v}_1,\mathbf{v}_2\}$ be the basis for R^2 given by $\mathbf{v}_1 = (-1,3)$ and $\mathbf{v}_2 = (2,1)$.
   (a) Find the dual basis $B^* = \{f_1,f_2\}$ of B.
   (b) Use Theorem 5-19 to express **(i)** the linear functional f with $f(x,y) = -x + y$ as a linear combination of B^*; and **(ii)** the vector $\mathbf{v} = (2,3)$ as a linear combination of B.
7. Prove Lemma 5-5.
8. Let F_0, $F_1{:}P(1) \to R$ be the evaluation functionals with $F_0(f) = f(0)$ and $F_1(f) = f(1)$ for $f \in P(1)$.
   (a) Show that $\{F_0,F_1\}$ is linearly independent by applying $c_0F_0 + c_1F_1 = 0$ to appropriate polynomials in $P(1)$.
   (b) Find the basis $B = \{f_0,f_1\}$ for $P(1)$ such that $B^* = \{F_0,F_1\}$ is the dual.
   (c) Express $Area{:}P(1) \to R$ (Example 5-14) as a linear combination of B^* using Theorem 5-19. Hence show that $Area(ax + b) = (a/2) + b$.
9. **(a)** Find the unique polynomial of degree at most 2 whose graph passes through the points (1,1), (2,2), and (3,3) in R^2.
   (b) Find the unique polynomial of degree at most 3 whose graph passes through the points (1,1), (2,1), (3,1), and (4,0) in R^2.

10. Let a_1, a_2, and a_3 be 3 distinct numbers and F_i:$P(2)\rightarrow R$ be the evaluation functionals as discussed in Example 5-42. Show that $\{F_1,F_2,F_3\}$ is linearly independent as follows.

(a) Suppose $c_1F_1 + c_2F_2 + c_3F_3 = 0$ for some scalars c_i. Apply this linear functional to the standard basis for $P(2)$ to obtain the system of equations below for the c_i.

$$\begin{aligned} c_1 \quad + c_2 \quad + c_3 \quad &= 0 \\ c_1a_1 + c_2a_2 + c_3a_3 &= 0 \\ c_1a_1^2 + c_2a_2^2 + c_3a_3^2 &= 0. \end{aligned}$$

(b) Show that all the c_i are 0 by proving that the coefficient matrix is invertible. This coefficient matrix is called a *Vandermonde matrix*.

11. Let T:$R^2\rightarrow R^3$ be the linear transformation defined by $T(x,y) = (2x - y, 2x + y, x - y)$.

(a) Find $T^t(g)$ where g is the linear functional on R^3 corresponding to the matrix $[a \quad b \quad c]$.

(b) Find $T^t([-1 \quad 0 \quad 2])$.

(c) Find $T^t([-1 \quad 0 \quad 2])(x,y)$.

12. Let F_1, F_2, and F_3 be the evaluation functionals at -1, 0, and 1 respectively on $P(2)$.

(a) Find the basis $B' = \{f_1,f_2,f_3\}$ for $P(2)$ so that $B'^* = \{F_1,F_2,F_3\}$ is the dual.

(b) Express $\{f_1,f_2,f_3\}$ in terms of the usual basis $B = \{1,x,x^2\}$ for $P(2)$ and find $[B/B']$. Express each of 1, x, and x^2 in terms of f_1, f_2, and f_3.

(c) Find $[B^*/B'^*]$. (Hint: Apply Theorem 5-20 to the identity map.)

13. Let D:$P(2)\rightarrow P(1)$ be the linear transformation defined by $D(ax^2 + bx + c) = 2ax + b$ (see Example 5-16). Let B and B' be the usual bases for $P(2)$ and $P(1)$ respectively.

(a) Find $[D^*;B'^*/B^*]$ and verify that it is the transpose of $[D;B/B']$.

(b) Find $D^*(Area)$. (Hint: Use the result of question 8.)

14. Justify the following proof of Theorem 5-20 by identifying the results used at each step. Let $B'^* = \{f'_1,\ldots,f'_m\}$. The ith column of $[T^t;B'^*/B^*] = [T^t(f'_i)]_{B^*} = [f'_iT]_{B^*} = [f'_iT;B/B(1)]^t = ([f'_i;B'/B(1)][T;B/B'])^t = [T;B/B']^t[f'_i]_{B^*} =$ the ith column of $[T;B/B']^t$.

15. Let V be finite dimensional. Show that the map d:$V\rightarrow(V^*)^*$ with $d(\mathbf{v})(f) = f(\mathbf{v})$ for $f \in V^*$ and $\mathbf{v} \in V$ is an isomorphism. (Hint: (i) Show that $d(\mathbf{v})$ is a linear functional on V^*; (ii) show d is linear; (iii) show d is 1-1; (iv) use $dim\ (V^*)^* = dim\ V$ to conclude that d is an isomorphism. The space $(V^*)^* = V^{**}$ is called the *bidual* or double dual of V. The result above shows that we cannot continue to generate *new* spaces by taking duals when V is finite dimensional.)

6 THE EIGENVALUE PROBLEM

6-1 Classification of Linear Operators
6-2 The Characteristic Polynomial
6-3 Diagonalization of Linear Operators
6-4 Computational Considerations II (Optional)
6-5 Application to Economics (Optional)

6-1 Classification of Linear Operators

In any scientific discipline, it is natural to attempt to classify (or enumerate) the distinct objects that arise in that discipline. In this chapter, we will classify (partially) linear operators on a finite dimensional vector space V.

Specifically, we will attempt to make a list of all possible linear operators T on V, starting with those that are in some sense "simplest". However, we will not complete the program of classification, but in considering it, we will develop numerous concepts which are important both for mathematics and for its applications.

The idea is simply to classify linear operators T by classifying their matrices: we shall say that T is a linear operator of a certain type iff for *some* basis B for V, $[T;B]$ has a certain special (and simple) form. And, by enumerating the various matrices with any given special form, we shall be enumerating the possible linear operators on V, since an operator is completely determined by its matrix with respect to a basis.

It is important to bear in mind that, since the matrix of T varies from basis to basis (Theorems 5-17 and 5-18), the main difficulty will be to determine (if possible) a basis B such that $[T;B]$ has the required special form.

To begin our program of classification, let us consider some "simple" matrices (admittedly, this is a somewhat subjective notion).

(1) Certainly the simplest $n \times n$ matrix (where $n = \dim V$) is the $n \times n$ zero matrix O. However, this is the matrix only of the zero operator 0 on V with respect to any basis B (Example 5-18(a)).

(2) Perhaps next in simplicity is the $n \times n$ identity matrix I. This is the matrix only of $1_V: V \to V$, the identity operator on V, with respect to any basis for V (Example 5-18(b)).

(3) Only slightly less simple than I is cI, where c is a scalar. The reader can verify that this is necessarily the matrix of $c1_V$ in any basis for V. These operators have the property that they simply multiply every vector in V by c, and are called **scalar transformations** (see question 4, Exercises 5-1 and Example 5-48)

Having dealt with these essentially trivial categories, we consider the one that is our principal interest. It is the category of operators whose matrices are *diagonal* in some basis B for V. This is the obvious generalization of category (3) above. Note that diagonal matrices are still relatively simple, particularly because their sums, products, and scalar products are also diagonal.

DEFINITION 6-1

Let T be a linear operator on a vector space V of dimension n. Then T is said to be **diagonalizable** iff there is a basis B for V such that $[T;B]$ is an $n \times n$ diagonal matrix.

Thus, T is diagonalizable iff there is a basis B such that

$$[T;B] = \begin{bmatrix} c_1 & 0 & \dots & 0 & 0 \\ 0 & c_2 & \dots & 0 & 0 \\ \dots & \dots & \dots & \dots & \dots \\ 0 & 0 & 0 & 0 & c_n \end{bmatrix} \tag{6-1}$$

where $c_1, c_2, \dots, c_n$ are scalars, which are not necessarily distinct.

EXAMPLE 6-1

Let $T{:}R^2 \to R^2$ be the linear operator defined by

$$T(x,y) = ((5/4)x + (9/4)y, (3/4)x - (1/4)y).$$

This operator was discussed in Examples 5-36 and 5-37. It was seen that if B is the standard basis for R^2 and $B' = \{\mathbf{v}'_1, \mathbf{v}'_2\}$, with $\mathbf{v}'_1 = (3,1)$ and $\mathbf{v}'_2 = (-1,1)$, then,

$$[T;B] = \begin{bmatrix} 5/4 & 9/4 \\ 3/4 & -1/4 \end{bmatrix}, [T;B'] = \begin{bmatrix} 2 & 0 \\ 0 & -1 \end{bmatrix}.$$

Thus, T is diagonalizable with $c_1 = 2$ and $c_2 = -1$, in the notation used in Definition 6-1. ❑

We see also, that given only the formula for T (or its matrix in the standard basis), it is not at all apparent that T is diagonalizable. One of our main goals in this chapter will be to determine the conditions for an operator to be diagonalizable. The fact that the matrices of a linear operator in different bases are all similar leads to the following terminology for the *diagonalizability of matrices*.

DEFINITION 6-2

A square matrix A is **diagonalizable** iff it is similar to a diagonal matrix.

Note that Definitions 6-1 and 6-2 are compatible in the following sense:

(1) A is diagonalizable iff the linear operator T_A is.
(2) $T{:}V \to V$ is diagonalizable iff $[T;B]$ is, for any choice of basis B for V. (Readers may prove this in question 5, Exercises 6-1.)

Thus, from the previous example, the matrix

$$\begin{bmatrix} 5/4 & 9/4 \\ 3/4 & -1/4 \end{bmatrix}$$

is diagonalizable.

In order to gain some insight into the nature of diagonalizable operators, let us consider again the operator T of Example 6-1. It has the property that

$$T(\mathbf{v}'_1) = 2\mathbf{v}'_1 \text{ and } T(\mathbf{v}'_2) = -1\mathbf{v}'_2.$$

In general, let $T:V \rightarrow V$ be a diagonalizable linear operator such that $[T;B]$ is diagonal as in equation 6-1, with $B = \{\mathbf{v}_1,\ldots,\mathbf{v}_n\}$. Then by definition,

$$T(\mathbf{v}_i) = c_i\mathbf{v}_i,\ 1 \leq i \leq n.$$

Thus, diagonalizable operators map the vectors of *some* basis to multiples of themselves. This important observation leads to the following terminology.

DEFINITION 6-3

A *nonzero* vector $\mathbf{v}$ is called an **eigenvector** of a linear operator T on V iff $T(\mathbf{v}) = c\mathbf{v}$ for some scalar c. In this case, c is called an **eigenvalue** of T **associated** with $\mathbf{v}$, and $\mathbf{v}$ is called an eigenvector of T **associated** with c.

Eigenvectors and eigenvalues are also called *characteristic vectors* and *characteristic values* respectively in some books.

EXAMPLE 6-2

Consider the linear operator T_A defined by

$$A = \begin{bmatrix} 2 & -1 \\ 1 & 0 \end{bmatrix}.$$

(a) Then, for example, the vector $\mathbf{v} = (1,1)$ is an eigenvector because

$$T_A(\mathbf{v}) = \begin{bmatrix} 2 & -1 \\ 1 & 0 \end{bmatrix}\begin{bmatrix} 1 \\ 1 \end{bmatrix} = \begin{bmatrix} 1 \\ 1 \end{bmatrix} = 1\mathbf{v}.$$

The associated eigenvalue is $c = 1$.

(b) On the other hand, for example, $\mathbf{v} = (1,2)$ is not an eigenvector because

$$T_A(\mathbf{v}) = \begin{bmatrix} 2 & -1 \\ 1 & 0 \end{bmatrix}\begin{bmatrix} 1 \\ 2 \end{bmatrix} = \begin{bmatrix} 0 \\ 1 \end{bmatrix} \neq c\mathbf{v},$$

for any scalar c. ❑

When $\mathbf{v}$ is an eigenvector of T, T acts simply as a scalar transformation by c on the 1-dimensional space *span* $\{\mathbf{v}\}$ generated by $\mathbf{v}$. For, if $a\mathbf{v}$ is any vector in it ($a \in R$),

$$T(a\mathbf{v}) = aT(\mathbf{v}) = a(c\mathbf{v}) = c(a\mathbf{v}).$$

In particular, we see that any nonzero multiple $a\mathbf{v}$ of an eigenvector associated with eigenvalue c is also an eigenvector associated with c. Of course, $T(\mathbf{0}) = \mathbf{0} = c\mathbf{0}$, but we *do not* call $\mathbf{0}$ an eigenvector.

It is important to realize that *a linear operator need not have any eigenvalues or eigenvectors.* We use the above geometrical thinking in terms of scalar transformations to illustrate this fact in the next example.

EXAMPLE 6-3

Let $T:R^2 \to R^2$ be the rotation of the plane through a counterclockwise angle of (say) 90°, as discussed in Example 5-7. Then T has no eigenvector $\mathbf{v}$ because if $\mathbf{v} \neq \mathbf{0}$, $T(\mathbf{v}) \neq c\mathbf{v}$ for any scalar c; if $T(\mathbf{v}) = c\mathbf{v}$, $T(\mathbf{v})$ and $\mathbf{v}$ would be parallel, contrary to the definition of T. Also, it follows that T is not diagonalizable. ❑

On the other hand, consider the next example.

EXAMPLE 6-4

Let $T:R^2 \to R^2$ be the rotation through 180° (reflection in the origin):

$$T(x,y) = (-x,-y) = (-1)(x,y).$$

Here *every* vector $\mathbf{v} = (x,y) \neq \mathbf{0}$ is an eigenvector of T associated with the eigenvalue -1. ❑

DEFINITION 6-4

If A is an $n \times n$ matrix, its eigenvalues and eigenvectors are those of the matrix transformation T_A.

Thus, for example,

$$A = \begin{bmatrix} 0 & -1 \\ 1 & 0 \end{bmatrix},$$

has no eigenvalues and eigenvectors, since T_A is the rotation of Example 6-3 above.

The matrices of Example 6-1,

$$A' = \begin{bmatrix} 5/4 & 9/4 \\ 3/4 & -1/4 \end{bmatrix} \text{ and } A'' = \begin{bmatrix} 2 & 0 \\ 0 & -1 \end{bmatrix},$$

are similar matrices and A'' clearly has eigenvalues 2 and -1. So, A' also has eigenvalues 2 and -1. At this point, we don't know whether or not they have any other eigenvalues.

The next result shows that the definitions of eigenvalues for linear operators and matrices are compatible in the way we would expect.

LEMMA 6-1

A scalar c is an eigenvalue of a linear operator T iff it is an eigenvalue of the matrix $[T;B]$ for any basis B of V.

PROOF c is an eigenvalue of T iff there is a nonzero vector $\mathbf{v}$ in V such that $T(\mathbf{v}) = c\mathbf{v}$. Taking matrices of both sides with respect to B, we see that this holds iff $[T(\mathbf{v})]_B = c[\mathbf{v}]_B$, or, using Theorem 5-7, iff $[T;B][\mathbf{v}]_B = c[\mathbf{v}]_B$. Since $[\mathbf{v}]_B \neq O$ iff $\mathbf{v} \neq \mathbf{0}$, this last equation is simply the requirement that c be an eigenvalue of $[T;B]$, thus completing the proof. ■

COROLLARY

Similar matrices have the same eigenvalues.

PROOF Let A and $A' = P^{-1}AP$ be similar. Then A and A' are the matrices of the linear operator T_A with respect to two (different) bases. By the lemma, their eigenvalues are those of T_A, and are therefore the same. (We shall see a different proof of this fact in the next section.) ■

We conclude this section by formulating the criterion for diagonalizability in terms of eigenvalues.

THEOREM 6-1

A linear operator T on a vector space V of dimension n is diagonalizable iff there is a basis B for V of eigenvectors of T.

PROOF By definition, if T is diagonalizable, there is a basis of eigenvectors. Conversely, if $B = \{\mathbf{v}_1, \ldots, \mathbf{v}_n\}$ is a basis for V of eigenvectors of T, then there are scalars $c_1, \ldots, c_n$ (not necessarily distinct) such that

$$T(\mathbf{v}_1) = c_1\mathbf{v}_1, \ldots, T(\mathbf{v}_n) = c_n\mathbf{v}_n.$$

Therefore,

$$[T;B] = \begin{bmatrix} c_1 & 0 & \dots & 0 & 0 \\ 0 & c_2 & \dots & 0 & 0 \\ \dots & \dots & \dots & \dots & \dots \\ 0 & 0 & 0 & 0 & c_n \end{bmatrix}$$

and we see that T is diagonalizable. ■

EXERCISES 6-1

1. Find all eigenvectors and eigenvalues of the following linear operators and state which (if any) are diagonalizable.

(a) The zero operator 0 on a vector space V (of finite dimension).

(b) The identity operator 1_V on V.

(c) The operator on V defined by $T(\mathbf{v}) = c\mathbf{v}$ for $\mathbf{v}$ in V, where c is a scalar.

(d) $T:R^2 \rightarrow R^2$ defined by the counterclockwise rotation through 30°.

(e) $T:R^2 \rightarrow R^2$ defined by the reflection in the x axis: $T(x,y) = (x, -y)$.

2. Given the following matrices A, determine whether or not the accompanying vectors $\mathbf{v}$ are eigenvectors of A. If they are, find the associated eigenvalues.

(a) $A = \begin{bmatrix} -1 & 1 \\ 1 & -1 \end{bmatrix}$

(i) $\mathbf{v} = (1,1)$ **(ii)** $\mathbf{v} = (-2,3)$ **(iii)** $\mathbf{v} = (-4,-4)$ **(iv)** $\mathbf{v} = (0,1)$ **(v)** $\mathbf{v} = (10,-10)$

(b) $A = \begin{bmatrix} 2 & -1 & 0 \\ 1 & 0 & 0 \\ 0 & 0 & 3 \end{bmatrix}$

(i) $\mathbf{v} = (0,0,1)$ **(ii)** $\mathbf{v} = (1,1,1)$ **(iii)** $\mathbf{v} = (1,1,0)$ **(iv)** $\mathbf{v} = (1,2,3)$ **(v)** $\mathbf{v} = (-2,-2,0)$

3. Let $T:P(1) \rightarrow P(1)$ be the linear operator defined by $T(ax + b) = 2bx - a + 3b$, for $ax + b$ in $P(1)$. Show that $x + 1$ and $2x + 1$ are eigenvectors, and find the associated eigenvalues of T. Show that $x - 1$ is not an eigenvector.

4. Given the matrix

$$A = \begin{bmatrix} -2 & -1 \\ 1 & 0 \end{bmatrix}.$$

(a) Show that $c = -1$ is an eigenvalue of A.

(b) Show that $c = 0$ is not an eigenvalue of A.

(c) Find all eigenvalues c of A by solving the equation $AX = cX$ for c.

5. **(a)** Prove that a square matrix A is diagonalizable iff the matrix operator defined by A is diagonalizable.
   (b) Prove that a linear operator $T:V \to V$ is diagonalizable iff $[T;B]$ is diagonalizable for any basis B for V.
6. Prove that $c = 0$ is an eigenvalue of a linear operator T (or a square matrix A) iff T (respectively, A) is singular.
7. Suppose that T is an invertible linear operator with eigenvalue c. Show that $1/c$ is an eigenvalue of T^{-1}. Prove the same fact for invertible matrices A. Also show that the operator and its inverse have the same eigenvectors in this case.

6-2 The Characteristic Polynomial

In this section, we develop a method for finding the eigenvalues and eigenvectors of a linear operator or square matrix. The idea is illustrated in the following example.

EXAMPLE 6-5

Let $T_A:R^2 \to R^2$ be the operator defined by

$$A = \begin{bmatrix} 5 & -4 \\ 3 & -2 \end{bmatrix}.$$

A scalar c is an eigenvalue of T_A (and of A) iff there is a (nonzero) vector $\mathbf{v} = (x,y)$ such that

$$T_A(\mathbf{v}) = T_A(x,y) = c\mathbf{v} = c(x,y),$$

that is,

$$\begin{bmatrix} 5 & -4 \\ 3 & -2 \end{bmatrix}\begin{bmatrix} x \\ y \end{bmatrix} = c\begin{bmatrix} x \\ y \end{bmatrix}. \tag{6-2}$$

Now, 6-2 may be written as the homogeneous system of equations in x and y:

$$\begin{aligned} (5 - c)x - \quad\quad\quad 4y &= 0 \\ 3x + (-2 - c)y &= 0. \end{aligned} \tag{6-3}$$

And, c is an eigenvalue iff this system has a *nontrivial* solution. By Theorem 4-13 (Corollary), the system has a nontrivial solution iff its coefficient matrix

$$\begin{bmatrix} 5 - c & -4 \\ 3 & -2 - c \end{bmatrix} = A - cI$$

is singular. This is the case iff $det\ (A - cI) = 0$ (Theorem 3-8). So, we find that c is an eigenvalue iff

$$0 = (5 - c)(-2 - c) + 12 = c^2 - 3c + 2 = (c - 1)(c - 2).$$

Therefore, the eigenvalues of T_A are $c = 1$ and $c = 2$, the roots of the equation $det\ (A - cI) = 0$. Notice that this is a polynomial equation of degree 2 in the variable c. ❑

The above example can be generalized to yield the next theorem.

THEOREM 6-2

Let T be a linear operator on a vector space V of dimension n, and let B be a basis for V. Then a scalar c is an eigenvalue of T iff $det\ (A - cI) = 0$, where $A = [T;B]$ and I is the $n \times n$ identity matrix.

PROOF First, suppose c is an eigenvalue of T. Then, as in the last example, there is a nonzero $\mathbf{v}$ in V such that

$$T(\mathbf{v}) = c\mathbf{v} = c1_V(\mathbf{v}),$$

where, as before, $1_V{:}V \to V$ is the identity operator on V. We can rewrite this as

$$(T - c1_V)(\mathbf{v}) = \mathbf{0},$$

from which it follows that the linear operator $T - c1_V$ is singular ($\mathbf{v} \neq \mathbf{0}$ is in its kernel). Consequently, its matrix with respect to any basis B is also singular. By Theorem 3-8, we get $det\ [T - c1_V;B] = 0$. But by Theorem 5-12,

$$[T - c1_V;B] = [T;B] - c[1_V;B] = A - cI,$$

so we see that if c is an eigenvalue, $det\ (A - cI) = 0$.

Conversely, if $det\ (A - cI) = 0$, reversing the steps above shows that $[T - c1_V;B]$ is singular, so $T - c1_V$ is singular, and there is therefore a vector $\mathbf{v} \neq \mathbf{0}$ in $Ker\ (T - c1_V)$. Thus, $(T - c1_V)(\mathbf{v}) = \mathbf{0}$, or, $T(\mathbf{v}) = c\mathbf{v}$, showing that c is an eigenvalue. ■

The corollary which follows is the analogous result for matrices.

COROLLARY

A scalar c is an eigenvalue of $n \times n$ matrix A iff $det\ (A - cI) = 0$.

PROOF By definition, c is an eigenvalue of A iff it is an eigenvalue of T_A. By the theorem, this is the case iff $det\ (A - cI) = 0$, because $[T_A] = A$. ■

Now, if $A = (a_{ij})$ is $n \times n$ and I is the $n \times n$ identity, then

$$det\ (A - cI) = det \begin{bmatrix} a_{11} - c & a_{12} & \dots & a_{1n} \\ a_{21} & a_{22} - c & \dots & a_{2n} \\ \dots & \dots & \dots & \dots \\ a_{n1} & a_{n2} & \dots & a_{nn} - c \end{bmatrix}. \tag{6-4}$$

Using the formula for *det* (Definition 3-4) and expanding, we see that $det\ (A - cI)$ is a polynomial of degree n in the indeterminate c. This is the case because the highest power of c will arise from the term that is the product of the diagonal elements in 6-4: off-diagonal elements don't contain c. Therefore, we can write 6-4 as

$$\begin{aligned} det\ (A - cI) &= (a_{11} - c)...(a_{nn} - c) + \{\text{lower powers of } c\} \\ &= (-1)^n c^n + \{\text{lower powers of } c\}. \end{aligned} \tag{6-5}$$

This shows that $det\ (A - cI)$ is a polynomial of degree n in the indeterminate c with leading coefficient $+1$ or -1. Since it is customary to use x as the indeterminate in a polynomial, we are led to the following definition.

DEFINITION 6-5

Let A be an $n \times n$ matrix. Then the **characteristic polynomial** f of A is the polynomial in x of degree n defined by $f = det\ (A - xI)$.

By the preceding discussion, *a scalar c is an eigenvalue of A iff it is a zero of the characteristic polynomial*. By Theorem B-5, Corollary, a polynomial of degree n has at most n real zeros. Since $\deg f = n$, it follows that an $n \times n$ matrix has at most n distinct eigenvalues.

If T is a linear operator on an n-dimensional space V, we have seen that the eigenvalues of T are those of $[T;B]$ for any choice of basis B. Thus we *define* the characteristic polynomial of T to be $f = det\ (A - cI)$ where $A = [T;B]$. It follows that T has at most n distinct eigenvalues. This definition is also justified by

LEMMA 6-2

Similar matrices have the same characteristic polynomial.

PROOF Let A and $A' = P^{-1}AP$ be similar. Then the characteristic polynomial f' of A' is

$$\begin{aligned} f' &= det\ (P^{-1}AP - xI) \\ &= det\ [P^{-1}(A - xI)P] \quad (\text{since } P^{-1}xIP = xI) \\ &= det\ P^{-1} det\ (A - xI) det\ P \\ &= det\ (A - xI) \\ &= f, \end{aligned}$$

which is the characteristic polynomial of A. (Note that the corollary of Lemma 6-1 also follows from this result.) ■

In many books, the Greek letter λ (lambda) is used instead of x in the characteristic polynomial f. Also, the equation $f(x) = 0$ is often called the **characteristic equation** of the matrix or operator.

EXAMPLE 6-6

We find the eigenvalues of the matrix

$$A = \begin{bmatrix} 4 & -3 \\ 3 & -2 \end{bmatrix}.$$

The characteristic polynomial is

$$\begin{aligned} f &= det\,(A - xI) \\ &= det \begin{bmatrix} 4 - x & -3 \\ 3 & -2 - x \end{bmatrix} \\ &= (4 - x)(-2 - x) + 9 = x^2 - 2x + 1 = (x - 1)^2. \end{aligned}$$

Thus, A has one eigenvalue $c = 1$. (See also question 16, Exercises 6-2.) ❑

EXAMPLE 6-7

Let $T{:}R^2 \to R^2$ be the rotation through 90° in a counterclockwise direction. We saw in Example 6-4 that T has no eigenvalues. We verify this using the characteristic polynomial f, which is

$$f = det\left\{\begin{bmatrix} 0 & -1 \\ 1 & 0 \end{bmatrix} - x\begin{bmatrix} 1 & 0 \\ 0 & 1 \end{bmatrix}\right\} = det\begin{bmatrix} -x & -1 \\ 1 & -x \end{bmatrix} = x^2 + 1.$$

This polynomial has no (real) zeros, so T has no eigenvalues. ❑

EXAMPLE 6-8

Consider

$$A = \begin{bmatrix} 5 & -2 & 0 \\ -2 & 6 & 2 \\ 0 & 2 & 7 \end{bmatrix}.$$

The characteristic polynomial is

$$
\begin{aligned}
f = det \begin{bmatrix} 5 - x & -2 & 0 \\ -2 & 6 - x & 2 \\ 0 & 2 & 7 - x \end{bmatrix} \\
&= (5 - x)[(6 - x)(7 - x) - 4] + 2[(-2)(7 - x) - 0] \\
&= (5 - x)(38 - 13x + x^2) - 28 + 4x \\
&= -x^3 + 18x^2 - 99x + 162.
\end{aligned}
$$

Factoring by the methods of Appendix B, we find that

$$
\begin{aligned}
f &= (x - 3)(-x^2 + 15x - 54) \\
&= -(x - 3)(x - 9)(x - 6).
\end{aligned}
$$

Thus A has eigenvalues $c = 3, 6$, and 9. ❑

As may be apparent from working the last example, finding the zeros of the characteristic polynomial is not an efficient or practical method for finding the eigenvalues of larger matrices. Indeed, there are more efficient ways of computing the characteristic polynomial itself than by finding $det\ (A - xI)$ (see question 3, Exercises 6-4). These numerical considerations are pursued in section 6-4.

We next consider the problem of finding eigenvectors associated with each eigenvalue. This is simply a problem of solving homogeneous systems of equations.

LEMMA 6-3

A vector $\mathbf{v} \neq \mathbf{0}$ is an eigenvector associated with an eigenvalue c of a linear operator T (or $n \times n$ matrix A) iff $\mathbf{v}$ is in $Ker\ (T - c1_V)$ (respectively for A, $\mathbf{v}$ is in $Ker\ (T_A - c1_{R^n})$).

PROOF This is just a restatement of the definition, since, $\mathbf{v} \neq \mathbf{0}$ is an eigenvector iff $T(\mathbf{v}) = c\mathbf{v}$, or, $(T - c1_V)(\mathbf{v}) = \mathbf{0}$, i.e., $\mathbf{v} \in Ker\ (T - c1_V)$ as stated. ■

DEFINITION 6-6

If c is an eigenvalue of T (or A), the **eigenspace** E_c of c is $Ker\ (T - c1_V)$ (respectively for A, $Ker\ (T_A - c1_{R^n})$).

This is the subspace of V (respectively, of R^n) consisting of all eigenvectors associated with c, together with $\mathbf{0}$. Now, if B is a basis for V, with $[T;B] = A$, $Ker\ (T - c1_V)$ consists of the vectors $\mathbf{v}$ such that $X = [\mathbf{v}]_B$ is a solution of the homogeneous system

$$(A - cI)X = O \tag{6-6}$$

(compare equation 6-3 in Example 6-5). Thus, to find the eigenvectors associated with c, we solve 6-6. The reader should review Theorem 4-14 and Example 4-31 for the systematic construction of bases for the solution spaces of such systems.

EXAMPLE 6-9

We find the eigenvectors for the matrix

$$A = \begin{bmatrix} 5 & -4 \\ 3 & -2 \end{bmatrix}.$$

The eigenvalues were found in Example 6-5 to be $c = 1$ and 2.

(a) For $c = 1$, $\mathbf{v} = (x,y)$ is an associated eigenvector iff

$$\left\{\begin{bmatrix} 5 & -4 \\ 3 & -2 \end{bmatrix} - 1\begin{bmatrix} 1 & 0 \\ 0 & 1 \end{bmatrix}\right\}\begin{bmatrix} x \\ y \end{bmatrix} = \begin{bmatrix} 0 \\ 0 \end{bmatrix}.$$

We solve this system by row-reducing the coefficient matrix:

$$\begin{bmatrix} 4 & -4 \\ 3 & -3 \end{bmatrix} \rightarrow \begin{bmatrix} 1 & -1 \\ 0 & 0 \end{bmatrix}.$$

The corresponding system of equations is $x - y = 0$, and the dimension of the solution space is $2 - 1 = 1$. Consequently, there is only one linearly independent eigenvector associated with $c = 1$ (so, $dim\ E_1 = 1$). For example, the vector $\mathbf{v}_1 = (1,1)$ is a basis. The reader should verify that

$$\begin{bmatrix} 5 & -4 \\ 3 & -2 \end{bmatrix}\begin{bmatrix} 1 \\ 1 \end{bmatrix} = 1\begin{bmatrix} 1 \\ 1 \end{bmatrix}.$$

(b) For $c = 2$, the coefficient matrix for the equations of the eigenvectors is

$$A - 2I = \begin{bmatrix} 3 & -4 \\ 3 & -4 \end{bmatrix},$$

which is row-reduced to

$$\begin{bmatrix} 1 & -4/3 \\ 0 & 0 \end{bmatrix}.$$

The corresponding system of equations is $x - (4/3)y = 0$. Thus, $dim\ E_2 = 1$ as well and a basis is, for example, $\mathbf{v}_2 = (4/3,1)$. ❑

Notice that the set $\{\mathbf{v}_1,\mathbf{v}_2\}$ in the last example is linearly independent, and therefore constitutes a basis B' for R^2. Thus, by Theorem 6-1, A is diagonalizable. We can let $P = [B'/B]$ where B is the standard basis, and obtain

$$P^{-1}AP = \begin{bmatrix} 1 & 0 \\ 0 & 2 \end{bmatrix},$$

a diagonal matrix. Of course, P is simply the matrix whose columns are the coordinate matrices of $\mathbf{v}_1$ and $\mathbf{v}_2$:

$$P = \begin{bmatrix} 1 & 4/3 \\ 1 & 1 \end{bmatrix}.$$

EXAMPLE 6-10

Let $T{:}P(2)\rightarrow P(2)$ be the linear operator defined by

$$T(a_0 + a_1x + a_2x^2) = (a_0 + a_1 + a_2) + (a_1 + a_2)x + a_2x^2.$$

Then the matrix of T with respect to the usual basis for $P(2)$ is easily seen to be

$$A = \begin{bmatrix} 1 & 1 & 1 \\ 0 & 1 & 1 \\ 0 & 0 & 1 \end{bmatrix}.$$

The characteristic polynomial is $f = -(x - 1)^3$ since A is upper triangular. So, T has only one ("repeated") eigenvalue $c = 1$. A polynomial function g with $g(x) = a_0 + a_1x + a_2x^2$ is an eigenvector iff

$$\left\{\begin{bmatrix} 1 & 1 & 1 \\ 0 & 1 & 1 \\ 0 & 0 & 1 \end{bmatrix} - 1\begin{bmatrix} 1 & 0 & 0 \\ 0 & 1 & 0 \\ 0 & 0 & 1 \end{bmatrix}\right\}\begin{bmatrix} a_0 \\ a_1 \\ a_2 \end{bmatrix} = \begin{bmatrix} 0 \\ 0 \\ 0 \end{bmatrix}.$$

We solve, as usual, by row-reducing the coefficient matrix:

$$\begin{bmatrix} 0 & 1 & 1 \\ 0 & 0 & 1 \\ 0 & 0 & 0 \end{bmatrix} \rightarrow \begin{bmatrix} 0 & 1 & 0 \\ 0 & 0 & 1 \\ 0 & 0 & 0 \end{bmatrix}.$$

The corresponding system of equations is $a_1 = 0$ and $a_2 = 0$. The dimension of the solution space is $3 - 2 = 1$, so there is only one linearly independent eigenvector for $c = 1$. For example, $\{g\}$ with $g(x) = 1$, for $x \in R$ $(a_0 = 1, a_1 = a_2 - 0)$ is a basis for the eigenspace. Thus T is not diagonalizable. ❑

THEOREM 6-3

Let $c_1, \ldots, c_k$ be the distinct eigenvalues of a linear operator T (or matrix A), and let $\mathbf{v}_i$ be an eigenvector associated with c_i, $i = 1, \ldots, k$. Then $\{\mathbf{v}_1, \ldots, \mathbf{v}_k\}$ is linearly independent.

PROOF We use induction on k. If $k = 1$, then $\{\mathbf{v}_1\}$ is certainly linearly independent because $\mathbf{v}_1 \neq \mathbf{0}$. Next, assuming the result is true for $k - 1$, we prove it for k. Suppose that

$$\sum_{j=1}^{k} a_j \mathbf{v}_j = \mathbf{0}, \tag{6-7}$$

for some scalars $a_1, \ldots, a_k$. Applying T to both sides yields

$$\sum_{j=1}^{k} a_j T(\mathbf{v}_j) = \sum_{j=1}^{k} a_j c_j \mathbf{v}_j = \mathbf{0}. \tag{6-8}$$

Subtracting c_k times equation 6-7 from the last part of 6-8 yields

$$\sum_{j=1}^{k-1} a_j (c_j - c_k) \mathbf{v}_j = \mathbf{0}, \tag{6-9}$$

since the term with $j = k$ drops out. By the induction hypothesis, $\mathbf{v}_1, \ldots, \mathbf{v}_{k-1}$ are linearly independent, so we conclude from 6-9 that

$$a_j(c_j - c_k) = 0, \text{ for } j = 1, \ldots, k - 1.$$

As the c_j are distinct, $c_j - c_k \neq 0$, and we get $a_1 = a_2 = \ldots = a_{k-1} = 0$. Substituting into 6-7 results in $a_k \mathbf{v}_k = \mathbf{0}$, so we see that all $a_j = 0$ ($\mathbf{v}_k \neq \mathbf{0}$). This completes the proof. ■

COROLLARY 1

If T is a linear operator on a space of dimension n and has n *distinct* eigenvalues, then T is diagonalizable. A similar result applies to $n \times n$ matrices. ■

COROLLARY 2

If the characteristic polynomial of a linear operator (or a square matrix) is a product of *distinct* linear factors, then the operator (or matrix) is diagonalizable. ■

The reader is warned that the converses are not true, as the next example shows.

EXAMPLE 6-11

Consider the matrix

$$A = \begin{bmatrix} 2 & 0 & 1 \\ 0 & 2 & -1 \\ 0 & 0 & 5 \end{bmatrix}.$$

The characteristic polynomial is $f = -(x - 2)^2(x - 5)$, so A has eigenvalues $c = 2, 5$. The eigenspace for $c = 2$ is the solution space of $(A - 2I)X = O$, and

$$A - 2I = \begin{bmatrix} 0 & 0 & 1 \\ 0 & 0 & -1 \\ 0 & 0 & 5 \end{bmatrix},$$

which has rank 1. Thus, there are $2 = 3 - 1$ linearly independent eigenvectors associated with $c = 2$ (for example, $\mathbf{v}_1 = (1,0,0)$ and $\mathbf{v}_2 = (0,1,0)$). Together with eigenvector associated with $c = 5$ (for example, $\mathbf{v}_3 = (1,-1,3)$), they constitute a basis for R^3 of eigenvectors of A, so A is diagonalizable. In fact,

$$P^{-1}AP = \begin{bmatrix} 2 & 0 & 0 \\ 0 & 2 & 0 \\ 0 & 0 & 5 \end{bmatrix}.$$

where P is the 3×3 matrix with columns $\mathbf{v}_1$, $\mathbf{v}_2$, and $\mathbf{v}_3$. Notice that the characteristic polynomial f is *not* a product of distinct linear factors. ❑

In the next section, we shall discuss the necessary and sufficient conditions for the diagonalizability of matrices and linear operators.

EXERCISES 6-2

In each of questions 1 through 5, find the characteristic polynomial and eigenvalues of the given matrix A.

1. $\begin{bmatrix} 4 & -2 \\ 1 & 1 \end{bmatrix}$ **2.** $\begin{bmatrix} 1 & 2 \\ 3 & 4 \end{bmatrix}$ **3.** $\begin{bmatrix} 0 & -2 & 3 \\ 1 & 0 & -4 \\ 0 & 0 & -1 \end{bmatrix}$

4. $\begin{bmatrix} 0 & -2 & -2 \\ 2 & 5 & 2 \\ -1 & -2 & 1 \end{bmatrix}$ **5.** $\begin{bmatrix} 5 & -4 & -1 & 3 \\ 3 & -2 & 4 & -5 \\ 0 & 0 & 1 & 1 \\ 0 & 0 & 1 & 1 \end{bmatrix}$

6. Use Lemma 6-2 to prove the corollary to Lemma 6-1.

7. Find a basis of eigenvectors for the eigenspaces of each of the matrices in questions 1 through 5. Which of these matrices are diagonalizable?

8. Prove that if A is $n \times n$, the constant term in its characteristic polynomial is *det* A.

9. Let $D:P(2) \rightarrow P(2)$ be the linear operator defined by

$$D(ax^2 + bx + c) = 2ax + b.$$

(a) Find the matrix A of D with respect to the usual basis for $P(2)$.

(b) Find the characteristic polynomial and eigenvalues of D.

(c) Find bases for the eigenspaces of D. Is D diagonalizable?

10. Let $T:R^3 \rightarrow R^3$ be defined by

$$T(x,y,z) = (2x + y + 4z, x + 2y - 2z, 2z).$$

(a) Find the characteristic polynomial and eigenvalues of T.

(b) Find a basis for R^3 of eigenvectors of T.

(c) Find an invertible 3×3 matrix P such that $P^{-1}AP$ is diagonal, where $A = [T]$.

11. Let A and B be $n \times n$ matrices. Show that AB and BA have the same eigenvalues. (Hint: Consider the cases where (i) c is a nonzero eigenvalue of AB and (ii) c is a zero eigenvalue of AB. Recall question 6, Exercises 6-1.)

12. **(a)** Prove that the characteristic polynomial of a triangular matrix is a product of linear factors.

(b) Prove that the eigenvalues of a triangular matrix are its diagonal entries.

13. **(a)** Consider the block-triangular matrix

$$A = \begin{bmatrix} B & C \\ O & D \end{bmatrix}.$$

Show that the characteristic polynomial of A is the product of the characteristic polynomials of B and D (see Theorem 3-10).

(b) State and prove a generalization of this result.

14. Show that a square matrix and its transpose have the same eigenvalues.

15. Find a basis for R^3 of eigenvectors of the matrix in Example 6-8.

16. Show that the matrix in Example 6-6 is not diagonalizable.

6-3 Diagonalization of Linear Operators

In the preceding section, we saw that the eigenvalues of linear operators and square matrices are determined as the zeros of a particular polynomial, namely, their characteristic polynomial. In fact, polynomials play a central role in the diagonalizability of linear operators and matrices. One reason for this is related to the following observation. Suppose that A is an $n \times n$ matrix. Since $M(n)(= M(n \times n))$ is a vector space of dimension n^2, the $(n^2 + 1)$ powers $I = A^0, A^1 = A, A^2, \ldots, A^{n^2}$ of A must be linearly dependent. Thus, there are scalars $a_0, \ldots, a_k$ (with $k = n^2$) such that

$$\begin{aligned} a_0 I + a_1 A + a_2 A^2 + \ldots + a_k A^k &= a_k A^k + a_{k-1} A^{k-1} + \ldots + a_0 I \\ &= O, \end{aligned} \tag{6-10}$$

where O is the $n \times n$ zero matrix, as usual. Let h be the polynomial in x of degree k,

$$h = a_k x^k + a_{k-1} x^{k-1} + \ldots + a_1 x + a_0.$$

(Readers unfamiliar with the concept of a polynomial should consult Appendix B.) We denote by $h(A)$ the result of replacing the indeterminate x by the matrix A everywhere in h. Note that in this substitution, we regard a_0 as $a_0 x^0$, so x^0 becomes $A^0 = I$. Then, equation 6-10 states that $h(A) = O$. We say that h **annihilates** the matrix A. It is important to note that $h(A)$ is also an $n \times n$ matrix.

A similar argument shows that if $T:V \rightarrow V$ is a linear operator on a space of dimension n, then there is a polynomial h of degree at most n^2 such that $h(T) = 0$, the zero linear operator on V.

In general, given *any* polynomial $f = a_n x^n + \ldots + a_0$ and any square matrix A, or any linear operator T on a vector space V, we can form a new matrix $f(A)$ of the same order as A and a new linear operator $f(T)$ on V, by setting

$$f(A) = a_n A^n + \ldots + a_0 I$$

and

$$f(T) = a_n T^n + \ldots + a_0 1_V$$

respectively. We have simply replaced x by A and by T in each case. Note that this is possible, because in both vector spaces $M(n)$ and $L(V)(= L(V,V))$, we can add and multiply vectors, so powers like A^k and T^k make sense for $k > 0$. (See Definitions 5-8 and 5-9, and Theorem 5-11.) Also, $M(n)$ and $L(V)$ have unique *multiplicative identities*:

(1) the $n \times n$ identity matrix $I \in M(n)$, which has the property that $AI = IA = A$ and $A^0 = I$ for any $A \in M(n)$;

(2) the identity operator $1_V \in L(V)$, which has the property that $T1_V = 1_V T = T$ and $T^0 = 1_V$ for $T \in L(V)$.

Vector spaces with these properties are called **linear algebras with identity** (see also Definition 5-10 and the discussion there). Of course, R itself is an important linear algebra, having the *number* 1 as identity.

EXAMPLE 6-12

(a) Let $f = 2x^2 - x + 3$ and A be the 2×2 matrix given by

$$A = \begin{bmatrix} 2 & -1 \\ 3 & 0 \end{bmatrix}.$$

Then, $f(A) = 2A^2 - A + 3I$ is the 2×2 matrix which we evaluate as

$$\begin{aligned} f(A) &= 2\begin{bmatrix} 2 & -1 \\ 3 & 0 \end{bmatrix}^2 - \begin{bmatrix} 2 & -1 \\ 3 & 0 \end{bmatrix} + 3\begin{bmatrix} 1 & 0 \\ 0 & 1 \end{bmatrix} \\ &= \begin{bmatrix} 2 & -4 \\ 12 & -6 \end{bmatrix} - \begin{bmatrix} 2 & -1 \\ 3 & 0 \end{bmatrix} + \begin{bmatrix} 3 & 0 \\ 0 & 3 \end{bmatrix} \\ &= \begin{bmatrix} 3 & -3 \\ 9 & -3 \end{bmatrix}. \end{aligned}$$

(b) With the same polynomial f as in (a), let $T{:}R^3 \to R^3$ be the linear operator defined by

$$T(x,y,z) = (x - y, 2z, y + z).$$

Then, $f(T) = 2T^2 - T + 3 1_V$, a linear operator on R^3 which we evaluate at any (x,y,z) in R^3 as

$$\begin{aligned} f(T)(x,y,z) &= 2T^2(x,y,z) - T(x,y,z) + 3(x,y,z) \\ &= 2T(x - y, 2z, y + z) - (x - y, 2z, y + z) + (3x, 3y, 3z) \\ &= 2(x - y - 2z, 2y + 2z, 2z + y + z) + (2x + y, 3y - 2z, 2z - y) \\ &= (4x - y - 4z, 7y + 2z, y + 8z). \;\square \end{aligned}$$

The important properties of substituting elements of linear algebras into polynomials are given in the next result.

LEMMA 6-4

Let $\alpha(T)$ denote the set of polynomials that annihilate a linear operator T (similarly, $\alpha(A)$ for a matrix A). Then $\alpha(T)$ is a subspace of $P(R)$ such that $h \in \alpha(T)$ and $g \in P(R)$ implies $gh = hg$ is also in $\alpha(T)$.

PROOF (Optional) We prove the result for $L = M(n)$. The proof for $L(V)$ is similar. Let $A \in M(n)$,

$$f = a_k x^k + \ldots + a_0 \text{ and } g = b_m x^m + \ldots + b_0, \text{ with } k \geq m.$$

Then,

$$cf + g = ca_k x^k + \ldots + ca_{m+1}x^{m+1} + (a_m + b_m)x^m + \ldots + (a_0 + b_0)x^0,$$

so that

$$\begin{aligned}(cf + g)(A) &= ca_k A^k + \ldots + ca_{m+1}A^{m+1} + (a_m + b_m)A^m + \ldots + (a_0 + b_0)A^0\\ &= c(a_k A^k + \ldots + a_0 I) + (b_m A^m + \ldots + b_0 I)\\ &= cf(A) + g(A),\end{aligned}$$

proving property (1).

To prove (2), we note that the coefficient of x^j in fg is

$$\sum_{i=0}^{j} a_i b_{j-i}.$$

Therefore,

$$fg = \sum_{j=0}^{k+m} \sum_{i=0}^{j} a_i b_{j-i} x^j.$$

So, substituting A for x we have,

$$\begin{aligned}fg(A) &= \sum_{j=0}^{k+m} \sum_{i=0}^{j} a_i b_{j-i} A^j\\ &= (a_k A^k + \ldots + a_0 I)(b_m A^m + \ldots + b_0 I)\\ &= f(A)g(A),\end{aligned}$$

as stated. ■

EXAMPLE 6-13

We consider the polynomials $f = x^2 - 1$ and $g = 2x + 1$. Let A be any square matrix. Then,

$$fg = 2x^3 + x^2 - 2x - 1,$$

so that

$$fg(A) = 2A^3 + A^2 - 2A - I = (A^2 - I)(2A + I) = f(A)g(A).$$

Similarly, with $c = -3$ (for example),

$$\begin{aligned}cf + g &= -3x^2 + 3 + 2x + 1\\ &= -3x^2 + 2x + 4.\end{aligned}$$

So,

$$\begin{aligned}(cf + g)(A) &= -3A^2 + 2A + 4I = (-3A^2 + 3I) + (2A + I)\\ &= -3f(A) + g(A). \;\square\end{aligned}$$

Polynomials h introduced at the beginning of this section, having the property that $h(A) = O$ or $h(T) = 0$, are called **annihilating polynomials**. Their special properties are summarized in the lemma below.

LEMMA 6-4

Let $\alpha(T)$ denote the set of polynomials that annihilate a linear operator T (similarly, $\alpha(A)$ for a matrix A). Then $\alpha(T)$ is a subspace of $P(R)$ such that $h \in \alpha(T)$ and $g \in P(R)$ implies $gh = hg$ is also in $\alpha(T)$.

(In algebra, sets with these properties are called **ideals** in $P(R)$.)

PROOF This is an immediate consequence of Theorem 6-4 above. We give the proof for linear operators T. Let $h, h' \in \alpha(T)$, c be a scalar, and $g \in P(R)$. Then,

$$(ch + h')(T) = ch(T) + h'(T) = c0 + 0 = 0,$$

showing that $\alpha(T)$ is a subspace of $P(R)$. And, since $h(T) = 0$,

$$(gh)(T) = g(T)h(T) = g(T)0 = 0,$$

so gh is in $\alpha(T)$, completing the proof. ■

Moreover, we next see that there is an unique monic polynomial p in $\alpha(T)$ such that every h in $\alpha(T)$ is simply a *multiple* of p, i.e., $h = gp$ for some g in $P(R)$.

In fact, since $\alpha(T)$ is nonempty (we know it contains at least a polynomial constructed as in equation 6-10), we may let p be a polynomial in it which is monic and has degree no greater than that of any other h in $\alpha(T)$. We say p has *minimal* (least) degree of all polynomials in $\alpha(T)$. Note that $\alpha(T)$ must contain monic polynomials, because if h is in $\alpha(T)$ with a leading coefficient a, then $(1/a)h$ is in $\alpha(T)$ and is monic.

Then, if h is any polynomial in $\alpha(T)$, we may divide h by p to get

$$h = qp + r,$$

where q is the quotient, and r is the remainder which is either zero or has degree less than that of p. Substituting T for x in this equation results in

$$0 = h(T) = q(T)p(T) + r(T) = q(T)0 + r(T) = r(T).$$

Thus, r annihilates T and is in $\alpha(T)$. It follows that $r = 0$, as otherwise r would have degree less than that of p, contrary to our definition of p. This shows that $h = qp$, so every polynomial in $\alpha(T)$ is a multiple of p. We leave it for the reader to prove the uniqueness of p in question 7, Exercises 6-3.

DEFINITION 6-7

Let T be a linear operator. The **minimal polynomial** p of T is the unique monic polynomial of least degree such that $p(T) = 0$. A similar description defines the minimal polynomial of an $n \times n$ matrix A.

The next theorem summarizes the preceding discussion.

THEOREM 6-5

The minimal polynomial divides any polynomial that annihilates a linear operator or an $n \times n$ matrix. ■

As we shall see, diagonalizability can be characterized by means of the minimal polynomial.

EXAMPLE 6-14

Recall that a square matrix A is called *nilpotent* iff $A^k = O$ for some positive integer k (question 17, Exercises 2-5). The *least* positive integer k with this property is called the **index** (of nilpotency) of A. (A similar definition applies to a linear operator T.) As an example, if

$$A = \begin{bmatrix} 0 & 1 \\ 0 & 0 \end{bmatrix},$$

then $A^2 = O$ (and also $A^k = O$ for $k \geq 2$), so A is nilpotent of index 2. It follows that the minimal polynomial of A is $p = x^2$. Similarly, if

$$B = \begin{bmatrix} 0 & 1 & 2 \\ 0 & 0 & 1 \\ 0 & 0 & 0 \end{bmatrix},$$

then B is nilpotent of index 3 since $B^3 = O$ while $B^2 \neq O$. So the minimal polynomial of B is $p = x^3$. In general, *if A is nilpotent of index k, its minimal polynomial is $p = x^k$*. ❑

For general matrices or operators, the minimal polynomial p is not as simple. To determine p, the next theorem is fundamental. We omit the proof here, but the steps are outlined for the reader in question 22, Exercises 6-3.

THEOREM 6-6 (Cayley-Hamilton Theorem)
Let T be a linear operator on a vector space of dimension n (or let A be an $n \times n$ matrix). Then $f(T) = 0$ (respectively, $f(A) = O$) where f is the characteristic polynomial. ■

The theorem is named after Arthur Cayley, cited earlier in connection with the invention of matrices, and William Rowan Hamilton, cited in Chapter 1 for his role in the invention of the concept of vectors.

COROLLARY
The minimal polynomial p divides the characteristic polynomial f.

PROOF This follows from Theorem 6-5, as f must be a multiple of p. ■

Consequently, any factor of p is a factor of f, and the zeros of p are eigenvalues. For, if $f = pq$ for some polynomial q and c is a zero of p, then

$$f(c) = p(c)q(c) = 0q(c) = 0,$$

showing that c is a zero of f, and hence an eigenvalue.

EXAMPLE 6-15

Let us again consider the nilpotent matrix

$$B = \begin{bmatrix} 0 & 1 & 2 \\ 0 & 0 & 1 \\ 0 & 0 & 0 \end{bmatrix},$$

for which we saw that the minimal polynomial is $p = x^3$. This is clearly also the characteristic polynomial f. Thus, in this case p and f have the same zeros. ❑

The result of the above example is actually true in general.

THEOREM 6-7
The minimal and characteristic polynomials have the same zeros.

PROOF In view of the preceding corollary and the remarks pursuant to it, we only have to show that every zero c of f (i.e., every eigenvalue) is a zero of p. We use the following lemma.

LEMMA 6-5

If c is an eigenvalue of T, and h is any polynomial, then $h(c)$ is an eigenvalue of $h(T)$. (A similar result holds for matrices.)

PROOF If c is an eigenvalue of T, then $T(\mathbf{v}) = c\mathbf{v}$ for some $\mathbf{v} \neq \mathbf{0}$ in V. Applying T to both sides of this equation yields

$$T^2(\mathbf{v}) = cT(\mathbf{v}) = c^2\mathbf{v},$$

showing that c^2 is an eigenvalue of T^2. More generally, it should be clear (by induction) that c^n is an eigenvalue of T^n, where n is any nonnegative integer. Thus, if

$$h = a_0 + a_1x + \ldots + a_kx^k$$

is any polynomial,

$$h(T)(\mathbf{v}) = (\sum_{i=0}^{h} a_iT^i)(\mathbf{v}) = \sum_{i=0}^{k} a_iT^i(\mathbf{v}) = \sum_{i=0}^{k} a_ic^i\mathbf{v} = h(c)\mathbf{v},$$

showing that $h(c)$ is an eigenvalue of $h(T)$. ■

To complete the proof of Theorem 6-7, we now let $h = p$ in the lemma, to get

$$\mathbf{0} = p(T)(\mathbf{v}) = p(c)\mathbf{v},$$

so that $p(c) = 0$ (as $\mathbf{v} \neq \mathbf{0}$). Thus, every eigenvalue is a zero of p, proving the theorem. ■

EXAMPLE 6-16

We use theorem 6-7 to compute the minimal polynomials of some of the matrices and operators from previous examples.

(a) (Example 6-5) The characteristic polynomial of

$$A = \begin{bmatrix} 5 & -4 \\ 3 & -2 \end{bmatrix}$$

was seen to be $f = (x - 1)(x - 2)$. Thus the minimal polynomial must be $p = f$.

(b) (Example 6-6) The characteristic polynomial of

$$A = \begin{bmatrix} 4 & -3 \\ 3 & -2 \end{bmatrix}$$

was seen to be $f = (x - 1)^2$. Thus, there are two possibilities for p: $p = x - 1$, or $p = f$. Now, substitution of A for x in $x - 1$ yields

$$A - I = \begin{bmatrix} 3 & -3 \\ 3 & -3 \end{bmatrix} \neq O,$$

so $x - 1$ is not the minimal polynomial. Thus, $p = f$.

(c) (Example 6-7) The characteristic polynomial of

$$A = \begin{bmatrix} 0 & -1 \\ 1 & 0 \end{bmatrix}$$

was seen to be $f = x^2 + 1$. Although f has no real zeros, it can be shown that f has two distinct complex zeros (see Appendix A). Therefore, by Theorem 6-7, $p = f$.

In general, it is true that any quadratic factor of f having *no* real zeros must be a factor of p. However, for this example, the reader may show directly (without using the stated fact) that $p = f$ (question 14, Exercises 6-3).

(d) (Example 6-8) Here again $p = f$.

(e) (Example 6-11) The characteristic polynomial of

$$A = \begin{bmatrix} 2 & 0 & 1 \\ 0 & 2 & -1 \\ 0 & 0 & 5 \end{bmatrix}$$

was seen to be $f = -(x - 2)^2(x - 5)$. The possibilities for p are therefore $(x - 2)(x - 5)$ and $-f$ (recall that p is monic). The reader may verify that substitution of A for x in the first possibility yields the zero matrix, so $p = (x - 2)(x - 5)$. ❑

Notice that among the examples considered above, the diagonalizable ones are precisely those in which the minimal polynomial is a product of *distinct, linear* factors. In general,

THEOREM 6-8

A linear operator T on an n-dimensional vector space V is diagonalizable iff its minimal polynomial p is a product of *distinct* linear factors:

$$p = (x - c_1)(x - c_2)...(x - c_k),$$

where $c_1, ..., c_k$ are distinct scalars. A similar result holds for $n \times n$ matrices.

PROOF We omit the proof, but the corresponding result for matrices is an easy corollary (see question 15, Exercises 6-3). ■

EXAMPLE 6-17

We determine all linear operators T on R^2 that satisfy the equation $T^2 - T = 0$. Let h be the polynomial $x^2 - x = x(x - 1)$. Then, $h(T) = 0$, so the minimal polynomial p of T divides h. Consequently, there are three possibilities for p: (i) x, or (ii) $x - 1$, or (iii) h. Now, in case (i), $T = 0$, while in case (ii), $T - I_{R^2} = 0$, so that $T = I_{R^2}$. In case (iii), T is diagonalizable with eigenvalues 1 and 0, so we can say that there is a basis B for R^2 such that

$$[T;B] = \begin{bmatrix} 1 & 0 \\ 0 & 0 \end{bmatrix}.$$

This exhausts the possibilities for T. ❑

We have seen that *not all linear operators T on a vector space V are diagonalizable*. Nevertheless, it can be shown that, depending on the characteristic polynomial of T, it may be possible to find a basis for V such that the matrix $[T;B]$ has a relatively simple (though not diagonal) special form called a **canonical form**. There are several such canonical forms. These will complete the classification of linear operators which we have discussed in part.

EXERCISES 6-3

1. For the matrix

$$A = \begin{bmatrix} 1 & -1 \\ 2 & 0 \end{bmatrix},$$

show by computation that its powers A^0, A, A^2, A^3, A^4 are linearly dependent. Hence construct a polynomial h of degree at most 4 such that $h(A) = O$.

2. Repeat the instructions of question 1 for the linear operator $T:P(1)\to P(1)$ defined by $T(1) = 1 + x$, $T(x) = 1 - x$.

3. Given the polynomial $f = x^3 - 2x^2 + 1$, the matrix

$$A = \begin{bmatrix} -2 & 3 \\ 2 & 4 \end{bmatrix}$$

and the linear operator $T:R^2 \to R^2$ defined by $T(x,y) = (2x + y, x - y)$, find $f(A)$ and $f(T)$.

4. Let $f = x^2 + x + 1$, $g = -x + 2$, $c = -3$, and $T(x,y) = (x + y, x - y)$.
   (a) Compute the following.
   (i) $[(cf + g)(T)](x,y)$ **(ii)** $[cf(T) + g(T)](x,y)$
   (iii) $[(fg)(T)](x,y)$ **(iv)** $[f(T)g(T)](x,y)$
   (b) Verify that **(i)** $(cf + g)(T) = cf(T) + g(T)$, and **(ii)** $(fg)(T) = f(T)g(T)$.
   (c) Verify that $g(T)f(T) = f(T)g(T)$, and generalize this result.

5. Let T be a linear operator on a finite dimensional vector space V having a basis B. Show that $[f(T);B] = f([T;B])$.

6. Prove Lemma 6-4 for $n \times n$ matrices A.

7. Prove that the minimal polynomial of a linear operator T (or a square matrix) is unique: it is the only monic polynomial of least degree in $\alpha(T)$ that divides every polynomial in $\alpha(T)$.

8. **(a)** Prove that if T is a linear operator on a finite dimensional vector space V with basis B, then the minimal polynomial of T is the minimal polynomial of $[T;B]$.
   (b) Use the result of **(a)** to show that similar matrices have the same minimal polynomial.

9. Compute the characteristic and minimal polynomials of the following matrices and linear operators.
   (a) $A = \begin{bmatrix} 0 & 0 \\ 1 & 0 \end{bmatrix}$ **(b)** $A = \begin{bmatrix} 0 & 0 & 0 \\ -1 & 0 & 0 \\ 3 & -2 & 0 \end{bmatrix}$
   (c) $T:R^3 \to R^3$ defined by $T(1,1,1) = (0,-1,1)$, $T^2(1,1,1) = (0,0,2)$, and $T^3(1,1,1) = (0,0,0)$. (Hint: First show that this rule defines a linear operator T.)

10. Compute the minimal polynomial of the matrices in questions 1 to 5 of Exercises 6-2.

11. Find the minimal polynomial of $D:P(2)\to P(2)$ defined in question 9, Exercises 6-2. Is D diagonalizable?

12. Suppose A and B are nilpotent $n \times n$ matrices such that $AB = BA$ (i.e., they *commute*). Show that $cA + B$ and AB are also nilpotent, where c is a scalar.

13. Compute the minimal polynomial of the linear operator defined in question 10, Exercises 6-2.

14. **(a)** Show (directly) that the minimal polynomial of

$$A = \begin{bmatrix} 0 & -1 \\ 1 & 0 \end{bmatrix}$$

is $p = x^2 + 1$.

(b) Compute the minimal polynomial of

$$B = \begin{bmatrix} 0 & 2 & 1 \\ -1 & 0 & 1 \\ 1 & 0 & 1 \end{bmatrix}.$$

(c) Are A and B diagonalizable?

15. Prove Theorem 6-8 for $n \times n$ matrices A. (Hint: It is a consequence of the result for linear operators.)

16. A vector space V is called the *sum* of subspaces $W_1, W_2, \ldots, W_k$ iff every vector $\mathbf{v}$ in V can be expressed (possibly in many ways) as

$$\mathbf{v} = \mathbf{w}_1 + \mathbf{w}_2 + \ldots + \mathbf{w}_k, \ \mathbf{w}_i \in W_i, \ i = 1, 2, \ldots, k.$$

In this case we write

$$V = W_1 + W_2 + \ldots + W_k = \sum_{i=1}^{k} W_i.$$

(See question 19, Exercises 4-2 for $k = 2$.)

(a) Show that $M(2)$ is the sum of subspaces W_1 and W_2 which consist of matrices of the form

$$\begin{bmatrix} a & b \\ b & c \end{bmatrix} \text{ and } \begin{bmatrix} a & b \\ c & a \end{bmatrix}$$

respectively, where $a, b, c \in R$.

(b) Find $W_1 \cap W_2$.

(c) Express the matrix

$$\begin{bmatrix} 1 & 2 \\ 3 & 4 \end{bmatrix}$$

as a sum $\mathbf{w}_1 + \mathbf{w}_2$, with $\mathbf{w}_i \in W_i$, in two different ways.

17. A vector space V is called the *direct sum* of subspaces $W_1, \ldots, W_k$ iff every vector $\mathbf{v}$ in V has an *unique* expression as $\mathbf{v} = \mathbf{w}_1 + \mathbf{w}_2 + \ldots + \mathbf{w}_k$, $\mathbf{w}_i \in W_i$. In this case we also say that the subspaces are *independent*.

(a) Show that R^2 is the direct sum of the x and y axes.

(b) Show that $M(2)$ is the direct sum of the subspaces of symmetric and skew-symmetric matrices.

18. Show that a finite dimensional vector space V is the direct sum of subspaces W_i, $i = 1, \ldots, k$, iff whenever B_i is a basis for W_i, then $B_1 \cup B_2 \cup \ldots \cup B_k$ is a basis for V.

19. Let T be a diagonalizable linear operator with characteristic polynomial $f = (x - c_1)^{n_1} \ldots (x - c_k)^{n_k}$, where $c_1, \ldots, c_k$ are distinct scalars. Show that n_i is the dimension of E_i, the eigenspace of c_i.

20. **(a)** Determine all linear operators T on R^2 such that $h(T) = 0$, where $h = x^2 - 5x + 6$.

(b) Find all diagonalizable 2×2 matrices A such that $A^2 = 2A - I$.

21. Prove that a square matrix and its transpose have the same minimal polynomial.

22. Use the steps below to prove the Cayley-Hamilton Theorem.

(a) Choose a basis B for V, let $A = [T;B]$, and define the matrix $F = A - xI$. F is a matrix with entries that are polynomials of degree ≤ 1. Show $det\ F = f$, the characteristic polynomial of T.

(b) Let $B = \{\mathbf{v}_1, \ldots, \mathbf{v}_n\}$. Show that

$$\sum_{i=1}^{n} F_{ij}(T)(\mathbf{v}_i) = \mathbf{0}$$

for any $j = 1, 2, \ldots, n$.

(c) Show that the entries of $adj\ F$ are polynomials of degree at most $n - 1$. Then $adj\ F(T)$ denotes the result of replacing x by T in $adj\ F$.

(d) Show that

$$\mathbf{0} = \sum_{j=1}^{n} \sum_{i=1}^{n} adj\ F(T)_{jk} F_{ij}(T)(\mathbf{v}_i) = f(T)(\mathbf{v}_k).$$

Hence conclude that $f(T)$ is the zero operator on V.

6-4 Computational Considerations II (Optional)

As noted earlier, eigenvalues are not normally computed by finding the roots of the characteristic polynomial. Instead, there are many numerical methods involving iterative procedures which may be used to *estimate* some or all of the eigenvalues and eigenvectors. The choice of procedure depends very much on the nature of the matrix or operator.

In this section, we describe a method which works well for matrices that are diagonalizable, contain many zeros (they are **sparse**), and have a **dominant eigenvalue**.

DEFINITION 6-8

Let A be an $n \times n$ matrix. An eigenvalue c_1 is called the **dominant eigenvalue** of A if its absolute value is larger than that of all other eigenvalues of A. An eigenvector $\mathbf{v}_1$ associated with c_1 is called a **dominant eigenvector** of A.

EXAMPLE 6-18

(a) If A is 4×4 and has eigenvalues $-5, 4, 0$, and 3, then -5 is the dominant eigenvalue of A.

(b) If A is 3×3 and has eigenvalues 2, 2, and 1 (i.e., these are the zeros of the characteristic polynomial), then A has no dominant eigenvalue.

(c) If A is 4×4 and has eigenvalues $-3, 2, 3$, and 0, then A has no dominant eigenvalue. ❑

The method discussed below (called the **power method**) can be used to estimate the dominant eigenvalue and eigenvector of diagonalizable matrices A. Of course, we do not know in advance whether or not A has a dominant eigenvalue; if it doesn't, then the method may *fail to converge*, that is, it may not yield an eigenvector (see question 4, Exercises 6-4).

We suppose that c_1 is the dominant eigenvalue of A and that the other eigenvalues $c_2, \ldots, c_n$ (not necessarily distinct) have been named so that

$$|c_1| > |c_2| \geq |c_3| \geq \ldots \geq |c_n|.$$

Let $B = \{\mathbf{v}_1, \ldots, \mathbf{v}_n\}$ be a corresponding basis for R^n of eigenvectors for A. Then, $\mathbf{v}_1$ is a dominant eigenvector. The method starts by arbitrarily choosing an approximation $\mathbf{x}_0$ for $\mathbf{v}_1$. Then, $\mathbf{x}_0$ may be expressed as a linear combination of the vectors of B:

$$\mathbf{x}_0 = \sum_{i=1}^{n} a_i \mathbf{v}_i. \qquad (6\text{-}11)$$

We may assume that $\mathbf{x}_0$ has been chosen so that $a_1 \neq 0$. Note that the equation above relates vectors in R^n, which we identify, as usual, with column vectors. Thus we may multiply equation 6-11 by A to obtain (as we shall see) a first approximation $\mathbf{x}_1$ for $\mathbf{v}_1$:

$$\mathbf{x}_1 = A\mathbf{x}_0 = A\left(\sum_{i=1}^{n} a_i \mathbf{v}_i\right) = \sum_{i=1}^{n} a_i A\mathbf{v}_i;$$

that is,

$$\mathbf{x}_1 = \sum_{i=1}^{n} a_i c_i \mathbf{v}_i, \qquad (6\text{-}12)$$

since the $\mathbf{v}_i$ are eigenvectors associated with the c_i.

Continuing in this way, the kth approximation $\mathbf{x}_k$ is obtained as

$$\mathbf{x}_k = A\mathbf{x}_{k-1} = A^k\mathbf{x}_0 = \sum_{i=1}^{n} a_i A^k \mathbf{v}_i = \sum_{i=1}^{n} a_i c_i^k \mathbf{v}_i,$$

or,

$$x_k = c_1^k[a_1\mathbf{v}_1 + a_2(c_2/c_1)^k\mathbf{v}_2 + \ldots + a_n(c_n/c_1)^k\mathbf{v}_n]$$

(since $c_1 \neq 0$ by dominance).

Now since $|c_j/c_1| < 1$ for $j = 2, \ldots, n$, it follows that $(c_j/c_1)^k$ can be made as close to 0 as we please by making k sufficiently large. So, for k sufficiently large,

$$\mathbf{x}_k = A^k\mathbf{x}_0 \approx c_1^k a_1 \mathbf{v}_1. \tag{6-13}$$

This shows that $\mathbf{x}_k$ is an approximation to a dominant eigenvector, because it is approximately equal to a scalar multiple of the dominant eigenvector $\mathbf{v}_1$.

Note that, if by chance $\mathbf{x}_0$ is an eigenvector for some other eigenvalue of A (so that $a_1 = 0$), the method fails (see question 9, Exercises 6-4).

It remains only to find an approximation for the dominant eigenvalue c_1. We do this as follows. Suppose that $\mathbf{v}_1 = (b_1,\ldots,b_n)$. As $\mathbf{v}_1 \neq \mathbf{0}$, some $b_i \neq 0$. Then, by equation 6-13, the ith component (entry) of $\mathbf{x}_k$ is approximately $c_1{}^k a_1 b_i$. Similarly, the ith entry of $\mathbf{x}_{k-1}$ is approximately $c_1{}^{k-1}a_1 b_i$. If we take the ratio $r(k,i)$ of these i^{th} entries, we obtain

$$r(k,i) = \frac{i^{\text{th}} \text{ entry of } \mathbf{x}_k}{i^{\text{th}} \text{ entry of } \mathbf{x}_{k-1}} \approx \frac{c_1{}^k a_1 b_i}{c_1{}^{k-1} a_1 b_i} = c_1$$

where i has been chosen so that the ith entry b_i of $\mathbf{v}_1$ is nonzero. Of course, we do not know this i in advance (since we are trying to find $\mathbf{v}_1$!), so in practice we could keep track of all the ratios $r(k,i)$, $i = 1, \ldots, n$. In fact, we shall modify the procedure after the next example to simplify the estimation of c_1.

EXAMPLE 6-19

We use the power method to estimate the dominant eigenvalue of

$$A = \begin{bmatrix} 3 & 1 \\ 4 & 3 \end{bmatrix}.$$

The reader may verify by computing the characteristic polynomial that the eigenvalues are actually 5 and 1. We start by arbitrarily choosing $x_0 = (1,1)$.

<u>$k=1$:</u>

$$\mathbf{x}_1 = A\mathbf{x}_0 = \begin{bmatrix} 3 & 1 \\ 4 & 3 \end{bmatrix}\begin{bmatrix} 1 \\ 1 \end{bmatrix} = \begin{bmatrix} 4 \\ 7 \end{bmatrix},$$

so $r(1,1) = 4/1 = 4$, $r(1,2) = 7/1 = 7$.

$\underline{k=2:}$

$$\mathbf{x}_2 = A\mathbf{x}_1 = \begin{bmatrix} 3 & 1 \\ 4 & 3 \end{bmatrix}\begin{bmatrix} 4 \\ 7 \end{bmatrix} = \begin{bmatrix} 19 \\ 37 \end{bmatrix},$$

so $r(2,1) = 19/4 = 4.750$, $r(2,2) = 37/7 \approx 5.286$ (to four significant figures).

$\underline{k=3:}$

$$\mathbf{x}_3 = A\mathbf{x}_2 = \begin{bmatrix} 3 & 1 \\ 4 & 3 \end{bmatrix}\begin{bmatrix} 19 \\ 37 \end{bmatrix} = \begin{bmatrix} 94 \\ 187 \end{bmatrix},$$

so $r(3,1) = 94/19 \approx 4.947$, $r(3,2) = 187/37 \approx 5.054$.

At this stage, it is already clear that $r(k,i)$ are approaching 5 as k increases, and that

$$\mathbf{x}_3 = \begin{bmatrix} 94 \\ 187 \end{bmatrix} = 94\begin{bmatrix} 1 \\ 1.989 \end{bmatrix} = 94(1, 1.989)$$

is an approximation to a dominant eigenvector. The reader may verify that (1,2) is a true dominant eigenvector. ❑

Notice that the entries in $\mathbf{x}_3$ are inconveniently large numbers (and they continue to increase as k does). Since a nonzero scalar multiple of an approximation to an eigenvector is still an approximation to an eigenvector, it is usual to modify the above procedure at each step (or "iteration") by **scaling** $\mathbf{x}_k$. This means that $\mathbf{x}_k$ is multiplied by $1/|L|$, where L is the entry of $\mathbf{x}_k$ with the largest absolute value. If we call the result $\mathbf{x}'_k$, then we may obtain the kth estimate, $c_1(k)$, for c_1 by solving

$$(\mathbf{x}_{k+1})_i = (A\mathbf{x}'_k)_i = c_1(k)(\mathbf{x}'_k)_i$$

for $c_1(k)$, where the ith entry of $\mathbf{x}'_k$ is $+1$ or -1.

EXAMPLE 6-20

We solve the previous example by the power method with scaling. As before,

$\underline{k=1:}$

$$\mathbf{x}_1 = A\mathbf{x}_0 = \begin{bmatrix} 4 \\ 7 \end{bmatrix}, \mathbf{x}'_1 = \begin{bmatrix} 4/7 \\ 7/7 \end{bmatrix} \approx \begin{bmatrix} 0.5714 \\ 1 \end{bmatrix}.$$

$\underline{k=2:}$

$$\mathbf{x}_2 = A\mathbf{x}'_1 = \begin{bmatrix} 3 & 1 \\ 4 & 3 \end{bmatrix}\begin{bmatrix} 0.5714 \\ 1 \end{bmatrix} \approx \begin{bmatrix} 2.714 \\ 5.286 \end{bmatrix}.$$

Thus, $(\mathbf{x}_2)_2 = 5.286(\mathbf{x}'_1)_2$, so we see that $c_1(1) = 5.286$ (the first approximation to c_1). Scaling, $\mathbf{x}'_2 = (1/5.286)\mathbf{x}_2 = (0.5134, 1)$.

$\underline{k=3:}$

$$\mathbf{x}_3 = A\mathbf{x}'_2 = \begin{bmatrix} 3 & 1 \\ 4 & 3 \end{bmatrix} \begin{bmatrix} 0.5134 \\ 1 \end{bmatrix} \approx \begin{bmatrix} 2.540 \\ 5.054 \end{bmatrix}.$$

Thus, $c_1(2) = 5.054$.

Continuing in this way yields

$\mathbf{x}'_3 = (1/5.054)(2.540,5.054) \approx (0.5026,1);$
$\mathbf{x}_4 = A\mathbf{x}'_3 \approx (2.508,5.010)$, so $c_1(3) = 5.010;$
$\mathbf{x}'_4 = (1/5.010)(2.508,5.010) \approx (0.5006,1);$
$\mathbf{x}_5 = A\mathbf{x}'_4 \approx (2.502,5.002)$, so $c_1(4) = 5.002;$
$\mathbf{x}'_5 = (1/5.002)(2.502,5.002) \approx (0.5002,1);$
$\mathbf{x}_6 = A\mathbf{x}'_5 \approx (2.501,5.001)$, so $c_1(5) = 5.001.$

Therefore, six iterations yield the dominant eigenvalue correct to two decimal places (since there was no change in the second decimal place in going from $c_1(4)$ to $c_1(5)$). ❑

Once the dominant eigenvalue and eigenvector have been found, the next step is to find the other eigenvalues if these are needed. They are often called the **subdominant eigenvalues**. The principal method here is known as **deflation**. It means the construction of a new matrix B from the original matrix A, such that B has eigenvalues $0, c_2, \ldots, c_n$, where A has eigenvalues $c_1, \ldots, c_n$ with $|c_1| > |c_2| \geq \ldots \geq |c_n|$. Then, if $|c_2|$ is strictly greater than $|c_3|$, c_2 is the dominant eigenvalue of B, and may be found by the power method.

Clearly, this procedure may be continued so long as the resulting matrices have dominant eigenvalues. However, the buildup of rounding errors limits the number of times the procedure may be repeated, but we shall not consider that problem here.

We consider one of many possible ways for constructing the new matrix B.

THEOREM 6-9

Let A be a diagonalizable $n \times n$ matrix with *distinct* eigenvalues $c_1, \ldots, c_k$ $(1 \leq k \leq n)$. Let $\mathbf{v}_1$ be an eigenvector associated with c_1 and assume that A^t has an eigenvector $\mathbf{w}_1$ associated with c_1 such that $\mathbf{w}_1{}^t\mathbf{v}_1 = 1$. Then

$$B = A - c_1\mathbf{v}_1\mathbf{w}_1{}^t$$

has eigenvalues $0, c_2, c_3, \ldots, c_k$ and the same eigenvectors as A.

PROOF We omit the proof, but remark that the result does *not* hold if A^t has no eigenvector $\mathbf{w}_1$ such that $\mathbf{w}_1{}^t\mathbf{v}_1 = 1$. This happens when all eigenvectors $\mathbf{w}_1$ of A^t associated with c_1 are *orthogonal* to $\mathbf{v}_1$, i.e., $\mathbf{w}_1{}^t\mathbf{v}_1 = 0$. If $\mathbf{w}_1 = (a_1,\ldots,a_n)$ and $\mathbf{v}_1 = (b_1,\ldots,b_n)$, this means that

$$\sum_{i=1}^{n} a_i b_i = 0$$

(see section 1-3 and Chapter 7). For example, the reader may verify that if

$$A = \begin{bmatrix} 4 & -3 & 0 \\ 3 & -2 & 0 \\ 0 & 0 & 1/2 \end{bmatrix},$$

then the eigenvectors of A and its transpose associated with eigenvalue 1 are orthogonal (question 6, Exercises 6-4). However, note that here A is not diagonalizable and 1 is not a dominant eigenvalue. ■

In view of the remarks stated above, Theorem 6-9 is usually applied to symmetric matrices A (where $A = A^t$), in which case the eigenvectors of A^t are of course the same as those of A. (We shall see in Chapter 7 that symmetric matrices are always diagonalizable.) In this case, we may set $\mathbf{w}_1 = (1/a)\mathbf{v}_1$, where $\mathbf{v}_1 = (a_1,\ldots,a_n)$ is an eigenvector associated with c_1 and

$$a = \mathbf{v}_1{}^t\mathbf{v}_1 = \sum_{i=1}^{n} a_i^2.$$

EXAMPLE 6-21

We use the power method with deflation to estimate the eigenvalues and eigenvectors of

$$A = \begin{bmatrix} 5 & -2 & 0 \\ -2 & 6 & 2 \\ 0 & 2 & 7 \end{bmatrix}.$$

We saw in Example 6-8 that the true eigenvalues are 3, 6, and 9. Eleven iterations of the power method applied to A yield

$$\mathbf{x}'_{10} \approx \begin{bmatrix} -0.47442 \\ 0.97446 \\ 1 \end{bmatrix}$$

and $c_1(10) \approx 8.9489$, which is an approximation for the true dominant eigenvalue 9. The calculation was done on a Burroughs B7900 with 11 significant figures, and then rounded to 5. (Note that this is not a particularly good approximation for $c_1 = 9$; 21 iterations yielded $c_1(20) \approx 8.9991$.)

For the deflation of A, we therefore choose $\mathbf{v}_1 = \mathbf{x}'_{10}$, and since $\mathbf{v}_1{}^t\mathbf{v}_1 = 2.1746$, we obtain $\mathbf{w}_1$ from

$$\mathbf{w}_1 = (1/2.1746)\mathbf{v}_1 = \begin{bmatrix} -0.21816 \\ 0.44811 \\ 0.45985 \end{bmatrix}$$

in order that $\mathbf{w}_1{}^t\mathbf{v}_1 = 1$.

Using the approximation $c_1(10)$ for c_1, we obtain the deflated matrix (Theorem 6-9)

$$B_1 = A - c_1\mathbf{v}_1\mathbf{w}_1{}^t \approx \begin{bmatrix} 4.0738 & -0.0976 & 1.9523 \\ -0.0976 & 2.0923 & -2.0101 \\ 1.9523 & -2.0101 & 2.8848 \end{bmatrix}.$$

Next, we apply the power method to B_1. In the same way as before, this produces

$$\mathbf{x}'_{10} = \begin{bmatrix} 1 \\ -0.51782 \\ 0.95993 \end{bmatrix}$$

and $c_1(10) \approx 5.9984$, which are approximations for an eigenvector $\mathbf{v}_2$ and for the true first subdominant eigenvalue $c_2 = 6$ of A (and the dominant eigenvalue of B_1).

Finally, we deflate B_1 using $\mathbf{v}_2 = \mathbf{x}'_{10}$ and $\mathbf{w}_2 = (1/2.1896)\mathbf{v}_2$ to get

$$B_2 = B_1 - c_2\mathbf{v}_2\mathbf{w}_2{}^t \approx \begin{bmatrix} 1.3343 & 1.3210 & -0.6774 \\ 1.3210 & 1.3578 & -0.6483 \\ -0.6774 & -0.6483 & 0.3605 \end{bmatrix}.$$

B_2 has the same eigenvalues as B_1 except c_2 is replaced by 0. Of course, B_2 should have eigenvalues 0, 0, 3, but the reader may check that, on account of rounding, the first is no longer 0 since

$$B_2\mathbf{v}_1 \approx \begin{bmatrix} 0.02 \\ 0.05 \\ 0.05 \end{bmatrix} \neq \mathbf{0}.$$

Applying the power method to B_2 results in

$$\mathbf{x}'_{10} \approx \begin{bmatrix} 0.99702 \\ 1 \\ -0.50149 \end{bmatrix}$$

and $c_3(10) \approx 3.0000$, completing the calculation. ❑

The reader is cautioned that the methods discussed in this section were included because of their relative simplicity and in order to give some idea of the numerical computation of eigenvalues and eigenvectors. In practice, however, numerous other more sophisticated techniques and variations are used to solve these types of problems more efficiently.

EXERCISES 6-4

1. **(a)** Use the power method with scaling to estimate the dominant eigenvalue and eigenvector of the following matrices. Start the procedure with $\mathbf{x}_0 = (1,1)$, round all computations to three significant figures, and use six iterations to get $\mathbf{x}_6$, $\mathbf{x}'_5$, and $c_1(5)$.

   (i) $A = \begin{bmatrix} 6 & 5 \\ 4 & 5 \end{bmatrix}$ **(ii)** $A = \begin{bmatrix} -2 & -4 \\ 3 & 5 \end{bmatrix}$ **(iii)** $A = \begin{bmatrix} -9 & 8 \\ -5 & 5 \end{bmatrix}$

   (b) Find the true dominant eigenvalue for each matrix A in (a) above.

2. The *percentage error* in an approximation z for a number c is defined as $|(z - c)/c| \times 100\%$. Find the percentage error in the approximations for the dominant eigenvalues of the matrices in question 1.

3. **(a)** Let A be the matrix

$$\begin{bmatrix} 1 & 0 & 0 & 1 \\ 0 & 0 & -3 & 0 \\ 0 & 2 & 0 & 1 \\ 1 & 1 & 0 & 1 \end{bmatrix},$$

   and suppose its characteristic polynomial is $f = x^4 + a_3x^3 + a_2x^2 + a_1x + a_0$. Use the Cayley-Hamilton Theorem to find the coefficients a_3, a_2, a_1, and a_0, and hence f. (Hint: $f(A) = O$, so multiplication of both sides by a suitable column vector X, e.g., $X = (1,0,0,0)$, yields a system of four equations in the unknowns a_3, a_2, a_1, and a_0:

$$A^4X + a_3A^3X + a_2A^2X + a_1AX + a_0X = O.$$

   For large matrices, this method is more efficient for finding f than the evaluation of $det\ (A - xI)$).

   (b) Generalize the ideas in (a) above to arbitrary $n \times n$ matrices.

4. Carry out six iterations of the power method on the matrix

$$A = \begin{bmatrix} 0 & -1 \\ 1 & 0 \end{bmatrix}$$

   to show that the estimates for a dominant eigenvector do not converge. Explain this result.

5. Let A be an $n \times n$ matrix with distinct eigenvalues $c_1, \ldots, c_k$ and associated eigenvectors $\mathbf{v}_1, \ldots, \mathbf{v}_k$. Let the corresponding eigenvectors of A^t be $\mathbf{w}_1, \ldots, \mathbf{w}_k$. Show that $\mathbf{w}_j{}^t\mathbf{v}_i = 0$ if $i \neq j$.

6. Consider the matrix

$$A = \begin{bmatrix} 4 & -3 & 0 \\ 3 & -2 & 0 \\ 0 & 0 & 1/2 \end{bmatrix}$$

discussed after Theorem 6-9 in the text. Show that

(a) $c = 1, 1/2$ are the eigenvalues;

(b) if $\mathbf{v}$ is an eigenvector associated with $c = 1$ and $\mathbf{w}$ is the associated eigenvector of A^t, then $\mathbf{w}^t \mathbf{v} = 0$;

(c) A is not diagonalizable.

7. The *estimated percentage error* after k iterations in estimating a number c by approximations $c(1), \ldots, c(k)$ is $|[c(k) - c(k-1)]/c(k)| \times 100\%$. How many iterations are required to find the dominant eigenvalues of the matrices in question 1 by the power method with scaling so that the estimated percentage error is less than 1%?

8. Use the power method (with scaling) and deflation to estimate the eigenvalues of the matrices A given in (a) and (b) below as follows:

(i) Apply the power method to A as in question 1.

(ii) Apply the power method in the same way to the deflated matrix B to obtain the subdominant eigenvalue.

(a) $A = \begin{bmatrix} 13 & 5 \\ 5 & 2 \end{bmatrix}$ **(b)** $A = \begin{bmatrix} 2 & 1 \\ 1 & 3 \end{bmatrix}$

9. Apply the power method with scaling to obtain the dominant eigenvalue of

$$A = \begin{bmatrix} 5 & -4 \\ 3 & -2 \end{bmatrix}.$$

Start with the initial estimate (2,1) and stop when the estimated percentage error is less than 1%. What happens if you start with the usual (1,1)? Explain this.

10. Use the power method with scaling followed by deflation to obtain estimates for the eigenvalues and eigenvectors of the following matrices. Round to three significant figures after each arithmetic operation and use a sufficient number of iterations of the power method so that the estimated percentage error is less than 2%.

(a) $\begin{bmatrix} 2 & 0 & 5 \\ 0 & -2 & 0 \\ 5 & 0 & 13 \end{bmatrix}$ **(b)** $\begin{bmatrix} -0.6 & -2.4 & -1.8 \\ -2.4 & -3.6 & -4.2 \\ -1.8 & -4.2 & -3.9 \end{bmatrix}$ **(c)** $\begin{bmatrix} 3 & -4 & 2 \\ -4 & -1 & 6 \\ 2 & 6 & -2 \end{bmatrix}$

11. **(a)** Suppose a matrix A has an eigenvalue which is smaller in absolute value than the other eigenvalues. Show that the reciprocal of this eigenvalue is the dominant eigenvalue of A^{-1}.

(b) Use the result in (a) above to estimate the smallest eigenvalues of the matrices in question 8.

6-5 Application to Economics (Optional)

Eigenvalues and eigenvectors are among the most important ideas used in the applications of linear algebra. However, many of these applications also require other mathematical machinery, such as calculus and differential equations. In this section, we examine an important application to economics which is relatively easy to understand. The application also illustrates some of the special mathematical considerations that arise in solving applied problems.

We consider a primitive economy with three industries: (1) fishing, (2) hunting, and (3) maize farming. For simplicity, the reader may assume that there is one fisherman who catches all the fish, one hunter supplying all the game, and one farmer producing all the maize. Each of these individuals buys goods from the others, and we also assume that they each pay for the part of their own commodity that they themselves consume. In the table below, called an *input-output table*, we summarize the fraction of each good consumed by each of the three individuals over some fixed period of time, say a year.

		Output (Income)		
		Fisherman	*Hunter*	*Farmer*
Input (Consumption)	*Fisherman*	0.25	0.25	0.125
	Hunter	0.5	0.5	0.25
	Farmer	0.25	0.25	0.625

As an example, we read from the table that the fisherman consumes 0.25 or 1/4 of the yearly catch of fish, the hunter consumes 0.5 of the catch, and the farmer consumes 0.25 of the catch. Alternatively, we see that the hunter's consumption makes an annual "input" of 0.5 of the fisherman's income, and so on. Notice that the sum of the entries in each column is 1, reflecting the fact that all goods are consumed within the society. For this reason, this model of the economy is said to be **closed**. Later we shall also examine an **open** model, in which there is sufficient production to meet an external demand.

To complete our description of this closed model, we further assume that the three individuals all pay the same price for a given commodity, and that their individual income and expenditures are exactly equal. The latter requirement is known as the **equilibrium** condition: it means that the individuals in the economy will survive. The problem then is to determine the prices of the goods (or the annual incomes) so that this condition will be met. (In fact, a solution of the weaker requirement that expenditures do not exceed income will necessarily make expenditures equal to income, although this is not obvious — see question 4, Exercises 6-5.)

We let x_1, x_2, and x_3 be the annual income (in dollars) of the fisherman, hunter, and farmer respectively. Then, for example, since the fisherman consumes 0.25 of the hunter's annual production, he must pay $0.25x_2$ dollars for it. Thus, the fisherman's annual expenditures are

$$0.25x_1 + 0.25x_2 + 0.125x_3.$$

To satisfy the equilibrium condition for the fisherman, we must therefore have

$$0.25x_1 + 0.25x_2 + 0.125x_3 = x_1.$$

In a similar way, we obtain the following equations for the hunter and the farmer respectively:

$$\begin{aligned} 0.5x_1 + 0.5x_2 + 0.25x_3 &= x_2, \\ 0.25x_1 + 0.25x_2 + 0.625x_3 &= x_3. \end{aligned}$$

We write the above three equations as the single matrix equation

$$AX = X, \tag{6-14}$$

where

$$A = \begin{bmatrix} 0.25 & 0.25 & 0.125 \\ 0.5 & 0.5 & 0.25 \\ 0.25 & 0.25 & 0.625 \end{bmatrix}$$

is the **input-output matrix** and

$$X = \begin{bmatrix} x_1 \\ x_2 \\ x_3 \end{bmatrix}$$

is the **price** (income) **vector**.

Note that if X is a (nontrivial) solution of equation 6-14, then it is an eigenvector of A associated with eigenvalue $c = 1$. We solve for X (if possible) by solving the homogeneous system of equations

$$(I - A)X = O.$$

The reader may verify that the solution space has dimension 1, with a basis vector $X = (1,2,2)$. Thus, the annual incomes are not determined absolutely, but only relative to one another. For example, they could be 10 000, 20 000 and 20 000 (dollars) respectively.

It is important to realize that we did not know in advance that the problem had a solution. However, as we shall see, the existence of a meaningful solution is assured by the special properties of input-output matrices.

Let us first formulate the **general closed Leontief model**, of which we have just seen an example. These models are named after Wassily Leontief, a contemporary American economist and Nobel laureate who has made many important contributions to the field of economics. The models are applied to a large part of the American economy.

In the general closed model, there are n industries (numbered 1 to n). We let a_{ij} be the fraction of the output of industry j consumed by industry i (in some fixed time period) and let x_i be the total income of industry i in that period of time (the price for the total output of industry i), $i, j = 1, \ldots, n$. We assume that the model is closed, so

$$a_{1j} + a_{2j} + \dots + a_{nj} = 1, \text{ for } j = 1, \dots, n.$$

We want to find $X = (x_i)$ so that (as before) $AX = X$, i.e., total incomes equal total expenditures for all industries. A is the input-output matrix and X is the price vector. Note that A is a special matrix with these properties:
(1) $a_{ij} \geq 0$ for all i and j; and
(2) the column sums of A are 1.

In the example of the closed model discussed above, the entries of A were all positive, but this is not necessary.

DEFINITION 6-9

A matrix A is said to be **nonnegative** (respectively, **positive**) if all its entries are nonnegative (respectively, positive). We write $A \geq O$ (respectively, $A > O$).

Thus, an input-output matrix is a nonnegative matrix. It is also called an **exchange matrix** (or column stochastic matrix) because of property (2). We may extend the notation in Definition 6-9 to write $A \geq B$ (or $A > B$) when A and B are matrices of the same order and $A - B \geq O$ (respectively, $A - B > O$).

Now, it is clear that the closed Leontief model always has a nontrivial solution X. For, since the column sums of A are 1, the row sums of A^t are 1. It follows that if $U = (1,1,\dots,1)$ is the $n \times 1$ vector with all entries equal to 1, then $A^tU = U$, since the ith row of A^tU is simply the sum of the entries in row i of A^t, and this is 1. This shows that U is an eigenvector of A^t associated with eigenvalue $c = 1$. Since A has the same eigenvalues as A^t, it follows that there is a nonzero vector X, such that $AX = 1X = X$, i.e., the model always has a nontrivial solution X, as stated. (For a different proof, see question 3, Exercises 6-5.)

However, a solution X will not be meaningful if some prices x_i are negative (in fact, if some $x_i = 0$, then industry i has zero income). Thus, a solution of the closed Leontief model is required to be a nonnegative X such that $AX = X$, i.e., X is an eigenvector of A associated with eigenvalue $c = 1$ and $X \geq O$.

While it can be shown without the theory of eigenvalues that the problem always has such a solution (see, for example, *The Theory of Linear Economic Models* by D. Gale, McGraw-Hill, 1960), we prefer to quote the fundamental theorem below regarding the eigenvalues and eigenvectors of nonnegative matrices. This theorem has other applications in many areas in addition to the one under consideration. In order to state this result in its most general form, we need the notion of an **irreducible matrix**. It is interesting that this concept may be defined using directed graphs. (See sections 2-3 and 4-5. Note that we do not assume the reader to be familiar with this material. The definitions should be clear from the figures and examples.)

DEFINITION 6-10

If A is an $n \times n$ matrix, its **digraph** is the directed graph $G(A)$ with n vertices $1, \ldots, n$ and a directed edge from i to j iff $a_{ij} > 0$. (Thus, there is a "loop" at vertex i if $a_{ii} > 0$.)

Some examples are shown in Figure 6-1, where a "*" in a matrix indicates a positive number. A is said to be **irreducible** if $G(A)$ is **strongly connected**, which means that for every pair of vertices i and j $(i \neq j)$, there is a path of edges from i to j on which the directions of the edges all lead from i to j. Intuitively, we may say that every vertex can "communicate" with every other vertex. If A is not irreducible, then it is said to be **reducible**.

Readers familiar with section 4-5 will recognize that a strongly connected graph is connected, but not necessarily conversely (Figure 6-1 (a), (d)). Of course, a positive matrix is irreducible, because there is then a directed edge from every i to every j.

FIGURE 6-1

(a)

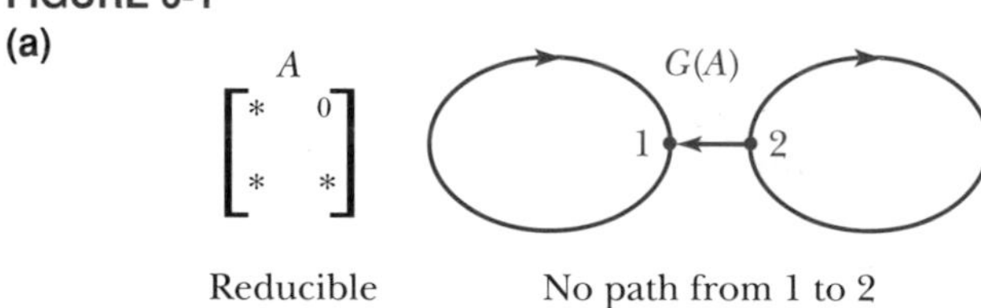

(b)

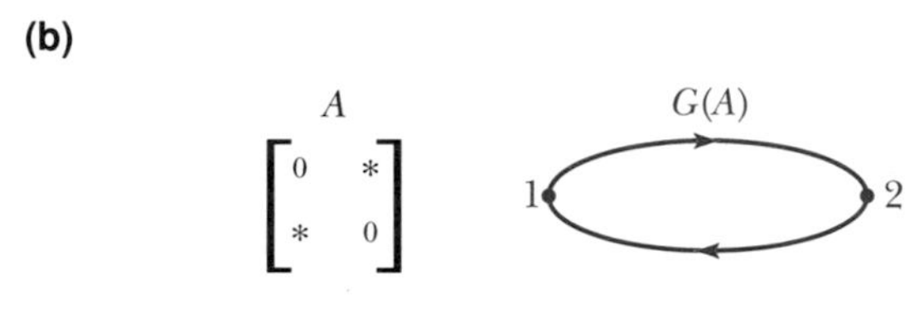

(c)

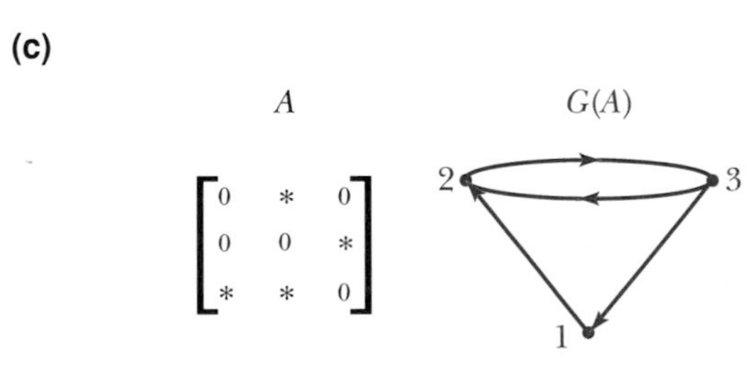

(d)

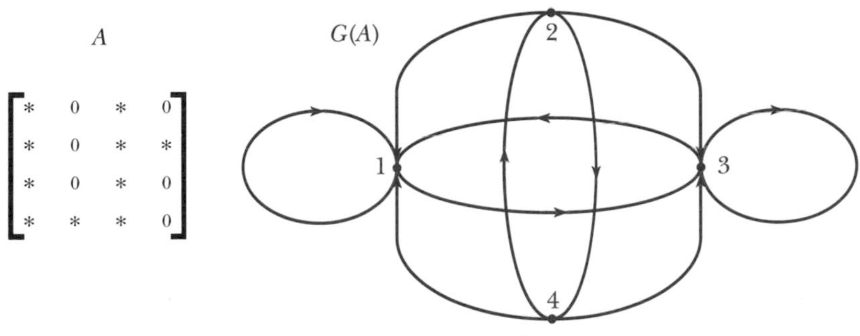

Reducible

No paths from 1 to 2, 1 to 4, 3 to 2, and 3 to 4

THEOREM 6-10 (Perron-Frobenius Theorem)

Let A be a nonnegative, irreducible $n \times n$ matrix. Then A has a positive eigenvalue c with an associated positive eigenvector $\mathbf{v}$ such that the following properties hold.

(1) If c' is any other eigenvalue of A, then $|c'| \le c$. If A is positive, then $|c'| < c$, so c is dominant.

(2) If $\mathbf{v}'$ is any positive eigenvector of A, then $\mathbf{v}'$ is necessarily a nonzero multiple of $\mathbf{v}$.

(3) If A' is a **principal** $k \times k$ **submatrix** $(k < n)$ of A obtained by deleting rows $i_1, \ldots, i_{n-k}$ and the *same* columns, then any eigenvalue c' of A' satisfies $|c'| < c$. ■

The proofs are beyond the scope of this book. (They may be found, for example, in *Applications of the Theory of Matrices* by F.R. Gantmacher, Interscience Publishing Co., 1959.) The theorem is named after the German mathematicians Oskar Perron (1880–1975) and Ferdinand Georg Frobenius (1849–1917). Perron made many contributions to the theory of differential and integral equations, while Frobenius is also noted for his discoveries in group theory.

EXAMPLE 6-22

(a) Consider the (exchange) matrix

$$A = \begin{bmatrix} 0 & 15/64 & 0 \\ 0 & 0 & 1 \\ 1 & 49/64 & 0 \end{bmatrix}.$$

The graph of A is shown in Figure 6-1(c) and is strongly connected, so that A is irreducible. The characteristic polynomial of A is

$$\begin{aligned} f &= -x^3 + (49/64)x + 15/64 \\ &= -(x-1)(x+3/8)(x+5/8). \end{aligned}$$

So, A has eigenvalues $-5/8$, $-3/8$, and 1. Therefore, $c_1 = 1$ is the largest eigenvalue of A. The reader may verify that an eigenvector is $\mathbf{v}_1 = (15,64,64)$, which is positive. Eigenvectors for $c_2 = -5/8$ and $c_3 = -3/8$ are $\mathbf{v}_2 = (3,-8,5)$ and $\mathbf{v}_3 = (5,-8,3)$ respectively. Notice that $\mathbf{v}_2$ and $\mathbf{v}_3$ are *not* positive, as predicted by the theorem (if they were, they would have to be multiples of $\mathbf{v}_1$, which is impossible as they are not associated with the eigenvalue 1).

To illustrate statement (3) in Theorem 6-10, let A' be the 2×2 principal submatrix obtained by deleting row 1 and column 1 of A:

$$A' = \begin{bmatrix} 0 & 1 \\ 49/64 & 0 \end{bmatrix}.$$

The characteristic polynomial is

$$f' = x^2 - 49/64 = (x - 7/8)(x + 7/8).$$

The eigenvalues have absolute values less than the largest eigenvalue 1 of A.

(b) Let A be the (exchange) matrix given by

$$A = \begin{bmatrix} 1/2 & 0 \\ 1/2 & 1 \end{bmatrix}.$$

Its graph is shown in Figure 6-1(a), so it is reducible. Its characteristic polynomial is

$$f = (x - 1/2)(x - 1),$$

so that A has eigenvalues 1/2 and 1. The reader may verify that the associated eigenvectors are $(1,-1)$ and $(0,1)$ respectively. Therefore, A has no positive eigenvectors. However, note that it does have a nonnegative eigenvector $(0,1)$. ❑

In fact, when a nonnegative matrix A is reducible, the following consequence of the Perron-Frobenius Theorem is still true. We omit the proof of this theorem, but will illustrate in Example 6-23 why it follows from the Perron-Frobenius Theorem.

THEOREM 6-11

A nonnegative matrix A has a nonnegative eigenvalue c with an associated nonnegative eigenvector $\mathbf{v}$. If c' is any other eigenvalue of A, then $|c'| \le c$. Furthermore, if A' is a $k\times k$ principal submatrix of A $(k < n)$, any eigenvalue c' of A' satisfies $|c'| \le c$. ■

This theorem guarantees a (nonnegative) solution for all closed Leontief input-output models. For, by question 4(b), Exercises 6-5, a nonnegative eigenvector of an exchange matrix must be associated with eigenvalue $c = 1$.

EXAMPLE 6-23

Let A be the matrix

$$A = \begin{bmatrix} 1 & 0 & 2 \\ 0 & 0 & 3 \\ 4 & 0 & 5 \end{bmatrix}.$$

Its graph $G(A)$ is shown in Figure 6-2, from which it is evident that A is reducible. However, vertices 1 and 3 can "communicate", so they constitute the vertices of a strongly connected **subgraph** G_1 (i.e., part) of $G(A)$, but of no larger strongly connected subgraph. We regard the vertex 2 as a subgraph G_2 of $G(A)$ (see Figure 6-3).

FIGURE 6-2

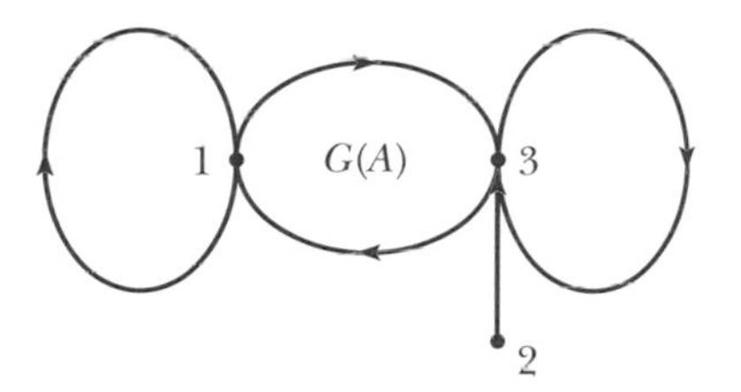

FIGURE 6-3

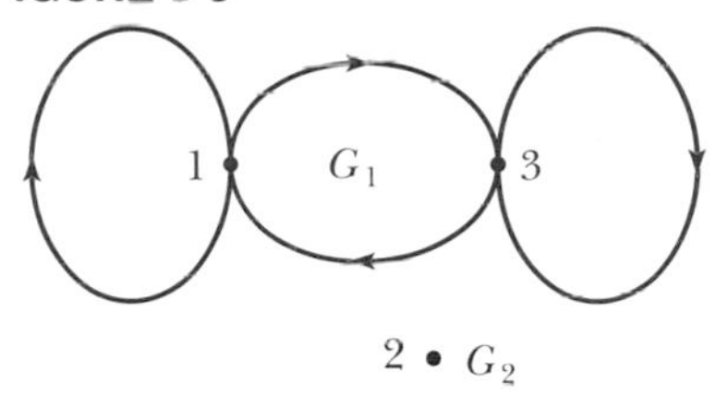

It is natural to relabel the vertices of $G(A)$ so that G_1 has vertices 1 and 2, while G_2 has vertex 3. So, we interchange vertices 2 and 3 to obtain the new graph $G(A')$ from $G(A)$ (Figure 6-4). A matrix A' corresponding to this graph is obtained from A by interchanging rows 2 and 3 and columns 2 and 3.

$$A' = \begin{bmatrix} 1 & 2 & 0 \\ 4 & 5 & 0 \\ 0 & 3 & 0 \end{bmatrix}.$$

FIGURE 6-4

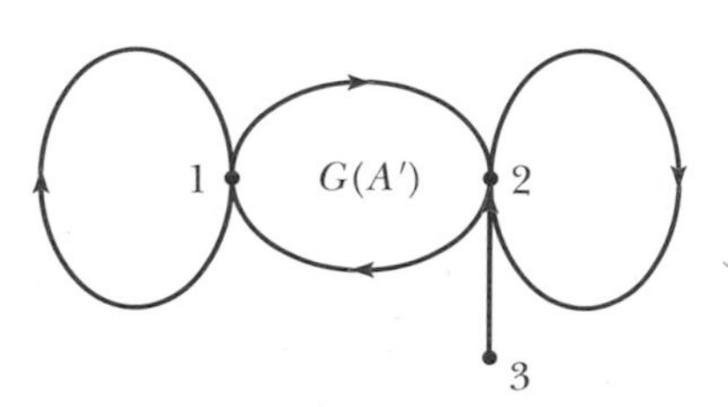

Let P be the 3×3 elementary matrix obtained from I by interchanging rows 2 and 3. Then, $A' = PAP^{-1}$, so A and A' have the same eigenvalues. Furthermore, A' has the block triangular form

$$A' = \begin{bmatrix} A_{11} & O \\ A_{21} & A_{22} \end{bmatrix},$$

where

$$A_{11} = \begin{bmatrix} 1 & 2 \\ 4 & 5 \end{bmatrix} \text{ and } A_{22} = [0]$$

are square matrices on the diagonal of A', and the graphs of A_{11} and A_{22} are G_1 and G_2 respectively. Also, note that A_{11} is irreducible, while A_{22} is a 1×1 matrix. Thus, by the Perron-Frobenius Theorem, A_{11} has a dominant positive eigenvalue. This is also a positive eigenvalue of A' and A (see question 13, Exercises 6-2). ❑

In general, any nonnegative $n \times n$ matrix A is similar to a block triangular matrix A' with irreducible square matrices or 1×1 matrices along its diagonal. Then, we may pick for the nonnegative eigenvalue c of A the largest nonnegative eigenvalue of the matrices on the diagonal of A. But, since the largest nonnegative eigenvalue may be 0, we can now only assert $c \geq 0$, not $c > 0$. For example, if

$$A = \begin{bmatrix} 0 & 1 & 2 \\ 0 & 0 & 3 \\ 0 & 0 & 0 \end{bmatrix},$$

then A is nonnegative, but its largest (and only) eigenvalue is 0.

We conclude this section with a discussion of the Leontief open model, also called a **linear production model**. It differs from the closed model in that the production from the various industries is not only distributed among the industries in the economy, but is also used to satisfy an external demand. Thus, in an open model, the prices are fixed, and we wish to find the dollar value x_i of the production of each industry (over some time period, say a year) necessary to satisfy both the needs of every industry and the external demand.

As before, we let A be the input-output matrix for the economy of n industries. In this context, it is often called the **consumption matrix**. Here $a_{ij} \geq 0$ is the dollar value of commodity i required to produce one dollar's worth of commodity j per year. In other words, a_{ij} is the dollar "input" of industry i required to produce one dollar of "output" of industry j. We assume commodity j is produced only by one industry, namely, the jth industry.

Next, $X = (x_i)$ is the $n \times 1$ **production vector**, where x_i is the annual dollar value of the production of the ith industry. Finally, $D = (d_i)$ is the **demand vector**, where d_i is the dollar value of the annual external demand for commodity i. Of course, A and D are nonnegative matrices, and we wish to find a nonnegative X such that both internal consumption and external demand are met.

Since the jth industry requires a_{ij} dollars of commodity i per year to produce one dollar's worth of its goods, it follows that $a_{ij}x_j$ dollars of commodity i are required to produce x_j dollars of j. Thus the total internal consumption of commodity i is

$$a_{i1}x_1 + a_{i2}x_2 + \dots + a_{in}x_n = \sum_{j=1}^{n} a_{ij}x_j,\ 1 \le i \le n.$$

The difference between the production x_i and the internal consumption is used to meet the external demand for the ith commodity, that is,

$$x_i - \sum_{j=1}^{n} a_{ij}x_j = d_i,$$

or, in matrix notation,

$$X - AX = (I - A)X = D. \tag{6-15}$$

Thus, in the open Leontief model we want a nonnegative vector X such that equation 6-15 holds.

Now if $(I - A)^{-1}$ exists and is nonnegative, then this problem has the unique solution $X = (I - A)^{-1}D$ for any demand vector D.

DEFINITION 6-11

An input-output matrix A is said to be **productive** if $(I - A)^{-1}$ exists and is nonnegative.

The next result, another consequence of the Perron-Frobenius Theorem, gives the necessary and sufficient conditions for an input-output matrix to be productive. The proof is due to G. Debreu, a contemporary American economist, and I.N. Herstein, one of the foremost contemporary algebraists. First we establish a lemma which is interesting in its own right.

LEMMA 6-6

Let A be an $n \times n$ nonnegative matrix whose largest nonnegative eigenvalue is c.

(1) If there is an $X > O$ such that $AX \le X$ (or, $\ge X$), then $c \le 1$ (respectively, $c \ge 1$).

(2) If there is an $X \ge O$ such that $AX < X$ (or, $> X$), then $c < 1$ (respectively, $c > 1$).

PROOF (Optional) The eigenvalue c referred to exists by the Perron-Frobenius Theorem, 6-10. We prove the first assertion and leave the proofs of the others, which are similar, for the reader.

Since A and A^t have the same eigenvalues, let $Z \geq O$ be an eigenvector of A^t associated with the eigenvalue c, as guaranteed by the Perron-Frobenius Theorem. Then,

$$A^tZ = cZ,$$

so

$$Z^tA = cZ^t.$$

Thus,

$$AX \leq X$$

implies

$$Z^t(AX) \leq Z^tX$$

and

$$(Z^tA)X \leq Z^tX,$$

or,

$$cZ^tX \leq Z^tX.$$

Since $Z^tX > O$, it follows that $c \leq 1$. ■

THEOREM 6-12

A nonnegative matrix A is productive iff its largest (nonnegative) eigenvalue $c < 1$.

PROOF (Optional) Suppose first that $c < 1$. Then 1 is not an eigenvalue of A, so $(I - A)^{-1}$ exists. We show that it is nonnegative as follows. Now, the existence of the inverse implies that the equation

$$(I - A)X = Y \tag{6-16}$$

has a solution

$$X = (I - A)^{-1}Y$$

for any vector $Y \in M(n \times 1)$. We first show that $Y \geq O$ implies $X \geq O$. This will imply that $(I - A)^{-1}$ is nonnegative because, by taking Y to be the standard basis vectors, we get for X the columns of $(I - A)^{-1}$. We proceed by contradiction: suppose that X is *not* nonnegative. Then, X has some negative entries. By permuting the rows of X, we may write

$$X = \begin{bmatrix} -X_1 \\ X_2 \end{bmatrix},$$

where $X_1 > O$ and $X_2 \geq O$. Then, after permuting the rows and columns of A, which leads to a similar matrix, which we again denote as A, we can rewrite 6-16 as

$$\begin{bmatrix} I - A_1 & -A'_1 \\ -A'_2 & I - A_2 \end{bmatrix} \begin{bmatrix} -X_1 \\ X_2 \end{bmatrix} = Y \geq O.$$

Therefore,

$$-(I - A_1)X_1 - A'_1 X_2 \geq O,$$

or,

$$-(I - A_1)X_1 \geq A'_1 X_2 \geq O,$$

or, $X_1 \leq A_1 X_1$, with $X_1 > O$. By applying Lemma 6-6 to A_1, we find that the largest eigenvalue c_1 of A_1 satisfies $c_1 > 1$. But, by Theorem 6-11 and the hypothesis, $c_1 \leq c < 1$. This contradiction means that $X \geq O$, and therefore $(I - A)^{-1} \geq O$.

To prove the converse, suppose $(I - A)^{-1} \geq O$ and let Y be any positive vector. Then,

$$X = (I - A)^{-1} Y \geq O,$$

so

$$(I - A)X = Y > O.$$

This means there is an $X \geq O$ such that $X > AX$. By Lemma 6-6, $c < 1$, thus completing the proof. ■

COROLLARY

(1) A is productive iff there is an $X \geq O$ such that $AX < X$.

(2) If the row sums of A are less than 1, then A is productive.

(3) If the column sums of A are less than 1, then A is productive.

PROOF The proofs for the above statements are left for the reader as questions 11, 12, and 13 of Exercises 6-5. ■

EXAMPLE 6-24

(a) If the input-output matrix for an economy with three industries is

$$A = \begin{bmatrix} 0.1 & 0 & 0.3 \\ 0 & 0.7 & 0.2 \\ 0.3 & 0 & 0.5 \end{bmatrix},$$

then A is reducible. It is productive by part (2) of the corollary to

Theorem 6-12. The reader may verify that the largest eigenvalue is $c = 0.7$ and

$$(I - A)^{-1} = \begin{bmatrix} 25/18 & 0 & 5/6 \\ 5/9 & 10/3 & 5/3 \\ 5/6 & 0 & 5/2 \end{bmatrix}.$$

Note that industry 3 is not profitable because, from column 3 of A, we see that the annual input in dollars from all industries to produce one dollar output of industry 3 is 1 (dollar).

(b) Suppose that the input-output matrix for an economy is

$$A = \begin{bmatrix} 0.2 & 0.5 & 0.55 \\ 0 & 0 & 0.2 \\ 0.2 & 0 & 0 \end{bmatrix}.$$

A is irreducible, and it is productive because the column sums are less than 1. The reader may verify that the largest (and only real) eigenvalue of A is $c = 0.5$, and that

$$(I - A)^{-1} = (1/67)\begin{bmatrix} 100 & 50 & 65 \\ 4 & 69 & 16 \\ 20 & 10 & 80 \end{bmatrix}. \square$$

EXAMPLE 6-25

Consider an economy with four industries: (1) coal, (2) steel, (3) shipbuilding, and (4) a railroad. Suppose the input-output matrix is

$$A = \begin{bmatrix} 0.1 & 0.4 & 0.4 & 0.2 \\ 0.1 & 0.1 & 0.4 & 0.2 \\ 0 & 0 & 0.1 & 0 \\ 0.3 & 0.1 & 0.1 & 0 \end{bmatrix}.$$

The reader may verify that A is reducible. Since neither the row sums nor the column sums of A is less than 1, we can't use parts 2 and 3 of the corollary to Theorem 6-12. However, with $X = (1, 1/2, 1/4, 1)$, we find that

$$AX = (0.1, 0.45, 0.025, 0.375) < X.$$

So, A is productive by part (1) of the same corollary. We calculate the largest eigenvalue of A to be $c \approx 0.495$ and (to three significant figures)

$$(I - A)^{-1} = \begin{bmatrix} 1.31 & 0.625 & 0.903 & 0.387 \\ 0.238 & 1.25 & 0.694 & 0.298 \\ 0 & 0 & 1.11 & 0 \\ 0.417 & 0.313 & 0.451 & 1.15 \end{bmatrix}.$$

As an example, if in a certain year the external demands for coal, steel, shipbuilding, and railroad transportation are 10, 20, 100, and 50 million dollars respectively, then the demand vector is (in millions):

$$D = \begin{bmatrix} 10 \\ 20 \\ 100 \\ 50 \end{bmatrix}.$$

The production vector must therefore be

$$X = (I - A)^{-1}D = \begin{bmatrix} 135.25 \\ 111.68 \\ 111.00 \\ 113.03 \end{bmatrix}.$$

Therefore the annual outputs of the four industries must be (in millions of dollars) 135.25, 111.68, 111.00, and 113.03 respectively. ❑

EXERCISES 6-5

1. For each of the following exchange matrices A, find a nonnegative price vector X satisfying $AX = X$.

(a) $\begin{bmatrix} 1/4 & 1/2 \\ 3/4 & 1/2 \end{bmatrix}$ (b) $\begin{bmatrix} 0 & 1/4 \\ 1 & 3/4 \end{bmatrix}$

(c) $\begin{bmatrix} 0.25 & 0 & 0.4 \\ 0.25 & 0.3 & 0.25 \\ 0.5 & 0.7 & 0.35 \end{bmatrix}$ (d) $\begin{bmatrix} 1 & 1/2 & 1/4 \\ 0 & 0 & 3/4 \\ 0 & 1/2 & 0 \end{bmatrix}$

2. Draw the graph $G(A)$ for each of the matrices A in question 1 and determine whether each A is reducible or irreducible.

3. (a) Show that if A is an $n \times n$ exchange matrix, then rank $(I - A) < n$. (Hint: Use the fact that the columns of A sum to 1.)

   (b) Use the result of (a) to conclude that $AX = X$ always has a nontrivial solution X.

4. (a) Show that if A is an $n \times n$ exchange matrix and X is a nonzero nonnegative $n \times 1$ vector, then $AX \leq X$ or $AX \geq X$ implies $AX = X$. (Hint: Let U be the $n \times 1$ vector with all entries 1. Then we have seen that $A^tU = U$. Use this to obtain a contradiction from the assumptions $AX < X$ or $AX > X$.)

   (b) Use the ideas in (a) above to show that if c is an eigenvalue of A associated with a nonnegative eigenvector X, then $c = 1$.

5. (a) Let A be the nonzero nonnegative matrix

$$A = \begin{bmatrix} a & b \\ c & d \end{bmatrix}.$$

Show that A has a nonnegative eigenvalue by considering the zeros of its characteristic polynomial.

(b) Determine the conditions on a, b, c, and d in order that A be irreducible.

6. Let A be the matrix

$$A = \begin{bmatrix} 0 & 1 & 0 \\ 1 & 1 & 1 \\ 0 & 1 & 2 \end{bmatrix}.$$

(a) Show that A is irreducible.

(b) Verify the Perron-Frobenius Theorem for A.

7. Consider the matrix

$$A = \begin{bmatrix} 1 & 0 & 1 & 0 \\ 1 & 0 & 1 & 1 \\ 1 & 0 & 1 & 0 \\ 1 & 1 & 1 & 0 \end{bmatrix}.$$

(a) Show that A is reducible.

(b) Use the ideas of Example 6-23 to obtain a matrix A' similar to A such that A' has a block triangular form. Hence find the largest eigenvalue of A and find a corresponding nonnegative eigenvector.

8. Let A be an irreducible nonnegative $n \times n$ matrix and consider the principal submatrix $A' = A(n/n)$ obtained by deleting row n and column n.

(a) Is A' necessarily irreducible? Give some examples.

(b) Suppose A' has a nonnegative eigenvalue c' with an associated nonnegative eigenvector X'. Show that $c' \leq c$, where c is the largest eigenvalue of A. (Hint: Let Y be the $n \times 1$ vector defined by

$$Y = \begin{bmatrix} X' \\ 0 \end{bmatrix},$$

and show that $AY \geq c'Y$.)

9. Determine which of the following input-output matrices are productive.

(a) $\begin{bmatrix} 1/8 & 1/2 & 0 \\ 0 & 1/2 & 0 \\ 7/8 & 0 & 1 \end{bmatrix}$ (b) $\begin{bmatrix} 1/2 & 1/3 & 0 \\ 1/2 & 1/3 & 0 \\ 0 & 1/3 & 1/3 \end{bmatrix}$

(c) $\begin{bmatrix} 0.2 & 0 & 0.1 & 0 \\ 0 & 0.5 & 0 & 0 \\ 0.2 & 0.6 & 0.3 & 0.2 \\ 0.3 & 0 & 0.2 & 0.3 \end{bmatrix}$

10. **(a)** Show that the input-output matrix

$$A = \begin{bmatrix} 1/2 & 0 & 1/4 \\ 1/8 & 1/2 & 1/4 \\ 1/8 & 0 & 1/8 \end{bmatrix}$$

is productive.

(b) Find a nonnegative production vector X for the economy described by A in (a) above, given that the external demand vector is

$$D = \begin{bmatrix} 125000 \\ 250000 \\ 90000 \end{bmatrix},$$

where the demands are in dollars.

11. Prove part (1) of the corollary to Theorem 6-12. (Hint: Use the proof of the theorem.)

12. Prove part (2) of the corollary to Theorem 6-12. (Hint: Consider the vector AU, where U is the vector with all entries equal to 1.)

13. Prove part (3) of the corollary to Theorem 6-12. (Hint: Consider the transpose of A and use the theorem.)

14. Show that a nonnegative 2×2 matrix A is productive iff $det\ (I - A) > 0$.

15. Show that if a matrix A is productive, then at least one of its row sums is less than 1.

7 INNER PRODUCTS

7-1 Inner Products
7-2 Inner Product Spaces
7-3 Linear Operators on Inner Product Spaces
7-4 Self-adjoint Operators
7-5 Bilinear and Quadratic Forms (Optional)
7-6 Complex Eigenvalues and Eigenvectors (Optional)

7-1 Inner Products

The object of this chapter is to generalize the notions of dot product, lengths, and angles (section 1-3) for arbitrary vector spaces. In a general vector space V, there is no natural choice for these notions as there is in R^3 (i.e., V does not have a distinguished "standard" basis as R^3 does). Thus, our approach will be to "abstract" the properties of the dot product (Theorem 1-5) in order to define the analogous concept (called an "inner product") on V. Of course, the dot product on R^3 will be an example of an inner product.

Vector spaces with the additional structure of an inner product (which are therefore called **inner product spaces**) are important in many applications because lengths and angles occur frequently as measurements of physical quantities.

We will also examine the behaviour of linear transformations and operators on inner product spaces. This will lead to some important properties of symmetric matrices and to an application in the geometry of quadratic equations.

DEFINITION 7-1

Let V be a vector space. An **inner product** on V is a function $<,>$ which assigns to every ordered pair of vectors **u** and **v** a number $<\mathbf{u},\mathbf{v}>$ such that

(1) $<\mathbf{u} + \mathbf{v},\mathbf{w}> = <\mathbf{u},\mathbf{w}> + <\mathbf{v},\mathbf{w}>$ for any **u**, **v**, and **w** in V;

(2) $<c\mathbf{u},\mathbf{v}> = c<\mathbf{u},\mathbf{v}>$ for any scalars c, and **u** and **v** in V;

(3) $<\mathbf{u},\mathbf{v}> = <\mathbf{v},\mathbf{u}>$ for all **u** and **v** in V;

(4) $<\mathbf{u},\mathbf{u}> > 0$ unless $\mathbf{u} = \mathbf{0}$.

The number $<\mathbf{u},\mathbf{v}>$ is called the **inner product** of **u** and **v**. It is also often denoted by $(\mathbf{u},\mathbf{v})$, $(\mathbf{u}|\mathbf{v})$, or $\mathbf{u}\cdot\mathbf{v}$.

Remarks

(1) Note that $<,>$ is a function of two arguments. Axioms (1) and (2) state that it is *linear* as a function of its left argument when the right one is fixed. Axiom (3) is called *symmetry*, and together with (1) and (2), it implies that $<,>$ is also linear in the *right* argument. For, if c is a scalar,

$$<\mathbf{u},c\mathbf{v} + \mathbf{w}> = <c\mathbf{v} + \mathbf{w},\mathbf{u}> = c<\mathbf{v},\mathbf{u}> + <\mathbf{w},\mathbf{u}>$$
$$= c<\mathbf{u},\mathbf{v}> + <\mathbf{u},\mathbf{w}>.$$

(2) Axiom (4) is called **positive definiteness**. In fact, together with the linearity, it implies another property:

$$(4') \ <\mathbf{u},\mathbf{u}> = 0 \text{ iff } \mathbf{u} = \mathbf{0}.$$

For,

$$<\mathbf{0},\mathbf{0}> \; = \; <\mathbf{0} + \mathbf{0},\mathbf{0}> \; = \; <\mathbf{0},\mathbf{0}> + <\mathbf{0},\mathbf{0}>,$$

so that $<\mathbf{0},\mathbf{0}> \; = \; 0$. More generally, a similar argument shows that $<\mathbf{u},\mathbf{0}> \; = \; <\mathbf{0},\mathbf{u}> \; = \; 0$ for any $\mathbf{u}$ in V (question 2, Exercises 7-1).

EXAMPLE 7-1

The most important example of an inner product is the dot product on R^n obtained by generalizing the dot product formula for R^3:

$$<\mathbf{u},\mathbf{v}> \; = \; u_1v_1 + \ldots + u_nv_n$$

for $\mathbf{u} = (u_1,\ldots,u_n)$ and $\mathbf{v} = (v_1,\ldots,v_n) \in R^n$. It is also often called the **standard**, **usual**, or **Euclidean inner product** on R^n and denoted by $\mathbf{u}\cdot\mathbf{v}$. We leave it for the reader to verify that $<\mathbf{u},\mathbf{v}>$ so defined is an inner product of R^n (question 3, Exercises 7-1). Note that we can write

$$<\mathbf{u},\mathbf{v}> \; = \; \mathbf{u}\cdot\mathbf{v} \; = \; [\mathbf{u}]_B{}^t[\mathbf{v}]_B,$$

where B is the standard basis for R^n. ❑

There are other inner products on R^n, as the next example shows.

EXAMPLE 7-2

We define $<,>$ on R^2 by

$$<\mathbf{u},\mathbf{v}> \; = \; u_2v_2 + 2u_1v_1 - u_1v_2 - u_2v_1,$$

where $\mathbf{u} = (u_1,u_2)$ and $\mathbf{v} = (v_1,v_2)$ are in R^2. Then, $<,>$ is an inner product on R^2. We verify the axioms given in Definition 7-1.

(1) $<\mathbf{u} + \mathbf{v},\mathbf{w}>$
$= (u_2 + v_2)w_2 + 2(u_1 + v_1)w_1 - (u_1 + v_1)w_2 - (u_2 + v_2)w_1$
$= (u_2w_2 + 2u_1w_1 - u_1w_2 - u_2w_1) + (v_2w_2 + 2v_1w_1 - v_1w_2 - v_2w_1)$
$= \; <\mathbf{u},\mathbf{w}> + <\mathbf{v},\mathbf{w}>$.

(2) $<c\mathbf{u},\mathbf{v}> \; = (cu_2)v_2 + 2(cu_1)v_1 - (cu_1)v_2 - (cu_2)v_1 = c<\mathbf{u},\mathbf{v}>$.

(3) $<\mathbf{v},\mathbf{u}> \; = v_2u_2 + 2v_1u_1 - v_1u_2 - v_2u_1 = \; <\mathbf{u},\mathbf{v}>$

(4) $<\mathbf{u},\mathbf{u}> \; = u_2u_2 + 2u_1u_1 - u_1u_2 - u_2u_1$
$= 2u_1{}^2 + u_2{}^2 - 2u_1u_2$
$= (u_1 - u_2)^2 + u_1{}^2 > 0$ if $\mathbf{u} \neq \mathbf{0}$. ❑

In general, it is true that a vector space V may have many different inner products. However, we shall see later that by choosing an appropriate basis B for V, $<\mathbf{u},\mathbf{v}>$ may be computed from $[\mathbf{u}]_B$ and $[\mathbf{v}]_B$ by a formula identical to the dot product formula.

EXAMPLE 7-3

Since each row of an $m \times n$ matrix A may be regarded as an n-tuple, we may regard A itself as an mn-tuple by writing its rows one after the other, in the order in which they appear in A. In this way we can use the dot product on R^{mn} to define an inner product (the **standard** inner product) on $M(m \times n)$ by

$$<A,B> = a_{11}b_{11} + a_{12}b_{12} + \ldots + a_{mn}b_{mn} \text{ } (mn \text{ terms})$$

where $A, B \in M(m \times n)$. The reader may verify that the axioms are satisfied (question 4, Exercises 7-1). As an example (with $m = n = 2$), if

$$A = \begin{bmatrix} 1 & 2 \\ -3 & 4 \end{bmatrix}, B = \begin{bmatrix} -2 & 3 \\ 2 & 1 \end{bmatrix},$$

then

$$<A,B> = (1)(-2) + (2)(3) + (-3)(2) + (4)(1) = 2.$$

Recall that the trace of a square matrix M, denoted trM, is the sum of the diagonal entries of M. Then, we may write $<A,B>$ more concisely as

$$<A,B> = tr\,(AB^t).$$

To justify this, note that AB^t is $m \times m$ and

$$tr\,(AB^t) = \sum_{i=1}^{m} (AB^t)_{ii} = \sum_{i=1}^{m} \sum_{j=1}^{n} A_{ij}B^t_{ji} = \sum_{i=1}^{m} \sum_{j=1}^{n} A_{ij}B_{ij} = <A,B>.$$

Using this alternative formula, the verification of the axioms for an inner product follows from the properties of the transpose (Theorem 2-5) and those of the trace (question 16, Exercises 4-2, and Example 5-43). ❑

EXAMPLE 7-4

In a similar way, we may define the standard inner product on $P(n)$ as follows. If $f(x) = a_0 + \ldots + a_nx^n$ and $g(x) = b_0 + \ldots + b_nx^n$ $(x \in R)$, we put

$$<f,g> = a_0b_0 + \ldots + a_nb_n.$$

Note that to compute this, we are simply using the dot product formula on the matrices of f and g with respect to the usual basis for $P(n)$. For example, in $P(2)$, if $f(x) = 2x + 3x^2$ and $g(x) = -2 - 4x - 5x^2$, then

$$<f,g> = 0(-2) + 2(-4) + 3(-5) = -23. \square$$

We have noted before that a vector space may have several different inner products. In fact, it has *infinitely* many. The next result shows how to construct them. (Readers will need the definition of "1-1" given in section 5-3.)

THEOREM 7-1

Let V and W be vector spaces of the same finite dimension, and suppose that $<,>$ is an inner product on W. Then, for any 1-1 linear transformation $T:V \to W$, there is an inner product $<,>_T$ on V defined by $<\mathbf{u},\mathbf{v}>_T = <T(\mathbf{u}),T(\mathbf{v})>$ for $\mathbf{u}$ and $\mathbf{v}$ in V. We call $<,>_T$ the inner product on V **induced** by T and the inner product $<,>$ on W.

PROOF $<,>_T$ is an inner product on V because T is linear and 1-1. For example, if $\mathbf{u}$, $\mathbf{v}$, and $\mathbf{w}$ are in V,

$$\begin{aligned} <\mathbf{u} + \mathbf{v},\mathbf{w}>_T &= <T(\mathbf{u} + \mathbf{v}), T(\mathbf{w})> \\ &= <T(\mathbf{u}),T(\mathbf{w})> + <T(\mathbf{v}),T(\mathbf{w})> \\ &= <\mathbf{u},\mathbf{w}>_T + <\mathbf{v},\mathbf{w}>_T, \end{aligned}$$

showing that $<,>_T$ satisfies axiom (1) for inner products.

Since T is 1-1, $T(\mathbf{u}) \neq \mathbf{0}$ when $\mathbf{u} \neq \mathbf{0}$. Therefore,

$$<\mathbf{u},\mathbf{u}>_T = <T(\mathbf{u}),T(\mathbf{u})> > 0,$$

unless $\mathbf{u} = \mathbf{0}$, so axiom (4) holds. We leave the verification of axioms (2) and (3) for the reader. ∎

If we take the dot product on $W = R^n$, then for every 1-1 linear transformation $T:V \to R^n$ (where $\dim V = n$), we obtain an inner product $<,>_T$ on V with

$$<\mathbf{u},\mathbf{v}>_T = T(\mathbf{u})\cdot T(\mathbf{v}),\ \mathbf{u},\ \mathbf{v} \in V.$$

EXAMPLE 7-5

(a) Let B be the standard basis for $P(n-1)$ and $T:P(n-1) \to R^n$, defined by $T(x^i) = \mathbf{e}_{i+1}$, $i = 0, \ldots, n-1$, be the unique-linear transformation that maps the vectors of B to $B(n)$, the standard basis for R^n. Then, the inner product on $P(n-1)$ discussed in Example 7-4 is induced by T and the dot product on R^n, because, using the notation of Example 7-4,

$$\begin{aligned} T(f)\cdot T(g) &= (a_0,\ldots,a_{n-1})\cdot(b_0,\ldots,b_{n-1}) \\ &= a_0b_0 + \ldots + a_{n-1}b_{n-1} \\ &= \langle f,g\rangle. \end{aligned}$$

(b) Let $T:R^2 \to R^2$ be the linear operator defined by

$$T(x,y) = (\sqrt{2}x - y/\sqrt{2}, y/\sqrt{2}).$$

The reader should verify that T is 1-1. If $\mathbf{u} = (u_1,u_2)$ and $\mathbf{v} = (v_1,v_2)$ are in R^2,

$$\begin{aligned} T(u)\cdot T(v) &= (\sqrt{2}u_1 - u_2/\sqrt{2}, u_2/\sqrt{2})\cdot(\sqrt{2}v_1 - v_2/\sqrt{2}, v_2/\sqrt{2}) \\ &= (\sqrt{2}u_1 - u_2/\sqrt{2})(\sqrt{2}v_1 - v_2/\sqrt{2}) + u_2v_2/2 \\ &= 2u_1v_1 - u_2v_1 - u_1v_2 + u_2v_2/2 + u_2v_2/2 \\ &= \langle \mathbf{u},\mathbf{v}\rangle, \end{aligned}$$

where $\langle , \rangle$ is the inner product of Example 7-2. Therefore, this inner product is induced by T and the dot product on R^2. ❑

In the next section, we shall see that *every inner product on an n-dimensional vector space V is induced by some 1-1 linear transformation $T:V \to R^n$ and the dot product.*

EXERCISES 7-1

1. Compute the inner products of the vectors below with respect to the inner product indicated.

(a) $\mathbf{u} = (4,-3)$, $\mathbf{v} = (-2,3)$ in R^2 with respect to **(i)** the dot product; and **(ii)** the inner product defined in Example 7-2.

(b) A and B in $M(2\times 3)$ given by

$$A = \begin{bmatrix} -1 & 2 & 0 \\ 3 & -4 & 2 \end{bmatrix}, B = \begin{bmatrix} 3 & -4 & 1 \\ 0 & 3 & 2 \end{bmatrix},$$

with respect to the standard inner product in $M(2\times 3)$.

(c) f and g in $P(2)$ with $f(x) = -2 + 3x + 4x^2$ and $g(x) = -6x + 2x^2$ for $x \in R$, with respect to the standard inner product in $P(2)$.

2. Show that $\langle \mathbf{u},\mathbf{0}\rangle = \langle \mathbf{0},\mathbf{u}\rangle = 0$ for $\mathbf{u}$ in any vector space with inner product $\langle , \rangle$.

3. Verify that the dot product on R^n satisfies the axioms for inner products.

4. Verify that the inner product defined on $M(m\times n)$ in Example 7-3 satisfies the axioms for inner products.

5. Which of the following functions define inner products on R^2? Justify your answers.
   (a) $<\mathbf{u},\mathbf{v}> = 3u_2v_2 - 4u_1v_1$
   (b) $<\mathbf{u},\mathbf{v}> = u_1v_1 + 4u_2v_2$
   (c) $<\mathbf{u},\mathbf{v}> = u_2v_2 + 2u_1v_1 + u_1v_2 - u_2v_1$
6. Let $T{:}R^2 \to R^2$ be the linear transformation defined by $T(x,y) = (2x - y, x + 2y)$.
   (a) Verify that T is 1-1.
   (b) Find the inner product $<,>_T$ induced on R^2 by T and the dot product by obtaining an expression for $<\mathbf{u},\mathbf{v}>_T$ in terms of the components of **u** and **v**.
7. Let f and g be the polynomial functions in $P(1)$ with $f(x) = ax + b$ and $g(x) = cx + d$ for x in R. Define $<,>$ by $<f,g> = ac/3 + (bc + ad)/2 + bd$. Show that this is an inner product on $P(1)$.
8. Let $<,>$ be an inner product on a vector space V and let k be a scalar. Show that $<\mathbf{u},\mathbf{v}>' = k<\mathbf{u},\mathbf{v}>$ for $\mathbf{u}, \mathbf{v} \in V$ also defines an inner product $<,>'$ on V iff $k > 0$.
9. Let $T{:}M(n) \to M(n)$ be the linear operator with $T(A) = A^t$ for A in $M(n)$. Find the inner product on $M(n)$ induced by T and the standard inner product on $M(n)$.
10. Suppose that $<\mathbf{u},\mathbf{v}> = 0$ for all $\mathbf{v}$ in V. Show that $\mathbf{u} = \mathbf{0}$.

7-2 Inner Product Spaces

In this section, we study the properties of a vector space V with a certain fixed inner product $<,>$ on it. Such a combination is called an **inner product space**. Note that if we choose a different inner product on the same vector space V, then a different inner product space results, even though the set of vectors V is the same. Whereas in the case of R^3, the definition of the dot product was motivated by the notions of lengths of vectors and angles (orthogonality), in a general inner product space, we can only define these notions *by means of* the inner product.

DEFINITION 7-2

Let V be an inner product space. If **u** is a vector in V, its **norm** (or length) is the number $|\mathbf{u}|$ (also denoted $\|\mathbf{u}\|$) defined by $|\mathbf{u}| = <\mathbf{u},\mathbf{u}>^{1/2}$.

Of course, in the inner product space R^3 with the dot product, this is just the familiar **Euclidean length** of a vector: $|\mathbf{u}| = (u_1^2 + u_2^2 + u_3^2)^{1/2}$.

EXAMPLE 7-6

(a) In $M(2)$ with the standard inner product, let

$$A = \begin{bmatrix} -1 & 2 \\ 3 & 0 \end{bmatrix}.$$

Then $|A| = \langle A,A\rangle^{1/2} = [(-1)^2 + 2^2 + 3^2 + 0]^{1/2} = \sqrt{14}$.

(b) In $P(2)$ with the standard inner product, let f be given by

$$f(x) = -3 + 2x - 4x^2.$$

Then, $|f| = [(-3)^2 + 2^2 + (-4)^2]^{1/2} = \sqrt{29}$.

(c) For $\mathbf{u} = (u_1,u_2) \in R^2$ with the inner product $\langle , \rangle$ of Example 7-2,

$$|\mathbf{u}| = (u_2{}^2 + 2u_1{}^2 - 2u_1u_2)^{1/2}.$$

As an example,

$$|\mathbf{e}_1| = \langle \mathbf{e}_1,\mathbf{e}_1\rangle^{1/2} = [0 + 2 - 2(0)]^{1/2} = \sqrt{2}.$$

While this result may be perplexing, it must be remembered that, as we saw in Example 7-5(b), in effect, $\langle \mathbf{e}_1,\mathbf{e}_1\rangle$ is the dot product $T(\mathbf{e}_1)\cdot T(\mathbf{e}_1)$ where T is the linear operator described in that example. Moreover, the next theorem shows that the norm of any inner product has the properties expected of the lengths of vectors. ❑

THEOREM 7-2 (Properties of the Norm)

Let V be an inner product space. Then the norm defined by the inner product has the following properties:

(1) $|c\mathbf{u}| = |c||\mathbf{u}|$ for $c \in R$ and $\mathbf{u} \in V$ (note that $|c|$ is the **absolute value** of c).

(2) $|\mathbf{u}| > 0$ unless $\mathbf{u} = \mathbf{0}$, $\mathbf{u} \in V$; and $|\mathbf{0}| = 0$.

(3) $|\langle \mathbf{u},\mathbf{v}\rangle| \leq |\mathbf{u}||\mathbf{v}|$ for all $\mathbf{u}$ and $\mathbf{v} \in V$ (the **Cauchy-Schwarz inequality**).

(4) $|\mathbf{u} + \mathbf{v}| \leq |\mathbf{u}| + |\mathbf{v}|$ (the **triangle inequality**).

PROOF We leave (2) as an exercise for the reader.

To prove (1), we have

$$|c\mathbf{u}| = \langle c\mathbf{u},c\mathbf{u}\rangle^{1/2} = (c^2)^{1/2}\langle \mathbf{u},\mathbf{u}\rangle^{1/2} = |c||\mathbf{u}|.$$

In order to prove (3), we will let

$$\mathbf{w} = \mathbf{v} - (\langle \mathbf{v},\mathbf{u}\rangle/\langle \mathbf{u},\mathbf{u}\rangle)\mathbf{u}.$$

Before we use this to prove the result, recall from Chapter 1 that when $V = R^2$ (or R^3), we know that $(\langle \mathbf{v},\mathbf{u}\rangle/\langle \mathbf{u},\mathbf{u}\rangle)\mathbf{u}$ is the **orthogonal projection** of **v** on **u**, as shown in Figure 7-1. Thus, **w** is a vector which is orthogonal to **u**, and **w** is **0** exactly when **u** and **v** are collinear (or, one of them is **0**); i.e., iff **u** and **v** are linearly dependent.

FIGURE 7-1

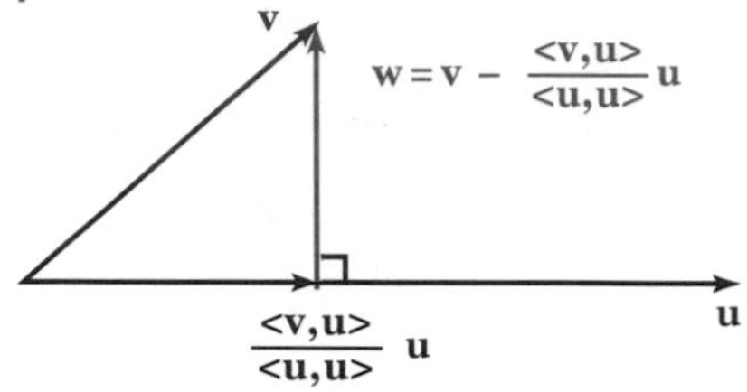

Now, in the general case, it is also true that $\langle \mathbf{w},\mathbf{u}\rangle = 0$. For,

$$\begin{aligned}
\langle \mathbf{w},\mathbf{u}\rangle &= \langle [\mathbf{v} - (\langle \mathbf{v},\mathbf{u}\rangle/\langle \mathbf{u},\mathbf{u}\rangle)\mathbf{u}],\mathbf{u}\rangle \\
&= \langle \mathbf{v},\mathbf{u}\rangle - (\langle \mathbf{v},\mathbf{u}\rangle/\langle \mathbf{u},\mathbf{u}\rangle)\langle \mathbf{u},\mathbf{u}\rangle \qquad \text{(by linearity)} \\
&= \langle \mathbf{v},\mathbf{u}\rangle - \langle \mathbf{v},\mathbf{u}\rangle \\
&= 0.
\end{aligned}$$

We complete the proof of (3) as follows.

$$\begin{aligned}
\langle \mathbf{w},\mathbf{w}\rangle &= \langle \mathbf{w},\mathbf{v} - (\langle \mathbf{v},\mathbf{u}\rangle/\langle \mathbf{u},\mathbf{u}\rangle)\mathbf{u}\rangle \\
&= \langle \mathbf{w},\mathbf{v}\rangle - (\langle \mathbf{v},\mathbf{u}\rangle/\langle \mathbf{u},\mathbf{u}\rangle)\langle \mathbf{w},\mathbf{u}\rangle \\
&= \langle \mathbf{w},\mathbf{v}\rangle \qquad (\text{as } \langle \mathbf{w},\mathbf{u}\rangle = 0) \\
&= \langle \mathbf{v} - (\langle \mathbf{v},\mathbf{u}\rangle/\langle \mathbf{u},\mathbf{u}\rangle)\mathbf{u},\mathbf{v}\rangle \\
&= \langle \mathbf{v},\mathbf{v}\rangle - (\langle \mathbf{v},\mathbf{u}\rangle/\langle \mathbf{u},\mathbf{u}\rangle)\langle \mathbf{u},\mathbf{v}\rangle \\
&= \langle \mathbf{v},\mathbf{v}\rangle - \langle \mathbf{u},\mathbf{v}\rangle^2/\langle \mathbf{u},\mathbf{u}\rangle.
\end{aligned}$$

But,

$$\langle \mathbf{w},\mathbf{w}\rangle \geq 0.$$

So,

$$\langle \mathbf{u},\mathbf{v}\rangle^2 \leq \langle \mathbf{u},\mathbf{u}\rangle\langle \mathbf{v},\mathbf{v}\rangle,$$

or, taking square roots,

$$|\langle \mathbf{u},\mathbf{v}\rangle| \leq |\mathbf{u}||\mathbf{v}|.$$

Finally, (4) is proved by considering

$$\begin{aligned}
|\mathbf{u} + \mathbf{v}|^2 &= \langle \mathbf{u} + \mathbf{v},\mathbf{u} + \mathbf{v}\rangle \\
&= \langle \mathbf{u},\mathbf{u}\rangle + 2\langle \mathbf{u},\mathbf{v}\rangle + \langle \mathbf{v},\mathbf{v}\rangle \\
&\leq \langle \mathbf{u},\mathbf{u}\rangle + 2|\langle \mathbf{u},\mathbf{v}\rangle| + \langle \mathbf{v},\mathbf{v}\rangle \\
&\leq \langle \mathbf{u},\mathbf{u}\rangle + 2|\mathbf{u}||\mathbf{v}| + \langle \mathbf{v},\mathbf{v}\rangle \\
&= |\mathbf{u}|^2 + 2|\mathbf{u}||\mathbf{v}| + |\mathbf{v}|^2 \\
&= (|\mathbf{u}| + |\mathbf{v}|)^2.
\end{aligned}$$

Therefore, $|\mathbf{u} + \mathbf{v}| \leq |\mathbf{u}| + |\mathbf{v}|$. ■

Remarks

(1) Inequality (3) is named after Augustin Louis Cauchy (1789–1857), an eminent French mathematician, and Hermann Amandus Schwarz (1843–1921), a German mathematician.

(2) Inequality (4) is called the triangle inequality because in R^2 (and R^3), it states the familiar fact that the length of any side of a triangle cannot exceed the sum of the lengths of the other two sides (Figure 7-2).

FIGURE 7-2

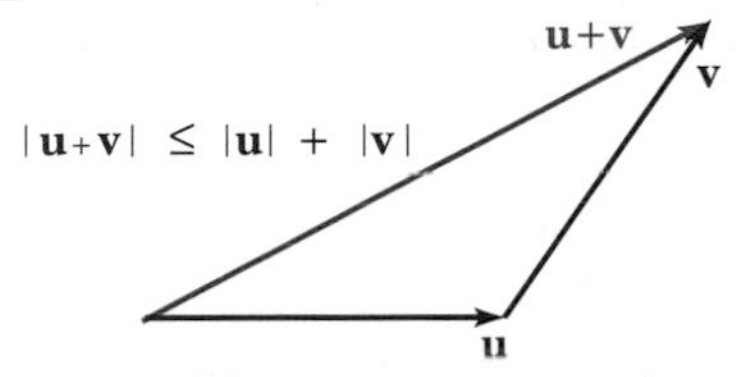

Both inequalities have important applications in other areas of mathematics.

EXAMPLE 7-7

In R^n with the dot product, let $\mathbf{u} = (u_1,...,u_n)$ and $\mathbf{v} = (v_1,...,v_n)$.

(a) Then, $<\mathbf{u},\mathbf{v}>^2 \le |\mathbf{u}|^2|\mathbf{v}|^2$ becomes

$$(\sum_{i=1}^{n} u_i v_i)^2 \le (\sum_{i=1}^{n} u_i^2)(\sum_{i=1}^{n} v_i^2).$$

In other words, "the square of the sum of products of n numbers cannot exceed the product of the sums of their squares". In particular, if we set $\mathbf{v} - (1,...,1)$, we get

$$(\sum_{i=1}^{n} u_i)^2 \le n\,(\sum_{i=1}^{n} u_i^2),$$

which is another useful inequality.

(b) By the triangle inequality,

$$(\sum_{i=1}^{n} (u_i + v_i)^2)^{1/2} \le (\sum_{i=1}^{n} u_i^2)^{1/2} + (\sum_{i=1}^{n} v_i^2)^{1/2}. \;\square$$

By virtue of the Cauchy-Schwarz inequality, we can define *angles* between vectors in arbitrary inner product spaces. For, if **u** and **v** are nonzero vectors in V, we may write property (3) in Theorem 7-2 as

$$\left|\frac{<\mathbf{u},\mathbf{v}>}{|\mathbf{u}||\mathbf{v}|}\right| \leq 1, \text{ or, } -1 \leq \frac{<\mathbf{u},\mathbf{v}>}{|\mathbf{u}||\mathbf{v}|} \leq 1.$$

Thus, there exists exactly one number α in the range $[0, \pi]$ such that

$$\cos \alpha = \frac{<\mathbf{u},\mathbf{v}>}{|\mathbf{u}||\mathbf{v}|}. \tag{7-1}$$

We call α the **angle between u and v**. Of course, if V is R^2 or R^3, this definition agrees with the notion of angles discussed in Chapter 1 (equation 1-3). Since the angle is $\pi/2$ (90°) iff $\cos \alpha = 0$, i.e., iff $<\mathbf{u},\mathbf{v}> = 0$ by equation 7-1, the following definition arises.

DEFINITION 7-3

Vectors **u** and **v** in an inner product space V are **orthogonal** (written $\mathbf{u} \perp \mathbf{v}$) iff $<\mathbf{u},\mathbf{v}> = 0$.

Notice that **u** and **v** need not be nonzero; the zero vector **0** is orthogonal to every vector by this definition.

EXAMPLE 7-8

(a) In R^5 with the dot product, let $\mathbf{u} = (-1,2,0,3,1)$ and $\mathbf{v} = (0,1,-2,4,-1)$. Then, $|\mathbf{u}| = \sqrt{15}$, $|\mathbf{v}| = \sqrt{22}$, $<\mathbf{u},\mathbf{v}> = 13$ and $\cos \alpha = 13/(\sqrt{15}\sqrt{22}) \approx 0.7156$. So, (from tables) $\alpha \approx 44.3°$.

(b) In $M(2)$ with the usual inner product, let

$$A = \begin{bmatrix} a & b \\ b & c \end{bmatrix}, \ B = \begin{bmatrix} 0 & d \\ -d & 0 \end{bmatrix}, \ C = \begin{bmatrix} 0 & e \\ -e & 0 \end{bmatrix}.$$

Notice that A is symmetric while B and C are skew-symmetric. Then,

$$<A,B> = bd - bd = 0,$$

so $A \perp B$. On the other hand,

$$<B,C> = de + de \neq 0,$$

unless B or $C = O$. Thus, nonzero skew-symmetric 2×2 matrices are not orthogonal. Intuitively, we see that this makes sense since the space of such matrices has dimension 1 (cf. question 8, Exercises 7-2).

(c) In R^n with the dot product, the angle α between the vector $\mathbf{u} = (1,\ldots,1)$ and any basis vector $\mathbf{e}_i$ is given by $\cos \alpha = <\mathbf{u},\mathbf{e}_i>/(|\mathbf{u}||\mathbf{e}_i|) = 1/\sqrt{n}$. ❑

DEFINITION 7-4

A nonempty set S of vectors in an inner product space V is **orthogonal** iff $\mathbf{u} \perp \mathbf{v}$ ($<\mathbf{u},\mathbf{v}> = 0$) for all $\mathbf{u}$ and $\mathbf{v} \in S$. It is **orthonormal** if, in addition, $|\mathbf{u}| = 1$ for all $\mathbf{u} \in S$. Vectors of length 1 are called **unit vectors**.

EXAMPLE 7-9

(a) The standard basis for R^n is an orthonormal set of n vectors with respect to the dot product, because $<\mathbf{e}_i,\mathbf{e}_j> = \delta_{ij}$ for all $i, j = 1, \ldots, n$.

(b) Similarly, the standard bases for $M(m \times n)$ and $P(n)$ are orthonormal sets with respect to the standard inner products in these spaces. For example, in $M(2)$, recall that the standard basis vectors are

$$E_1 = \begin{bmatrix} 1 & 0 \\ 0 & 0 \end{bmatrix}, E_2 = \begin{bmatrix} 0 & 1 \\ 0 & 0 \end{bmatrix}, E_3 = \begin{bmatrix} 0 & 0 \\ 1 & 0 \end{bmatrix}, E_4 = \begin{bmatrix} 0 & 0 \\ 0 & 1 \end{bmatrix}.$$

This is clearly an orthonormal set.

(c) In R^3, the set $\{\mathbf{v}_1,\mathbf{v}_2,\mathbf{v}_3\}$ with $\mathbf{v}_1 = (1/\sqrt{3})(1,1,1)$, $\mathbf{v}_2 = (1/\sqrt{6})(1,1,-2)$, and $\mathbf{v}_3 = (1/\sqrt{2})(-1,1,0)$ is an orthonormal basis with respect to the dot product. ❑

THEOREM 7-3

Let V be an inner product space.

(1) Any orthogonal set of nonzero vectors in V is linearly independent.

(2) If $B = \{\mathbf{v}_1,\ldots,\mathbf{v}_n\}$ is an orthonormal basis for V, then any $\mathbf{u} \in V$ may be written as

$$\mathbf{u} = \sum_{i=1}^{n} <\mathbf{u},\mathbf{v}_i>\mathbf{v}_i.$$

In other words, the **components** of $\mathbf{u}$ with respect to the basis B may be computed simply as inner products: $[\mathbf{u}]_B = (<\mathbf{u},\mathbf{v}_i>)$ (an $n \times 1$ matrix).

PROOF (1) Let S be an orthogonal set of vectors in V and suppose that

$$\sum_{i=1}^{k} c_i\mathbf{u}_i = \mathbf{0},$$

for some $\mathbf{u}_1,\ldots,\mathbf{u}_k$ in S and scalars $c_i \in R$. As both sides of the equation are vectors, we may take their inner products with $\mathbf{u}_j$ to get

$$<\sum_{i=1}^{k} c_i\mathbf{u}_i,\mathbf{u}_j> = <\mathbf{0},\mathbf{u}_j> = 0,$$

or,

$$\sum_{i=1}^{k} c_i<\mathbf{u}_i,\mathbf{u}_j> = 0.$$

Since S is orthogonal, $<\mathbf{u}_i,\mathbf{u}_j> = 0$ when $i \neq j$, and equals $|\mathbf{u}_j|^2$ when $i = j$. So, this equation becomes

$$c_j|\mathbf{u}_j|^2 = 0.$$

Since $\mathbf{u}_j \neq \mathbf{0}$, it follows that $c_j = 0$ for $j = 1, \ldots, k$. Therefore, S is linearly independent.

(2) Certainly any $\mathbf{u} \in V$ is a linear combination

$$\mathbf{u} = \sum_{i=1}^{n} c_i\mathbf{v}_i, c_i \in R.$$

As in part (1), we take the inner product of both sides of the equation with $\mathbf{v}_j$ to get

$$<\mathbf{u},\mathbf{v}_j> = <\sum_{i=1}^{n} c_i\mathbf{v}_i,\mathbf{v}_j> = c_j<\mathbf{v}_j,\mathbf{v}_j> = c_j.$$

Thus,

$$\mathbf{u} = \sum_{i=1}^{n} c_i\mathbf{v}_i = \sum_{i=1}^{n} <\mathbf{u},\mathbf{v}_i>\mathbf{v}_i,$$

completing the proof. ■

COROLLARY

If $\mathbf{v}$ is orthogonal to every vector of an orthonormal basis for V, then $\mathbf{v} = \mathbf{0}$. ■

It follows from property (1) in the theorem that any set of n nonzero orthogonal vectors in an inner product space V of dimension n is a basis for V. Consequently, *the dimension of an inner product space is the number of "orthogonal directions" in it.*

EXAMPLE 7-10

We compute the matrix of $\mathbf{v} = (1,2,3)$ with respect to the orthonormal basis B of Example 7-9(c):

$$<\mathbf{v},\mathbf{v}_1> = (1/\sqrt{3})(1 + 2 + 3) = 6/\sqrt{3};$$

$$<\mathbf{v},\mathbf{v}_2> = (1/\sqrt{6})(1 + 2 - 6) = -3/\sqrt{6};$$

$$<\mathbf{v},\mathbf{v}_3> = (1/\sqrt{2})(-1 + 2 + 0) = 1/\sqrt{2}.$$

Thus,

$$[\mathbf{v}]_B = \begin{bmatrix} 6/\sqrt{3} \\ -3/\sqrt{6} \\ 1/\sqrt{2} \end{bmatrix}. \square$$

Orthonormal bases are important not only because of their computational simplicity, but also because in physical processes, it is natural to make measurements of quantities along orthogonal directions. For this reason, the following result is fundamental.

THEOREM 7-4 (Gram-Schmidt Orthogonalization Process)
Let V be an inner product space with basis $B = \{\mathbf{u}_1,\ldots,\mathbf{u}_n\}$. Then there is an orthogonal basis $B' = \{\mathbf{v}_1,\ldots,\mathbf{v}_n\}$ for V such that *span* $\{\mathbf{v}_1,\ldots,\mathbf{v}_k\}$ = *span* $\{\mathbf{u}_1,\ldots,\mathbf{u}_k\}$ for $k = 1, \ldots, n$. Furthermore, $B'' = \{\mathbf{w}_1,\ldots,\mathbf{w}_n\}$ with $\mathbf{w}_i = \mathbf{v}_i/|\mathbf{v}_i|$ is an **orthonormal** basis for V with the same property. The bases B' and B'' are constructed from B.

PROOF The proof is based on the idea used earlier in proving the Cauchy-Schwarz inequality. We define $\mathbf{v}_1 = \mathbf{u}_1$ and

$$\mathbf{v}_2 = \mathbf{u}_2 - \frac{<\mathbf{u}_2,\mathbf{v}_1>}{<\mathbf{v}_1,\mathbf{v}_1>}\mathbf{v}_1, \tag{7-2}$$

i.e., $\mathbf{v}_2 = \mathbf{u}_2$ − orthogonal projection of $\mathbf{u}_2$ on $\mathbf{v}_1$ (see Figure 7-1, where "$\mathbf{u}$" = $\mathbf{v}_1$, "$\mathbf{v}$" = $\mathbf{u}_2$). Then, clearly $\mathbf{v}_2 \perp \mathbf{v}_1$ and in this case, $\mathbf{v}_2 \neq \mathbf{0}$, because if it were, equation 7-2 would yield

$$\mathbf{u}_2 = (<\mathbf{u}_2,\mathbf{v}_1>/<\mathbf{v}_1,\mathbf{v}_1>)\mathbf{u}_1 \qquad (\text{as } \mathbf{v}_1 = \mathbf{u}_1),$$

which contradicts the fact that B is a basis for V (so $\mathbf{u}_1$ and $\mathbf{u}_2$ are linearly independent). Also, as equation 7-2 expresses $\mathbf{v}_2$ as a linear combination of $\mathbf{u}_2$ and $\mathbf{v}_1$, where $\mathbf{u}_1 = \mathbf{v}_1$,

$$\textit{span}\ \{\mathbf{v}_1,\mathbf{v}_2\} = \textit{span}\ \{\mathbf{u}_1,\mathbf{u}_2\}.$$

Next, we would put

$$\mathbf{v}_3 = \mathbf{u}_3 - \frac{<\mathbf{u}_3,\mathbf{v}_1>}{<\mathbf{v}_1,\mathbf{v}_1>}\mathbf{v}_1 - \frac{<\mathbf{u}_3,\mathbf{v}_2>}{<\mathbf{v}_2,\mathbf{v}_2>}\mathbf{v}_2 \qquad (7\text{-}3)$$

and show that $\mathbf{v}_3 \perp \mathbf{v}_1$ and $\mathbf{v}_2$, $\mathbf{v}_3 \neq \mathbf{0}$, and *span* $\{\mathbf{v}_1,\mathbf{v}_2,\mathbf{v}_3\}$ = *span* $\{\mathbf{u}_1,\mathbf{u}_2,\mathbf{u}_3\}$. However, let us proceed by induction and assume that we have constructed $\mathbf{v}_1, \ldots, \mathbf{v}_k$ $(k \geq 2)$ such that

(i) $\{\mathbf{v}_1,\ldots,\mathbf{v}_k\}$ is an orthogonal set of nonzero vectors and
(ii) *span* $\{\mathbf{v}_1,\ldots,\mathbf{v}_k\}$ = *span* $\{\mathbf{u}_1,\ldots,\mathbf{u}_k\}$.

We define

$$\mathbf{v}_{k+1} = \mathbf{u}_{k+1} - \sum_{i=1}^{k} \frac{<\mathbf{u}_{k+1},\mathbf{v}_i>}{<\mathbf{v}_i,\mathbf{v}_i>}\mathbf{v}_i. \qquad (7\text{-}4)$$

First, $\mathbf{v}_{k+1} \neq \mathbf{0}$, as, otherwise equation 7-4 would yield a nontrivial linear combination of $\mathbf{u}_1, \ldots, \mathbf{u}_{k+1}$ equal to the zero vector, contradicting the fact that these vectors are linearly independent. Second, equation 7-4 can be used to express $\mathbf{v}_{k+1}$ as a linear combination of $\mathbf{u}_1, \ldots, \mathbf{u}_{k+1}$ because, by assumption, each $\mathbf{v}_i$ $(i \leq k)$ is a linear combination of only $\mathbf{u}_1, \ldots, \mathbf{u}_i$. Thus, $\mathbf{v}_{k+1}$ is in *span* $\{\mathbf{u}_1,\ldots,\mathbf{u}_{k+1}\}$, so *span* $\{\mathbf{v}_1,\ldots,\mathbf{v}_{k+1}\}$ is a subspace of *span* $\{\mathbf{u}_1,\ldots,\mathbf{u}_{k+1}\}$ (but, we see below that they are equal).

Finally, $\mathbf{v}_{k+1} \perp \mathbf{v}_j$ for $j \leq k$ because

$$\begin{aligned} <\mathbf{v}_{k+1},\mathbf{v}_j> &= <\mathbf{u}_{k+1},\mathbf{v}_j> - \sum_{i=1}^{k} \frac{<\mathbf{u}_{k+1},\mathbf{v}_i>}{<\mathbf{v}_i,\mathbf{v}_i>} <\mathbf{v}_i,\mathbf{v}_j> \\ &= <\mathbf{u}_{k+1},\mathbf{v}_j> - <\mathbf{u}_{k+1},\mathbf{v}_j> = 0, \end{aligned}$$

(since $<\mathbf{v}_i,\mathbf{v}_j> = 0$ unless $i = j$ for $i, j \leq k$, by (i)). So, $\{\mathbf{v}_1,\ldots,\mathbf{v}_{k+1}\}$ is linearly independent, from which it follows that

$$span\ \{\mathbf{v}_1,\ldots,\mathbf{v}_{k+1}\} = span\ \{\mathbf{u}_1,\ldots,\mathbf{u}_{k+1}\}.$$

This completes the induction process and the proof. Note that since the vectors of B'' are just scalar multiples of those of B', B'' has the same properties noted for B'. ■

COROLLARY 1

Any finite dimensional inner product space has an orthonormal basis. ■

Remarks

(1) Theorem 7-4 is named after Jorgen Pederson Gram (1850–1916), a Danish mathematician and actuary, and Erhardt Schmidt (1876–1959), a German mathematician.

(2) In the Gram-Schmidt process, it should be clear that different choices for the "starting" basis B will generally result in different orthonormal bases B'' for V. Of course, if B is orthonormal itself, then $B'' = B$.

(3) The vector

$$\sum_{i=1}^{k} \frac{<\mathbf{u}_{k+1},\mathbf{v}_i>}{<\mathbf{v}_i,\mathbf{v}_i>}\mathbf{v}_i$$

is in the subspace $W(k) = span\ \{\mathbf{v}_1,\ldots,\mathbf{v}_k\}$. Generalizing the terminology of the case $k = 1$, we call this vector the **orthogonal projection** of $\mathbf{u}_{k+1}$ on $W(k)$. More generally, we have the following definition.

DEFINITION 7-5

If $\mathbf{v}_1, \ldots, \mathbf{v}_k$ are any **orthonormal** vectors in V and $\mathbf{u}$ is an arbitrary vector, the vector

$$\sum_{i=1}^{k} <\mathbf{u},\mathbf{v}_i>\mathbf{v}_i$$

is called the **orthogonal projection** of $\mathbf{u}$ on $W = span\ \{\mathbf{v}_1,\ldots,\mathbf{v}_k\}$. It depends only on W, and not on the $\mathbf{v}$'s. (See question 18, Exercises 7-2.)

EXAMPLE 7-11

We apply the Gram-Schmidt process to construct an orthonormal basis B'' for R^3 (with the dot product) from the basis $B = \{\mathbf{u}_1,\mathbf{u}_2,\mathbf{u}_3\}$, where $\mathbf{u}_1 = (-1,1,1)$, $\mathbf{u}_2 = (2,1,0)$, and $\mathbf{u}_3 = (-1,2,2)$. First, we construct an orthogonal basis $B' = \{\mathbf{v}_1,\mathbf{v}_2,\mathbf{v}_3\}$ with $\mathbf{v}_1 = \mathbf{u}_1$ and

$$\mathbf{v}_2 = \mathbf{u}_2 - (<\mathbf{u}_2,\mathbf{v}_1>/<\mathbf{v}_1,\mathbf{v}_1>)\mathbf{v}_1$$

(equation 7-2). Since $<\mathbf{u}_2,\mathbf{v}_1> = <\mathbf{u}_2,\mathbf{u}_1> = -2 + 1 + 0 = -1$ and $<\mathbf{v}_1,\mathbf{v}_1> = 3$, we obtain

$$\mathbf{v}_2 = (2,1,0) - (-1/3)(-1,1,1) = (5/3,4/3,1/3).$$

To compute $\mathbf{v}_3$ (using equation 7-3 or 7-4), we need

$$\begin{aligned} <\mathbf{u}_3,\mathbf{v}_1> &= 1 + 2 + 2 = 5, \\ <\mathbf{u}_3,\mathbf{v}_2> &= -5/3 + 8/3 + 2/3 = 5/3, \\ <\mathbf{v}_2,\mathbf{v}_2> &= 25/9 + 16/9 + 1/9 = 42/9. \end{aligned}$$

So,

$$\begin{aligned} \mathbf{v}_3 &= (-1,2,2) - (5/3)(-1,1,1) - (5/3)(9/42)(5/3,4/3,1/3) \\ &= (2/3,1/3,1/3) - (5/42)(5,4,1) \\ &= (1/14,-2/14,3/14). \end{aligned}$$

Finally, $B'' = \{\mathbf{w}_1,\mathbf{w}_2,\mathbf{w}_3\}$ is obtained by "normalizing" B':

$$\mathbf{w}_1 = \mathbf{v}_1/|\mathbf{v}_1| = (1/\sqrt{3})(-1,1,1),$$
$$\mathbf{w}_2 = \mathbf{v}_2/|\mathbf{v}_2| = (3/\sqrt{42})(5/3,4/3,1/3) = (1/\sqrt{42})(5,4,1),$$
$$\mathbf{w}_3 = \mathbf{v}_3/|\mathbf{v}_3| = (1/\sqrt{14})(1,-2,3). \square$$

EXAMPLE 7-12

Consider the space V of 2×2 symmetric matrices. A basis for this 3-dimensional space is $B = \{\mathbf{u}_1,\mathbf{u}_2,\mathbf{u}_3\}$ with

$$\mathbf{u}_1 = \begin{bmatrix} 1 & 0 \\ 0 & 0 \end{bmatrix}, \mathbf{u}_2 = \begin{bmatrix} 0 & 1 \\ 1 & 0 \end{bmatrix}, \mathbf{u}_3 = \begin{bmatrix} 0 & 0 \\ 0 & 1 \end{bmatrix}.$$

In fact, B is already orthogonal, so application of the Gram-Schmidt process to it would yield $B' = B$ and $B'' = \{\mathbf{w}_1,\mathbf{w}_2,\mathbf{w}_3\}$ with $\mathbf{w}_1 = \mathbf{u}_1$, $\mathbf{w}_3 = \mathbf{u}_3$, and (by normalizing)

$$\mathbf{w}_2 = 1/\sqrt{2}\begin{bmatrix} 0 & 1 \\ 1 & 0 \end{bmatrix}. \square$$

The following corollaries of the Gram-Schmidt process show that in any inner product space V, the computation of inner products is essentially a dot product, provided we represent vectors by their matrices with respect to an orthonormal basis for V.

COROLLARY 2

Let $B = \{\mathbf{v}_1,\ldots,\mathbf{v}_n\}$ be an orthonormal basis for an inner product space V. Then, if $\mathbf{u}$ and $\mathbf{v}$ are in V, $<\mathbf{u},\mathbf{v}> = [\mathbf{u}]_B\cdot[\mathbf{v}]_B$ (the dot product of column vectors).

PROOF By Theorem 7-3,

$$\mathbf{u} = \sum_{i=1}^{n} <\mathbf{u},\mathbf{v}_i>\mathbf{v}_i \text{ and } \mathbf{v} = \sum_{j=1}^{n} <\mathbf{v},\mathbf{v}_j>\mathbf{v}_j.$$

Thus,

$$<\mathbf{u},\mathbf{v}> = \sum_{i=1}^{n}\sum_{j=1}^{n} <\mathbf{u},\mathbf{v}_i><\mathbf{v},\mathbf{v}_j><\mathbf{v}_i,\mathbf{v}_j> = \sum_{i=1}^{n} <\mathbf{u},\mathbf{v}_i><\mathbf{v},\mathbf{v}_i>$$
$$= [\mathbf{u}]_B\cdot[\mathbf{v}]_B. \blacksquare$$

COROLLARY 3

Any inner product on an n-dimensional vector space V is induced by the dot product on R^n and some 1-1 linear transformation $T{:}V \to R^n$.

PROOF Let B be an orthonormal basis for V with respect to the given inner product on V, and let $B(n)$ be the standard basis for R^n. By Theorem 5-2, there is then a unique linear transformation $T{:}V \to R^n$ such that T maps B to $B(n)$. In fact,

$$T(\mathbf{v}) = [\mathbf{v}]_B.$$

T is easily seen to be 1-1 (readers who have covered section 5-3 will recognize T as the isomorphism $mat{:}V \to R^n$). Therefore, when $\mathbf{u}$ and $\mathbf{v}$ are in V,

$$<\mathbf{u},\mathbf{v}> = [\mathbf{u}]_B \cdot [\mathbf{v}]_B = T(\mathbf{u}) \cdot T(\mathbf{v}),$$

so that $<,>$ is the inner product on V induced by T and the dot product. ■

The ideas discussed below will be needed in the next section.

DEFINITION 7-6

Let W be a subspace of an inner product space V. The **orthogonal complement** $W\perp$ (read "W perp") of W in V is the set of vectors in V that are orthogonal to all vectors of W:

$$W\perp = \{\mathbf{v} \in V \mid <\mathbf{w},\mathbf{v}> = 0 \text{ for all } \mathbf{w} \text{ in } W\}.$$

FIGURE 7-3

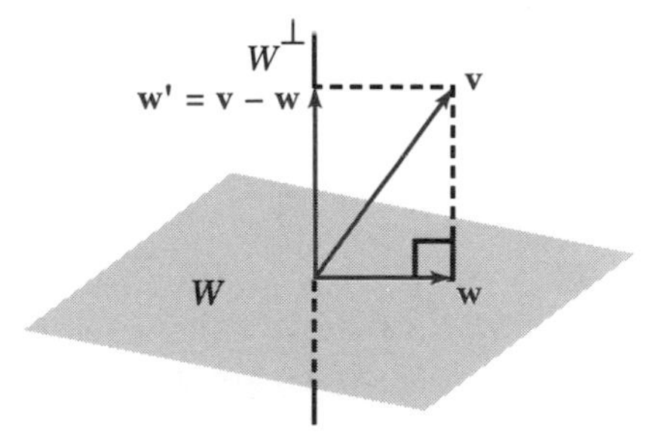

Thus, for example, $\{\mathbf{0}\}\perp = V$, $V\perp = \{\mathbf{0}\}$, and, intuitively, we see that if W is a plane through the origin in R^3 (with the dot product), then $W\perp$ is the unique line through the origin and perpendicular to this plane, as shown in Figure 7-3. Notice that in this case, $W\perp$ is a subspace, and every vector $\mathbf{v}$ in

R^3 can be expressed in exactly one way as $\mathbf{v} = \mathbf{w} + \mathbf{w}'$ where $\mathbf{w}$ is the orthogonal projection of $\mathbf{v}$ on W and $\mathbf{w}' = \mathbf{v} - \mathbf{w}$.

More generally,

THEOREM 7-5

If W is a subspace of a finite dimensional inner product space V, then so is $W\perp$, and *every* vector $\mathbf{v}$ in V may be *uniquely* expressed as $\mathbf{v} = \mathbf{w} + \mathbf{w}'$, with $\mathbf{w}$ in W and $\mathbf{w}'$ in $W\perp$.

We denote this fact by writing $V = W \oplus W\perp$, read "V is the direct sum of W and W perp" (see question 17, Exercises 6-3).

PROOF We leave the proof that $W\perp$ is a subspace of V for the reader (question 17(a), Exercises 7-2). Let B be an orthonormal basis for W, $B = \{\mathbf{v}_1,\ldots,\mathbf{v}_k\}$, and let $\mathbf{v}$ be an arbitrary vector in V. We set $\mathbf{w}$ equal to the orthogonal projection of $\mathbf{v}$ on W (Definition 7-5):

$$\mathbf{w} = \sum_{i=1}^{k} \langle \mathbf{v},\mathbf{v}_i \rangle \mathbf{v}_i.$$

As in the example above, let $\mathbf{w}' = \mathbf{v} - \mathbf{w}$. Then, $\mathbf{w}'$ is in $W\perp$. For,

$$\begin{aligned} \langle \mathbf{w}',\mathbf{v}_j \rangle &= \langle \mathbf{v} - \mathbf{w},\mathbf{v}_j \rangle \\ &= \langle \mathbf{v},\mathbf{v}_j \rangle - \sum_{i=1}^{k} \langle \mathbf{v},\mathbf{v}_i \rangle \langle \mathbf{v}_i,\mathbf{v}_j \rangle \\ &= \langle \mathbf{v},\mathbf{v}_j \rangle - \langle \mathbf{v},\mathbf{v}_j \rangle \quad (\text{as } \langle \mathbf{v}_i,\mathbf{v}_j \rangle = \delta_{ij}) \\ &= 0. \end{aligned}$$

So, $\mathbf{w}'$ is orthogonal to the basis B, which implies that it is in $W\perp$ (question 17(b), Exercises 7-2).

Consequently, $\mathbf{v} = \mathbf{w} + \mathbf{w}'$ is an expression of the required type. If $\mathbf{v} = \mathbf{u} + \mathbf{u}'$ were another such expression (with $\mathbf{u} \in W$ and $\mathbf{u}' \in W\perp$), then,

$$\mathbf{u} + \mathbf{u}' = \mathbf{w} + \mathbf{w}'$$

would imply

$$\mathbf{u} - \mathbf{w} = \mathbf{w}' - \mathbf{u}'.$$

As $\mathbf{u} - \mathbf{w}$ is in W while $\mathbf{w}' - \mathbf{u}'$ is in $W\perp$ (why?), it follows that $\mathbf{u} - \mathbf{w}$ and $\mathbf{w}' - \mathbf{u}'$ are in $W \cap W\perp$. But we can show that $W \cap W\perp = \{\mathbf{0}\}$ (question 17(c), Exercises 7-2), so we get $\mathbf{u} = \mathbf{w}$ and $\mathbf{u}' = \mathbf{w}'$, completing the proof. ■

COROLLARY

dim $W\perp$ = *dim* V − *dim* W.

PROOF This follows from questions 17 and 18, Exercises 6-3; it is left as an exercise (question 17(d)). ■

EXAMPLE 7-13

Let W be the subspace of R^3 (with the dot product) generated by $\mathbf{v}_1 = (1,1,0)$ and $\mathbf{v}_2 = (1,2,1)$. Then, $\mathbf{v} = (x,y,z)$ is in $W\perp$ iff $<\mathbf{v},\mathbf{v}_i> = \mathbf{v}\cdot\mathbf{v}_i = 0$ for $i = 1, 2$. This leads to the system of equations

$$\begin{aligned} x + y \quad &= 0 \qquad (i = 1) \\ x + 2y + z &= 0 \qquad (i = 2), \end{aligned}$$

or, after solving,

$$\begin{aligned} x \quad - z &= 0 \\ y + z &= 0. \end{aligned}$$

An example of a basis for the solution space is $\mathbf{u} = (1,-1,1)$, and $W\perp = span\,\{\mathbf{u}\}$. In other words, $W\perp$ is the line containing the vector $(1,-1,1)$. ❑

EXERCISES 7-2

1. Find $|\mathbf{u}|$ for each $\mathbf{u} \in R^5$ with the dot product.
   (a) $\mathbf{u} = (-1,2,0,3,-3)$ **(b)** $\mathbf{u} = (3,-2,4,6,1)$
2. Find $|\mathbf{u}|$ for each $\mathbf{u} \in R^2$ with the inner product of Example 7-2.
   (a) $\mathbf{u} = (2,3)$ **(b)** $\mathbf{u} = (0,1)$
3. **(a)** Find the norm of the matrix

   $$\begin{bmatrix} -1 & 3 & 2 \\ 4 & 0 & 5 \end{bmatrix}$$

   with respect to the usual inner product on $M(2\times 3)$.

   (b) Find $|f|$, where f in $P(4)$ with the standard inner product is given by $f(x) = -2x^4 + 3x^2 - 5$.
4. Prove property (2) in Theorem 7-2.
5. Show that equality holds in the Cauchy-Schwarz inequality iff $\mathbf{u}$ and $\mathbf{v}$ are linearly dependent.
6. Find the cosine of the angle between the following pairs of vectors; all spaces have their standard inner products.

   (a) $\mathbf{u} = (-1,2,0,4,1)$, $\mathbf{v} = (0,0,2,3,3)$ in R^5.

   (b) In $M(2)$,

   $$A = \begin{bmatrix} 1 & 2 \\ 3 & 4 \end{bmatrix}, B = \begin{bmatrix} -4 & 6 \\ 2 & 1 \end{bmatrix}.$$

(c) f and g in $P(3)$ with $f(x) = 2x^3 - x^2 + 3$ and $g(x) = 3x^2 + 1/2$, $x \in R$.

7. Find the cosine of the angle between the standard basis vectors in R^2, with the inner product of Example 7-2.

8. **(a)** Construct examples of two 3×3 skew-symmetric matrices that are orthogonal.

(b) What is the largest number of elements possible in an orthogonal set of nonzero, 3×3 skew-symmetric matrices?

9. Find the values of the number c such that

$$A = \begin{bmatrix} c & 8 \\ -3 & 1 \end{bmatrix} \text{ and } B = \begin{bmatrix} 2c & c \\ c & -3 \end{bmatrix}$$

are orthogonal.

10. Prove the identity

$$|\mathbf{u} + \mathbf{v}|^2 + |\mathbf{u} - \mathbf{v}|^2 = 2|\mathbf{u}|^2 + 2|\mathbf{v}|^2,$$

when **u** and **v** are in an inner product space. This is called the *parallelogram law*.

11. Prove the *polarization identity*

$$<\mathbf{u},\mathbf{v}> = (1/4)|\mathbf{u} + \mathbf{v}|^2 - (1/4)|\mathbf{u} - \mathbf{v}|^2,$$

for **u** and **v** in any inner product space V. It shows that the inner product is determined if the norm function is known on V.

In questions 12 through 15, use the Gram-Schmidt process to construct an orthonormal basis B'' for the given inner product space V from the given basis B. Also use Theorem 7-3 to find $[\mathbf{v}]_{B''}$ for the given vector **v**.

12. $V = R^2$ with the dot product; $B = \{\mathbf{u}_1,\mathbf{u}_2\}$ with $\mathbf{u}_1 = (1,2)$ and $\mathbf{u}_2 = (1,1)$; $\mathbf{v} = (-2,3)$.

13. $V = R^3$ with the dot product; $B = \{\mathbf{u}_1,\mathbf{u}_2,\mathbf{u}_3\}$ where $\mathbf{u}_1 = (-1,1,1)$, $\mathbf{u}_2 = (1,-2,1)$, and $\mathbf{u}_3 = (1,2,-3)$; $\mathbf{v} = (-2,3,-1)$.

14. V is the space of 2×2 symmetric matrices with the standard inner product; $B = \{\mathbf{u}_1,\mathbf{u}_2,\mathbf{u}_2\}$ and **v** are given by

$$\mathbf{u}_1 = \begin{bmatrix} 1 & 1 \\ 1 & 1 \end{bmatrix}, \mathbf{u}_2 = \begin{bmatrix} 1 & 1 \\ 1 & 0 \end{bmatrix}, \mathbf{u}_3 = \begin{bmatrix} 0 & 1 \\ 1 & 1 \end{bmatrix}, \mathbf{v} = \begin{bmatrix} 1 & 2 \\ 2 & 3 \end{bmatrix}.$$

15. $V = R^3$ with the inner product induced by the dot product and the linear transformation $T{:}V \to R^3$, given by $T(x,y,z) = (x + y, y + z, x + z)$; B is the standard basis for R^3; $\mathbf{v} = (-1,2,-3)$.

16. Let W be the subspace of R^5 with the dot product, having basis $B = \{\mathbf{u}_1,\mathbf{u}_2,\mathbf{u}_3\}$ where $\mathbf{u}_1 = (-1,0,1,1,0)$, $\mathbf{u}_2 = (1,-1,0,1,1)$, and $\mathbf{u}_3 = (0,0,1,1,0)$.

(a) Construct an orthonormal basis for W from B.

(b) Determine which of the following vectors $\mathbf{v}$ are in W using property (2) in Theorem 7-3. (Hint: The orthogonal projection of $\mathbf{v}$ on W must be $\mathbf{v}$.)
(i) $\mathbf{v} = (0,-1,2,3,1)$ **(ii)** $\mathbf{v} = (0,1,0,0,0)$ **(iii)** $\mathbf{v} = (0,0,0,0,1)$

(c) Find an orthonormal basis for R^5 containing B. (Hint: Use the results of (b).)

17. **(a)** Let W be a subspace of an inner product space V. Show that $W\perp$ is a subspace of V.

(b) Show that if $\mathbf{v}$ in V is orthogonal to every vector in a basis for W, then $\mathbf{v}$ is in $W\perp$.

(c) Show that $W \cap W\perp = \{\mathbf{0}\}$.

(d) If V has finite dimension, show that $\dim W\perp = \dim V - \dim W$. (Hint: Choose an orthonormal basis for W and extend it to a basis for V.)

(e) Show that if V is finite dimensional, then $(W\perp)\perp = W$.

18. Use Theorem 7-5 to show that the orthogonal projection $\mathbf{w}$ of a vector $\mathbf{v}$ on a subspace W of a finite dimensional vector space V is independent of the orthonormal basis for W used to compute it (Definition 7-5).

19. Let W be the subspace of symmetric matrices in $M(2)$ with the standard inner product. Find $W\perp$.

20. **(a)** Let W be the subspace of R^3 generated by $(1,1,1)$. Find a basis for $W\perp$ and the equation of $W\perp$ as a plane.

(b) Let W be the plane in R^3 given by the equation $2x + 3y + z = 0$. Find a basis for $W\perp$ and the parametric equations for $W\perp$ as a line.

21. Prove the following "Pythagoras' Theorem" for any inner product space V: If $\mathbf{u}$, $\mathbf{v}$ are orthogonal vectors in V, then $|\mathbf{u} + \mathbf{v}|^2 = |\mathbf{u}|^2 + |\mathbf{v}|^2$.

22. Let W be a subspace of an inner product space V, and let $\mathbf{v}$ be in V. Show that the orthogonal projection $\mathbf{w}$ of $\mathbf{v}$ on W is the "best approximation" to $\mathbf{v}$ in W in the sense that $|\mathbf{v} - \mathbf{w}| \leq |\mathbf{v} - \mathbf{w}'|$ for any vector $\mathbf{w}' \in W$. (Hint: Write $\mathbf{v} - \mathbf{w}' = (\mathbf{v} - \mathbf{w}) + (\mathbf{w} - \mathbf{w}')$ and use question 21.)

23. **(a)** Find an orthonormal basis for R^2 with the inner product of Example 7-2. Start the Gram-Schmidt process with the standard basis for R^2.

(b) Find an orthonormal basis for $P(1)$ with the inner product of question 7, Exercises 7-1. Start the process with the standard basis for $P(1)$.

24. Let V be an inner product space. Define the *distance* $d(\mathbf{u},\mathbf{v})$ between vectors $\mathbf{u}$ and $\mathbf{v}$ in V by $d(\mathbf{u},\mathbf{v}) = |\mathbf{u} - \mathbf{v}|$.

(a) In R^3 with the dot product, find $d(\mathbf{u},\mathbf{v})$ where $\mathbf{u} = (-1,2,3)$ and $\mathbf{v} = (-2,-2,5)$.

(b) In R^n with the dot product, show that

$$d(\mathbf{u},\mathbf{v}) = \{\sum_{i=1}^{n} (u_i - v_i)^2\}^{1/2}.$$

(c) If V is any inner product space, show that **(i)** $d(\mathbf{u},\mathbf{v}) \geq 0$, **(ii)** $d(\mathbf{u},\mathbf{v}) = 0$ iff $\mathbf{u} = \mathbf{v}$, **(iii)** $d(\mathbf{u},\mathbf{v}) = d(\mathbf{v},\mathbf{u})$, and **(iv)** $d(\mathbf{u},\mathbf{v}) \leq d(\mathbf{u},\mathbf{w}) + d(\mathbf{w},\mathbf{v})$, where **u**, **v**, and **w** are any vectors in V. Interpret (iv), the "triangle inequality", using a sketch for the case $V = R^2$.

7-3 Linear Operators on Inner Product Spaces

We now consider the properties of linear operators on an inner product space V. We shall see that, given a linear operator T on V, there is a natural way of obtaining a *new* operator T^* on V, called the **adjoint** of T. The various possible relationships between T and T^* are important in the classification of linear operators on V.

We begin by recalling that if $T{:}V \to W$ is a 1-1 linear transformation of inner product spaces, then it induces a new inner product $<,>_T$ on V by the rule $<\mathbf{u},\mathbf{v}>_T = <T(\mathbf{u}),T(\mathbf{v})>$ for **u** and **v** in V. Note that the right-hand inner product is the one on W. We now consider those linear transformations T for which this induced inner product *coincides* with the specified inner product $<,>$ on V. This happens iff $<\mathbf{u},\mathbf{v}> = <T(\mathbf{u}),T(\mathbf{v})>$ for all **u** and **v** in V. We say that "T preserves inner products". This leads to the definition below.

DEFINITION 7-7

A linear transformation $T{:}V \to W$ of inner product spaces is **isometric** if

$$<T(\mathbf{u}),T(\mathbf{v})> = <\mathbf{u},\mathbf{v}> \qquad (7\text{-}5)$$

for all **u** and **v** in V. If $V = W$ (so that T is a linear operator on V), then T is called an **orthogonal operator** when equation 7-5 holds.

EXAMPLE 7-14

Let $V = P(2)$ and W be the space of 2×2 symmetric matrices, both with their standard inner products. We define $T{:}V \to W$ by

$$T(a + bx + cx^2) = \begin{bmatrix} a & b/\sqrt{2} \\ b/\sqrt{2} & c \end{bmatrix}.$$

The reader may verify that T is linear. T is isometric since

$$<T(a + bx + cx^2), T(a' + b'x + c'x^2)>$$

$$= <\begin{bmatrix} a & b/\sqrt{2} \\ b/\sqrt{2} & c \end{bmatrix}, \begin{bmatrix} a' & b'/\sqrt{2} \\ b'/\sqrt{2} & c' \end{bmatrix}>$$

$$= aa' + bb' + cc'$$

$$= <a + bx + cx^2, a' + b'x + c'x^2>. \square$$

Notice that, by equation 7-5, an isometric linear transformation necessarily "preserves norms" of vectors, i.e., if $\mathbf{u}$ is in V, then

$$|T(\mathbf{u})|^2 = <T(\mathbf{u}),T(\mathbf{u})> = <\mathbf{u},\mathbf{u}> = |\mathbf{u}|^2.$$

Also, by Theorem 5-8, T must be 1-1 since $\mathbf{u} \neq \mathbf{0}$ implies $T(\mathbf{u}) \neq \mathbf{0}$. And, by equation 7-1, T also preserves angles between vectors.

THEOREM 7-6

A linear transformation $T:V \to W$ of inner product spaces is isometric iff $|T(\mathbf{v})| = |\mathbf{v}|$ for all $\mathbf{v} \in V$.

PROOF We have just seen that an isometric map preserves norms of vectors. To prove that T is isometric if it preserves norms, recall the polarization identity (question 11, Exercises 7-2):

$$<\mathbf{u},\mathbf{v}> = (1/4)|\mathbf{u} + \mathbf{v}|^2 - (1/4)|\mathbf{u} - \mathbf{v}|^2,$$

for $\mathbf{u}$ and $\mathbf{v}$ in any inner product space. Replacing $\mathbf{u}$ and $\mathbf{v}$ in this identity by $T(\mathbf{u})$ and $T(\mathbf{v})$, we obtain

$$\begin{aligned} <T(\mathbf{u}),T(\mathbf{v})> &= (1/4)|T(\mathbf{u}) + T(\mathbf{v})|^2 - (1/4)|T(\mathbf{u}) - T(\mathbf{v})|^2 \\ &= (1/4)|T(\mathbf{u} + \mathbf{v})|^2 - (1/4)|T(\mathbf{u} - \mathbf{v})|^2 \quad \text{(by linearity)} \\ &= (1/4)|\mathbf{u} + \mathbf{v}|^2 - (1/4)|\mathbf{u} - \mathbf{v}|^2 \quad \text{(since } T \text{ preserves norms)} \\ &= <\mathbf{u},\mathbf{v}>, \end{aligned}$$

by the polarization identity in V. Thus T is isometric. ∎

COROLLARY

If c is an eigenvalue of an orthogonal operator T, then $|c| = 1$.

PROOF Let $\mathbf{v}$ be an eigenvector associated with c. Then, $T(\mathbf{v}) = c\mathbf{v}$, so

$$|\mathbf{v}| = |T(\mathbf{v})| = |c\mathbf{v}| = |c||\mathbf{v}|,$$

by the theorem and property (1) of the norm (Theorem 7-2). Therefore, $|c| = 1$, since $\mathbf{v} \neq \mathbf{0}$. ■

EXAMPLE 7-15

(a) By Theorem 7-6, it is clear that any rotation of the plane is isometric (i.e., it is an orthogonal linear operator on R^2), since rotations preserve lengths.

(b) Similarly, we see intuitively that a linear operator that reflects vectors in a fixed line through the origin of R^2 is an orthogonal operator on R^2. For example, $T(x,y) = (y,x)$ is a reflection in the line $y = x$. The reader may verify that it preserves the dot product. ❑

In order to continue our study of linear operators on finite dimensional inner product spaces, we make use of the following important property of such spaces.

THEOREM 7-7

Let V be a finite dimensional inner product space. Then every linear map $f{:}V \to R$ may be expressed as $f(\mathbf{u}) = <\mathbf{u},\mathbf{v}>$ for $\mathbf{u} \in V$, where $\mathbf{v}$ is an *unique* fixed vector in V, depending only on f.

PROOF We first note that for every vector $\mathbf{v}$ in V, the function $f_{\mathbf{v}}{:}V \to R$ with $f_{\mathbf{v}}(\mathbf{u}) = <\mathbf{u},\mathbf{v}>$ for $\mathbf{u}$ in V is certainly linear. We must show that *every* linear $f{:}V \to R$ is of this type. Let $B = \{\mathbf{v}_1,\ldots,\mathbf{v}_n\}$ be an orthonormal basis for V. We define the vector $\mathbf{v}$ by

$$\mathbf{v} = \sum_{i=1}^{n} f(\mathbf{v}_i)\mathbf{v}_i.$$

Then,

$$f_{\mathbf{v}}(\mathbf{v}_j) = <\mathbf{v}_j,\mathbf{v}> = \sum_{i=1}^{n} f(\mathbf{v}_i)<\mathbf{v}_j,\mathbf{v}_i> = f(\mathbf{v}_j).$$

Therefore, $f_{\mathbf{v}}$ and f agree on the basis B, and so are equal (Theorem 5-2). Finally, if $\mathbf{w}$ were another vector such that $f = f_{\mathbf{w}}$, then we would have $f_{\mathbf{v}} = f_{\mathbf{w}}$ and so $<\mathbf{u},\mathbf{v}> = <\mathbf{u},\mathbf{w}>$ for all $\mathbf{u}$ in V. Alternatively, $<\mathbf{u},\mathbf{v} - \mathbf{w}> = 0$ for all $\mathbf{u} \in V$, so $\mathbf{v} - \mathbf{w}$ is orthogonal to all vectors in V. Thus, $\mathbf{v} - \mathbf{w} = \mathbf{0}$ by Theorem 7-3, Corollary, i.e., $\mathbf{v} = \mathbf{w}$. ■

Using this theorem, we can now prove the existence and uniqueness of the **cross product** in R^3.

COROLLARY (Theorem 3-14 The Cross Product in R^3)
Let $\mathbf{u} = (u_1,u_2,u_3)$ and $\mathbf{v} = (v_1,v_2,v_3)$ be arbitrary vectors in R^3. Then there is an unique vector $\mathbf{u}\times\mathbf{v}$ in R^3 which is completely characterized by the equation

$$(\mathbf{u}\times\mathbf{v})\cdot\mathbf{w} = det\begin{bmatrix} u_1 & u_2 & u_3 \\ v_1 & v_2 & v_3 \\ w_1 & w_2 & w_3 \end{bmatrix} \tag{7-6}$$

for any vector $\mathbf{w} = (w_1,w_2,w_3)$ in R^3.

PROOF (Optional) Let us define $D:R^3 \to R$ by $D(\mathbf{w})$ = the determinant in equation 7-6 above for any $\mathbf{w} \in R^3$. Then, the properties of the determinant function (Theorems 3-4 and 3-5) show that D is linear. We elaborate on what this means in the present context. We first show that if $\mathbf{w} = (w_1,w_2,w_3)$ and $\mathbf{w}' = (w'_1,w'_2,w'_3)$ are any two vectors in R^3, and c is a scalar, then

$$D(c\mathbf{w} + \mathbf{w}') = cD(\mathbf{w}) + D(\mathbf{w}').$$

By our definition of D,

$$D(c\mathbf{w} + \mathbf{w}') = det\begin{bmatrix} u_1 & u_2 & u_3 \\ v_1 & v_2 & v_3 \\ cw_1 + w'_1 & cw_2 + w'_2 & cw_3 + w'_3 \end{bmatrix}$$

$$= c\,det\begin{bmatrix} u_1 & u_2 & u_3 \\ v_1 & v_2 & v_3 \\ w_1 & w_2 & w_3 \end{bmatrix} + det\begin{bmatrix} u_1 & u_2 & u_3 \\ v_1 & v_2 & v_3 \\ w'_1 & w'_2 & w'_3 \end{bmatrix}$$

$$= cD(\mathbf{w}) + D(\mathbf{w}'),$$

as stated (we have used Theorems 3-4 and 3-5 in breaking up the above determinant).

Next, we apply Theorem 7-7 to $f = D$ on R^3 with the dot product, to conclude that there is an unique vector $\mathbf{p} \in R^3$ such that

$$f(\mathbf{w}) = D(\mathbf{w}) = \langle\mathbf{w},\mathbf{p}\rangle = \mathbf{w}\cdot\mathbf{p}, \text{ for all } \mathbf{w} \in R^3.$$

Of course, since $\mathbf{p}$ depends on the fixed vectors $\mathbf{u}$ and $\mathbf{v}$, we denote it by $\mathbf{u}\times\mathbf{v}$. Therefore,

$$D(\mathbf{w}) = \mathbf{w}\cdot(\mathbf{u}\times\mathbf{v}) = (\mathbf{u}\times\mathbf{v})\cdot\mathbf{w}, \text{ for all } \mathbf{w} \in R^3. \blacksquare$$

The interested reader is alerted to question 9, Exercises 7-3, in which this result is used to generalize the cross product of vectors in R^3 to a cross product in R^n.

The next example shows how to compute the vector $\mathbf{v}$ in Theorem 7-7.

EXAMPLE 7-16

Let $f{:}R^3 \to R$ be given by

$$f(x,y,z) = x - 2y + 3z.$$

Then, $f(\mathbf{e}_1) = 1$, $f(\mathbf{e}_2) = -2$, and $f(\mathbf{e}_3) = 3$. Therefore, we set $\mathbf{v} = (1,-2,3)$. If $\mathbf{u} = (x,y,z)$,

$$\mathbf{u}\cdot\mathbf{v} = x - 2y + 3z = f(\mathbf{u}),$$

showing that $f = f_{\mathbf{v}}$. ❑

Now let $T{:}V \to V$ be a linear operator and $\mathbf{v}$ be any vector in V. Then, the function $T'{:}V \to R$ with $T'(\mathbf{u}) = <T(\mathbf{u}),\mathbf{v}>$ for $\mathbf{u} \in V$ is clearly linear. By Theorem 7-7, there is an unique vector $\mathbf{w}$ in V such that

$$T'(\mathbf{u}) = <T(\mathbf{u}),\mathbf{v}> = f_{\mathbf{w}}(\mathbf{u}) = <\mathbf{u},\mathbf{w}> \text{ for all } \mathbf{u} \in V.$$

We *define* the function $T^*{:}V \to V$ by $T^*(\mathbf{v}) = \mathbf{w}$. Then, we can write the above as

$$<T(\mathbf{u}),\mathbf{v}> = <\mathbf{u},T^*(\mathbf{v})>, \text{ for all } \mathbf{u} \text{ and } \mathbf{v} \in V. \tag{7-7}$$

This is legitimate, since $\mathbf{w}$ is uniquely determined by $\mathbf{v}$ and T. In fact, T^* so defined is linear as justified below. If $\mathbf{v}$, $\mathbf{v}'$ are in V and c is a scalar, then for any vector $\mathbf{u}$,

$$\begin{aligned} <\mathbf{u},T^*(c\mathbf{v} + \mathbf{v}')> &= <T(\mathbf{u}),c\mathbf{v} + \mathbf{v}'> \\ &= c<T(\mathbf{u}),\mathbf{v}> + <T(\mathbf{u}),\mathbf{v}'> \\ &= c<\mathbf{u},T^*(\mathbf{v})> + <\mathbf{u},T^*(\mathbf{v}')> \\ &= <\mathbf{u},cT^*(\mathbf{v}) + T^*(\mathbf{v}')>. \end{aligned}$$

Since this is true for any $\mathbf{u}$ in V, it follows that $T^*(c\mathbf{v} + \mathbf{v}') = cT^*(\mathbf{v}) + T^*(\mathbf{v}')$, showing that T^* is linear.

Furthermore, T^* is the unique linear operator satisfying equation 7-7. For, if S were another linear operator such that $<T(\mathbf{u}),\mathbf{v}> = <\mathbf{u},S(\mathbf{v})>$ for all $\mathbf{u}$ and $\mathbf{v}$, then we would have $<\mathbf{u},S(\mathbf{v})> = <\mathbf{u},T^*(\mathbf{v})>$ for all $\mathbf{u}$ and $\mathbf{v}$. So, $S(\mathbf{v}) = T^*(\mathbf{v})$ for all $\mathbf{v}$, and we conclude that $S = T$.

DEFINITION 7-8

The linear operator T^* on V is called the **adjoint** of T. It is uniquely characterized by equation 7-7.

(Readers who have covered sections 5-3 and 5-5 will find a more elegant equivalent definition of T^* in question 10, Exercises 7-3.)

EXAMPLE 7-17

Let $T:R^2 \to R^2$ be the linear operator defined by $T(x,y) = (x + y, 2x - y)$ for (x,y) in R^2. Then,

$$[T] = \begin{bmatrix} 1 & 1 \\ 2 & -1 \end{bmatrix}.$$

We find T^* using appropriate choices for $\mathbf{u}$ and $\mathbf{v}$ (above) and the dot product for $<,>$. Since the standard basis is orthonormal, by property (2) in Theorem 7-3 and by equation 7-7,

$$\begin{aligned} T^*(\mathbf{e}_i) &= <T^*(\mathbf{e}_i),\mathbf{e}_1>\mathbf{e}_1 + <T^*(\mathbf{e}_i),\mathbf{e}_2>\mathbf{e}_2 \\ &= <\mathbf{e}_1,T^*(\mathbf{e}_i)>\mathbf{e}_1 + <\mathbf{e}_2,T^*(\mathbf{e}_i)>\mathbf{e}_2 \quad \text{(symmetry)} \\ &= <T(\mathbf{e}_1),\mathbf{e}_i>\mathbf{e}_1 + <T(\mathbf{e}_2),\mathbf{e}_i>\mathbf{e}_2, \end{aligned} \tag{7-8}$$

for $i = 1$ and 2.

Now $T(\mathbf{e}_1) = (1,2)$ and $T(\mathbf{e}_2) = (1,-1)$, so $<T(\mathbf{e}_1),\mathbf{e}_1> = 1$, $<T(\mathbf{e}_1),\mathbf{e}_2> = 2$, $<T(\mathbf{e}_2),\mathbf{e}_1> = 1$, and $<T(\mathbf{e}_2),\mathbf{e}_2> = -1$. Therefore, substitution into 7-8 with $i = 1$ and $i = 2$ yields

$$\begin{aligned} T^*(\mathbf{e}_1) &= 1\mathbf{e}_1 + 1\mathbf{e}_2 = (1,1), \\ T^*(\mathbf{e}_2) &= 2\mathbf{e}_1 - 1\mathbf{e}_2 = (2,-1). \end{aligned}$$

So,

$$[T^*] = \begin{bmatrix} 1 & 2 \\ 1 & -1 \end{bmatrix},$$

and $T^*(x,y) = (x + 2y, x - y)$ for (x,y) in R^2. ❑

It is no coincidence that $[T^*] = [T]^t$ in this example. In fact, the argument used in the example can be generalized to prove the next theorem.

THEOREM 7-8

Let V be a finite dimensional inner product space and T a linear operator on V. Then

$$[T^*;B] = [T;B]^t$$

provided B is an orthonormal basis for V.

PROOF Question 13(a), Exercises 7-3. ■

COROLLARY

Let A be an $n \times n$ matrix. Then $T_A{}^* = T_{A^t}$.

PROOF Question 14, Exercises 7-3. ■

EXAMPLE 7-18

Let $T{:}R^2 \to R^2$ be the rotation through an angle θ in a counterclockwise direction. Then, as seen from Examples 5-7 and 5-15(b),

$$[T] = \begin{bmatrix} \cos\theta & -\sin\theta \\ \sin\theta & \cos\theta \end{bmatrix}.$$

So (with the dot product on R^2),

$$[T^*] = \begin{bmatrix} \cos\theta & \sin\theta \\ -\sin\theta & \cos\theta \end{bmatrix} = \begin{bmatrix} \cos(-\theta) & -\sin(-\theta) \\ \sin(-\theta) & \cos(-\theta) \end{bmatrix}.$$

Therefore, T^* is a rotation through θ in a *clockwise* direction. ❑

We also see that, in this case, T^* is the inverse T^{-1} of T (see Definition 5-11 and Example 5-32(a)). In order to make the following discussion relatively self-contained (for the benefit of those who have not read section 5-3), we note that the inverse linear operator T^{-1} of a linear operator T on V exists iff T is 1-1 (*or* onto), and is completely characterized by the fact that

$$TT^{-1} = T^{-1}T = 1_V.$$

Therefore, we may write for the orthogonal operator (rotation) in the preceding example that $T^* = T^{-1}$ (or, $T^*T = TT^* = 1_V$), since a rotation through θ followed by a rotation through $-\theta$ has the same values as the identity operator. The next result establishes this equation as a criterion for orthogonality of *any* linear operator.

THEOREM 7-9

A linear operator T on a finite dimensional inner product space is orthogonal iff $T^*T = TT^* = 1_V$.

PROOF First suppose that T is orthogonal. Then, as we have seen, T is 1-1, since it preserves norms. Thus T^{-1} exists and $TT^{-1} = T^{-1}T = 1_V$. For any $\mathbf{u}$ and $\mathbf{v}$ in V, we then have

$$<T(\mathbf{u}),\mathbf{v}> = <T(\mathbf{u}),T(T^{-1}(\mathbf{v}))> = <\mathbf{u},T^{-1}(\mathbf{v})>$$

(as T preserves inner products). Thus T^{-1} satisfies equation 7-7 for T^*, and by the uniqueness of T^*, we conclude that $T^* = T^{-1}$, so $T^*T = TT^* = I_V$.

Conversely, if the condition holds, then

$$<T(\mathbf{u}),T(\mathbf{v})> = <\mathbf{u},T^*T(\mathbf{v})> = <\mathbf{u},\mathbf{v}>$$

for any $\mathbf{u}$ and $\mathbf{v} \in V$, so T is orthogonal. ■

COROLLARY

T is orthogonal iff its matrix A with respect to some (and hence any*) orthonormal basis B for V satisfies $AA^t = A^tA = I$, the identity matrix (i.e., $A^{-1} = A^t$).*

PROOF The stated result follows at once when we take matrices. Details are left for the reader. ■

DEFINITION 7-9

An $n \times n$ matrix A is **orthogonal** iff $AA^t = I$ (iff $A^tA = I$).

So, the corollary above states that a linear operator on a finite dimensional inner product space is orthogonal iff its matrix with respect to any orthonormal basis is orthogonal. In particular, a matrix transformation T_A on R^n is orthogonal iff A is.

EXAMPLE 7-19

We can verify by multiplication of A and its transpose that the following matrices A are orthogonal.

(a) $\begin{bmatrix} \cos\theta & -\sin\theta \\ \sin\theta & \cos\theta \end{bmatrix}$ **(b)** $\begin{bmatrix} 0 & 1 \\ 1 & 0 \end{bmatrix}$ **(c)** $\begin{bmatrix} 1/\sqrt{2} & 1/\sqrt{2} \\ -1/\sqrt{2} & 1/\sqrt{2} \end{bmatrix}$

(d) $\begin{bmatrix} 1/\sqrt{3} & 1/\sqrt{3} & 1/\sqrt{3} \\ -1/\sqrt{6} & -1/\sqrt{6} & 2/\sqrt{6} \\ -1/\sqrt{2} & 1/\sqrt{2} & 0 \end{bmatrix}$ **(e)** I

Of course, (a) and (c) are the matrices of rotations, while (b) is a reflection in the line $y = x$. Notice that in all cases, the rows (and columns) constitute an orthonormal set of vectors when regarded as vectors in R^n (with the appropriate value of n). ❑

More generally,

LEMMA 7-1

An $n \times n$ matrix A is orthogonal iff its rows and columns are orthonormal sets of n vectors in R^n.

PROOF We prove the result for rows. The result for columns then follows from the fact that A is orthogonal iff its transpose is. Now, suppose that A is orthogonal. The ith row $A_{[i]}$ of A has entries a_{ik}, $k = 1, \ldots, n$, and letting $A^t = (a^t_{ij})$,

$$\begin{aligned} <A_{[i]}, A_{[j]}> &= A_{[i]} \cdot A_{[j]} \\ &= \sum_{k=1}^{n} a_{ik} a_{jk} = \sum_{k=1}^{n} a_{ik} a^t_{kj} = (AA^t)_{ij} = \delta_{ij}, \end{aligned}$$

so the rows of A form an orthonormal set.

Conversely, if the rows of A form an orthonormal set, the above equations show that A is orthogonal. ■

Note that by Theorem 7-6, Corollary, the eigenvalues of an orthogonal matrix have absolute value 1. Of course, an orthogonal matrix may have no eigenvalues at all (e.g., a suitable rotation). Another useful fact is that when A is orthogonal, $det\, A = \pm 1$. The reader may verify this for the matrices in Example 7-19. The general proof is left as an exercise. However, the converse is *not* true, as may be seen from

$$A = \begin{bmatrix} 2 & 1 \\ 1 & 1 \end{bmatrix},$$

which is not an orthogonal matrix.

EXERCISES 7-3

1. Let V be the space of 2×2 symmetric matrices and W be the space of 3×3 skew-symmetric matrices, both with the standard inner product. Show that $T{:}V \rightarrow W$ defined by

$$T(\begin{bmatrix} a & b \\ b & c \end{bmatrix}) = \begin{bmatrix} 0 & a/\sqrt{2} & b \\ -a/\sqrt{2} & 0 & c/\sqrt{2} \\ -b & -c/\sqrt{2} & 0 \end{bmatrix}$$

is an isometric linear map.

2. Let V be an n-dimensional inner product space and B be an orthonormal basis for V. Show that $mat{:}V \to R^n$ with $mat(\mathbf{v}) = [\mathbf{v}]_B$ is isometric, where R^n has the dot product.

3. Let $T{:}R^3 \to R^3$ be the linear operator defined by $T(\mathbf{e}_1) = \mathbf{e}_2$, $T(\mathbf{e}_2) = -\mathbf{e}_1$, and $T(\mathbf{e}_3) = -\mathbf{e}_3$. Show that T is orthogonal.

4. Which of the following linear operators on R^2 with the dot product are orthogonal? Justify your answer.

(a) reflection in the origin: $T(x,y) = (-x,-y)$

(b) projection onto the x axis: $T(x,y) = (x,0)$

(c) a scalar transformation

5. **(a)** Let $V = R^2$ with the inner product of Example 7-2 and $W = R^2$ with the dot product. Is $T{:}V \to W$ as defined in Example 7-5(b) an isometric linear map?

(b) Is a rotation of R^2 through 90° in a counterclockwise direction an orthogonal operator if R^2 has the inner product of Example 7-2?

6. Let $T{:}V \to W$ be a linear transformation of inner product spaces of the same finite dimension. Show that T is isometric iff T maps orthonormal bases for V to orthonormal bases for W.

7. **(a)** Let $f{:}R^3 \to R$ be the linear map (functional) defined by $f(x,y,z) = -x + 3y - 4z$. Find a vector $\mathbf{v}$ in R^3 such that $f(\mathbf{u}) = \mathbf{u} \cdot \mathbf{v}$ for all $\mathbf{u} \in R^3$.

(b) Define $E{:}P(2) \to R$ by $E(g) = g(1)$ for g in $P(2)$ (see Example 5-42). Find a polynomial function p in $P(2)$ such that $E(h) = <h,p>$ for h in $P(2)$, where $<,>$ is the standard inner product on $P(2)$.

8. (For readers who have covered sections 5-3 and 5-5) Let V be a finite dimensional inner product space. Show that the function $*{:}V \to V^*$, defined by $*(\mathbf{v}) = f_\mathbf{v}$ where $f_\mathbf{v}(\mathbf{u}) = <\mathbf{u},\mathbf{v}>$ for $\mathbf{u}$ and $\mathbf{v} \in V$, is an isomorphism.

9. Let R^n have the dot product.

(a) Given $n - 1$ vectors $\mathbf{u}_1, \ldots, \mathbf{u}_{n-1}$ in R^n, show that there is an unique vector $\mathbf{v}$ in R^n such that for all $\mathbf{u} \in R^n$,

$$<\mathbf{u},\mathbf{v}> = det \begin{bmatrix} \mathbf{u}_1 \\ \cdots \\ \mathbf{u}_{n-1} \\ \mathbf{u} \end{bmatrix}.$$

The vector $\mathbf{v}$ is called the *cross product* of $\mathbf{u}_1, \ldots, \mathbf{u}_{n-1}$ and is denoted by $\mathbf{u}_1 \times \ldots \times \mathbf{u}_{n-1}$.

(b) Show each of the following.

(i) If σ is a permutation of $\{1,\ldots,n-1\}$ then $\mathbf{u}_{\sigma(1)} \times \ldots \times \mathbf{u}_{\sigma(n-1)} = (sgn\ \sigma)\mathbf{u}_1 \times \ldots \times \mathbf{u}_{n-1}$ (skew-symmetry).

(ii) $\mathbf{u}_1 \times \ldots \times (c\mathbf{u}_i + \mathbf{u}'_i) \times \ldots \times \mathbf{u}_{n-1}$
$= c(\mathbf{u}_1 \times \ldots \times \mathbf{u}_i \times \ldots \times \mathbf{u}_{n-1}) + (\mathbf{u}_1 \times \ldots \times \mathbf{u}'_i \times \ldots \times \mathbf{u}_{n-1})$ for $i = 1, \ldots, n-1$ (linearity).

10. (For readers who have read sections 5-3 and 5-5) With the notation of question 8, let $T: V \to V$ be a linear operator. Show that $T^* = *^{-1}T^t*$ (composition of maps).

11. Use the method of Example 7-17 to compute T^*, where T on R^2 is given by $T(x,y) = (2x - y, x + 3y)$ and R^2 has the dot product.

12. **(a)** Find 0^* and 1_V^*.

(b) Construct an example to show that $[T^*;B] \neq [T;B]^t$ when B is not orthonormal.

13. **(a)** Prove Theorem 7-8 using property (2) in Theorem 7-3 as follows: If $B = \{\mathbf{v}_1, \ldots, \mathbf{v}_n\}$ is an orthonormal basis for V, let $[T;B] = A = (a_{ij})$. Then,

$$T^*(\mathbf{v}_i) = \sum_{j=1}^{n} <T^*(\mathbf{v}_i), \mathbf{v}_j> \mathbf{v}_j = \sum_{j=1}^{n} <\mathbf{v}_i, T(\mathbf{v}_j)> \mathbf{v}_j.$$

Now substitute

$$T(\mathbf{v}_j) = \sum_{k=1}^{n} a_{kj} \mathbf{v}_k.$$

(b) (For readers who have read sections 5-3 and 5-5) Prove Theorem 7-8 using the equivalent definition of T^* given in question 10.

14. Prove the corollary of Theorem 7-8.

15. (For readers who have read section 5-3) Let S and T be linear operators on a finite dimensional inner product space V and let c be a scalar. Show the following.

(a) $(S + T)^* = S^* + T^*$ **(b)** $(cS)^* = cS^*$
(c) $(ST)^* = T^*S^*$ **(d)** $(T^*)^* = T$
(e) If T is invertible, $(T^*)^{-1} = (T^{-1})^*$.

16. Which of the following matrices are orthogonal?

(a) $\begin{bmatrix} 1 & 0 \\ 0 & 0 \end{bmatrix}$ **(b)** $\begin{bmatrix} 1/\sqrt{5} & 2/\sqrt{5} \\ -2/\sqrt{5} & 1/\sqrt{5} \end{bmatrix}$ **(c)** $\begin{bmatrix} 0 & 1 & 0 \\ 1 & 0 & 0 \\ 0 & 0 & 1 \end{bmatrix}$

(d) $\begin{bmatrix} 1/\sqrt{2} & 0 & 1/\sqrt{6} & 1/\sqrt{3} \\ 0 & 1 & 0 & 0 \\ -1/\sqrt{3} & 1/\sqrt{3} & 1/\sqrt{3} & 0 \\ 0 & 0 & -2/\sqrt{6} & 1/\sqrt{3} \end{bmatrix}$ **(e)** $1/2\begin{bmatrix} 1 & -1 & 1 & -1 \\ -1 & 1 & -1 & 1 \\ 1 & 1 & 1 & 1 \\ -1 & -1 & 1 & 1 \end{bmatrix}$

17. Show that if A is orthogonal, $det\, A = +1$ or -1.

18. Use question 6 to show that if B and B' are orthonormal bases for an inner product space V, then $[B'/B]$ is orthogonal.

19. Prove that products and inverses of orthogonal matrices are also orthogonal.

20. Construct an orthogonal 3×3 matrix whose first row is $(1/\sqrt{2}, 1/\sqrt{2}, 0)$.

21. Prove that every 2×2 orthogonal matrix A with $det\, A = +1$ is the matrix of a rotation of the plane. (Hint: Show that A may be written as

$$\begin{bmatrix} \cos\theta & -\sin\theta \\ \sin\theta & \cos\theta \end{bmatrix}$$

for some number θ, using the given conditions on A.)

22. In this question we indicate how any nonsingular $n \times n$ matrix A may be factored by the Gram-Schmidt process as $A = QR$, where Q is an $n \times n$ orthogonal matrix and R is an $n \times n$ upper triangular matrix. This factorization is called the *QR factorization* of A. It is important in numerical methods for solving the system of equations $AX = B$. For simplicity, we illustrate it for the 3×3 case. Let the columns of A be $\mathbf{u}_1$, $\mathbf{u}_2$, and $\mathbf{u}_3$. Then we may write

$$A^t = \begin{bmatrix} \mathbf{u}_1 \\ \mathbf{u}_2 \\ \mathbf{u}_3 \end{bmatrix}.$$

(a) With the notation of Theorem 7-4, show that $\mathbf{u}_1$, $\mathbf{u}_2$, and $\mathbf{u}_3$ may be expressed in terms of the orthonormal basis vectors $\mathbf{w}_1$, $\mathbf{w}_2$, and $\mathbf{w}_3$ constructed by the Gram-Schmidt process as follows:

$$\begin{aligned} \mathbf{u}_1 &= |\mathbf{v}_1|\mathbf{w}_1 \\ \mathbf{u}_2 &= <\mathbf{u}_2, \mathbf{w}_1>\mathbf{w}_1 + |\mathbf{v}_2|\mathbf{w}_2 \\ \mathbf{u}_3 &= <\mathbf{u}_3, \mathbf{w}_1>\mathbf{w}_1 + <\mathbf{u}_3, \mathbf{w}_2>\mathbf{w}_2 + |\mathbf{v}_3|\mathbf{w}_3. \end{aligned}$$

(Hint: Use equations 7-2 and 7-3.)

(b) Conclude that $A^t = R^t Q^t$, where

$$R^t = \begin{bmatrix} |\mathbf{v}_1| & 0 & 0 \\ <\mathbf{u}_2, \mathbf{w}_1> & |\mathbf{v}_2| & 0 \\ <\mathbf{u}_3, \mathbf{w}_1> & <\mathbf{u}_3, \mathbf{w}_2> & |\mathbf{v}_3| \end{bmatrix}; \; Q^t = \begin{bmatrix} \mathbf{w}_1 \\ \mathbf{w}_2 \\ \mathbf{w}_3 \end{bmatrix}.$$

Thus, $A = QR$, Q is orthogonal, and R is upper triangular.

(c) Generalize the method for the $n \times n$ case.

23. Use question 22 to find the QR factorization of the matrix

$$A = \begin{bmatrix} -1 & 2 & -1 \\ 1 & 1 & 2 \\ 1 & 0 & 2 \end{bmatrix}.$$

7-4 Self-adjoint Operators

We next examine the diagonalization problem for linear operators on finite dimensional inner product spaces. While we already have numerous criteria for diagonalization (Chapter 6), it is of interest in many applications (especially in physics and engineering) to know whether or not there is an orthonormal basis of eigenvectors for a linear operator on an inner product space. Briefly, the reason for this is that in physical problems, it is natural to make measurements of phenomena in mutually orthogonal directions.

Our main goal in this section will be to show that the linear operators for which such a basis exists are the "self-adjoint" ones.

DEFINITION 7-10

A linear operator T on an inner product space V is **self-adjoint** iff $T = T^*$.

Since $[T^*;B] = [T;B]^t$ when B is orthonormal (Theorem 7-8), it follows that the self-adjoint operators are those whose matrices with respect to any orthonormal basis are symmetric. Thus, the theorems of this section deal with symmetric matrices. In particular, a matrix transformation $T_A:R^n \to R^n$ is self-adjoint iff A is a symmetric matrix.

The following example illustrates an important special case.

EXAMPLE 7-20

Consider first the simple orthogonal projection $T:R^2 \to R^2$ onto the x axis, with $T(x,y) = (x,0)$. Then,

$$[T] = \begin{bmatrix} 1 & 0 \\ 0 & 0 \end{bmatrix},$$

so, certainly T is self-adjoint. ❑

We can generalize this example as follows. Recall (Definition 7-5) that if W is a subspace of an n-dimensional inner product space V, and $B = \{\mathbf{v}_1,\ldots,\mathbf{v}_k\}$ is an orthonormal basis for W, then the orthogonal projection on W of a vector $\mathbf{v}$ in V is

$$\sum_{i=1}^{k} <\mathbf{v},\mathbf{v}_i>\mathbf{v}_i.$$

We can therefore define a function $T:V \to V$ by

$$T(\mathbf{v}) = \sum_{i=1}^{k} <\mathbf{v},\mathbf{v}_i>\mathbf{v}_i,\ \mathbf{v} \in V.$$

We leave it for the reader to prove that T so defined is linear and that $Im\ T = W$, $Ker\ T = W^{\perp}$. T is called the **orthogonal projection operator** onto W.

To see that T is self-adjoint, we may extend the basis B for W to an orthonormal basis $B' = \{\mathbf{v}_1,\ldots,\mathbf{v}_k,\ldots,\mathbf{v}_n\}$ for V ($k \le n$). Then,

$$T(\mathbf{v}_i) = \begin{cases} \mathbf{v}_i, & 1 \le i \le k, \\ \mathbf{0}, & k < i \le n. \end{cases}$$

For, if $\mathbf{v}$ is in W,

$$\mathbf{v} = \sum_{i=1}^{k} <\mathbf{v},\mathbf{v}_i>\mathbf{v}_i = T(\mathbf{v})$$

(Theorem 7-3), while if $\mathbf{v}$ is not in W, $T(\mathbf{v}) = \mathbf{0}$. Therefore,

$$[T;B'] = \begin{bmatrix} I & O \\ O & O \end{bmatrix},$$

where I is the $k \times k$ identity matrix and the O's are zero matrices. In particular, we see that $[T;B']$ is symmetric, so T is self-adjoint.

Our first objective is to show that every self-adjoint linear operator T on an n-dimensional inner product space (or any $n \times n$ symmetric matrix A) has at least one eigenvalue. When n is odd, this follows from Theorem B-5, Corollary, since the characteristic polynomial f then has at least one real zero. However, the general proof of this fact (n even or odd) requires one of several possible theorems of advanced mathematics, which we must quote without proof. The route we follow below is intuitively plausible and introduces some interesting new concepts. The reader who is primarily interested in computing orthonormal bases of eigenvectors for symmetric matrices may skip to Theorem 7-13 and the examples following it without loss of continuity.

The theorem we need to quote without proof (it requires calculus) is stated below.

THEOREM 7-10

Let A be an $n \times n$ matrix. There is a unit vector $X \in M(n \times 1)$, i.e., $|X| = 1$, with the dot product such that

$$|AY| \le |AX| \tag{7-9}$$

for all unit vectors $Y \in M(n \times 1)$. ■

Note that since $|AX|^2 = AX \cdot AX$, equation 7-9 may also be written as

$$AY \cdot AY \le AX \cdot AX, \text{ for all unit } Y \in M(n \times 1).$$

It is interesting to observe that the set S of unit vectors in $M(n \times 1) = R^n$ is the "unit sphere" in R^n with center at the origin. For example, with $n = 3$, let $X = (x,y,z)$. Then,

$$|X|^2 = x^2 + y^2 + z^2 = 1$$

iff X is a point on the sphere S in R^3 with center at the origin and radius 1. In R^2, S is, of course, the unit circle with center at (0,0). So, the theorem states that the function $|AY|$, $Y \in S$, takes on a maximum value at some X *in* S.

DEFINITION 7-11

We call the vector(s) X whose existence is asserted by Theorem 7-10 **maximal vector(s)** for A.

EXAMPLE 7-21

Consider the matrix

$$A = \begin{bmatrix} 2 & 0 \\ 0 & 1 \end{bmatrix}.$$

Let $X = (x,y)$ with $|X|^2 = x^2 + y^2 = 1$. Then $AX = (2x,y)$ and

$$AX \cdot AX = 4x^2 + y^2 = 3x^2 + 1.$$

This expression assumes its largest value when $x = +1$ or -1 (as $-1 \leq x \leq +1$). So, the vectors (1,0) and $(-1,0)$ are maximal vectors for A. Notice that they are also eigenvectors for A. ❑

DEFINITION 7-12

If A is an $n \times n$ matrix, its (matrix) **norm** $\|A\|$ is defined by $\|A\| = |AX|$, where X is a maximal vector for A.

So, for the matrix of the preceding example, $\|A\| = 2$. Note that $\|A\| \neq |A|$, the norm of A as a vector in $M(n)$ with the usual inner product.

LEMMA 7-2

If Y is any vector in $M(n \times 1)$, $|AY| \leq \|A\| \, |Y|$.

PROOF As $(1/|Y|)Y$ is a unit vector,

$$|A(1/|Y|)Y| \leq \|A\|.$$

But, by properties of norms of vectors,

$$|A(1/|Y|)Y| = (1/|Y|)|AY|.$$

Therefore, $(1/|Y|)|AY| \leq \|A\|$, or, $|AY| \leq \|A\|\,|Y|$ as stated. ■

The result we want is a corollary of the next theorem.

> **THEOREM 7-11**
>
> Let A be an $n \times n$ symmetric matrix and X be a maximal vector for A. Then X is an eigenvector of A^2 associated with the eigenvalue $\|A\|^2$.

PROOF We first note that if A is nonzero and symmetric, then A^2 is nonzero and symmetric (question 7, Exercises 7-4). Next,

$$\begin{aligned}
\|A\|^2 &= |AX|^2 \\
&= AX \cdot AX \\
&= (T_A X)\cdot(T_A X) \\
&= (T_A{}^2 X)\cdot X \quad \text{(since } T_A \text{ is self-adjoint)} \\
&= (A^2 X)\cdot X \\
&\leq |A^2 X|\,|X| \quad \text{(by Cauchy-Schwarz)} \\
&= |A^2 X| \quad \text{(since } |X| = 1\text{)} \\
&= |A(AX)| \\
&\leq \|A\|\,|AX| \quad \text{(by Lemma 7-2)} \\
&= \|A\|^2 \quad \text{(since } |AX| = \|A\|\text{)}.
\end{aligned}$$

In summary, we have

$$\|A\|^2 = (A^2 X)\cdot X \leq |A^2 X|\,|X| \leq \|A\|^2. \qquad (7\text{-}10)$$

Since the two ends of the inequalities above are the same, they must all be equalities, so that

$$(A^2 X)\cdot X = |A^2 X|\,|X|.$$

It follows that A^2X and X must be linearly dependent: $A^2X = cX$ for some scalar c (question 5, Exercises 7-2). This shows that X is an eigenvector of A^2 associated with eigenvalue c. To find c, we take the dot product of both sides of this equation with X:

$$A^2 X \cdot X = (cX)\cdot X = c,$$

as $|X|^2 = 1$. Therefore, $c = \|A\|^2$ by 7-10, completing the proof. ■

COROLLARY 1

A has eigenvalue $+\|A\|$ or $-\|A\|$.

PROOF If X is a maximal eigenvector for A, then, by the theorem, $A^2X = \|A\|^2X$, or,

$$(A^2 - \|A\|^2I)X = O.$$

We may factorize the left side of this equation to yield

$$(A + \|A\|I)(A - \|A\|I)X = O. \qquad (7\text{-}11)$$

If $(A - \|A\|I)X = O$, then X is an eigenvector of A associated with the eigenvalue $\|A\|$. Otherwise, by equation 7-11, $(A - \|A\|I)X$ is an eigenvector of A associated with eigenvalue $-\|A\|$. ∎

(The preceding proof follows material in *Tensor Geometry* by C.T.J. Dodson and T. Poston, Pitman Publishing Ltd., 1977.)

COROLLARY 2

If T is a self-adjoint operator on a finite dimensional inner product space, then T has at least one eigenvalue.

PROOF Let B be an orthonormal basis for A. Then, $A = [T;B]$ is symmetric and, by Corollary 1, has an eigenvalue. This is also an eigenvalue of T. ∎

EXAMPLE 7-22

Consider the matrix A of Example 7-21,

$$A^2 = \begin{bmatrix} 4 & 0 \\ 0 & 1 \end{bmatrix},$$

which has eigenvalues 4 and 1. By Theorem 7-11, a maximal vector X for A is an eigenvector of A^2, so X must be associated with eigenvalue 4 or 1. We see that $X = (1,0)$ is a unit eigenvector associated with 4, while $X' = (0,1)$ is a unit eigenvector associated with 1. Since $|AX| = 2$ while $|AX'| = 1$, $X = (1,0)$ must be a maximal vector for A and $\|A\| = \sqrt{4} = 2$. This agrees with our findings in Example 7-21. ❑

Notice that for the above matrix A, the eigenvectors are orthogonal. The next result generalizes this observation.

THEOREM 7-12

If T is a self-adjoint linear operator on an inner product space V, then eigenvectors associated with *distinct* eigenvalues are orthogonal.

PROOF Let $\mathbf{v}$ and $\mathbf{v}'$ be eigenvectors of T associated with distinct eigenvalues c and c' respectively. Then,

$$T(\mathbf{v}) = c\mathbf{v},\ T(\mathbf{v}') = c'\mathbf{v}',$$

and

$$\begin{aligned} c<\mathbf{v},\mathbf{v}'> &= <c\mathbf{v},\mathbf{v}'> = <T(\mathbf{v}),\mathbf{v}'> = <\mathbf{v},T(\mathbf{v}')> \\ &= <\mathbf{v},c'\mathbf{v}'> = c'<\mathbf{v},\mathbf{v}'>. \end{aligned}$$

Consequently, $(c - c')<\mathbf{v},\mathbf{v}'> = 0$. Since $c \neq c'$, it follows that $<\mathbf{v},\mathbf{v}'> = 0$, so $\mathbf{v}$ and $\mathbf{v}'$ are orthogonal. ■

We now state the main idea of this chapter.

THEOREM 7-13

If T is a self-adjoint linear operator on an n-dimensional inner product space V, then there is an orthonormal basis for V consisting of eigenvectors of T.

PROOF (Outline) By Corollary 2 of Theorem 7-11, we know that T has an eigenvector $\mathbf{v}$. Let $\mathbf{v}' = (1/|\mathbf{v}|)\mathbf{v}$, so $\mathbf{v}'$ is a unit eigenvector of T. If V has dimension $n = 1$, the proof is complete. If not, let $W = span\ \{\mathbf{v}'\}$, a 1-dimensional subspace of V. Then, by Theorem 7-5, $V = W \oplus W^{\perp}$, and because T is self-adjoint, it induces (by restriction of its domain to $W^{\perp}$) a self-adjoint linear operator T' on $W^{\perp}$. As $dim\ W^{\perp} = n - 1$, by induction on the dimension of V, the theorem is true for T', yielding an orthonormal basis for $W^{\perp}$ of eigenvectors of T', which are also eigenvectors of T. Together with $\mathbf{v}'$, these yield the stated orthonormal basis for V. Hints for the complete proof are given in question 8, Exercises 7-4. ■

COROLLARY 1

Let A be an $n \times n$ symmetric matrix. There is an orthogonal matrix Q such that $Q^t AQ$ is a diagonal matrix D having the eigenvalues of A as entries.

PROOF By the theorem, there is an orthonormal basis B' for R^n of eigenvectors of the self-adjoint matrix transformation $T = T_A$.

So, $D = [T;B']$ is diagonal with the eigenvalues of A (and T) as entries. By Theorem 5-17,

$$D = [T;B'] = [B'/B]^{-1}[T;B][B'/B] = Q^{-1}AQ,$$

where B is the standard basis for R^n and $Q = [B'/B]$. But, by question 18, Exercises 7-3, Q is an orthogonal matrix, so that $Q^{-1} = Q^t$, completing the proof. ■

COROLLARY 2

Let T be a linear operator on a finite dimensional inner product space V. There is an orthonormal basis for V consisting of eigenvectors of T iff T is self-adjoint.

PROOF Question 9, Exercises 7-4. ■

DEFINITION 7-13

An $n \times n$ matrix A is **orthogonally similar** to an $n \times n$ matrix B iff there is an orthogonal matrix Q such that $A = Q^tBQ$.

The next corollary results from Corollaries 1 and 2.

COROLLARY 3

An $n \times n$ matrix A is orthogonally similar to a diagonal matrix iff A is symmetric. ■

In summary, a self-adjoint linear operator may be thought of geometrically as operating by scalar transformations (with scalars equal to the eigenvalues of T) in mutually orthogonal directions.

EXAMPLE 7-23

Consider the linear operator $T:R^2 \rightarrow R^2$ defined by

$$T(x,y) = (x + \sqrt{3}y, \sqrt{3}x - y).$$

We find an orthonormal basis for R^2 consisting of eigenvectors for T. Now,

$$A = [T] = \begin{bmatrix} 1 & \sqrt{3} \\ \sqrt{3} & -1 \end{bmatrix},$$

so the characteristic polynomial is

$$f = -1 + x^2 - 3 = x^2 - 4 = (x - 2)(x + 2).$$

Therefore, the eigenvalues are $+2$ and -2. An eigenvector for $c = 2$ is easily seen to be $\mathbf{u}_1 = (3, \sqrt{3})$, while an eigenvector for $c = -2$ is $\mathbf{u}_2 = (-\sqrt{3}, 3)$. Note that $\mathbf{u}_1 \perp \mathbf{u}_2$ (Theorem 7-12). We normalize to obtain the required orthonormal basis for R^2:

$$\mathbf{v}_1 = (1/\sqrt{12})(3, \sqrt{3}) = (1/2)(\sqrt{3}, 1),$$
$$\mathbf{v}_2 = (1/\sqrt{12})(-\sqrt{3}, 3) = (1/2)(-1, \sqrt{3}).$$

If we let

$$Q = \begin{bmatrix} \sqrt{3}/2 & -1/2 \\ 1/2 & \sqrt{3}/2 \end{bmatrix},$$

it follows that Q is orthogonal and

$$Q^tAQ = \begin{bmatrix} 2 & 0 \\ 0 & -2 \end{bmatrix},$$

which is a diagonal matrix. ❑

In general, it is clear that we may construct an orthonormal basis to diagonalize any self-adjoint linear operator T (or symmetric matrix A) by following these steps:

(1) Find bases for the eigenspaces of T (respectively, A). Their union is a basis B for the space in question.

(2) Construct orthonormal bases for the eigenspaces using the Gram-Schmidt procedure. Their union is the required orthonormal basis B'.

(3) Let Q be the transition matrix $[B'/B]$. Then Q^tAQ is diagonal.

EXAMPLE 7-24

We orthogonally diagonalize the matrix

$$A = \begin{bmatrix} 1 & 2 & -4 \\ 2 & -2 & -2 \\ -4 & -2 & 1 \end{bmatrix}.$$

The characteristic polynomial is seen to be

$$f = -x^3 + 27x + 54 = -(x + 3)^2(x - 6).$$

The eigenvalues are -3 and 6, and the reader may verify that two linearly independent eigenvectors for $c = -3$ are $\mathbf{u}_1 = (1,0,1)$ and $\mathbf{u}_2 = (1,-2,0)$. Applying the Gram-Schmidt process to these vectors yields $\mathbf{v}_1 = (1/\sqrt{2})(1,0,1)$ and $\mathbf{v}_2 = (\sqrt{2}/3)(1/2,-2,-1/2)$.

An eigenvector X associated with $c = 6$ is a solution of

$$(A - 6I)X = O,$$

which has coefficient matrix

$$\begin{bmatrix} -5 & 2 & -4 \\ 2 & -8 & -2 \\ -4 & -2 & -5 \end{bmatrix}.$$

This matrix is reduced to

$$\begin{bmatrix} 1 & 0 & 1 \\ 0 & 1 & 1/2 \\ 0 & 0 & 0 \end{bmatrix}.$$

So, an eigenvector for $c = 6$ is $\mathbf{u}_3 = (2,1,-2)$, which is normalized to $\mathbf{v}_3 = (1/3)(2,1,-2)$. Therefore $\{\mathbf{v}_1,\mathbf{v}_2,\mathbf{v}_3\}$ constitutes an orthonormal basis B' for R^3 of eigenvectors of A. When we set

$$Q = \begin{bmatrix} 1/\sqrt{2} & \sqrt{2}/6 & 2/3 \\ 0 & -2\sqrt{2}/3 & 1/3 \\ 1/\sqrt{2} & -\sqrt{2}/6 & -2/3 \end{bmatrix},$$

we find that

$$Q^tAQ = \begin{bmatrix} -3 & 0 & 0 \\ 0 & -3 & 0 \\ 0 & 0 & 6 \end{bmatrix}. \square$$

EXERCISES 7-4

1. Which of the following linear operators are self-adjoint? All spaces have the standard inner products.

(a) $T:R^2 \rightarrow R^2$ with $T(x,y) = (2x + 3y, 3x - y)$

(b) $T:R^2 \rightarrow R^2$ with $T(x,y) = (x + y, -x + y)$

(c) $T:R^3 \rightarrow R^3$ with $T(x,y,z) = (x + 2y + 3z, 2x + z, 3x + z)$

(d) $T:M(n) \rightarrow M(n)$ with $T(A) = A^t$

(e) $T:P(2) \rightarrow P(2)$ with $T(ax^2 + bx + c) = 2ax + b$

2. (a) Can a nonzero square matrix be both symmetric and nilpotent? Explain this.

(b) Can a square matrix be both orthogonal and symmetric? Explain this and characterize such matrices, if they exist.

3. Show that for a self-adjoint linear operator T, $T^2(\mathbf{v}) = \mathbf{0}$ implies $T(\mathbf{v}) = \mathbf{0}$.

4. Show that the orthogonal projection operator $T:V \rightarrow V$ (see definition after Example 7-20) is linear.

5. (a) If S and T are self-adjoint linear operators, show that ST is self-adjoint iff $ST = TS$.

(b) If A and B are $n \times n$ symmetric matrices, show that AB is symmetric iff $AB = BA$.

6. Find the maximal vectors and $\|A\|$ for the matrix

$$A = \begin{bmatrix} 2 & 1 \\ 1 & 2 \end{bmatrix}.$$

7. Prove that if A is a nonzero symmetric matrix, then $A^2 \neq O$.

8. (a) Let W be a subspace of a finite dimensional inner product space V such that W is *invariant* under a linear operator T on V (this means that $T(\mathbf{w}) \in W$ whenever $\mathbf{w} \in W$). Show that $W^{\perp}$ is invariant under T^*.

(b) Show that if $W \subseteq V$ is invariant under a linear operator T, then $T':W \rightarrow W$, defined by $T'(\mathbf{w}) = T(\mathbf{w})$, $\mathbf{w} \in W$, defines a linear operator T' on W, called the *restriction* of T to W.

(c) Use (a) and (b) above to complete the proof of Theorem 7-13.

9. Prove Corollary 2 of Theorem 7-13.

10. Show that the minimal polynomial of a symmetric matrix is a product of distinct linear factors.

11. For each of the following matrices A, find an orthogonal matrix Q such that $Q^t AQ = D$ is diagonal, and find D.

(a) $\begin{bmatrix} 1 & 2 \\ 2 & 1 \end{bmatrix}$ **(b)** $\begin{bmatrix} 2 & 6 \\ 6 & 18 \end{bmatrix}$

(c) $\begin{bmatrix} -1 & 4 & -8 \\ 4 & -7 & -4 \\ -8 & -4 & -1 \end{bmatrix}$ **(d)** $\begin{bmatrix} 1 & 2 & 2 \\ 2 & 4 & 4 \\ 2 & 4 & 4 \end{bmatrix}$

(e) $\begin{bmatrix} -1 & 1 & 1 & 1 \\ 1 & -1 & 1 & 1 \\ 1 & 1 & -1 & 1 \\ 1 & 1 & 1 & -1 \end{bmatrix}$

12. Find an orthogonal matrix Q such that the following matrices A are orthogonally diagonalized.

(a) $\begin{bmatrix} 1 & 1 & 1 \\ 1 & 1 & 1 \\ 1 & 1 & 1 \end{bmatrix}$ **(b)** $\begin{bmatrix} k & k & k \\ k & k & k \\ k & k & k \end{bmatrix}$

13. Let $T:M(2) \to M(2)$ be the linear operator that takes any matrix to its transpose. Construct an orthonormal basis for $M(2)$ of eigenvectors for T.

14. A self-adjoint operator on an inner product space V is *positive definite* iff $<T(\mathbf{v}),\mathbf{v}> > 0$ for all nonzero $\mathbf{v} \in V$.

(a) Show that T is positive definite iff the rule $<\mathbf{u},\mathbf{v}>' = <T(\mathbf{u}),\mathbf{v}>$ for $\mathbf{u}, \mathbf{v} \in V$ defines an inner product $<,>'$ on V.

(b) Show that if V is a finite dimensional inner product space with inner product $<,>$, then any other inner product $<,>'$ on V is given by the formula in (a) for some positive definite linear operator T. (Hint: Use Theorem 7-7 to define T.)

15. Let T be a linear operator on a finite dimensional inner product space V. Show that the following conditions on T are equivalent by proving (1) $\Rightarrow$ (2) $\Rightarrow$ (3) $\Rightarrow$ (4) $\Rightarrow$ (1).

(1) T is positive definite (question 14).

(2) T is self-adjoint and all its eigenvalues are positive numbers.

(3) $T = S^2$ for some self-adjoint invertible linear operator S on V.

(4) $T = S^*S$ for some invertible linear operator S on V.

16. Use question 15 to find an invertible matrix B such that $A = B^2$, where A is the matrix in question 6 above (B is called a *square root* of A).

7-5 Bilinear and Quadratic Forms (Optional)

In this section, we consider a natural generalization (called a **bilinear form**) of an inner product which is especially important in the sciences and engineering. It has numerous applications, including the theory of conic sections in the plane. The latter will be discussed at the end of this section (after Example 7-37); it is relatively independent of the intervening material, as long as the reader is willing to accept the relevant theorem.

Suppose V is a vector space. If we omit axioms (3) and (4) for inner products on V, we obtain the definition of a bilinear form on V.

DEFINITION 7-14

A **bilinear form** F on a vector space V is a function that assigns a scalar $F(\mathbf{u},\mathbf{v})$ to every pair of vectors, $\mathbf{u}$ and $\mathbf{v}$, from V such that

(1) $F(c\mathbf{u} + \mathbf{u}',\mathbf{v}) = cF(\mathbf{u},\mathbf{v}) + F(\mathbf{u}',\mathbf{v})$,

(2) $F(\mathbf{u},c\mathbf{v} + \mathbf{v}') = cF(\mathbf{u},\mathbf{v}) + F(\mathbf{u},\mathbf{v}')$,

for all $\mathbf{u}$, $\mathbf{u}'$, $\mathbf{v}$, $\mathbf{v}' \in V$ and scalars $c \in R$.

As in the case of inner products, the two axioms above mean that F is linear in each argument when the other argument is fixed. This leads to the name bilinear form. In fact, bilinear forms are particular examples of **tensors**, functions of k vector arguments ($k \geq 1$) which are linear in each argument. Tensors are fundamental concepts in physics and relativity theory.

Of course, any inner product is a bilinear form. The following are examples of bilinear forms that are not inner products.

EXAMPLE 7-25

Consider the matrix

$$A = \begin{bmatrix} 2 & 1 \\ 3 & 2 \end{bmatrix}.$$

We define the bilinear form F on R^2 by

$$F(\mathbf{u},\mathbf{v}) = [\mathbf{u}]^t A[\mathbf{v}]. \tag{7-12}$$

Let $\mathbf{u} = (u_1,u_2)$ and $\mathbf{v} = (v_1,v_2)$. Then,

$$F(\mathbf{u},\mathbf{v}) = \begin{bmatrix} u_1 & u_2 \end{bmatrix} \begin{bmatrix} 2 & 1 \\ 3 & 2 \end{bmatrix} \begin{bmatrix} v_1 \\ v_2 \end{bmatrix} = 2u_1v_1 + u_1v_2 + 3u_2v_1 + 2u_2v_2.$$

The fact that F is bilinear is evident from its defining equation 7-12 and from properties of matrix multiplication. For example,

$$\begin{aligned} F(c\mathbf{u} + \mathbf{u}',\mathbf{v}) &= [c\mathbf{u} + \mathbf{u}']^t A[\mathbf{v}] \\ &= c[\mathbf{u}]^t A[\mathbf{v}] + [\mathbf{u}']^t A[\mathbf{v}] \\ &= cF(\mathbf{u},\mathbf{v}) + F(\mathbf{u}',\mathbf{v}). \end{aligned}$$

Note that F is not an inner product because it does not satisfy axiom (3) for inner products. For example, if $\mathbf{u} = (1,0)$ and $\mathbf{v} = (1,1)$, then $F(\mathbf{u},\mathbf{v}) = 3 \neq 5 = F(\mathbf{v},\mathbf{u})$. However, the reader may verify that (here) F is positive definite: $F(\mathbf{u},\mathbf{u}) > 0$ unless $\mathbf{u} = \mathbf{0}$.

On the other hand, if we define the bilinear form G by equation 7-12 but using the matrix

$$A' = \begin{bmatrix} -1 & 0 \\ 0 & 0 \end{bmatrix},$$

then the reader may verify that G satisfies axiom (3) but not axiom (4) for inner products. ❑

More generally, if V is an arbitrary n-dimensional vector space with a basis B, and A is an $n \times n$ matrix, we may define a bilinear form F by

$$F(\mathbf{u},\mathbf{v}) = [\mathbf{u}]_B{}^t A[\mathbf{v}]_B. \tag{7-13}$$

We shall see shortly that *every* bilinear form on V may be described in this way.

EXAMPLE 7-26

Let V be a vector space and suppose that $f,g:V \to R$ are linear (i.e., f and g are linear functionals; see section 5-5). We define a bilinear form denoted $f \otimes g$ on V by

$$f \otimes g(\mathbf{u},\mathbf{v}) = f(\mathbf{u})g(\mathbf{v}) \text{ for } \mathbf{u}, \mathbf{v} \in V.$$

The reader may show that $f \otimes g$ is bilinear. It is called the **tensor product** of f and g.

For example, with $V = P(1)$, let $E_c:P(1) \to R$ be the evaluation at c:

$$E_c(ax + b) = ac + b, \; c \in R.$$

E_c is linear (see Example 5-42). And, for example,

$$E_{-2}(ax + b) = -2a + b \quad (c = -2);$$
$$E_1(ax + b) = a + b \quad (c = 1).$$

So,

$$E_{-2} \otimes E_1(ax + b, a'x + b') = (-2a + b)(a' + b')$$
$$= -2aa' + a'b - 2ab' + bb'.$$

Note that $E_{-2} \otimes E_1$ is not an inner product because, for example,

$$E_{-2} \otimes E_1(x,x) = (-2)(1) < 0. ❑$$

Tensor products of linear functionals are important because they can be used to construct a basis for the vector space of bilinear forms. These ideas are pursued in the exercises.

DEFINITION 7-15

Let F be a bilinear form on an n-dimensional vector space V with (ordered) basis $B = \{\mathbf{v}_1,\ldots,\mathbf{v}_n\}$. The **matrix** of F with respect to B is the $n \times n$ matrix $[F;B]$ with (i,j) entry $F(\mathbf{v}_i,\mathbf{v}_j)$, for $i, j = 1, \ldots, n$.

With the notation of the Definition 7-15, we get

THEOREM 7-14

For any vectors $\mathbf{u}$ and $\mathbf{v} \in V$,

$$F(\mathbf{u},\mathbf{v}) = [\mathbf{u}]_B{}^t[F;B][\mathbf{v}]_B.$$

PROOF Let

$$\mathbf{u} = \sum_{i=1}^{n} a_i\mathbf{v}_i \text{ and } \mathbf{v} = \sum_{j=1}^{n} b_j\mathbf{v}_j,$$

so that $[\mathbf{u}]_B = (a_i)$ and $[\mathbf{v}]_B = (b_j)$. Also, let $a_{ij} = F(\mathbf{v}_i,\mathbf{v}_j)$, so that $[F;B] = (a_{ij})$. Then,

$$F(\mathbf{u},\mathbf{v}) = \sum_{i-1}^{n}\sum_{j-1}^{n} a_ib_jF(\mathbf{v}_i,\mathbf{v}_j) = \sum_{i=1}^{n} a_i\,(\sum_{j=1}^{n} a_{ij}b_j) = [\mathbf{u}]_B{}^t[F;B][\mathbf{v}]_B$$

as stated. ■

It follows that every bilinear form on V may be said to be of the type considered in Example 7-25 (equation 7-13). In particular, if A is an $n \times n$ matrix and F is the bilinear form on R^n given by $F(X,Y) = X^tAY$, then $[F;B] = A$, where B is the standard basis for R^n. (Note that we are again identifying R^n and $M(n \times 1)$.)

EXAMPLE 7-27

We compute the matrix A (with respect to the standard basis) of the bilinear form of Example 7-26, $F = E_{-2}\otimes E_1$ defined on $P(1)$. Recall that the standard (ordered) basis for $P(1)$ is $B = \{1,x\}$. Then,

$$a_{11} = F(1,1) = (1)(1) = 1, a_{12} = F(1,x) = (1)(1) = 1,$$
$$a_{21} = F(x,1) = (-2)(1) = -2, a_{22} = F(x,x) = (-2)(1) = -2.$$

Therefore,

$$A = [F;B] = \begin{bmatrix} 1 & 1 \\ -2 & -2 \end{bmatrix}. \square$$

Of course, the matrix of F depends on the choice of B. The relationship between $[F;B]$ and $[F;B']$ for two choices of bases introduces a new and important relation among square matrices.

THEOREM 7-15

Let B, B' be two bases for an n-dimensional vector space V. If F is a bilinear form on V, then there is a nonsingular matrix Q such that

$$[F;B'] = Q^t[F;B]Q.$$

In fact, $Q = [B'/B]$.

PROOF This theorem is proved by a simple substitution using Theorem 5-16; details are left for the reader. ■

DEFINITION 7-16

Two $n \times n$ matrices A and B are **congruent** iff $B = Q^tAQ$ for an invertible matrix Q.

COROLLARY

Two $n \times n$ matrices A and B are congruent iff they are the matrices of some bilinear form F on R^n with respect to two bases for R^n.

PROOF We only need to show that if A and B are congruent, then they are the matrices of a bilinear form F on R^n. Define F by

$$F(X,Y) = X^tAY, \text{ for } X, Y \in R^n.$$

Then, $[F;B'] = A$ where B' (here) is the standard basis for R^n. We let B'' be the basis for R^n such that $[B''/B'] = Q$, where Q arises in $B = Q^tAQ$. Then, it follows that $[F;B''] = B$ as required. ■

Readers will no doubt have noted the close parallel between congruence and similarity of matrices (Theorem 5-18). Furthermore,

LEMMA 7-3

Congruent matrices have the same rank.

PROOF Question 8, Exercises 7-5. ■

As a consequence of Lemma 7-3 and Theorem 7-15, Corollary, the following definition arises.

> **DEFINITION 7-17**
>
> If F is a bilinear form on a finite dimensional vector space V, its **rank** is the rank of $[F;B]$ for any choice of basis B for V.

EXAMPLE 7-28

(a) If F is the bilinear form of Example 7-27, we see from its matrix A that it has rank 1.

(b) Let $<,>$ be an inner product on a vector space V of dimension n. Then $<,>$ is a bilinear form on V of rank n because, if $B = \{\mathbf{v}_1,\ldots,\mathbf{v}_n\}$ is an orthonormal basis for V, the matrix A of $<,>$ has entries

$$a_{ij} = <\mathbf{v}_i,\mathbf{v}_j> = \delta_{ij};$$

i.e., $A = I$, which has rank n. ❑

A bilinear form F on an n-dimensional vector space is said to be **nondegenerate** if its rank is n. Such bilinear forms share many of the properties of inner products and are important in numerous applications.

EXAMPLE 7-29

In Einstein's special theory of relativity, an observer describes (or models) the universe by considering not only the spatial location of objects in relation to himself, but also their temporal relation to himself. There is no "absolute" time (i.e., single clock) valid for all observers. The "points" in the observer's universe are thus called "events", and are represented by 4-tuples $\mathbf{u} = (x,y,z,t)$, where x, y, and z locate the event in space, and t locates it in the observer's time. In other words, an observer represents the universe by R^4, called "space-time".

Furthermore, the geometry of space-time is not determined by the Euclidean geometry of the dot product, but by the nondegenerate bilinear form g on R^4 whose matrix with respect to the standard basis for R^4 is

$$A = \begin{bmatrix} 1 & 0 & 0 & 0 \\ 0 & 1 & 0 & 0 \\ 0 & 0 & 1 & 0 \\ 0 & 0 & 0 & -1 \end{bmatrix}.$$

Clearly, g is nondegenerate; it is called the **Lorentz metric**. For example, the "distance" (see question 24, Exercises 7-2), or **space-time separation**, d, of an event $\mathbf{u}$ from the observer $\mathbf{0} = (0,0,0,0)$ is then given by $d = |g(\mathbf{u},\mathbf{u})|^{1/2}$ where

$$g(\mathbf{u},\mathbf{u}) = [x \quad y \quad z \quad t]\begin{bmatrix} 1 & 0 & 0 & 0 \\ 0 & 1 & 0 & 0 \\ 0 & 0 & 1 & 0 \\ 0 & 0 & 0 & -1 \end{bmatrix}\begin{bmatrix} x \\ y \\ z \\ t \end{bmatrix}$$

$$= x^2 + y^2 + z^2 - t^2.$$

(The absolute value in the formula is necessary because $g(\mathbf{u},\mathbf{u})$ may be negative.) ❑

The bilinear form in the above example has an additional special property called **symmetry**.

DEFINITION 7-18

A bilinear form F on a vector space V is **symmetric** iff $F(\mathbf{u},\mathbf{v}) = F(\mathbf{v},\mathbf{u})$ for all $\mathbf{u}$ and $\mathbf{v}$ in V.

EXAMPLE 7-30

Let F be *any* bilinear form on V. We define F' on V by $F'(\mathbf{u},\mathbf{v}) = (1/2)[F(\mathbf{u},\mathbf{v}) + F(\mathbf{v},\mathbf{u})]$ for $\mathbf{u}, \mathbf{v} \in V$. Then F' is a symmetric bilinear form on V. ❑

THEOREM 7-16

A bilinear form F on a finite dimensional space V is symmetric iff $[F;B]$ is a symmetric matrix for some basis B for V (and hence for any basis).

PROOF Suppose F is symmetric and let $B = \{\mathbf{v}_1,\ldots,\mathbf{v}_n\}$ be a basis for V. Since

$$F(\mathbf{v}_i,\mathbf{v}_j) = F(\mathbf{v}_j,\mathbf{v}_i), \ 1 \leq i, j \leq n,$$

it follows that $[F;B]$ is symmetric. Conversely, if $[F;B]$ is symmetric, and $\mathbf{u}$ and $\mathbf{v}$ are in V,

$$
\begin{aligned}
F(\mathbf{u},\mathbf{v}) &= [\mathbf{u}]_B{}^t[F;B][\mathbf{v}]_B \\
&= ([\mathbf{u}]_B{}^t[F;B][\mathbf{v}]_B)^t \quad \text{(as any } 1 \times 1 \text{ matrix is symmetric)} \\
&= [\mathbf{v}]_B{}^t[F;B][\mathbf{u}]_B \quad \text{(as } [F;B] = [F;B]^t\text{)} \\
&= F(\mathbf{v},\mathbf{u}),
\end{aligned}
$$

which means that F is symmetric. ■

COROLLARY 1

A bilinear form F on a finite dimensional vector space V is symmetric iff there is a basis B for V such that $[F;B]$ is diagonal.

PROOF If $[F;B]$ is diagonal for some B, then certainly F is symmetric. Conversely, if F is symmetric, $A = [F;B]$ is symmetric for any choice of basis (by the theorem). But, by Theorem 7-13, Corollary 1, there is an orthogonal matrix Q such that Q^tAQ is diagonal. Then, by Theorem 7-15, if we let B' be the basis such that $[B'/B] = Q$, we find that

$$[F;B'] = Q^t[F;B]Q = Q^tAQ$$

is diagonal, which completes the proof. ■

Note that neither B nor B' is orthonormal since V here is not assumed to have an inner product, and the notion has no meaning without an inner product.

In fact, we can state the stronger result that F is symmetric iff its matrix $[F;B]$ is **orthogonally similar** to a diagonal matrix D, for any basis B. For, as noted, the matrix Q above is orthogonal. Note that the matrix D has as diagonal entries the eigenvalues of $[F;B]$, but that *different* choices for B lead to *different* $[F;B]$ and diagonal matrices D (question 11, Exercises 7-5).

COROLLARY 2

If V is an inner product space, there is an orthonormal basis for V such that $[F;B]$ is diagonal. ■

It is also important to note that Corollary 1 may be proved without using the result about orthogonal diagonalization of symmetric matrices described earlier. Moreover, given a symmetric matrix A, there is an algorithm for computing a nonsingular matrix Q and a diagonal matrix D such that $Q^tAQ = D$. The method uses only elementary row and column operations; the matrix Q is *not* in general orthogonal. The steps are illustrated in the following example.

EXAMPLE 7-31

Let A be given by

$$A = \begin{bmatrix} 1 & 2 & -4 \\ 2 & -2 & -2 \\ -4 & -2 & 1 \end{bmatrix}.$$

We diagonalize A by performing successive elementary operations to the rows of A and the *same* operations to the columns of A (i.e., elementary column operations). The matrix Q^t is obtained by performing these same row operations (without the column operations) to the identity matrix I, as in the method for finding A^{-1}. We can do these at the same time by using the augmented matrix $(A{:}I)$:

$$(A{:}I) = \begin{bmatrix} 1 & 2 & -4 & 1 & 0 & 0 \\ 2 & -2 & -2 & 0 & 1 & 0 \\ -4 & -2 & 1 & 0 & 0 & 1 \end{bmatrix}$$

$$\to \begin{bmatrix} 1 & 2 & -4 & 1 & 0 & 0 \\ 0 & -6 & 6 & -2 & 1 & 0 \\ -4 & -2 & 1 & 0 & 0 & 1 \end{bmatrix} \to \begin{bmatrix} 1 & 0 & -4 & 1 & 0 & 0 \\ 0 & -6 & 6 & -2 & 1 & 0 \\ -4 & 6 & 1 & 0 & 0 & 1 \end{bmatrix}$$

(we replaced row 2 by row 2 $-$ 2$\times$row 1 in $(A{:}I)$ and column 2 by column 2 $-$ 2$\times$column 1)

$$\to \begin{bmatrix} 1 & 0 & -4 & 1 & 0 & 0 \\ 0 & -6 & 6 & -2 & 1 & 0 \\ 0 & 6 & -15 & 4 & 0 & 1 \end{bmatrix} \to \begin{bmatrix} 1 & 0 & 0 & 1 & 0 & 0 \\ 0 & -6 & 6 & -2 & 1 & 0 \\ 0 & 6 & -15 & 4 & 0 & 1 \end{bmatrix}$$

(we replaced row 3 by row 3 + 4$\times$row 1 and then column 3 by column 3 + 4$\times$column 1)

$$\to \begin{bmatrix} 1 & 0 & 0 & 1 & 0 & 0 \\ 0 & -6 & 6 & -2 & 1 & 0 \\ 0 & 0 & -9 & 2 & 1 & 1 \end{bmatrix} \to \begin{bmatrix} 1 & 0 & 0 & 1 & 0 & 0 \\ 0 & -6 & 0 & -2 & 1 & 0 \\ 0 & 0 & -9 & 2 & 1 & 1 \end{bmatrix}$$

(we replaced row 3 by row 3 + row 2 and column 3 by column 3 + column 2).

From the last matrix, we read that

$$D = \begin{bmatrix} 1 & 0 & 0 \\ 0 & -6 & 0 \\ 0 & 0 & -9 \end{bmatrix}, Q^t = \begin{bmatrix} 1 & 0 & 0 \\ -2 & 1 & 0 \\ 2 & 1 & 1 \end{bmatrix}. \square$$

The following result is important in many applications.

COROLLARY 3

If F is a symmetric bilinear form of rank r on an n-dimensional vector space V, then there is a basis B' for V such that $[F;B']$ is diagonal with entries $+1$, -1, and 0 only:

$$[F;B] = \begin{bmatrix} 1 & & & & & & & & \\ & \cdot & & & & & & \text{O} & \\ & & 1 & & & & & & \\ & & & \cdot & & & & & \\ & & & & -1 & & & & \\ & & & & & \cdot & & & \\ & & & & & & -1 & & \\ & & & & & & & \cdot & \\ & \text{O} & & & & & & 0 & \\ & & & & & & & & \cdot \\ & & & & & & & & & 0 \end{bmatrix}.$$

The (total) number of nonzero entries is r.

PROOF By Corollary 1, there is a basis $B = \{\mathbf{v}_1,\ldots,\mathbf{v}_n\}$ for V such that $[F;B] = D$ is diagonal. Since congruent matrices have the same rank, $D = (d_{ij})$ has r nonzero entries. By reordering the elements of B (if necessary), we can arrange to have

$$\begin{aligned} &d_{ii} = 0 \text{ for } i > r, \\ &d_{ii} > 0 \text{ for } i \le k \qquad (\text{for some } 1 \le k \le r), \text{ and} \\ &d_{ii} < 0 \text{ for } k < i \le r \end{aligned}$$

(recall $d_{ii} = F(\mathbf{v}_i,\mathbf{v}_i)$). We "scale" B to obtain the new basis $B' = \{\mathbf{v}'_1,\ldots,\mathbf{v}'_n\}$ where

$$\mathbf{v}'_i = \begin{cases} (1/\sqrt{d_{ii}})\mathbf{v}_i, & 1 \le i \le k, \\ (1/\sqrt{-d_{ii}})\mathbf{v}_i, & k < i \le r, \\ \mathbf{v}_i, & i > r. \end{cases}$$

Then $[F;B']$ has the required form. ∎

EXAMPLE 7-32

Let F be the symmetric bilinear form on R^3 with matrix

$$A = \begin{bmatrix} 1 & 2 & -4 \\ 2 & -2 & -2 \\ -4 & -2 & 1 \end{bmatrix}$$

with respect to the standard basis. In Example 7-31, we showed that $Q^t A Q = D$, where D has diagonal entries $d_{11} = 1$, $d_{22} = -6$ and $d_{33} = -9$. Let $\mathbf{v}_1$, $\mathbf{v}_2$, and $\mathbf{v}_3$ be the columns of Q (i.e., the rows of Q^t, which form a basis for R^3 with respect to which the matrix of F is D). Then, following the proof of the corollary,

$$\begin{aligned} \mathbf{v}'_1 &= (1/1)\mathbf{v}_1 = (1,0,0), \\ \mathbf{v}'_2 &= (1/\sqrt{6})\mathbf{v}_2 = (1/\sqrt{6})(-2,1,0), \\ \mathbf{v}'_3 &= (1/\sqrt{9})\mathbf{v}_3 = (1/3)(2,1,1). \end{aligned}$$

And $B' = \{\mathbf{v}'_1, \mathbf{v}'_2, \mathbf{v}'_3\}$ has the property stated in Corollary 3:

$$[F;B'] = \begin{bmatrix} 1 & 0 & 0 \\ 0 & -1 & 0 \\ 0 & 0 & -1 \end{bmatrix}.$$

(Alternatively, we could use the matrices D and Q constructed in Example 7-24.) ❑

In fact, it can be shown that the number k of $+1$'s in the above matrix of F is independent of the choice of the basis B' for which $[F;B']$ has the form of Corollary 3; it is called the **index** of F. As a result,

$$\begin{aligned} tr[F;B'] &= k(+1) + (r - k)(-1) \\ &= (\text{number of } +1\text{'s}) - (\text{number of } -1\text{'s}) \\ &= 2k - r, \end{aligned}$$

is also independent of B' (but see question 14, Exercises 7-5). This number is called the **signature** of F. The matrix $[F;B']$ is sometimes called the **Sylvester Canonical form** for the matrix of F.

EXAMPLE 7-33

The rank, index, and signature of each bilinear form with matrix A below are as indicated.

(a) $A = \begin{bmatrix} 1 & 0 & 0 \\ 0 & -1 & 0 \\ 0 & 0 & -1 \end{bmatrix}$ rank $= 3$, index $= 1$, signature $= -1$.

(b) $A = \begin{bmatrix} 1 & 0 & 0 \\ 0 & -1 & 0 \\ 0 & 0 & 0 \end{bmatrix}$ rank $= 2$, index $= 1$, signature $= 0$.

(c) $A = \begin{bmatrix} 1 & 0 & 0 & 0 \\ 0 & 1 & 0 & 0 \\ 0 & 0 & -1 & 0 \\ 0 & 0 & 0 & 0 \end{bmatrix}$ rank = 3, index = 2, signature = 1.

(d) $A = \begin{bmatrix} 1 & 0 & 0 & 0 \\ 0 & 1 & 0 & 0 \\ 0 & 0 & 1 & 0 \\ 0 & 0 & 0 & -1 \end{bmatrix}$ rank = 4, index = 3, signature = 2. ❑

Note that any inner product on an n-dimensional space is a nondegenerate, symmetric bilinear form. It has rank, index, and signature equal to n.

We turn now to the notion of **quadratic forms**. This is similar to the norm function associated with an inner product.

> **DEFINITION 7-19**
>
> Let F be a symmetric bilinear form on a vector space V. The **quadratic form** q associated with F is the function $q: V \to R$ defined by $q(\mathbf{v}) = F(\mathbf{v},\mathbf{v})$ for $\mathbf{v} \in V$.

Note that q is *not* a linear function on V.

EXAMPLE 7-34

(a) Let F be the symmetric bilinear form on R^2 having matrix

$$A = \begin{bmatrix} 1 & -2 \\ -2 & 1 \end{bmatrix}$$

with respect to the standard basis. If $\mathbf{v} = (x,y)$ is in R^2,

$$q(\mathbf{v}) = q(x,y) = [x \quad y]\begin{bmatrix} 1 & -2 \\ -2 & 1 \end{bmatrix}\begin{bmatrix} x \\ y \end{bmatrix} = x^2 - 4xy + y^2.$$

Note that $q(x,y)$ is a homogeneous polynomial in x and y of degree 2 (this means that the sum of the degrees in every term is 2).

(b) Let g be the Lorentz metric on R^4 (Example 7-29). The associated quadratic form q is defined by $q(x,y,z,t) = x^2 + y^2 + z^2 - t^2$. ❑

In general, if F is a symmetric bilinear form on an n-dimensional vector space

V with basis B, the associated quadratic form q may be computed from

$$q(\mathbf{v}) = F(\mathbf{v},\mathbf{v}) = [\mathbf{v}]_B{}^t[F;B][\mathbf{v}]_B = \sum_{i=1}^{n}\sum_{j=1}^{n} a_{ij}x_ix_j$$

$$= \sum_{i=1}^{n} a_{ii}x_i{}^2 + 2\sum_{i<j} a_{ij}x_ix_j \qquad (7\text{-}14)$$

where $[\mathbf{v}]_B = (x_i)$, $[F;B] = (a_{ij}) = A$, and the second term in the last expression of 7-14 is the sum over all $i, j = 1, \ldots, n$ with $i < j$. We have used the symmetry of A to group a_{ji} with a_{ij} when $i \neq j$. The terms in the second sum in the last expression of 7-14 are called **cross-product** (or **rectangular**) **terms** because they contain products x_ix_j with $i \neq j$.

We see that given a quadratic form q by an equation such as 7-14, we obtain the matrix $A = (a_{ij}) = [F;B]$ of the associated symmetric bilinear form F as

$$a_{ij} = \begin{cases} \text{coefficient of } x_i{}^2, \text{ if } i = j, \\ (1/2) \text{ coefficient of } x_ix_j, \text{ if } i \neq j. \end{cases}$$

In fact, the reader may show (question 15, Exercises 7-5) that q determines F through the identity

$$F(\mathbf{u},\mathbf{v}) = (1/2)[q(\mathbf{u} + \mathbf{v}) - q(\mathbf{u}) - q(\mathbf{v})],\ \mathbf{u},\ \mathbf{v} \in V.$$

EXAMPLE 7-35

(a) The matrix of the bilinear form on R^2 associated with the quadratic form q given by $q(x,y) = -2x^2 + 6xy + y^2$ for (x,y) in R^2 is

$$A = \begin{bmatrix} -2 & 3 \\ 3 & 1 \end{bmatrix}$$

(with respect to the standard basis for R^2).

(b) The matrix of the symmetric bilinear form on R^3 associated with $q(x,y,z) = y^2 - 3z^2 + 2xy - 3xz + 4yz$ is

$$A = \begin{bmatrix} 0 & 1 & -3/2 \\ 1 & 1 & 2 \\ -3/2 & 2 & -3 \end{bmatrix}$$

(with respect to the standard basis for R^3). ❑

From Corollary 3 of Theorem 7-16, we obtain the next theorem.

THEOREM 7-17

When q is a quadratic form on a vector space V of dimension n, there is a basis B' for V such that

$$q(\mathbf{v}) = x_1^2 + \dots + x_k^2 - x_{k+1}^2 - \dots - x_r^2$$

if $[\mathbf{v}]_{B'} = (x_i)$, where k is the index of the associated symmetric bilinear form and r is its rank. ■

EXAMPLE 7-36

Let q be the quadratic form associated with the symmetric bilinear form F of Example 7-32. Then,

$$q(x_1,x_2,x_3) = x_1^2 - 2x_2^2 + x_3^2 + 4x_1x_2 - 8x_1x_3 - 4x_2x_3,$$

for $(x_1,x_2,x_3) \in R^3$. If we take the basis constructed in that example for B', it follows that if $\mathbf{v} = x_1\mathbf{v}'_1 + x_2\mathbf{v}'_2 + x_3\mathbf{v}'_3$ is in R^3, then

$$q(\mathbf{v}) = x_1^2 - x_2^2 - x_3^2. \square$$

Thus, by an appropriate choice of basis for a vector space V, the cross-product terms may be removed from the expression for a quadratic form on V. When V is a finite dimensional inner product space, we may, in fact, choose the new basis to be orthonormal.

THEOREM 7-18

Let q be a quadratic form on an inner product space V of dimen sion n. Then there is an orthonormal basis B for V such that if $[\mathbf{v}]_B = (x_i)$,

$$q(\mathbf{v}) = c_1x_1^2 + c_2x_2^2 + \dots + c_nx_n^2$$

for some scalars $c_1, \dots, c_n$. In fact, the scalars c_i are the eigenvalues of the (symmetric) matrix $A = [F;B]$ of the symmetric bilinear form F associated with q, and B consists of eigenvectors of A.

PROOF This theorem is an immediate consequence of Theorem 7-16, Corollary 2, and Theorem 7-13, Corollary 1 (orthogonal diagonalization of symmetric matrices). ■

EXAMPLE 7-37

Let q be the quadratic form on R^3 defined in the preceding example: q is associated with the symmetric bilinear form having matrix A considered earlier in Example 7-24, where A was orthogonally diagonalized. Let B be the orthonormal basis for R^3 constructed in that example. Then, if $\mathbf{v} = x_1\mathbf{v}_1 + x_2\mathbf{v}_2 + x_3\mathbf{v}_3$, it follows that $q(\mathbf{v}) = -3x_1^2 - 3x_2^2 + 6x_3^2$. ❑

We conclude this section by applying the preceding ideas to the problem of identifying the graphs of quadratic equations in two variables x and y. These are equations of the form

$$ax^2 + 2bxy + cy^2 + dx + ey + f = 0 \qquad (7\text{-}15)$$

where a, b, c, d, e, and f are constants (not all 0) and the factor 2 is included for convenience. It can be shown that the graph of this equation must represent one of the following, depending on the specific values of the constants:

(1) a **nondegenerate conic** (ellipse, circle, hyperbola, or parabola);
(2) a **degenerate conic** (a pair of lines or a single point);
(3) the empty set (if there is *no* graph, sometimes called an **imaginary conic**).

The nondegenerate conics in **standard position** are the curves in the plane illustrated in Figures 7-4, 7-5, and 7-6.

FIGURE 7-4

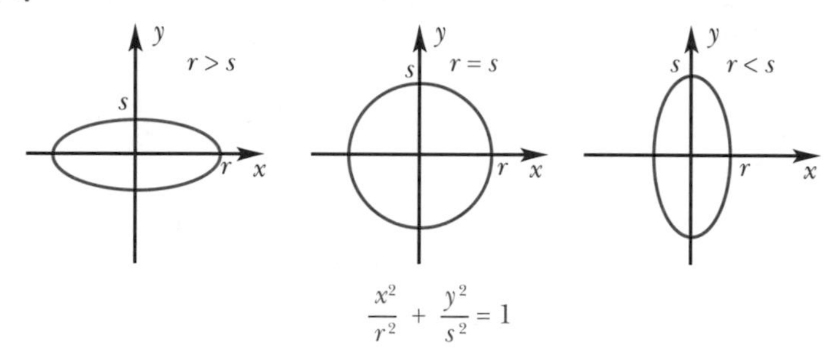

$$\frac{x^2}{r^2} + \frac{y^2}{s^2} = 1$$

For example, the ellipse in standard position (Figure 7-4) may be written as

$$s^2x^2 + r^2y^2 = r^2s^2,$$

from which we see that it has the form of equation 7-15 with $a = s^2$, $b = 0$, $c = r^2$, $d = e = 0$, and $f = -r^2s^2$.

On the other hand, if in equation 7-15, we set $a = c = 1$ and all other constants equal to 0, the resulting equation $x^2 + y^2 = 0$ has a graph consisting of the single point (0,0). Finally, if we set $a = 1 = f$ and the other constants equal to 0, we get the equation $x^2 + 1 = 0$ which has no graph.

FIGURE 7-5

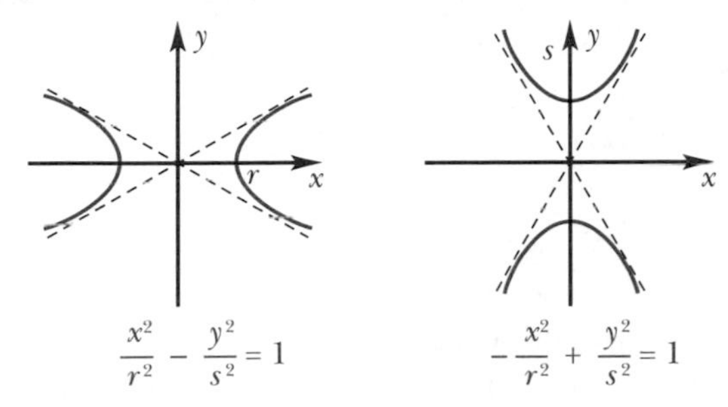

FIGURE 7-6

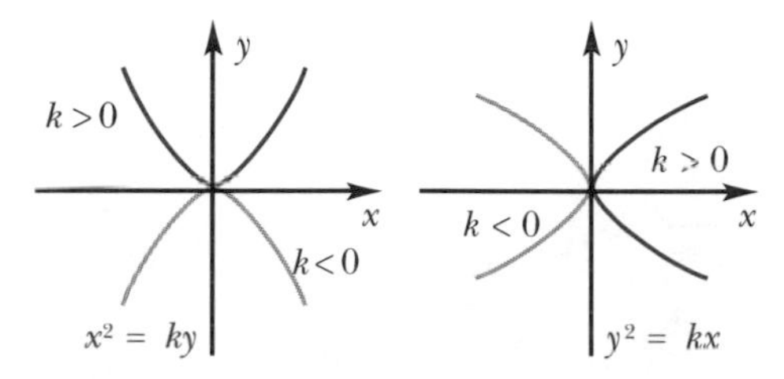

Identification of the graph of equation 7-15 is achieved by the following steps:

(1) *Rotating the axes* to eliminate the cross-product term $2bxy$. This means, as we shall see, choosing a new orthonormal basis for R^2 so that with respect to the axes defined by this basis, the equation of the curve contains no xy term.

(2) *Translating the axes*, if necessary, so that the equation becomes that of one of the conics in standard position (or that of a recognizable degenerate conic, or of the empty set).

A *translation* of the plane is a mapping $Z:R^2 \to R^2$ of the form

$$Z(x,y) = (x + k, y + k'),$$

where k and $k' \in R$. This is *not* a linear map of R^2 as a vector space, since, for example, $Z(0,0) = (k,k') \neq (0,0)$ unless $k = k' = 0$. In other words, Z moves $\mathbf{0} = (0,0)$ to (k,k'). However, readers may verify that Z maps lines in R^2 to lines in R^2.

In practice, the translation that moves a conic to standard position is determined by *completing the square* in its equation (after the cross-product term has been removed). For example, consider the quadratic equation

$$x^2 + 4y^2 + 4x - 8y + 4 = 0. \tag{7-16}$$

We may write the equation as

$$(x^2 + 4x + 4) - 4 + 4(y^2 - 2y + 1) - 4 + 4 = 0,$$

having regrouped the terms and having added and subtracted the quantities necessary to make the terms in x and those in y *perfect squares*. Therefore, equation 7-16 may be rewritten as

$$(x + 2)^2 + 4(y - 1)^2 = 4,$$

or,

$$\frac{(x + 2)^2}{4} + \frac{(y - 1)^2}{1} = 1,$$

which is an ellipse in standard position with respect to the axes $x' = x + 2$, $y' = y - 1$ (Figure 7-7). These axes are obtained by the translation

$$Z(x,y) = (x',y') = (x + 2, y - 1).$$

FIGURE 7-7

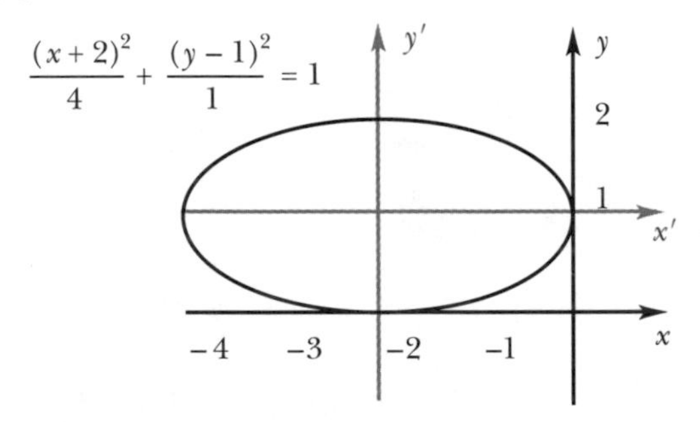

In order to eliminate the xy term from the quadratic equation 7-15, we note that this equation contains the quadratic form q on R^2 defined by $q(x,y) = ax^2 + 2bxy + cy^2$ which is associated with the symmetric bilinear form F on R^2 having matrix

$$A = \begin{bmatrix} a & b \\ b & c \end{bmatrix}.$$

This quadratic form q is sometimes called the quadratic form *associated with* the quadratic equation 7-15. In fact, we may write the latter more concisely in matrix notation as

$$X^tAX + LX + f = 0, \tag{7-17}$$

where

$$X = \begin{bmatrix} x \\ y \end{bmatrix} \text{ and } L = \begin{bmatrix} d & e \end{bmatrix}.$$

LX is called the **linear part** of equation 7-17.

Now, by Theorem 7-18, there is an orthonormal basis $B' = \{\mathbf{v}'_1, \mathbf{v}'_2\}$ (denoted B in the theorem) for R^2 such that

$$q(\mathbf{v}) = c_1x'^2 + c_2y'^2,$$

when $\mathbf{v} = x'\mathbf{v}'_1 + y'\mathbf{v}'_2$ is in R^2. Recall that this is the case because A is orthogonally diagonalizable. Thus,

(1) B' is an orthonormal basis for R^2 of eigenvectors for A;

(2) c_1 and c_2 are the eigenvalues of A;

(3) the matrix $Q = [B'/B]$ has columns $\mathbf{v}'_1$ and $\mathbf{v}'_2$ and is orthogonal, and $Q^tAQ = D$ where

$$D = \begin{bmatrix} c_1 & 0 \\ 0 & c_2 \end{bmatrix}.$$

Moreover, if we let

$$X = [\mathbf{v}]_B = \begin{bmatrix} x \\ y \end{bmatrix} \text{ and } X' = [\mathbf{v}]_{B'} = \begin{bmatrix} x' \\ y' \end{bmatrix}$$

(where B is the standard basis for R^2), then by Theorem 5-16,

$$X = QX'.$$

Substituting this equation into equation 7-17, we obtain for the equation of the quadratic (in the new coordinates x', y' defined by B'):

$$(QX')^tA(QX') + L(QX') + f = 0,$$

or,

$$X'^t(Q^tAQ)X' + L'X' + f = 0,$$

(where $L' = LQ$). This simplifies to

$$X'^tDX' + L'X' + f = 0,$$

or

$$c_1x'^2 + c_2y'^2 + d'x' + e'y' + f = 0, \tag{7-18}$$

which has no xy term.

It remains only to be shown that we may choose Q to be the matrix of a rotation since $\det Q = +1$ or -1 (see question 17, Exercises 7-3). If $\det Q = -1$, it is not the matrix of a rotation (see question 21, Exercises 7-3). However, by interchanging two columns of Q (i.e., the elements of B'), we obtain a new orthogonal matrix Q' such that $Q'^tAQ' = D'$, where

$$D' = \begin{bmatrix} c_2 & 0 \\ 0 & c_1 \end{bmatrix}.$$

Furthermore, $\det Q' = -\det Q = +1$, so Q' is the matrix of a rotation of the plane.

We may therefore summarize by saying that for any quadratic equation 7-15, we may find a rotation of the plane to new axes, such that equation 7-15 becomes equation 7-18, in which the cross product term has been eliminated. This result is known as the **Principal Axis Theorem** for the plane. It can be generalized to higher dimensions using Theorem 7-18.

We illustrate the procedure by identifying the conic with equation

$$6x^2 + 24xy - y^2 - 96x - 22y + 35 = 0. \tag{7-19}$$

The associated quadratic form is defined by the matrix

$$A = \begin{bmatrix} 6 & 12 \\ 12 & -1 \end{bmatrix}.$$

The characteristic polynomial is

$$\begin{aligned} det\,(A - xI) &= (6 - x)(-1 - x) - 144 \\ &= x^2 - 5x - 150 \\ &= (x - 15)(x + 10). \end{aligned}$$

So, the eigenvalues are $c_1 = 15$ and $c_2 = -10$. An eigenvector for $c_1 = 15$ is found by solving the system $(A - 15I)X = O$, or,

$$\begin{aligned} -9x + 12y &= 0 \\ 12x - 16y &= 0. \end{aligned}$$

Clearly, (4,3) is a solution; we normalize it to get unit eigenvector $\mathbf{v}'_1 = (1/5)(4,3)$. Similarly, a unit eigenvector for $c_2 = -10$ is easily seen to be $\mathbf{v}'_2 = (1/5)(3,-4)$. Therefore, the matrix Q is given by

$$Q = (1/5)\begin{bmatrix} 4 & 3 \\ 3 & -4 \end{bmatrix}.$$

However, this has determinant -1, so we interchange the columns and take

$$Q = (1/5)\begin{bmatrix} 3 & 4 \\ -4 & 3 \end{bmatrix}.$$

Of course, this is equivalent to letting the (ordered) orthonormal basis be $B' = \{\mathbf{v}'_2, \mathbf{v}'_1\}$. The rotation with matrix Q then changes equation 7-19 to

$$-10x'^2 + 15y'^2 + [-96 \quad -22]\begin{bmatrix} 3/5 & 4/5 \\ -4/5 & 3/5 \end{bmatrix}\begin{bmatrix} x' \\ y' \end{bmatrix} + 35 = 0,$$

or,

$$-10x'^2 + 15y'^2 - 40x' - 90y' + 35 = 0.$$

Completing the squares on the left, we obtain

$$-10(x' + 2)^2 + 40 + 15(y' - 3)^2 - 135 + 35 = 0,$$

or,

$$-10(x' + 2)^2 + 15(y' - 3)^2 = 60,$$

or,

$$\frac{-(x' + 2)^2}{6} + \frac{(y' - 3)^2}{4} = 1.$$

Translating with $x'' = x' + 2$ and $y'' = y' - 3$, we obtain the hyperbola in standard position having equation

$$\frac{-x''^2}{6} + \frac{y''^2}{4} = 1. \qquad (7\text{-}20)$$

The graph of the quadratic equation 7-19 and the various axes used to transform it to 7-20 are shown in Figure 7-8.

FIGURE 7-8

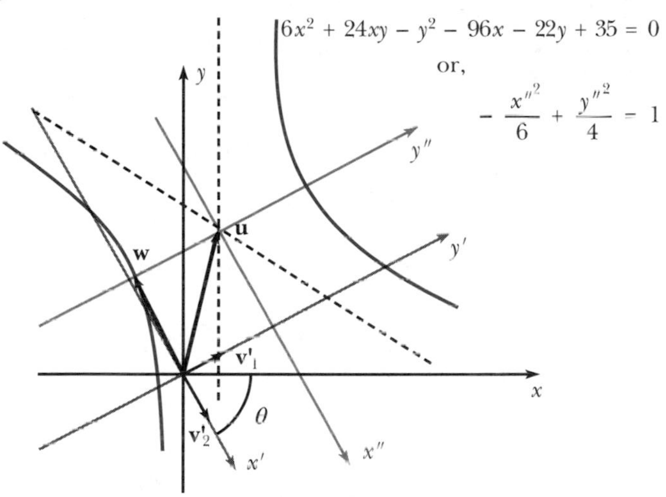

It is interesting to note that the angle of rotation is the angle θ between $\mathbf{v}'_2$ (which defines the x' axis) and $\mathbf{e}_1$ (which defines the x axis). This is given by $\cos\theta = \mathbf{e}_1 \cdot \mathbf{v}'_2 = 3/5 = 0.6$. So, the rotation is through an angle whose cosine is 0.6 and in a clockwise direction, as shown in the figure.

The center $\mathbf{u}$ of the hyperbola (shown in the figure) is the origin of the x'', y'' system, and has coordinates $x' = -2$, $y' = 3$ in the x', y' system. Therefore,

$$[\mathbf{u}]_{B'} = \begin{bmatrix} -2 \\ 3 \end{bmatrix},$$

so that if B denotes the standard basis for R^2,

$$[\mathbf{u}]_B = Q[\mathbf{u}]_{B'} = (1/5)\begin{bmatrix} 3 & 4 \\ -4 & 3 \end{bmatrix}\begin{bmatrix} -2 \\ 3 \end{bmatrix} = \begin{bmatrix} 6/5 \\ 17/5 \end{bmatrix}.$$

In other words, the center $\mathbf{u}$ has x, y coordinates $(6/5,17/5)$.

In a similar way, we may locate the intersection $\mathbf{w}$ of the hyperbola (as shown in the figure) with its axes. Now, in the x'', y'' system, it has coordinates $(0,-2)$ by equation 7-20. So, it has x', y' coordinates $(-2,1)$, and

$$[\mathbf{w}]_B = (1/5)\begin{bmatrix} 3 & 4 \\ -4 & 3 \end{bmatrix}\begin{bmatrix} -2 \\ 1 \end{bmatrix} = \begin{bmatrix} -2/5 \\ 11/5 \end{bmatrix}.$$

The reader may verify by substitution that these values satisfy the original equation 7-19.

EXERCISES 7-5

1. Determine which of the following functions F are bilinear forms on the spaces V indicated. Justify your answers.

(a) $V = R^2$, $F((x,y),(x',y')) = xx' - 4xy' + 3yy'$ for (x,y) and $(x',y') \in R^2$.

(b) $V = R^2$, $F((x,y),(x',y')) = -xx' + y + 4xy' - x'$ for (x,y) and $(x',y') \in R^2$.

(c) $V = M(n)$, $F(A,B) = tr(AB)$ for A, $B \in M(n)$.

(d) $V = M(n)$, $F(A,B) = (tr\,A)(tr\,B)$ for A, $B \in M(n)$.

(e) $V = M(2\times 1)$, $F(X,Y) = (AX)(BY)$ for X, $Y \in M(2\times 1)$, and $A = [-1 \quad 2]$, $B = [3 \quad -2]$.

(f) $V = R^2$, $F(\mathbf{u}_1,\mathbf{u}_2) = det\,U$, where U is the 2×2 matrix with rows $\mathbf{u}_1$, $\mathbf{u}_2 \in R^2$.

(g) $V = P(2)$, $F(f,g) = (fg)(0)$ for f and $g \in P(2)$.

(h) V is *any* inner product space, $F(\mathbf{u},\mathbf{v}) = d(\mathbf{u},\mathbf{v})$, where $d(\mathbf{u},\mathbf{v}) = |\mathbf{u} - \mathbf{v}|$, $\mathbf{u}$, $\mathbf{v} \in V$ (cf. question 24, Exercises 7-2).

(i) $V = M(2)$, $F(A,B) = 2tr(AB) - (tr\,A)(tr\,B)$.

(j) $V = M(m\times n)$, $F(A,B) = a_{ij}b_{kl}$ for some $1 \le i, k \le m$, $1 \le j, l \le n$.

2. Prove that if F is any bilinear form on V, then $F(\mathbf{u},\mathbf{0}) = F(\mathbf{0},\mathbf{u}) = 0$ for $\mathbf{u} \in V$.

3. **(a)** Let f and g be 1×2 matrices given by $f = [-2 \quad 1]$, $g = [3 \quad 4]$.

(i) Show that $f\otimes g$ is a bilinear form on $M(2\times 1)$, where $f\otimes g(X,Y) = (fX)(gY)$ for X, Y in $M(2\times 1)$ (cf. Example 7-26).

(ii) Compute $f \otimes g\ (X,Y)$ when

$$X = \begin{bmatrix} -1 \\ 1 \end{bmatrix}, Y = \begin{bmatrix} -2 \\ 3 \end{bmatrix}.$$

(b) Let V be any vector space and $f,g:V \to R$ be linear maps. Show that $f \otimes g$ is bilinear (Example 7-26).

4. Let V be any vector space. Prove that the set $L(V,V;R)$ of all bilinear forms on V is a vector space with respect to the addition and scalar multiplication defined by $(F + G)(\mathbf{u},\mathbf{v}) = F(\mathbf{u},\mathbf{v}) + G(\mathbf{u},\mathbf{v})$, $(cF)(\mathbf{u},\mathbf{v}) = cF(\mathbf{u},\mathbf{v})$, for $\mathbf{u}, \mathbf{v} \in V$ and $c \in R$.

5. (For readers who have covered sections 5-3 and 5-5) Let V be a vector space with basis $B = \{\mathbf{v}_1,\ldots,\mathbf{v}_n\}$ and dual basis $B^* = \{f_1,\ldots,f_n\}$.

(a) Show that $\{f_i \otimes f_j \mid 1 \le i,j \le n\}$ is a basis for $L(V,V;R)$ by showing that any F may be expressed as

$$F = \sum_{i=1}^{n} \sum_{j=1}^{n} F(\mathbf{v}_i,\mathbf{v}_j) f_i \otimes f_j.$$

(b) Hence show that when V is finite dimensional, $dim\ L(V,V;R) = (dim\ V)^2$.

(c) Show that the map $F \to [F;B]$ is an isomorphism of $L(V,V;R)$ with $M(n)$ when $dim\ V = n$.

6. For each of the bilinear forms F in questions 1 (a), (c), (d), (e), (f), (g), and (i), compute $[F;B]$, where B is the standard basis for the vector space V in question (use $n = 2$ for (c) and (d)).

7. Show that congruence of $n \times n$ matrices is an equivalence relation on $M(n)$.

8. Follow the steps below to prove that congruent $n \times n$ matrices have the same rank.

(a) Show that $rank\ QA = rank\ A$ when Q is invertible. (Hint. Q is a product of elementary matrices.)

(b) Use (a) to conclude that $rank\ AQ = rank\ A^t = rank\ A$.

(c) Use (a) and (b) to prove the result.

9. Prove that a bilinear form F on an n-dimensional vector space V is nondegenerate iff **(a)** for any $\mathbf{u} \neq \mathbf{0}$ in V, there is a $\mathbf{v} \in V$ such that $F(\mathbf{u},\mathbf{v}) \neq 0$, or **(b)** for any $\mathbf{v} \neq \mathbf{0}$ in V, there is a $\mathbf{u} \in V$ such that $F(\mathbf{u},\mathbf{v}) \neq 0$.

10. Let F be a nondegenerate symmetric bilinear form on a finite dimensional vector space V and W be a subspace of V. Define $W^\perp = \{\mathbf{v} \in V \mid F(\mathbf{u},\mathbf{v}) = 0 \text{ for all } \mathbf{u} \in W\}$.

(a) Show that $W^\perp$ is a subspace of V.

(b) Show that $W \cap W^\perp$ need not be $\{\mathbf{0}\}$. (Hint: Use Example 7-29.)

11. Construct an example of two 2×2 matrices that are congruent but not similar. Hence show that if F is a symmetric bilinear form on a vector space V of dimension n, and $[F;B]$, $[F;B']$ are diagonal for some bases B and B', these matrices may have different diagonal entries.

12. Use the method of Example 7-31 to find nonsingular matrices Q such that $Q^tAQ = D$ is diagonal for the following matrices A.

(a) $\begin{bmatrix} 1 & -2 \\ -2 & 3 \end{bmatrix}$ **(b)** $\begin{bmatrix} 1 & 2 & 2 \\ 2 & 3 & 0 \\ 2 & 0 & -1 \end{bmatrix}$

13. **(a)** Let F be the bilinear form on R^2 with matrix A of question 12 (a), with respect to the standard basis. Find a basis B for R^2 such that $[F;B]$ is diagonal with entries $+1$, -1, or 0 only.

(b) Repeat part (a) using the bilinear form F on R^3 whose matrix is A in question 12 (b).

(c) What are the rank, index, and signature of the bilinear forms in (a) and (b) above?

14. Show that congruent matrices need not have the same trace. (Recall that similar matrices do: question 16 in Exercises 5-4.)

15. Let q be the quadratic form associated with a symmetric bilinear form F on a vector space V. Show that $F(\mathbf{u},\mathbf{v}) = (1/2)[q(\mathbf{u} + \mathbf{v}) - q(\mathbf{u}) - q(\mathbf{v})]$.

16. **(a)** Find the symmetric bilinear form F on $V = R^2$ or R^3 associated with the following quadratic forms.

(i) $V = R^2,\ q(x,y) = -x^2 + 2xy + 3y^2$

(ii) $V = R^2,\ q(x,y) = 2x^2 + 4xy + 2y^2$

(iii) $V = R^3,\ q(x,y,z) = z^2 + 3xy - 4xz$

(iv) $V = R^3,\ q(x,y,z) = -x^2 + 2y^2 + 3z^2 - xy + 2yz$

(b) What is the rank, index, and signature of each F above?

17. For each of the quadratic forms q in question 16, construct a basis B' so that $q(\mathbf{v})$ has the form stated in Theorem 7-17 for any $\mathbf{v} \in V$.

18. For each of the quadratic forms q in (i) and (ii) of question 16(a), construct an orthonormal basis B such that $q(\mathbf{v})$ has the form stated in Theorem 7-18 for any $\mathbf{v} \in V$.

19. Identify each of the following quadratic equations using suitable rotations and/or translations of the plane in order to bring it to standard position. Sketch the graphs, showing all coordinate axes used.

(a) $x^2 + 4y^2 + 4x - 8y - 8 = 0$ **(b)** $2x^2 - 6x + 4y + 10 = 0$

(c) $2x^2 - 6xy + 2y^2 + 5 = 0$

(d) $6x^2 + 24xy - y^2 - 48x + 54y - 99 = 0$

(e) $25x^2 - 30xy + 9y^2 - 100 = 0$

(f) $x^2 + 4xy + y^2 - 4x + 8y - 16 = 0$

7-6 Complex Eigenvalues and Eigenvectors (Optional)

In this section, we consider the eigenvalue problem for complex matrices. In fact, all the definitions and the methods for computing eigenvalues and eigenvectors discussed in Chapter 6 remain valid for linear operators T on complex vector spaces V. For example, a complex number c is an eigenvalue of T if there is a vector $\mathbf{v} \in V$, $\mathbf{v} \neq \mathbf{0}$, such that $T(\mathbf{v}) = c\mathbf{v}$. However, we will concentrate on eigenvalues of complex matrix operators T_A, where A is a complex $n \times n$ matrix. Thus, c is an eigenvalue of $A \in M_n(C)$ if there is a nonzero $\mathbf{v} \in M_{n\times 1}(C)$ such that $A\mathbf{v} = c\mathbf{v}$. As in the real case, we will identify $n \times 1$ complex matrices $\mathbf{v} \in M_{n\times 1}(C)$ with complex n-tuples $\mathbf{v} \in C^n$. The characteristic polynomial $f = det\ (A - xI)$ of A is now a complex polynomial of degree n. By the Fundamental Theorem of Algebra (Appendix A), f can be expressed as

$$f = \pm(x - c_1)(x - c_2)...(x - c_n),$$

where the $c_i \in C$ are the eigenvalues of A, which are not necessarily distinct. In particular, any complex matrix has at least one eigenvalue.

The main result of Chapter 6 still holds in this setting, which is that a matrix A is diagonalizable iff its minimal polynomial p is a product of distinct linear factors.

EXAMPLE 7-38

Consider the matrix

$$A = \begin{bmatrix} 0 & -1 \\ 1 & 0 \end{bmatrix}$$

which is the matrix of the 90° counterclockwise rotation of the plane R^2. We saw that the characteristic polynomial of A is $f = x^2 + 1$, so that A has no real eigenvalues. However, we may regard A as a complex matrix (and T_A as a linear operator on C^2). Then, f can be factorized as

$$f = (x - i)(x + i),$$

so that A has two distinct complex eigenvalues $+i$ and $-i$. Since the minimal polynomial p has the same zeros as f, f is diagonalizable. Therefore, A is similar to the diagonal matrix

$$D = \begin{bmatrix} i & 0 \\ 0 & -i \end{bmatrix}.$$

In order to find the matrix P such that $P^{-1}AP = D$, we find the eigenvectors of A. As before, $\mathbf{v} = (v_1,v_2) \in C^2 = M_{2\times 1}(C)$ is an eigenvector associated with $c_1 = i$ iff $(A - iI)\mathbf{v} = \mathbf{0} = (0,0)$. The coefficient matrix of this homogeneous system of equations for v_1 and v_2 is

$$\begin{bmatrix} -i & 1 \\ 1 & -i \end{bmatrix},$$

which is reduced to

$$\begin{bmatrix} 1 & -i \\ 0 & 0 \end{bmatrix}.$$

Thus, the corresponding system of equations is simply $v_1 = iv_2$ and an eigenvector for $c_1 = i$ is, for example, $\mathbf{v}_1 = (-1,i)$. We may verify this by noting that

$$\begin{bmatrix} 0 & -1 \\ 1 & 0 \end{bmatrix}\begin{bmatrix} -1 \\ i \end{bmatrix} = \begin{bmatrix} -i \\ -1 \end{bmatrix} = i\begin{bmatrix} -1 \\ i \end{bmatrix}.$$

Readers may verify that an eigenvector for $c_2 = -i$ is $\mathbf{v}_2 = (-i,1)$. Therefore, the matrix P is

$$P = \begin{bmatrix} -1 & -i \\ i & 1 \end{bmatrix}. \square$$

Now, in section 7-3, we determined a special class of real matrices which are always diagonalizable. These are the symmetric matrices A for which $A = A^t$. Recall that they arise naturally as the matrices of self-adjoint linear operators on real inner product spaces and yield an orthonormal basis of eigenvectors for R^n.

In fact, complex symmetric matrices are *not* generally diagonalizable, as can be seen from the example

$$A = \begin{bmatrix} 2 & i \\ i & 0 \end{bmatrix}.$$

The characteristic polynomial is $f = (x - 1)^2$, and this is also seen to be the minimal polynomial (alternatively, it is easy to see that the eigenspace of the single eigenvalue $c = 1$ has dimension 1).

In order to generalize these ideas for the case of complex vector spaces and matrices, we must first define the notion of an inner product on a complex vector space (cf. Definition 7-1).

DEFINITION 7-20

Let V be a complex vector space. An **inner product** on V is a function $<,>$ which assigns to every ordered pair of vectors $\mathbf{u}$ and $\mathbf{v}$ a **complex** number $<\mathbf{u},\mathbf{v}>$ such that

(1) $<\mathbf{u} + \mathbf{v},\mathbf{w}> = <\mathbf{u},\mathbf{w}> + <\mathbf{v},\mathbf{w}>$ for any $\mathbf{u}$, $\mathbf{v}$, and $\mathbf{w}$ in V;

(2) $<z\mathbf{u},\mathbf{v}> = z<\mathbf{u},\mathbf{v}>$ for any scalars $z \in C$, and $\mathbf{u}$ and $\mathbf{v}$ in V;

(3) $<\mathbf{u},\mathbf{v}> = \overline{<\mathbf{v},\mathbf{u}>}$ for all $\mathbf{u}$ and $\mathbf{v}$ in V;

(4) $<\mathbf{u},\mathbf{u}> > 0$ unless $\mathbf{u} = \mathbf{0}$.

Note that axiom (3) implies that a (complex) inner product is *not* symmetric. Also, it implies that axiom (4) makes sense, since $<\mathbf{u},\mathbf{u}> = \overline{<\mathbf{u},\mathbf{u}>}$ means that $<\mathbf{u},\mathbf{u}>$ is real, and therefore can be compared in size to 0. As in Definition 7-1, axiom (4) is also called **positive definiteness.**

In addition, the following properties are immediate consequences.

(1) $<\mathbf{u},\mathbf{v} + \mathbf{w}> = <\mathbf{u},\mathbf{v}> + <\mathbf{u},\mathbf{w}>$.

(2) $<\mathbf{u},\mathbf{0}> = <\mathbf{0},\mathbf{u}> = 0$.

(3) $<\mathbf{u},z\mathbf{v}> = \bar{z}<\mathbf{u},\mathbf{v}>$.

(We leave the proofs for the reader.) These properties imply that the inner product is *not* linear in the right-hand argument.

As in the case of real inner products, a **complex inner product space** (or **unitary space**) V is a complex vector space with a particular inner product $<,>$ defined on V. In addition, we can again define the **norm** of a vector $\mathbf{u} \in V$ as $|\mathbf{u}| = <\mathbf{u},\mathbf{u}>^{1/2}$. Then $\mathbf{u}$ is a unit vector if $|\mathbf{u}| = 1$. In this connection, Theorem 7-2 (properties of the norm) is valid without change, except, of course, that $c \in C$ and $|c|$ means the **modulus** of c. Therefore, for any $\mathbf{u}$, $\mathbf{u}/|\mathbf{u}|$ is a unit vector.

The notions of orthogonal vectors and orthogonal and orthonormal sets of vectors also apply to complex inner product spaces V. Thus, vectors $\mathbf{u}$ and $\mathbf{v}$ are orthogonal iff $<\mathbf{u},\mathbf{v}> = 0$. However, these concepts no longer have an easy geometrical interpretation as they do in the case of, for example, R^2.

EXAMPLE 7-40

Let $\mathbf{u} = (u_1,\ldots,u_n)$ and $\mathbf{v} = (v_1,\ldots,v_n) \in C^n$. The **standard inner product** on C^n is defined by

$$<\mathbf{u},\mathbf{v}> = u_1\bar{v}_1 + u_2\bar{v}_2 + \ldots + u_n\bar{v}_n.$$

It is also often denoted $\mathbf{u}\cdot\mathbf{v}$ and called the **dot product** on C^n. The proof that $<\mathbf{u},\mathbf{v}>$ defines an inner product on C^n is similar to the real case and is left as an exercise.

Note that to calculate the dot product $<\mathbf{u},\mathbf{v}>$, we must first find the **complex conjugate** $\overline{v}_i$ of each component v_i of $\mathbf{v}$. In particular, the formula for the norm may be written as

$$|\mathbf{u}| = \sqrt{|u_1|^2 + |u_2|^2 + \dots + |u_n|^2}.$$

For example, let $\mathbf{u}_1 = (-2 - i, 3 + 2i, 4 - 2i)$, $\mathbf{u}_2 = (-i, -1 - 5i, 3 + 2i)$, and $\mathbf{u}_3 = (10 - 10i, 20 + 10i, -16 + 3i) \in C^3$. Then,

$$\begin{aligned}
<\mathbf{u}_1,\mathbf{u}_2> &= (-2 - i)(i) + (3 + 2i)(-1 + 5i) + (4 - 2i)(3 - 2i) \\
&= -4 - 3i, \\
<\mathbf{u}_2,\mathbf{u}_1> &= (-i)(-2 + i) + (-1 - 5i)(3 - 2i) + (3 + 2i)(4 + 2i) \\
&= -4 + 3i = \overline{<\mathbf{u}_1,\mathbf{u}_2>}, \\
<\mathbf{u}_1,\mathbf{u}_3> &= (-2 - i)(10 + 10i) + (3 + 2i)(20 - 10i) \\
&\quad + (4 - 2i)(-16 - 3i) \\
&= 0 = <\mathbf{u}_3,\mathbf{u}_1>, \\
<\mathbf{u}_2,\mathbf{u}_3> &= (-i)(10 + 10i) + (-1 - 5i)(20 - 10i) \\
&\quad + (3 + 2i)(-16 - 3i) \\
&= -102 - 141i.
\end{aligned}$$

Thus, only $\mathbf{u}_1$ and $\mathbf{u}_3$ are orthogonal. Also, for example,

$$\begin{aligned}
|\mathbf{u}_1| &= <\mathbf{u}_1,\mathbf{u}_1>^{1/2} \\
&= [(-2 - i)(-2 + i) + (3 + 2i)(3 - 2i) + (4 - 2i)(4 + 2i)]^{1/2} \\
&= \sqrt{38}. \ \square
\end{aligned}$$

More examples of unitary spaces (such as $M_{m\times n}(C)$, $P_n(C)$) which are analogous to real inner product spaces are given in the exercises.

The next example shows how to compute orthonormal bases for C^n (with the dot product) using the Gram-Schmidt process (Theorem 7-4).

EXAMPLE 7-41

Let $\mathbf{u}_1$, $\mathbf{u}_2$, and $\mathbf{u}_3$ be the vectors defined in the preceding example. The reader may show that these vectors are linearly independent, and so constitute a basis B for C^3. To start the Gram-Schmidt process, we construct an orthogonal basis $B' = \{\mathbf{v}_1,\mathbf{v}_2,\mathbf{v}_3\}$ from B. Then, $\mathbf{v}_1 = \mathbf{u}_1$ and, using equation 7-2 (section 7-2) and the results of Example 7-40, we get

$$\begin{aligned}
\mathbf{v}_2 &= \mathbf{u}_2 - \frac{<\mathbf{u}_2,\mathbf{v}_1>}{<\mathbf{v}_1,\mathbf{v}_1>}\mathbf{v}_1 \\
&= (-i, -1 - 5i, 3 + 2i) - \frac{-4 + 3i}{38}(-2 - i, 3 + 2i, 4 - 2i) \\
&= \frac{1}{38}(-11 - 36i, -20 - 191i, 124 + 56i).
\end{aligned}$$

Next, we use equation 7-3 to calculate $\mathbf{v}_3$. We first need to calculate the following:

$$\begin{aligned}
<\mathbf{v}_1,\mathbf{v}_1> &= <\mathbf{u}_1,\mathbf{u}_1> = 38,\\
<\mathbf{u}_3,\mathbf{v}_1> &= <\mathbf{u}_3,\mathbf{u}_1> = 0 \quad \text{(from Example 7-40)},\\
<\mathbf{u}_3,\mathbf{v}_2> &= -102 + 141i,\\
<\mathbf{v}_2,\mathbf{v}_2> &= \frac{1495}{38}.
\end{aligned}$$

Therefore,

$$\begin{aligned}
\mathbf{v}_3 &= \mathbf{u}_3 - \frac{0}{38}\mathbf{v}_1 - \frac{38}{1495}(-102 + 141i)\mathbf{v}_2\\
&= \frac{1}{1495}(8752 - 17071i, 929 - 1712i, 3376 - 7287i).
\end{aligned}$$

Finally, we normalize to obtain the orthonormal basis $B'' = \{\mathbf{w}_1,\mathbf{w}_2,\mathbf{w}_3\}$. Since

$$<\mathbf{v}_3,\mathbf{v}_3> = \frac{58369}{299}$$

and $\mathbf{w}_i = \dfrac{1}{\sqrt{<\mathbf{v}_i,\mathbf{v}_i>}}\mathbf{v}_i, i = 1, 2, 3,$

we get

$$\mathbf{w}_1 = \frac{1}{\sqrt{38}}(-2 - i, 3 + 2i, 4 - 2i),$$

$$\mathbf{w}_2 = \frac{1}{\sqrt{38 \times 1495}}(-11 - 36i, -20 - 191i, 124 + 56i),$$

$$\mathbf{w}_3 = \frac{1}{5\sqrt{299 \times 58369}}(8752 - 17071i, 929 - 1712i, -3376 - 7287i). \ \square$$

To continue our discussion of the complex analogue of the orthogonal diagonalization of symmetric matrices, we need the next definition.

DEFINITION 7-21

Let $A = (a_{ij}) \in M_{m\times n}(C)$. The **conjugate transpose** of A is the matrix $A^* = \overline{A}^t = (\overline{a}_{ji})$.

Recall that in the above notation, $\overline{A}$ denotes the matrix whose entries are the complex conjugates of those of A. Note that A^* is $n \times m$.

EXAMPLE 7-42

Let

$$A = \begin{bmatrix} 1-i & 2+3i & 4-3i \\ -2-i & i & 5 \end{bmatrix}.$$

Then,

$$A^* = \begin{bmatrix} 1+i & -2+i \\ 2-3i & -i \\ 4+3i & 5 \end{bmatrix}. \square$$

A^* is the analogue in $M_{m\times n}(C)$ of the transpose of a real matrix and has the properties summarized in the next result.

THEOREM 7-19 (Properties of the Conjugate Transpose Operation) Let A and B be complex matrices and $z \in C$. Then, provided the expressions are defined, the following properties hold:

(1) $(A + B)^* = A^* + B^*$.

(2) $(zA)^* = \bar{z}A^*$.

(3) $(A^*)^* = A$.

(4) $(AB)^* = B^*A^*$.

PROOF The proof is similar to that of Theorem 2-5 and is left as an exercise (question 15, Exercises 7-6). ■

DEFINITION 7-22

A complex $n\times n$ matrix $A \in M_n(C)$ is **normal** if $A^*A = AA^*$.

EXAMPLE 7-43

The following matrices are normal:

(a) $A = \begin{bmatrix} 1 & i \\ i & 1 \end{bmatrix}$.

For,

$$A^*A = \begin{bmatrix} 1 & -i \\ -i & 1 \end{bmatrix}\begin{bmatrix} 1 & i \\ i & 1 \end{bmatrix} = \begin{bmatrix} 2 & 0 \\ 0 & 2 \end{bmatrix} = AA^*.$$

(b) $A = \begin{bmatrix} 1+i & -1-i \\ 1+i & 1+i \end{bmatrix}$.

We leave the verification for the reader.

(c) Let A be any *real* symmetric matrix. Then A is normal, since $A = A^t = A^*$ (as $A = \overline{A}$) implies $A^*A = AA = AA^*$. However, a symmetric complex matrix is *not* in general normal. For example, the reader may verify that

$$\begin{bmatrix} 1 & i \\ i & 2 \end{bmatrix}$$

is not normal, although it is symmetric. ❑

The appropriate complex analogue of a real symmetric matrix is a **Hermitian matrix**:

DEFINITION 7-23

A complex $n \times n$ matrix $A \in M_n(C)$ is called a **Hermitian matrix** if $A = A^*$.

Clearly, any Hermitian matrix is normal (but the converse is not true, as can be seen from the previous example). Hermitian matrices are important in many physical applications. They are named after the French mathematician Charles Hermite (1822–1901). Note that since $a_{ii} = \overline{a}_{ii}$, it follows that the diagonal entries of a Hermitian matrix are real. For example, the following matrix is Hermitian:

$$\begin{bmatrix} 2 & 1-3i & 2+i \\ 1+3i & -5 & -i \\ 2-i & i & 0 \end{bmatrix}.$$

We can now state (without proof) the complex analogue of Theorem 7-13, Corollary 3. Recall that the latter states, in effect, that if A is a real $n \times n$ matrix, then there is an orthonormal basis for R^n consisting of eigenvectors of A iff A is symmetric.

THEOREM 7-20

Let A be a complex $n \times n$ matrix. Then there is an orthonormal basis B for C^n consisting of eigenvectors of A iff A is normal. ■

As in the real case, this means that if P is the matrix having the orthonormal basis of eigenvectors as its columns, then $P^{-1}AP = D$ is diagonal with the eigenvalues of A along the diagonal. Moreover, P has a special property analogous to the orthogonality for real matrices:

$$P^*P = PP^* = I.$$

To see this, let $\mathbf{v}_j$ be the jth column of P. Of course, this is the jth vector of B. Then, the ith row of P^* is $\mathbf{v}_i{}^* = (\overline{v}_{i,k})$ when $\mathbf{v}_i = (v_{i,k})$. Therefore, the (i,j) entry of P^*P is the matrix product ($1 \times n$ by $n \times 1$)

$$\begin{aligned} \mathbf{v}_i{}^*\mathbf{v}_j &= \sum_{k=1}^{n} \overline{v}_{i,k} v_{j,k} \\ &= \langle \mathbf{v}_j, \mathbf{v}_i \rangle \qquad \text{(the dot product)} \\ &= \delta_{ji}, \end{aligned}$$

showing that $P^*P = I$. The proof that $PP^* = I$ is similar and is left for the reader.

The complex analogue of Definition 7-9 is stated below.

DEFINITION 7-24

A complex $n \times n$ matrix $P \in M_n(C)$ is **unitary** if $P^*P = PP^* = I$.

It follows that P is unitary iff $P^{-1} = P^*$. For example, the matrix

$$\begin{bmatrix} 1/\sqrt{2} & -i/\sqrt{2} \\ -i/\sqrt{2} & 1/\sqrt{2} \end{bmatrix}$$

is unitary because the columns are an orthonormal set of vectors in C^2.

The analogue of Definition 7-13 is

DEFINITION 7-25

A matrix $A' \in M_n(C)$ is **unitarily similar** to A iff there is a unitary matrix P such that $A' = P^*AP$.

Readers may show that this defines an equivalence relation on $M_n(C)$.

We can now restate Theorem 7-20 as

COROLLARY

A complex $n \times n$ matrix $A \in M_n(C)$ is unitarily similar to a diagonal matrix iff A is normal.

PROOF We have seen from the theorem that if A is normal, then $P^*AP = D$ is diagonal for some unitary matrix P. Therefore, A is unitarily similar to a diagonal matrix.

Conversely, if $P^*AP = D$, with P unitary, then $A = PDP^*$, so $A^* = PD^*P^*$ and

$$\begin{aligned} AA^* &= (PDP^*)(PD^*P^*) \\ &= PDD^*P^* \\ &= PD^*DP^* \qquad \text{(since } D \text{ is diagonal)} \\ &= (PD^*P^*)(PDP^*) \\ &= A^*A, \end{aligned}$$

showing that A is normal. ■

Given a normal matrix A, we may diagonalize it using the same procedure as for real symmetric matrices. We may do so because it can be shown that eigenvectors corresponding to distinct eigenvalues are orthogonal and hence linearly independent. The next example illustrates the method.

EXAMPLE 7-44

We diagonalize the matrix

$$A = \begin{bmatrix} 4 & -2i & 4i \\ 2i & 1 & -2 \\ -4i & -2 & 4 \end{bmatrix}.$$

A is Hermitian, so we know it is normal. The characteristic polynomial is

$$f = \det \begin{bmatrix} 4-x & -2i & 4i \\ 2i & 1-x & -2 \\ -4i & -2 & 4-x \end{bmatrix} = -x^2(x-9),$$

so the eigenvalues are $c_1 = 0$ and $c_2 = 9$.

To find the eigenvectors $\mathbf{v} = (v_1,v_2,v_3) \in C^3$ for $c_1 = 0$, we solve the homogeneous system $(A - 0I)\mathbf{v} = \mathbf{0} = (0,0,0)$. The coefficient matrix is A itself, which is reduced by Gauss-Jordan reduction to

$$\begin{bmatrix} 1 & -i/2 & i \\ 0 & 0 & 0 \\ 0 & 0 & 0 \end{bmatrix}.$$

The corresponding system of equations is

$$v_1 = (i/2)v_2 - iv_3,$$

so the eigenspace of $c_1 = 0$ has dimension 2, and an example of a basis is $\{\mathbf{u}_1,\mathbf{u}_2\}$, where $\mathbf{u}_1 = (i/2,1,0)$ and $\mathbf{u}_2 = (-i,0,1)$. This basis is not orthogonal, so we apply the Gram-Schmidt process to it. We set $\mathbf{v}_1 = \mathbf{u}_1$ and

$$\begin{aligned}\mathbf{v}_2 &= \mathbf{u}_2 - \frac{<\mathbf{u}_2,\mathbf{v}_1>}{<\mathbf{v}_1,\mathbf{v}_1>}\mathbf{v}_1 \\ &= (-i,0,1) - \frac{-1/2}{5/4}(i/2,1,0) \\ &= (-4i/5,2/5,1).\end{aligned}$$

Normalizing, we get $\mathbf{w}_1 = \dfrac{2}{\sqrt{5}}(i/2,1,0)$ and $\mathbf{w}_2 = \dfrac{\sqrt{5}}{3}(-4i/5,2/5,1)$.

Similarly, to find the eigenvectors corresponding to $c_2 = 3$, we solve the system $(A - 9I)\mathbf{v} = (0,0,0)$. The coefficient matrix is

$$\begin{bmatrix} -5 & -2i & 4i \\ 2i & -8 & -2 \\ -4i & -2 & -5 \end{bmatrix},$$

which is reduced to

$$\begin{bmatrix} 1 & 0 & -1 \\ 0 & 1 & 1/2 \\ 0 & 0 & 0 \end{bmatrix}.$$

Therefore, the corresponding system of equations is

$$\begin{aligned} v_1 &= \quad\quad v_3 \\ v_2 &= (-1/2)v_3, \end{aligned}$$

and an example of a basis for the 1-dimensional eigenspace is $\mathbf{u}_3 = (1,-1/2,1)$. Normalizing, we get $\mathbf{w}_3 = (2/3)(1,-1/2,1)$. Note that $\mathbf{w}_3$ is guaranteed to be orthogonal to $\mathbf{w}_1$ and $\mathbf{w}_2$.

Thus, an orthonormal basis for C^3 consisting of eigenvectors of A is $B = \{\mathbf{w}_1,\mathbf{w}_2,\mathbf{w}_3\}$. Therefore, $P^*AP = D$ is diagonal, where P has $\mathbf{w}_1,\mathbf{w}_2$, and $\mathbf{w}_3$ as its columns and D has the eigenvalues on its diagonal:

$$P = \begin{bmatrix} \frac{i}{\sqrt{5}} & \frac{-4i}{3\sqrt{5}} & \frac{2}{3} \\ \frac{2}{\sqrt{5}} & \frac{2}{3\sqrt{5}} & \frac{-1}{3} \\ 0 & \frac{\sqrt{5}}{3} & \frac{2}{3} \end{bmatrix}, D = \begin{bmatrix} 0 & 0 & 0 \\ 0 & 0 & 0 \\ 0 & 0 & 9 \end{bmatrix}. \square$$

Note that the eigenvalues of A above are *real*. This is true for *any* Hermitian matrix, but of course, not for all normal matrices. The proof is left for the reader (question 26(a), Exercises 7-6).

EXERCISES 7-6

1. Find the dot product of the following vectors in C^2.

(a) $\mathbf{u} = (1,i)$, $\mathbf{v} = (1,-i)$

(b) $\mathbf{u} = (1 + 2i,-2 + 3i)$, $\mathbf{v} = (-4 - 2i,5 + 2i)$

(c) $\mathbf{u} = (-3 + 4i,4 - 3i)$, $\mathbf{v} = (2 - i,-2 + i)$

2. Find the dot product of the following vectors in C^3.

(a) $\mathbf{u} = (1,-i,i)$, $\mathbf{v} = (i,i,i)$

(b) $\mathbf{u} = (2 + i,i,4 - 2i)$, $\mathbf{v} = (-2i,0,3 + i)$

(c) $\mathbf{u} = (1 - i,1 + i,i)$, $\mathbf{v} = (1 + i,1 - i,i)$

3. **(a)** Find the norms of the vectors **u** and **v** in question 1.

(b) Find the norms of the vectors **u** and **v** in question 2.

4. Prove the following properties of an inner product $<,>$ on a complex vector space V, where $\mathbf{u}, \mathbf{v} \in V$ and $z \in C$.

(a) $<\mathbf{u},\mathbf{v} + \mathbf{w}> = <\mathbf{u},\mathbf{v}> + <\mathbf{u},\mathbf{w}>$

(b) $<\mathbf{u},\mathbf{0}> = <\mathbf{0},\mathbf{u}> = 0$

(c) $<\mathbf{u},z\mathbf{v}> = \bar{z}<\mathbf{u},\mathbf{v}>$

5. Which of the following define inner products on C^2? For each given inner product, either verify that all axioms hold, or list those that fail. Let $\mathbf{u} = (u_1,u_2)$ and $\mathbf{v} = (v_1,v_2) \in C^2$.

(a) $<\mathbf{u},\mathbf{v}> = u_1v_1 - u_2v_2$

(b) $<\mathbf{u},\mathbf{v}> = \bar{u}_1v_1 + \bar{u}_2v_2$

(c) $<\mathbf{u},\mathbf{v}> = u_1\bar{v}_1 + 2u_2\bar{v}_2$

(d) $<\mathbf{u},\mathbf{v}> = u_1\bar{v}_1 - u_2\bar{v}_2 + iu_1\bar{v}_1 - 2u_2\bar{v}_2$

6. Let **u** and **v** be vectors in C^n with the dot product. Use the Cauchy-Schwarz inequality to show that

$$\left|\sum_{i=1}^{n} u_i\bar{v}_i\right| \leq \sqrt{\sum_{i=1}^{n} |u_i|^2}\sqrt{\sum_{j=1}^{n} |v_j|^2}.$$

7. **(a)** Show that $M_{m\times n}(C)$ is an inner product space with $<,>$ defined by $<A,B> = tr\ AB^*$ (cf. Example 7-3 and Definition 7-21).

(b) Compute $<A,B>$, $|A|$, and $|B|$ for

$$A = \begin{bmatrix} i & 1 - i \\ 1 + i & 2 \end{bmatrix}, B = \begin{bmatrix} 2 - i & 1 \\ i & 1 + i \end{bmatrix},$$

where $<,>$ is the inner product on $M_2(C)$ defined in (a).

8. **(a)** Show that $P_n(C)$ is an inner product space with $<,>$ defined by

$$<f,g> = \sum_{i=0}^{n} a_i\bar{b}_i,$$

when

$$f(z) = \sum_{i=0}^{n} a_i z^i, g(z) = \sum_{j=0}^{n} b_j z^j, z \in C.$$

(b) Compute $<f,g>$, $|f|$, and $|g|$ for f and $g \in P_2(C)$ defined by

$$f(z) = 1 + i - (2 - 3i)z + iz^2,$$
$$g(z) = 1 - 3i + iz - (2 - i)z^2, z \in C,$$

where $<,>$ is the inner product defined in (a).

9. Let V be a complex inner product space. Show that any finite set of orthogonal vectors is linearly independent.

10. Which of the following sets of vectors in C^3 (with the dot product) are orthogonal? Which are orthonormal?

(a) $(i/\sqrt{3},i/\sqrt{3},-i/\sqrt{3})$, $(i,-i,0)$, $(-i,0,-i)$

(b) $(i/\sqrt{3},i/\sqrt{3},-i/\sqrt{3})$, $(i/\sqrt{2},0,i/\sqrt{2})$, $(i/\sqrt{2},0,-i/\sqrt{2})$

(c) $(2/3,-2i/3,i/3)$, $(1/3,2i/3,2i/3)$, $(2/3,i/3,-2i/3)$

11. Show that each of the following sets of vectors $\{\mathbf{u}_1,\mathbf{u}_2\}$ is a basis for C^2. Then use the Gram-Schmidt process to construct an orthonormal basis from each set.

(a) $\mathbf{u}_1 = (1,i)$, $\mathbf{u}_2 = (i,i)$

(b) $\mathbf{u}_1 = (1 + i,1 - i)$, $\mathbf{u}_2 = (i,-i)$

12. Show that each of the following sets of vectors $\{\mathbf{u}_1,\mathbf{u}_2,\mathbf{u}_3\}$ is a basis for C^3. Then use the Gram-Schmidt process to construct an orthonormal basis from each set.

(a) $\mathbf{u}_1 = (0,1,i)$, $\mathbf{u}_2 = (i,i,0)$, $\mathbf{u}_3 = (i,1,0)$

(b) $\mathbf{u}_1 = (1 + i,1 - i,i)$, $\mathbf{u}_2 = (0,i,-i)$, $\mathbf{u}_3 = (i,i,i)$

13. Find A^* for each of the following matrices.

(a) $\begin{bmatrix} i & 1-i \\ 5 & -i \end{bmatrix}$ **(b)** $\begin{bmatrix} 3-2i & 4+5i \end{bmatrix}$

(c) $\begin{bmatrix} i & i & -2i \\ 1+4i & 0 & 2 \end{bmatrix}$

14. Show that if $A \in M_n(C)$ is invertible, then $(A^{-1})^* = (A^*)^{-1}$.

15. Prove Theorem 7-19.

16. **(a)** For $A \in M_n(C)$, show that $det\,\overline{A} = \overline{det\,A}$.

(b) Use the result of (a) to show $det\,A^* = \overline{det\,A}$.

17. Which of the following matrices are normal? Which are Hermitian?

(a) $\begin{bmatrix} 0 & -1 \\ 1 & 0 \end{bmatrix}$ (b) $\begin{bmatrix} -i & i \\ i & i \end{bmatrix}$

(c) $\begin{bmatrix} 1-i & 1+i \\ 1-i & 1+i \end{bmatrix}$ (d) $\begin{bmatrix} 1 & -i & 2 \\ i & i & 2+i \\ 2 & 2-i & 0 \end{bmatrix}$

(e) $\begin{bmatrix} 1 & i \\ i & 2 \end{bmatrix}$ (f) $\begin{bmatrix} 1 & 2+3i & i \\ 2-3i & -2 & 0 \\ -i & 0 & 3 \end{bmatrix}$

(g) $\begin{bmatrix} i & -i & i \\ i & -i & i \\ -i & -i & i \end{bmatrix}$

18. Show that the set of $n \times n$ Hermitian matrices is not a subspace of $M_n(C)$ (compare with the symmetric matrices).

19. Which of the following matrices are unitary? Which are also Hermitian?

(a) $\begin{bmatrix} 1/\sqrt{2} & -1/\sqrt{2} \\ 1/\sqrt{2} & 1/\sqrt{2} \end{bmatrix}$ (b) $\begin{bmatrix} i/\sqrt{2} & -1/\sqrt{2} \\ -i/\sqrt{2} & 1/\sqrt{2} \end{bmatrix}$

(c) $\begin{bmatrix} 1/\sqrt{3} & (1-i)/\sqrt{3} \\ (1+i)/\sqrt{3} & -1/\sqrt{3} \end{bmatrix}$ (d) $\begin{bmatrix} i & 0 & 0 \\ 0 & 1-i & 0 \\ 0 & 0 & 1+2i \end{bmatrix}$

(e) $\begin{bmatrix} 2/3 & 1/3 & 2/3 \\ -2i/3 & 2i/3 & i/3 \\ i/3 & 2i/3 & -2i/3 \end{bmatrix}$

In questions 20 to 24, find the eigenvalues and eigenvectors of the matrix A and a unitary matrix P such that P^*AP is diagonal.

20. $A = \begin{bmatrix} -1 & 1-i \\ 1+i & -2 \end{bmatrix}$

21. $A = \begin{bmatrix} 0 & i \\ -i & 0 \end{bmatrix}$

22. $A = \begin{bmatrix} 1+i & -1-i \\ 1+i & 1+i \end{bmatrix}$

23. $A = \begin{bmatrix} 0 & 1 & i \\ 1 & 1 & 0 \\ -i & 0 & 1 \end{bmatrix}$

24. $A = \begin{bmatrix} 0 & -2i & 2i \\ 2i & -1 & 0 \\ -2i & 0 & 1 \end{bmatrix}$

25. Show that the rotation matrix of Example 7-38 is normal and find a unitary matrix P which diagonalizes it.

26. **(a)** Show that a Hermitian matrix A has real eigenvalues.

(b) Show that $det\ A$ is real if A is Hermitian.

(c) Show that $|det\ A| = 1$ if A is unitary. Is $det\ A$ necessarily real? Illustrate with an example.

27. Prove that a real $n \times n$ matrix is normal and has real eigenvalues iff it is symmetric.

A COMPLEX NUMBERS

This appendix introduces readers to the basic concepts of complex numbers to enable them to study the optional topics in complex linear algebra in sections 4-6 and 7-6.

Complex numbers were introduced in the eighteenth century in order that polynomial equations such as $x^2 + 1 = 0$ would have a solution. More generally, readers familiar with the general quadratic equation

$$ax^2 + bx + c = 0,\ a,\ b,\ c \in R,$$

will recall that this equation has no real roots (i.e., solutions) iff the discriminant $b^2 - 4ac < 0$. Thus, in a sense, the real number system is incomplete in that it does not contain the solutions to certain polynomial equations having real coefficients (for the definition of a polynomial, see Appendix B). For this reason, the set of real numbers was extended by adding a new number denoted by the symbol i and having the property

$$i^2 = -1.$$

Then, by substitution, we see that i is a solution of the equation $x^2 + 1 = 0$. The symbol i was introduced by the great Swiss mathematician Leonhard Euler (1707–1783), who made many contributions to number theory and analysis.

More generally, expressions of the form

$$a + bi,$$

with a and $b \in R$, were called **complex numbers** and used according to the familiar rules of arithmetic, together with the property $i^2 = -1$. For example,

$$\begin{aligned}(-3 + 2i) + (5 - 3i) &= -3 + 5 + 2i - 3i = 2 - i,\\ (-3 + 2i)(5 - 3i) &= -15 + 10i + 9i - 6i^2\\ &= -15 + 19i + 6 \quad (\text{since } i^2 = -1)\\ &= -9 + 19i.\end{aligned}$$

This informal use of the symbol i is unsatisfactory because it does not indicate what set the symbol i belongs to. We can remedy this by recognizing that $a + bi$ is completely determined by the ordered pair (a,b).

> **DEFINITION A-1**
> A **complex number** is an ordered pair of real numbers (a,b). In this notation, a is called the **real part** of (a,b) and b is called the **imaginary part**.

For example, $(-3,2)$ is equivalent to the informal notation $-3 + 2i$; -3 is the real part and 2 is the imaginary part. Complex numbers of the form $(a,0) = a + 0i$ are called **real numbers**, while those of the form $(0,b) = 0 + bi$ are called **pure imaginary**. They are usually abbreviated to a and bi respectively. For example,

$$\begin{aligned} 2 &= (2,0) = 2 + 0i, \\ -3i &= (0,-3) = 0 - 3i, \\ 0 &= (0,0) = 0 + 0i. \end{aligned}$$

Moreover, with this notation, we have located the symbol $i = 0 + 1i = (0,1)$ as a point in the plane R^2. It is also common to denote complex numbers by a single symbol such as

$$z = a + bi,$$

in which case we may write

$$\begin{aligned} a &= Re(z), \text{ the real part of } z; \\ b &= Im(z), \text{ the imaginary part of } z. \end{aligned}$$

For example, if $z = -2 + 3i$, $Re(z) = -2$ and $Im(z) = 3$.

So, for the moment, the set of complex numbers is simply R^2. For this reason, we may represent complex numbers as points or vectors in the plane, as shown in Figures A-1 and A-2 respectively. Figure A-1 is often called an **Argand diagram**.

FIGURE A-1

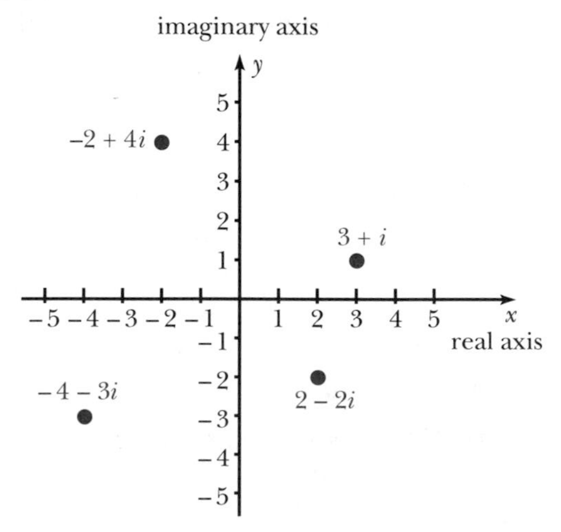

FIGURE A-2

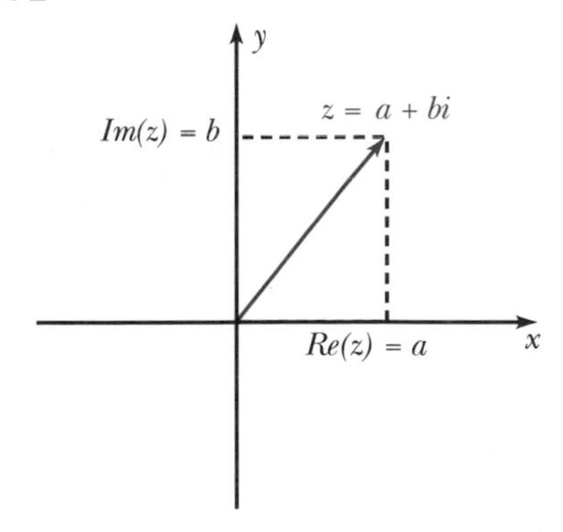

When the plane is construed in these ways, it is called the **complex plane**, and the x and y axes are referred to as the **real** and **imaginary** axes respectively. As we can see, the real numbers $a = (a,0)$ lie along the real axis and the pure imaginary numbers $(0,b)$ lie along the imaginary axis.

We shall introduce operations (addition, subtraction, multiplication, and division) on R^2 such that all the familiar properties of arithmetic for the reals are true and $i^2 = -1$. The resulting set, together with these operations will be denoted C and called the **field of complex numbers**.

> **DEFINITION A-2**
>
> Two complex numbers (a,b) and $(c,d) \in C$ are **equal**, written
>
> $$(a,b) = (c,d)$$
>
> iff
>
> $$a = c \text{ and } b = d.$$

In the informal notation, this definition states that $a + bi = c + di$ iff $a = c$ and $b = d$.

EXAMPLE A-1

Suppose $-2 + 4yi = 2x - 8i$. This means $-2 = 2x$ and $4y = -8$, so $x = -1$ and $y = -2$. ❑

The definitions of **addition** and **subtraction** in C are actually the usual vector definitions in the plane.

DEFINITION A-3
If (a,b) and $(c,d) \in C$, their **sum** is

$$(a,b) + (c,d) = (a + c, b + d),$$

and their **difference** is

$$(a,b) - (c,d) = (a - c, b - d).$$

In other words,

$$(a + bi) \pm (c + di) = (a \pm c) + (b \pm d)i.$$

Then (cf. Chapter 1), $0 = (0,0)$ is the **additive identity**, since

$$(a,b) + (0,0) = (0,0) + (a,b) = (a,b).$$

Furthermore, addition is **commutative** and **associative**:

$$\begin{aligned}(a,b) + (c,d) &= (c,d) + (a,b),\\ (a,b) + [(c,d) + (e,f)] &= [(a,b) + (c,d)] + (e,f).\end{aligned}$$

And, every complex number $z = (a,b)$ has an unique **additive inverse** (or **negative**) denoted $-z = (-a,-b)$, since

$$z + (-z) = (a,b) + (-a,-b) = 0 = (0,0).$$

EXAMPLE A-2

Let $z_1 = -2 + 3i$, $z_2 = 7 - 4i$, and $z_3 = 1 - 2i$. Then,

$$\begin{aligned}z_1 + z_2 &= (-2 + 7) + (3 - 4)i = 5 - i\\ &= z_2 + z_1,\\ z_1 - z_2 &= (-2 - 7) + [3 - (-4)]i = -9 + 7i,\\ -z_1 &= 2 - 3i,\\ z_1 + (-z_1) &= (-2 + 3i) + (2 - 3i) = 0,\\ z_2 + z_3 &= (7 - 4i) + (1 - 2i) = 8 - 6i,\\ (z_1 + z_2) + z_3 &= (5 - i) + (1 - 2i) = 6 - 3i = (-2 + 3i) + (8 - 6i)\\ &= z_1 + (z_2 + z_3).\ \square\end{aligned}$$

To obtain the formal definition of the **product** of complex numbers, recall that we want the usual rules of arithmetic to hold and $i^2 = -1$. By expanding the product, we obtain

$$\begin{aligned}(a + bi)(c + di) &= ac + bdi^2 + (ad + bc)i\\ &= (ac - bd) + (ad + bc)i.\end{aligned}$$

Therefore,

> **DEFINITION A-4**
> If (a,b) and $(c,d) \in C$, their **product** is
> $$(a,b)(c,d) = (ac - bd, ad + bc).$$

EXAMPLE A-3

Let $z_1 = -2 + 3i$, $z_2 = 7 - 4i$, and $z_3 = 1 - 2i$ (as in Example A-2). Then,

$$\begin{aligned}
z_1z_2 &= (-2 + 3i)(7 - 4i) = (-14 + 12) + (8 + 21)i = -2 + 29i,\\
z_2z_1 &= (7 - 4i)(-2 + 3i) = (-14 + 12) + (21 + 8)i = -2 + 29i\\
&= z_1z_2,\\
z_2z_3 &= (7 - 4i)(1 - 2i) = (7 - 8) + (-14 - 4)i = -1 - 18i,\\
z_1z_3 &= (-2 + 3i)(1 - 2i) = (-2 + 6) + (4 + 3)i = 4 + 7i,\\
(z_1z_2)z_3 &= (-2 + 29i)(1 - 2i) = (-2 + 58) + (4 + 29)i = 56 + 33i,\\
z_1(z_2z_3) &= (-2 + 3i)(-1 - 18i) = (2 + 54) + (36 - 3)i = 56 + 33i\\
&= (z_1z_2)z_3,\\
z_1(z_2 + z_3) &= (-2 + 3i)(8 - 6i) = (-16 + 18) + (12 + 24)i = 2 + 36i,\\
z_1z_2 + z_1z_3 &= (-2 + 29i) + (4 + 7i) = 2 + 36i\\
&= z_1(z_2 + z_3). \quad \square
\end{aligned}$$

The above example illustrates the familiar properties from real arithmetic:

$$\begin{aligned}
z_1z_2 &= z_2z_1 \quad \text{(commutative law of multiplication)},\\
(z_1z_2)z_3 &= z_1(z_2z_3) \quad \text{(associative law of multiplication)},\\
z_1(z_2 + z_3) &= z_1z_2 + z_1z_3 \quad \text{(distributive law)}.
\end{aligned}$$

Note also that, *using the definition*, we obtain

$$\begin{aligned}
i^2 &= (0 + i)(0 + i) = (0 - 1) + (0 + 0)i = -1,\\
1z &= (1 + 0i)(a + bi) = (a + 0) + (b + 0)i = a + bi = z,
\end{aligned}$$

as we would expect.

Before discussing division of complex numbers, we first introduce the following concept.

> **DEFINITION A-5**
> If $z = a + bi \in C$, the (complex) **conjugate** of z is $\bar{z} = a - bi$.

As an example, if $z = -3 - 4i$, $\bar{z} = -3 + 4i$. Also, $\bar{i} = -i$ and $\bar{1} = 1$. Note that $\bar{z}$ is the mirror image of z in the real axis, as shown in Figure A-3.

FIGURE A-3

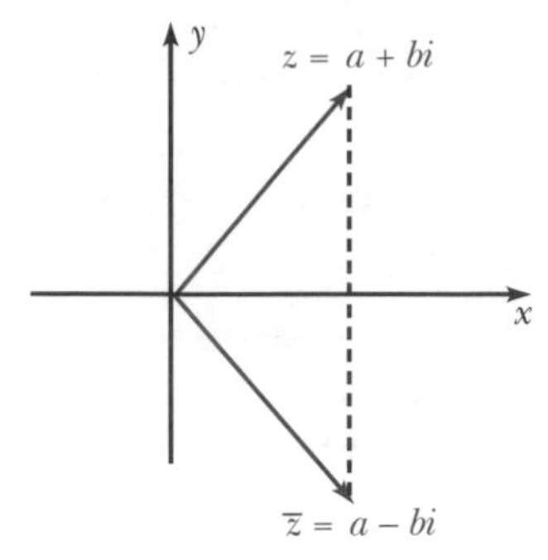

We see that, as vectors in the plane, z and $\bar{z}$ have the same **length**. This gives rise to the next definition.

> **DEFINITION A-6**
> If $z = a + bi \in C$, the **modulus** of z is $|z| = \sqrt{a^2 + b^2}$.

Therefore, $|z| = |\bar{z}|$. This is also called the **absolute value**, or magnitude of z. For example, if $z = -3 - 4i$, $|z| = \sqrt{9 + 16} = \sqrt{25} = 5$. Also, $|i| = 1 = |1|$. When $z = a + 0i$, $|z| = \sqrt{a^2 + 0} = |a|$, the absolute value of the real number a. Note that $|z| = \sqrt{a^2 + b^2} = 0$ iff $a = b = 0$, i.e., iff $z = 0$.

> **THEOREM A-1**
> If $z = a + bi \in C$, $z\bar{z} = |z|^2$.

PROOF Let $z = a + bi$. Then,

$$z\bar{z} = (a + bi)(a - bi) = a^2 + b^2 = |z|^2. \blacksquare$$

We next show that division by $z \neq 0$ is possible. We first show there is an unique $z' \in C$ such that

$$zz' = 1.$$

To show that z' exists, let $z = a + bi$ and $z' = x + yi$. We must show that real numbers x and y can be found so that

$$(a + bi)(x + yi) = 1,$$

or,

$$(ax - by) + (ay + bx)i = 1 + 0i.$$

This last equation is true iff the system of equations

$$\begin{aligned} ax - by &= 1 \\ bx + ay &= 0 \end{aligned}$$

has a solution for x and y. Since the coefficient matrix has determinant $(a^2 + b^2) = |z|^2 \neq 0$ when $z \neq 0$, an unique solution exists which is given by

$$x = \frac{a}{a^2 + b^2}, \quad y = \frac{-b}{a^2 + b^2}.$$

Therefore, we have proved that z' exists and is unique; we denote it by $\frac{1}{z}$. So,

THEOREM A-2

If $z = a + bi \neq 0 \in C$,

$$\frac{1}{z} = \frac{a}{a^2 + b^2} + \frac{-b}{a^2 + b^2}i. \blacksquare$$

Alternatively, we may write this as

$$\frac{1}{z} = \frac{\bar{z}}{|z|^2}.$$

We use this fact to define **division**.

DEFINITION A-7

If $z_1, z_2 \in C$ and $z_2 \neq 0$,

$$\frac{z_1}{z_2} = z_1 \frac{1}{z_2} = \frac{z_1 \bar{z}_2}{|z_2|^2}.$$

Consequently, *C has the same arithmetical properties as the field of real numbers* R: addition, subtraction, multiplication, and division. Furthermore, *R may be regarded as a subset of C*, so C does indeed "extend" R.

EXAMPLE A-4

(a) If $z = -3 - 4i$, then $\bar{z} = -3 + 4i$ and $|z| = 5$, so

$$\frac{1}{z} = \frac{-3 + 4i}{25} = \frac{-3}{25} + \frac{4}{25}i.$$

(b) $$\frac{-2+3i}{3-5i} = \frac{-2+3i}{3-5i}\left(\frac{3+5i}{3+5i}\right) = \frac{(-2+3i)(3+5i)}{(3-5i)(3+5i)} = \frac{-21-i}{34} = -\frac{21}{34} - \frac{1}{34}i. \square$$

This last example illustrates an easy way to remember division: *multiply numerator and denominator by the conjugate of the denominator.*

Readers are alerted to the following important properties of $\bar{z}$, $Re(z)$, $Im(z)$, and $|z|$. The proofs are left as exercises (question 9).

THEOREM A-3

If $z, z_1, z_2 \in C$, the following properties hold:

(1) $\overline{z_1 \pm z_2} = \bar{z}_1 \pm \bar{z}_2$.

(2) $\overline{z_1 z_2} = \bar{z}_1 \bar{z}_2$.

(3) $\overline{z_1/z_2} = \bar{z}_1/\bar{z}_2$ $(z_2 \neq 0)$.

(4) $\bar{\bar{z}} = z$.

(5) $Re(z) = \frac{1}{2}(z + \bar{z})$.

(6) $Im(z) = \frac{-i}{2}(z - \bar{z})$.

(7) $Re(z_1 \pm z_2) = Re(z_1) \pm Re(z_2)$.

(8) $Im(z_1 \pm z_2) = Im(z_1) \pm Im(z_2)$.

(9) $|z_1 + z_2| \leq |z_1| + |z_2|$. ■

We now exploit the geometry of the plane to obtain another formulation of complex numbers. In Figure A-4, we see that if θ is the angle between the vector $z = x + yi = x + iy$ and the positive real axis (i.e., the vector $1 = 1 + 0i$), then

$$x = r\cos\theta, y = r\sin\theta, \text{ where } r = |z| = \sqrt{x^2 + y^2}.$$

Therefore, we may write

$$z = r(\cos\theta + i\sin\theta).$$

This is called the **polar form** of the complex number z. Conversely, for any

FIGURE A-4

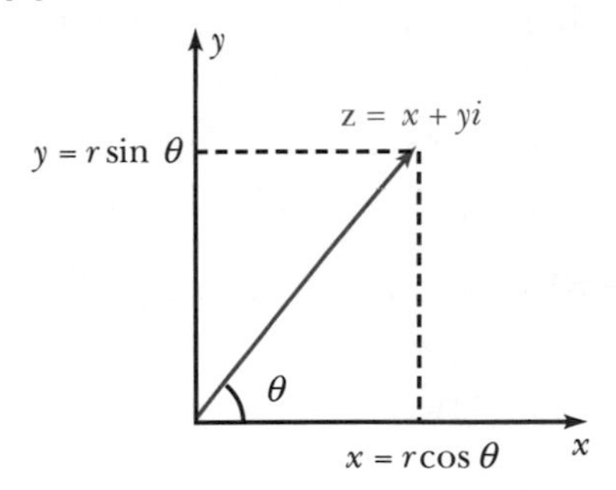

$r \geq 0$ and $\theta \in R$, the above expression represents a complex number and $r = 0$ iff $z = 0$.

The number θ is called an **argument** of z and is often denoted by *arg z*. In fact, the polar forms

$$r\,[\cos(\theta + 2k\pi) + i\sin(\theta + 2k\pi)],\ k = 0, \pm 1, \pm 2, \ldots$$

all represent the same complex number $z = r(\cos\theta + i\sin\theta)$, where $-\pi < \theta \leq \pi$, so *arg z* actually has infinitely many values. To avoid this ambiguity, we define the **principal argument** *Arg z* of z to be the unique value of θ satisfying $-\pi < \theta \leq \pi$. (Note the capital "*A*" in the notation to distinguish the principal argument from *arg z*.)

EXAMPLE A-5

(a) Let $z = 5 + 5i$. Then, $x = 5$ and $y = 5$, so,

$$r = \sqrt{5^2 + 5^2} = 5\sqrt{2},$$

$$x = 5 = r\cos\theta = 5\sqrt{2}\cos\theta,$$

$$y = 5 = r\sin\theta = 5\sqrt{2}\sin\theta.$$

Thus, $\cos\theta = 1/\sqrt{2} = \sin\theta$. The unique angle θ in the range $-\pi < \theta \leq \pi$ corresponding to these values is $\theta = \pi/4$ (= 45°). Therefore, $Arg\, z = \pi/4$ and the polar form of z is

$$z = 5\sqrt{2}(\cos\pi/4 + i\sin\pi/4).$$

(b) Let $z = -\sqrt{3} - i$. Then, $x = -\sqrt{3}$ and $y = -1$, so

$$r = \sqrt{(-\sqrt{3})^2 + (-1)^2} = 2,$$

$$x = -\sqrt{3} = r\cos\theta = 2\cos\theta,$$

$$y = -1 = r\sin\theta = 2\sin\theta.$$

Thus, $\cos\theta = -\sqrt{3}/2$, $\sin\theta = -1/2$. The unique angle θ in the range $-\pi < \theta \le \pi$ corresponding to these values is $\theta = -5\pi/6$ ($= -150°$). Therefore, $Arg\, z = -5\pi/6$ and the polar form of z is

$$z = 2[\cos(-5\pi/6) + i\sin(-5\pi/6)].$$

(c) If a complex number z has polar form

$$z = 3[\cos(-\pi/3) + i\sin(-\pi/3)],$$

then $r = 3$, so

$$x = 3\cos(-\pi/3) = 3(1/2) = 3/2,$$

$$y = 3\sin(-\pi/3) = 3(-\sqrt{3}/2) = -3\sqrt{3}/2.$$

Thus, $z = 3/2 - (3\sqrt{3}/2)\,i$. ❑

The next theorem gives the expressions for multiplication and division of complex numbers in polar form. We will use them to obtain a geometric interpretation of those operations.

THEOREM A-4

If $z_1 = r_1(\cos\theta_1 + i\sin\theta_1)$ and $z_2 = r_2(\cos\theta_2 + i\sin\theta_2)$, then

$$z_1z_2 = r_1r_2[\cos(\theta_1+\theta_2) + i\sin(\theta_1+\theta_2)],$$

$$\frac{z_1}{z_2} = \frac{r_1}{r_2}[\cos(\theta_1-\theta_2) + i\sin(\theta_1-\theta_2)], \text{ for } z_2 \ne 0.$$

PROOF

$$\begin{aligned} z_1z_2 &= [r_1(\cos\theta_1 + i\sin\theta_1)][r_2(\cos\theta_2 + i\sin\theta_2)] \\ &= r_1r_2[(\cos\theta_1\cos\theta_2 - \sin\theta_1\sin\theta_2) + i(\sin\theta_1\cos\theta_2 \\ &\quad + \cos\theta_1\sin\theta_2)] \\ &= r_1r_2[\cos(\theta_1+\theta_2) + i\sin(\theta_1+\theta_2)], \end{aligned}$$

by the well-known trigonometric identities for sums of angles.

Next, since $\bar{z}_2 = r_2(\cos\theta_2 - i\sin\theta_2)$ and $|z_2|^2 = r_2^2$, then, for $z_2 \ne 0$, we have

$$\begin{aligned} \frac{z_1}{z_2} &= \frac{z_1\bar{z}_2}{|z_2|^2} \\ &= \frac{[r_1(\cos\theta_1 + i\sin\theta_1)][r_2(\cos\theta_2 - i\sin\theta_2)]}{r_2^2} \\ &= \frac{r_1}{r_2}[(\cos\theta_1\cos\theta_2 + \sin\theta_1\sin\theta_2) + i(\sin\theta_1\cos\theta_2 - \cos\theta_1\sin\theta_2)] \\ &= \frac{r_1}{r_2}[\cos(\theta_1-\theta_2) + i\sin(\theta_1-\theta_2)], \end{aligned}$$

by the trigonometric identities for differences of angles. ■

COROLLARY

If z_1 and $z_2 \in C$, then

$$|z_1 z_2| = |z_1||z_2|,$$

$$\left|\frac{z_1}{z_2}\right| = \frac{|z_1|}{|z_2|}, z_2 \neq 0,$$

$$arg\ z_1 z_2 = arg\ z_1 + arg\ z_2,$$

$$arg\frac{z_1}{z_2} = arg\ z_1 - arg\ z_2, z_2 \neq 0. \blacksquare$$

We may restate Theorem A-4 as follows:

(1) The product of two complex numbers is obtained by multiplying their moduli and adding their arguments.

(2) The quotient of two complex numbers is obtained by dividing their moduli and subtracting their arguments.

EXAMPLE A-6

Let

$$z_1 = 5 + 5i = 5\sqrt{2}(\cos \pi/4 + i \sin \pi/4)$$

and

$$z_2 = -\sqrt{3} - i = 2[\cos(-5\pi/6) + i \sin(-5\pi/6)]$$

(Example A-5). Then,

$$\begin{aligned} z_1 z_2 &= r_1 r_2 [\cos(\theta_1 + \theta_2) + i \sin(\theta_1 + \theta_2)] \\ &= (5\sqrt{2})(2)[\cos(\pi/4 - 5\pi/6) + i \sin(\pi/4 - 5\pi/6)] \\ &= 10\sqrt{2}[\cos(-7\pi/12) + i \sin(-7\pi/12)] \\ &= 10\sqrt{2}(-\cos 75° - i \sin 75°) \\ &= -10\sqrt{2}\left(\frac{\sqrt{3} - 1}{2\sqrt{2}} - \frac{\sqrt{3} + 1}{2\sqrt{2}}i\right) \\ &= (-5\sqrt{3} + 5) - (5\sqrt{3} + 5)i. \end{aligned}$$

And,

$$\begin{aligned}\frac{z_1}{z_2} &= \frac{r_1}{r_2}[\cos(\theta_1 - \theta_2) + i \sin(\theta_1 - \theta_2)] \\ &= \frac{5\sqrt{2}}{2}\{\cos[\pi/4 - (-5\pi/6)] + i \sin[\pi/4 - (-5\pi/6)]\} \\ &= \frac{5\sqrt{2}}{2}(\cos 13\pi/12 + i \sin 13\pi/12) \\ &= \frac{5\sqrt{2}}{2}(-\cos 15° - i \sin 15°) \\ &= -\frac{5\sqrt{2}}{2}(\frac{\sqrt{3}+1}{2\sqrt{2}} + \frac{\sqrt{3}-1}{2\sqrt{2}}i) \\ &= (-5\sqrt{3}/4 - 5/4) - (5\sqrt{3}/4 - 5/4)i. \ \square\end{aligned}$$

EXAMPLE A-7

Let z be any complex number. Since $i = \cos \pi/2 + i \sin \pi/2$, $arg\ i = \pi/2$ and $|i| = 1$. Therefore,

$$|zi| = |z||i| = |z|$$

and

$$arg\ (zi) = arg\ z + arg\ i = arg\ z + \pi/2.$$

This means that the result of multiplying the vector z by i is the vector zi obtained by rotating z counterclockwise through an angle of $\pi/2$.

Similarly, for example, the complex number $z' = (1/\sqrt{2})(1 + i)$ has modulus 1 and argument $\pi/4$, so for any $z \in C$, zz' is obtained by rotating z counter-clockwise through an angle of $\pi/4$.

In general, *counterclockwise rotation of* z *through* θ $(\theta > 0)$ *is achieved by the multiplication* zz'', *where* $z'' = \cos\theta + i \sin\theta$. $\square$

We use the polar form of a complex number to obtain the following important result.

THEOREM A-5

If $z = r(\cos\theta + i \sin\theta)$ and n is an integer, then

$$z^n = r^n(\cos n\theta + i \sin n\theta).$$

PROOF When $n = 0$ and 1, the result is trivially true. For $n > 1$, we proceed by induction and repeated application of the multiplication formula with $\theta_1 = \theta_2 = \theta$. For the case $n < 0$, the result follows from the fact when $n > 0$, $z^{-n} = 1/z^n$, and this has argument $(0 - n\theta) = -n\theta$. ■

The case in which $r = 1$ is called **de Moivre's Formula**, after Abraham de Moivre (1667–1754), a French mathematician who lived most of his life in England and made significant contributions to the theories of probability and statistics.

EXAMPLE A-8

We compute $(1 + i)^{10}$ using Theorem A-5. Since

$$1 + i = \sqrt{2}(\cos \pi/4 + i \sin \pi/4),$$

$$\begin{aligned}(1 + i)^{10} &= (\sqrt{2})^{10}[\cos 10(\pi/4) + i \sin 10(\pi/4)] \\ &= 32(\cos \pi/2 + i \sin \pi/2) \\ &= 32(0 + i) \\ &= 32i. \ \square\end{aligned}$$

Next, we illustrate the use of the formula in Theorem A-5 to find roots of complex numbers. We say that u is an nth **root** of a complex number z if $u^n = z$. In this case, we denote u by $z^{1/n}$. Let $z = r(\cos \theta + i \sin \theta)$ $(\neq 0)$ and $u = s(\cos \mu + i \sin \mu)$. Then the equation $u^n = z$ and Theorem A-5 imply $r = |z| = |u^n| = s^n$ and $arg\ z = arg\ u^n$. So,

$$s = \sqrt[n]{r}$$

and

$$n\mu = \theta + 2k\pi, k = 0, \pm 1, \pm 2, \ldots$$

or,

$$\mu = \frac{\theta}{n} + \frac{2k\pi}{n}, k = 0, \pm 1, \pm 2, \ldots$$

Therefore,

$$u = \sqrt[n]{r}\left[\cos\left(\frac{\theta}{n} + \frac{2k\pi}{n}\right) + i \sin\left(\frac{\theta}{n} + \frac{2k\pi}{n}\right)\right], k = 0, \pm 1, \pm 2, \ldots$$

Now, it can be shown that the n numbers produced by setting $k = 0, 1, \ldots, n - 1$ are *distinct* and include *all* the values obtained for other k. Thus, there are n **distinct** n^{th} roots of z given by

$$z^{1/n} = \sqrt[n]{r}\,[\cos\left(\frac{\theta}{n} + \frac{2k\pi}{n}\right) + i \sin\left(\frac{\theta}{n} + \frac{2k\pi}{n}\right)],\ k = 0, 1, \ldots, n - 1.$$

Notice that since they all have the same modulus, they lie on a *circle* in the complex plane with center at the origin and radius $\sqrt[n]{r}$.

EXAMPLE A-9

(a) We calculate the three cube (3rd) roots of $-i = \cos(-\pi/2) + i \sin(-\pi/2.)$ From the previous equation (with $z = -i$ and $n = 3$),

$$(-i)^{1/3} = \sqrt[3]{1}\,[\cos\left(\frac{-\pi/2}{3} + \frac{2k\pi}{3}\right) + i \sin\left(\frac{-\pi/2}{3} + \frac{2k\pi}{3}\right)],\ k = 0, 1, 2.$$

Therefore, the three 3rd roots of $-i$ are

(i) $(k = 0)$: $\cos\frac{-\pi/2}{3} + i \sin\frac{-\pi/2}{3}$

$$= \cos\frac{-\pi}{6} + i \sin\frac{-\pi}{6}$$

$$= \sqrt{3}/2 - i/2;$$

(ii) $(k = 1)$: $\cos\left(\frac{-\pi/2}{3} + \frac{2\pi}{3}\right) + i \sin\left(\frac{-\pi/2}{3} + \frac{2\pi}{3}\right)$

$$= \cos\frac{3\pi}{6} + i \sin\frac{3\pi}{6}$$

$$= \cos\frac{\pi}{2} + i \sin\frac{\pi}{2}$$

$$= i;$$

(iii) $(k = 2)$: $\cos\left(\frac{-\pi/2}{3} + \frac{4\pi}{3}\right) + i \sin\left(\frac{-\pi/2}{3} + \frac{4\pi}{3}\right)$

$$= \cos\frac{7\pi}{6} + i \sin\frac{7\pi}{6}$$

$$= -\sqrt{3}/2 - i/2.$$

These roots are sketched in Figure A-5. They lie on a circle of radius 1 and are spaced $2\pi/3 = 120°$ apart.

FIGURE A-5

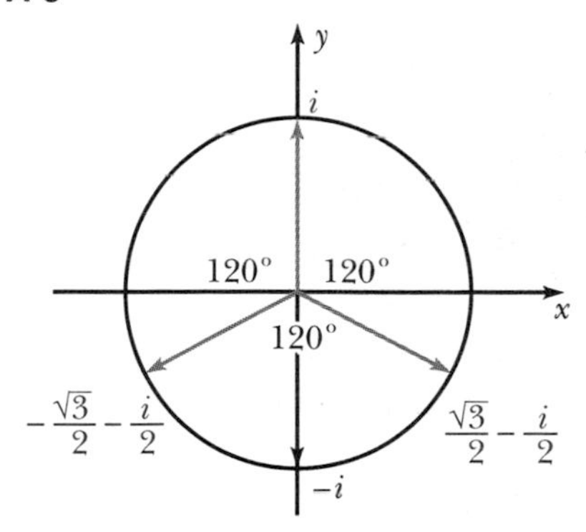

(b) We find the five 5th roots of $1 = \cos 0 + i \sin 0$. By the general formula with $z = 1$ and $n = 5$,

$$1^{1/5} = \sqrt[5]{1}\,[\cos(\frac{0}{5} + \frac{2k\pi}{5}) + i \sin(\frac{0}{5} + \frac{2k\pi}{5})],$$

$$= \cos(\frac{2k\pi}{5}) + i \sin(\frac{2k\pi}{5}), k = 0, 1, 2, 3, 4.$$

Therefore, these five roots are

(i) $(k = 0)$: $\cos 0 + i \sin 0 = 1$;

(ii) $(k = 1)$: $\cos(\frac{2\pi}{5}) + i \sin(\frac{2\pi}{5})$

$$= \cos 72° + i \sin 72°$$

$$= \frac{\sqrt{5} - 1}{4} + \frac{\sqrt{10 + 2\sqrt{5}}}{4} i;$$

(iii) $(k = 2)$: $\cos(\frac{4\pi}{5}) + i \sin(\frac{4\pi}{5})$

$$= \cos 144° + i \sin 144°$$
$$= -\cos 36° + i \sin 36°$$
$$= -\frac{\sqrt{5} - 1}{4} + \frac{\sqrt{10 - 2\sqrt{5}}}{4} i;$$

(iv) $(k = 3)$: $\cos(\frac{6\pi}{5}) + i \sin(\frac{6\pi}{5})$

$$= \cos 216° + i \sin 216°$$
$$= -\cos 36° - i \sin 36°$$
$$= \frac{-\sqrt{5} - 1}{4} - \frac{\sqrt{10 - 2\sqrt{5}}}{4} i;$$

(v) $(k = 4)$: $\cos\left(\frac{8\pi}{5}\right) + i\sin\left(\frac{8\pi}{5}\right)$

$$= \cos 288° + i \sin 288°$$

$$= -\cos 72° - i \sin 72°$$

$$= \frac{\sqrt{5}-1}{4} - \frac{\sqrt{10+2\sqrt{5}}}{4}i.$$

The roots are sketched in Figure A-6, where w_k represents the root corresponding to $k = 0, 1, 2, 3,$ and 4 above. Note that they are spaced $2\pi/5 = 72°$ apart. ❑

FIGURE A-6

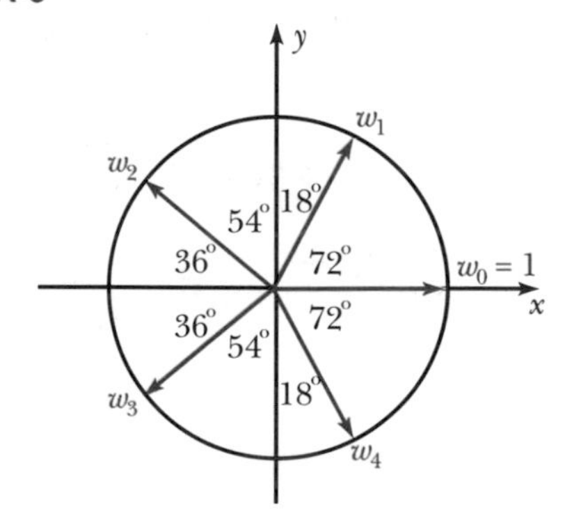

On the basis of the above discussion, we note that the roots of any quadratic equation

$$az^2 + bz + c = 0,$$

where a, b, and $c \in C$, are the complex numbers given by the well-known quadratic formula

$$z = \frac{-b \pm \sqrt{b^2 - 4ac}}{2a}.$$

When a, b, and c are complex, $u = b^2 - 4ac$ may, of course, be complex. If we represent u in polar form, $u = r(\cos\theta + i\sin\theta)$, as before, we see that the square roots of u are

$$u^{1/2} = \sqrt[2]{r}\left[\cos\left(\frac{\theta}{2} + \frac{2k\pi}{2}\right) + i\sin\left(\frac{\theta}{2} + \frac{2k\pi}{2}\right)\right], k = 0, 1,$$

or,

$$\sqrt[2]{r}\left[\cos\frac{\theta}{2} + i\sin\frac{\theta}{2}\right] \text{ and } \sqrt[2]{r}\left[\cos\left(\frac{\theta}{2} + \pi\right) + i\sin\left(\frac{\theta}{2} + \pi\right)\right],$$

which are *negatives* of each other. So, a quadratic equation continues to have two roots, as we would expect. In particular, the quadratic polynomial

$$az^2 + bz + c$$

can be factored as the product of two *linear* polynomials

$$a(z - z_1)(z - z_2),$$

where z_1 and z_2 are the roots of the corresponding quadratic equation.

More generally, we state the following important results without proof. For a discussion of polynomials with real coefficients, see Appendix B.

THEOREM A-6 (Fundamental Theorem of Algebra)
If $a_i \in C$, $a_n \neq 0$, the polynomial $f = a_n z^n + a_{n-1} z^{n-1} + \ldots + a_1 z + a_0$ has a zero in C. ∎

COROLLARY
If $a_i \in C$, $a_n \neq 0$, the polynomial $f = a_n z^n + a_{n-1} z^{n-1} + \ldots + a_1 z + a_0$ can be factored as

$$f = a_n(z - z_1)(z - z_2)\ldots(z - z_n),$$

where $z_n \in C$, not necessarily distinct, are the zeros of f. ∎

It follows that any polynomial equation $f(z) = 0$ has all its roots in C. Therefore, C cannot be further extended (as R was) by adding roots of polynomial equations.

EXERCISES

1. Express each of the following complex numbers z as an ordered pair of real numbers. Plot each number as a vector and find $Re(z)$ and $Im(z)$.
   (a) $-2i$ **(b)** $4 - 3i$ **(c)** $-1 - i$ **(d)** $1 + 3i$
2. Simplify each of the following expressions.
   (a) $(3 - 3i) - (4 + 2i) + 5i$ **(b)** $2i - (-2 - 4i) - (-i)$
   (c) $(-3 + i) + (-1 - i) + 2 - (3 - 3i)$
3. Given $(2 - 3yi) + (2 - 2i) = 2x - 5i$, find x and y.
4. Evaluate the following products.
   (a) $(-5 + 3i)(2i)$ **(b)** $(-1 + 2i)(1 + i)$
   (c) $(4 - 5i)(-2 - 3i)$

5. Find $\bar{z}$ and $|z|$ for each of the following.
(a) i **(b)** $-3 + 2i$ **(c)** 3 **(d)** $1 - i$ **(e)** $4 + 7i$

6. Given $z_1 = -2 - 3i$ and $z_2 = 1 - 2i$, express the following in the form $a + bi$.
(a) $(z_1 + 2i)z_2$ **(b)** $z_2 - (1 - i)z_1z_2$
(c) $z_1{}^2\bar{z}_2$ **(d)** $(z_2 + \bar{z}_2)^2 + iz_2$

7. Graph $(1 + i)$ and $(1 + i)(-2 + i)$.

8. Express the following in the form $a + bi$.
(a) $[(-1 - 3i) + (2 - 2i)](-1 - i)$
(b) $(-3 - i)\overline{(4 + 5i)} - (3 + 2i)^2$ **(c)** $\overline{(1 - 5i)}\overline{(2 + 2i)}\overline{(5i)}$
(d) $(-1 - i)^2 + (2 + i)\overline{(1 + i)}$ **(e)** $(1 + i)^8$.

9. Prove Theorem A-3.

10. Solve the quadratic equations.
(a) $z^2 + z + 1 = 0$ **(b)** $2z^2 + z + 2 = 0$

11. **(a)** Prove that $z_1z_2 = 0$ iff $z_1 = 0$ or $z_2 = 0$.
(b) Prove that $zz_1 = zz_2$, $z \neq 0$, implies $z_1 = z_2$.

12. Find $1/z$ for the following.
(a) $z = -i$ **(b)** $z = 1 - i$ **(c)** $z = 2 - 3i$

13. Express the following in the form $a + bi$.
(a) $\dfrac{1 + i}{1 - i}$ **(b)** $\dfrac{2 - 3i}{(2 + i)(-1 - i)}$ **(c)** $\dfrac{\overline{(1 - 5i)}}{\overline{(2 + 2i)}(5i)}$

(d) $\dfrac{2 - 4i}{3 + i} - \overline{(2 + 2i)}\overline{(1 - 2i)}$ **(e)** $\dfrac{i}{(1 - i)^2} + \dfrac{1 + 3i}{-i}$

14. **(a)** Prove that z is real iff $z = \bar{z}$.
(b) Prove that z is pure imaginary iff $z = -\bar{z}$.

15. Show that if z_1 is a root of $z^2 + 2z + 3 = 0$, then so is $\bar{z}_1$. Generalize this observation to an arbitrary quadratic equation with real coefficients.

16. Find the principal argument of each of the following complex numbers z.
(a) -2 **(b)** $3i$ **(c)** $1 + i$ **(d)** $-1 - \sqrt{3}i$ **(e)** $-\sqrt{2} + \sqrt{2}i$

17. Express each of the complex numbers z in question 16 in polar form.

18. Express the following complex numbers z in polar form.
(a) $-i$ **(b)** $\sqrt{2} - \sqrt{2}i$ **(c)** $2 - 2\sqrt{3}i$ **(d)** $-\sqrt{5} - 2i$ **(e)** $3 - 3i$

19. Calculate the following by first expressing the numerator and denominator in polar form.
(a) $\dfrac{1 - i}{\sqrt{2} + \sqrt{2}i}$ **(b)** $\dfrac{\overline{(\sqrt{3} + i)}}{\overline{(1 + i)}(i)}$

20. Given $z = \cos\theta + i\sin\theta$, show that

(a) $|z| = 1$, **(b)** $z^2 = \cos 2\theta + i\sin 2\theta$, **(c)** $\dfrac{1}{z} = \cos\theta - i\sin\theta$.

21. Use Theorem A-5 to calculate the following.
(a) $(1 - i)^{-3}$ **(b)** $(\sqrt{3} + i)^4$ **(c)** $(1/2 + i/2)^8$
(d) $(\sqrt{5} + 2i)^3$ **(e)** $(2 - i)^{-3}$

22. Find all roots of the following complex numbers and sketch them in the complex plane.
(a) $i^{1/3}$ **(b)** $(-16)^{1/4}$ **(c)** $(\sqrt{3} + i)^{1/2}$ **(d)** $1^{-1/5}$ **(e)** $(8i)^{1/3}$

23. Identify the integers n such that $(1 - i)^n$ is real.

24. Find the zeros of the following polynomials f and hence factorize each as a product of linear factors.
(a) $f = z^2 + z + 1$ **(b)** $f = z^2 + 2z - \sqrt{3}i$ **(c)** $f = z^2 + 2iz - 1$

B POLYNOMIALS

Polynomials arise in a natural way in the study of linear operators and matrices, notably in the study of eigenvalues of matrices and linear transformations (Chapter 6). For this reason, we review in this appendix some basic concepts of factorization and zeros of polynomials with real coefficients. However, the basic definitions also apply to polynomials with complex coefficients, unless otherwise stated.

DEFINITION B-1

A **polynomial** f of **degree** n is an expression of the form

$$f = a_n x^n + a_{n-1} x^{n-1} + \dots + a_1 x^1 + a_0 \qquad \text{(B-1)}$$

where $a_n \neq 0$ and $a_n, a_{n-1}, \dots, a_0$ are real numbers called the **coefficients** of f.

The letter x in the expression is regarded simply as a symbol to locate the coefficients and is often called an **indeterminate** in this context. The equation $f = 0$ is often called the corresponding **polynomial equation**.

The following are examples of polynomials.

(1) $f = 2x^4 - x^3 + x - 3$ is a polynomial of degree $n = 4$ with $a_4 = 2$, $a_3 = -1$, $a_2 = 0$, $a_1 = 1$, and $a_0 = 3$. The corresponding polynomial equation is

$$2x^4 - x^3 + x - 3 = 0.$$

(2) $g = x - 1$ is a polynomial of degree 1. The corresponding polynomial equation is $x - 1 = 0$.

(3) $h = 5 = 5x^0$ is a polynomial of degree 0.

Readers are no doubt struck (and possibly confused!) by the close similarity between this definition of a polynomial and the notion of a polynomial function (Chapter 4). In fact, for our purposes the two concepts are "essentially" the same, although, in the most general setting, they are quite different as described below.

(1) A polynomial is a linear combination of powers on the "indeterminate" x.

(2) A polynomial function is a function $f:R \rightarrow R$ having the special form

$$f(x) = a_n x^n + a_{n-1}x^{n-1} + \ldots + a_1 x^1 + a_0, \textit{ for every number } x \in R.$$

Now, we can use any polynomial f to construct a polynomial function by letting the indeterminate x take on real values and computing $f(x)$ using equation B-1. Conversely, any polynomial function yields a polynomial f when we regard x in the formula for $f(x)$ as an indeterminate. When the coefficients are real or complex numbers, it can be shown that there is a one-to-one correspondence between the set of polynomials and the set of polynomial functions. For this reason, we will *identify* polynomials with polynomial functions. Nevertheless, the reader is cautioned that in more advanced treatments, the coefficients may be taken to be elements of sets different from the real and complex numbers. It can then happen that different polynomials produce the same polynomial function, so the one-to-one correspondence breaks down.

One reason for distinguishing the notion of polynomial from that of polynomial function is that in the eigenvalue problem (Chapter 6), we want to substitute square matrices A (and linear operators T) for the indeterminate x to obtain a matrix $f(A)$ (respectively, a linear operator $f(T)$). On the other hand, if f is a polynomial function, it makes no sense to evaluate it at x other than numbers. With the identification proposed, it follows that the set of polynomials of degree at most n is a **vector space** with the obvious definitions of addition and scalar multiplication:

If $f = a_n x^n + \ldots + a_0$ and $g = b_m x^m + \ldots + b_0$, with $n \geq m$ (say), then

$$f + g = a_n x^n + \ldots + a_{m+1}x^{m+1} + (a_m + b_m)x^m + \ldots + (a_0 + b_0).$$

And if c is a scalar,

$$cf = ca_n x^n + \ldots + ca_0.$$

The **zero polynomial** 0 (all coefficients 0) will be identified with the number 0; we *do not* assign it a degree. Two polynomials f and g are **equal** iff coefficients of all like powers of x are equal.

In addition to these operations, we can also form **products** of polynomials (just as for polynomial functions). In practice, this is called **long multiplication**, with which the reader is no doubt familiar.

EXAMPLE B-1

The product of $f = 3x^3 + 2x^2 - 2$ and $g = -x^2 + 2x + 1$ is the polynomial of degree 5 given by

$$fg = -3x^5 + 4x^4 + 7x^3 + 4x^2 - 4x - 2.$$

This may be obtained by the familiar process shown in Figure B-1. ❑

FIGURE B-1

$$\begin{array}{rrrrrr}
 & & 3x^3 & + 2x^2 & & - 2 \\
 & \times & -x^2 & + 2x & & + 1 \\
\hline
-3x^5 & - 2x^4 & & + 2x^2 & & \\
 & 6x^4 & + 4x^3 & & - 4x & \\
 & & 3x^3 & + 2x^2 & & - 2 \\
\hline
-3x^5 & + 4x^4 & + 7x^3 & + 4x^2 & - 4x & - 2
\end{array}$$

Long Multiplication

In regard to products of polynomials, the following facts are fundamental (recall *deg f* means the degree of f):

> **THEOREM B-1**
>
> Let f and g be polynomials.
>
> (1) If $fg = 0$, either $f = 0$ or $g = 0$ (or both).
>
> (2) If f and g are nonzero, $deg\, fg = deg\, f + deg\, g$. ■

The proof is left as an exercise for the reader (question 2). It depends on the corresponding fact for real numbers that a product $ab = 0$ iff $a = 0$ or $b = 0$ (or both). Note that Example B-1 preceding the theorem illustrates property (2):

$$def\, fg = 5 = 3 + 2 = deg\, f + deg\, g.$$

The reverse process of multiplication is **division**: given a polynomial f (called the **dividend**) and a polynomial d (called the **divisor**), we find a polynomial q (called the **quotient**) such that

$$f = qd.$$

When this is possible (it isn't always), we say that d **divides** f (evenly), that q and d are **factors** of f, and that f is a **multiple** of d and q.

EXAMPLE B-2

Let h be the product of f and g in Example B-1: $h = fg$. Then both f and g divide h, and are factors of h. But, if we put $k = h + 1$, then, since

$$k = fg + 1,$$

f and g do not divide k evenly. Instead, the result of the division leaves a **remainder** of 1. ❑

The process used for determining whether or not one polynomial divides another is called **long division** and is no doubt also familiar to the reader. The fact that division (with or without a remainder) is always possible is stated in the next theorem. We omit the proof, but the idea is illustrated in Example B-3.

> **THEOREM B-2** (Division Algorithm)
>
> Let f and $d \neq 0$ be polynomials. Then there are polynomials q and r such that
>
> $$f = dq + r$$
>
> and either $r = 0$ (in which case d divides f) or $deg\, r < deg\, d$ (in which case r is called the remainder of the division of f by d). ■

We write this symbolically in the long division scheme as shown below.

$$\begin{array}{r|l} & q = \text{quotient} \\ \hline d = \text{divisor} & f = \text{dividend} \\ & \cdots\cdots\cdots\cdots\cdots \\ \hline & r = \text{remainder} \end{array}$$

EXAMPLE B-3

We illustrate the division algorithm using $d = 2x^2 - x + 2$ and $f = 6x^5 - x^3 + 3x + 1$ in Figure B-2.

FIGURE B-2

Long Division

$$\begin{array}{r|l}
 & 3x^3 + (3/2)x^2 - (11/4)x - 23/8 \\ \hline
2x^2 - x + 2 & 6x^5 - x^3 + 3x + 1 \\
 & 6x^5 - 3x^4 + 6x^3 \\ \hline
 & 3x^4 - 7x^3 + 3x + 1 \\
 & 3x^4 - (3/2)x^3 + 3x^2 \\ \hline
 & -(11/2)x^3 - 3x^2 + 3x + 1 \\
 & -(11/2)x^3 + (11/4)x^2 - (11/2)x \\ \hline
 & -(23/4)x^2 + (17/2)x + 1 \\
 & -(23/4)x^2 + (23/8)x - 23/4 \\ \hline
 & \text{remainder } r = (45/8)x + 27/4
\end{array}$$

We see that

$$6x^5 - x^3 + 3x + 1 = (2x^2 - x + 2)[3x^3 + (3/2)x^2 - (11/4)x - 23/8] + [(45/8)x + 27/4],$$

so

$$q = 3x^3 + (3/2)x^2 - (11/4)x - 23/8 \text{ and } r = (45/8)x + 27/4. \square$$

We shall be interested mainly in the case where the divisor d is a **monic** linear polynomial: $d = x - c$ for some scalar c. In general, *a polynomial is monic iff the leading coefficient (i.e., that of the highest power of* x*) is 1*. For the divisor $d = x - c$, there is an easy criterion for deciding whether or not it divides a given polynomial f. For, by the division algorithm, we may write

$$f = (x - c)q + r, \tag{B-2}$$

for some polynomials q and r. Also, either $r = 0$, or $deg\ r < deg(x - c) = 1$. Thus, either $r = 0$ or r is a constant. As noted earlier, we may regard equation B-2 as an equation relating polynomial functions. Let us evaluate both sides at $x = c$ to obtain

$$f(c) = (c - c)q(c) + r(c) = r(c).$$

And $r(c)$ is actually independent of c (the same for all c), since r is a constant. We see from the last equation that if $x - c$ divides f, so that $r(c) = 0$, then we must have $f(c) = 0$. Conversely, if $f(c) = 0$, then $r(c) = r = 0$. Thus, we have proved the following theorems.

THEOREM B-3 (Factor Theorem)

A polynomial $f \neq 0$ contains $x - c$ as a factor iff $f(c) = 0$. ■

THEOREM B-4 (Remainder Theorem)

If a polynomial $f \neq 0$ is divided by $x - c$, then the remainder r is $f(c)$. ■

DEFINITION B-2

A scalar c is called a **zero** of a polynomial f and a **root** of the polynomial equation $f(x) = 0$ iff $f(c) = 0$.

It should be emphasized that in writing $f(c)$, we are regarding the polynomial f as a polynomial function and evaluating it at $x = c$. This value can often be computed much more quickly than by direct substitution using an abbreviated version of long division called **synthetic division**.

EXAMPLE B-4

We find the value $f(2)$ for the polynomial $f = 2x^4 - 3x^2 + x - 3$. The long division of f by $x - 2$ to produce the remainder $f(2) = 19$ is shown in Figure B-3, together with the abbreviated synthetic division scheme. The latter simply keeps track of the important numbers in the long division process (shown in color in the figure). ❑

The steps for the synthetic division of $f = a_n x^n + \ldots + a_0$ by $x - c$ are summarized as follows:

(1) Write down the coefficients of f in order of *descending* powers of x and include 0's for missing powers.

(2) Write c in the box to the right of these coefficients. Note that for division by $(x - c)$, we write c rather than $-c$, which allows us to convert the subtraction of long division to the simpler addition in synthetic division.

(3) Leave a blank line and draw a horizontal addition line.

(4) Rewrite a_n in the first space below the addition line; multiply it by c and enter the product $a_n c$ $(= 2 \times 2 = 4$ in the example) directly below a_{n-1} $(=0$ in the example) in the blank line above the addition line.

(5) Add a_{n-1} and $a_n c$ $(=0 + 4 = 4$ in the example), and write the sum below the addition line.

(6) Repeat steps (4) and (5) using the new numbers generated below the addition line. Then the last number produced is $f(c)$ and the others in the same line are the coefficients of the quotient.

FIGURE B-3

Long Division

versus

Synthetic Division

$$
\begin{array}{r|llllllll}
 & 2x^3 & + & 4x^2 & + & 5x & + & 11 & \\
\hline
x - 2 & 2x^4 & + & 0x^3 & - & 3x^2 & + & x & - 3 \\
 & 2x^4 & - & 4x^3 & & & & & \\
\hline
 & & & 4x^3 & + & -3x^2 & + & x & - 3 \\
 & & & 4x^3 & - & 8x^2 & & & \\
\hline
 & & & & & 5x^2 & + & 1x & - 3 \\
 & & & & & 5x^2 & - & 10x & \\
\hline
 & & & & & & & 11x & + -3 \\
 & & & & & & & 11x & - 22 \\
\hline
 & & & & & & & & 19 = f(2)
\end{array}
$$

$$
\begin{array}{rrrrr|l}
2 & 0 & -3 & 1 & -3 & \underline{2} \\
 & 4 & 8 & 10 & 22 & \\
\hline
2 & 4 & 5 & 11 & 19 & = f(2)
\end{array}
$$

$$f(x) = (x - 2)(2x^3 + 4x^2 + 5x + 11) + 19$$

EXAMPLE B-5

We use synthetic division to calculate the remainder and quotient resulting from the division of $f = 5x^4 - x + 2$ by $x + 3$.

$$\begin{array}{lrrrrrl} x = -3: & 5 & 0 & 0 & -1 & 2 & \underline{|\ -3} \\ & & -15 & 45 & -135 & 408 & \\ \hline & 5 & -15 & 45 & -136 & 410 & = f(-3) \end{array}$$

Thus, $f = (x + 3)(5x^3 - 15x^2 + 45x - 136) + 410$. ❑

As noted earlier in this section, polynomials arise naturally in certain considerations in linear algebra, and their zeros have a special significance (see Chapter 6). We are therefore interested in finding all zeros of a given polynomial f, or, equivalently, the linear factors $(x - c)$ of f. We shall discuss several theorems that will limit the possibilities for the zeros of f and allow the actual zeros to be determined in certain cases.

First, a polynomial of degree 1 has (of course) exactly one zero: if $f = ax + b$, the unique real zero is $x = -b/a$.

Next, recall that the two roots of a quadratic equation

$$f(x) = ax^2 + bx + c = 0$$

are given by the quadratic formula

$$x = \frac{-b \pm \sqrt{b^2 - 4ac}}{2a}.$$

Furthermore, if the discriminant $b^2 - 4ac < 0$, we say that there are **no real roots**, or that the zeros of f are **complex** (see also Appendix A). The next example illustrates these possibilities.

EXAMPLE B-6

(a) Let $f = x^2 + 2x + 1$. Then there are two equal ("repeated") zeros $x = -1$, so we write

$$f = (x + 1)(x + 1)$$

by the factor theorem.

(b) Let $f = x^2 - 6x - 1$. The roots of $f(x) = 0$ are

$$x = \frac{6 \pm \sqrt{36 + 4}}{2} = 3 + \sqrt{10} \text{ and } 3 - \sqrt{10},$$

so we can factorize f as

$$f = (x - 3 - \sqrt{10})(x - 3 + \sqrt{10}).$$

(c) Let $f = x^2 - x + 1$. Then, $f(x) = 0$ has no real roots. ❑

The next theorem tells us what to expect for the zeros of an arbitrary polynomial with real coefficients. It is a consequence of the corollary of the Fundamental Theorem of Algebra stated in Appendix A (Theorem A-6).

THEOREM B-5

Every polynomial f of degree at least one having real coefficients can be factorized as a product of linear and quadratic factors with real coefficients. The quadratic factors have no real zeros. ■

COROLLARY

A polynomial f with real coefficients and of degree n has at most n real zeros. Furthermore, if n is odd, it has at least one real zero.

PROOF Suppose f is factorized according to the theorem as

$$f = (x - c_1) \dots (x - c_r)q_1 \dots q_s,$$

where $c_1, \dots, c_r$ are real numbers (not necessarily different) and q_i are quadratic polynomials with no real zeros. Then, using the fact that the degree of a product is the sum of the degrees of the factors, we see that $n = r + 2s$. So, the number of real zeros of f is $0 \leq r \leq n$. And, if n is odd, r must be odd since $2s$ is even. Thus, r must be at least one. ■

EXAMPLE B-7

Consider $f = x^5 - x$. By inspection, we may factorize it as

$$f = x(x^4 - 1) = x(x - 1)(x + 1)(x^2 + 1).$$

So, f has three real zeros: $x = 0$, 1, and -1. ❑

The question may have occurred to the reader that, since there is an explicit formula for the zeros of a quadratic polynomial, perhaps there is a similar formula for the zeros of an arbitrary polynomial of degree n. In fact, such formulae do exist for $n = 3$ and 4, although they are more complicated than the quadratic case. For $n \geq 5$, however, one of the great discoveries in algebra in the nineteenth century was that no such formula can exist. Therefore, to determine the zeros of such polynomials, other methods must be used.

The next theorem will permit us to find all **rational zeros** of a polynomial with *rational* coefficients. We shall omit the proof.

THEOREM B-6

If $f = a_n x^n + a_{n-1}x^{n-1} + \dots + a_1 x + a_0$ is a polynomial with *integral* coefficients, and $x = p/q$ is a rational zero (expressed in "lowest terms": i.e., p and q are integers with no common factors except 1), then p must be a factor of a_0 (the constant term) and q must be a factor of a_n (the leading coefficient). ■

Note that if f has rational coefficients, we can make them integral by multiplying f by a suitable integer.

EXAMPLE B-8

We factorize the polynomial $f = 6x^4 + x^3 + 11x^2 + 2x - 2$. By the theorem, any rational zero p/q is such that p divides -2 and q divides 6. Thus, the possibilities are as follows:

$$\begin{array}{rl}
\text{for } p: & +1, -1, +2, -2; \\
\text{for } q: & +1, -1, +2, -2, +3, -3, +6, -6; \\
\text{for } p/q: & +1, -1, +2, -2, +1/2, -1/2, +1/3, -1/3, \\
 & +2/3, -2/3, +1/6, -1/6.
\end{array}$$

We must now test whether these 12 possibilities are zeros by checking whether $f(x) = 0$ (for example, using synthetic division). Note that it may happen that *none* of them is a zero; Theorem B-6 does not preclude this. As a part of the procedure, each time a zero c of f is found, it should be factored out as $f = (x - c)q$. The procedure is then continued with the factor q since every zero of q must be a zero of f.

Some sample synthetic divisions are shown below.

(a) $p/q = 2$:

$$\begin{array}{ccccc|c}
6 & 1 & 11 & 2 & -2 & 2 \\
 & 12 & 26 & 74 & 152 & \\
\hline
6 & 13 & 37 & 76 & 150 & = f(2)
\end{array}$$

Thus, $x = 2$ is not a zero.

(b) $p/q = 1/3$:

$$\begin{array}{ccccc|c}
6 & 1 & 11 & 2 & -2 & 1/3 \\
 & 2 & 1 & 4 & 2 & \\
\hline
6 & 3 & 12 & 6 & 0 & = f(1/3)
\end{array}$$

So, $x = 1/3$ is a zero and

$$\begin{aligned} f &= (x - 1/3)(6x^3 + 3x^2 + 12x + 6) \\ &= 3(x - 1/3)(2x^3 + x^2 + 4x + 2). \end{aligned}$$

Applying the procedure to $q = 2x^3 + x^2 + 4x + 2$, we see that its possible rational zeros are $+1, -1, +2, -2, +1/2$, and $-1/2$. Of course we have already established in (a) that $x = 2$ is not a zero. Testing the others, we find that only $x = -1/2$ is a zero, so the complete factorization of f is

$$f = 3(x - 1/3)(x + 1/2)(2x^2 + 4) = 6(x - 1/3)(x + 1/2)(x^2 + 2). \square$$

Notice that while there were initially 12 possibilities for the rational zeros of f in the preceding example, of course we knew by Theorem B-5, Corollary, that at most 4 of these could be actual zeros of f ($deg\, f = 4$).

The next theorem is also useful in further limiting the number of real roots we may expect. But first, let $f = a_n x^n + \ldots + a_0$, as before. We say that f has a **variation in sign** whenever two successive terms in the sum differ in sign, ignoring powers of x with 0 coefficient (i.e., those that are "missing").

EXAMPLE B-9

(a) **(i)** $f = 2x^3 - 3x + 1$ has two variations in sign;

(ii) $g = 6x^5 - x^4 + 2x^2 - 3x + 4$ has four variations in sign;

(iii) $h = -4x^3 - 3x^2 - x + 2$ has only one variation in sign.

We shall also need to consider variations in sign in the polynomial $f(-x)$ obtained by replacing x by $-x$ in the expression for f. Using the above examples, we see

(b) **(i)** $f(-x) = -2x^3 + 3x + 1$ has one variation in sign;

(ii) $g(-x) = -6x^5 - x^4 + 2x^2 + 3x + 4$ has one variation in sign;

(iii) $h(-x) = 4x^3 - 3x^2 + x + 2$ has two variations in sign. ❑

THEOREM B-7 (Descartes' Rule of Signs)

The number of *positive real* zeros of a polynomial f with real coefficients is either equal to the number of variations in sign of f, or to the number of variations in sign decreased by an even number. The number of *negative* zeros of f is the number of positive zeros of $f(-x)$. ■

This theorem is attributed to René Descartes (1596–1650), one of the greatest philosophers of all time and the father of analytic geometry.

EXAMPLE B-10

We use the polynomials in Example B-9 to illustrate the use of Theorem B-7.

(i) f has either 2 or 0 positive real zeros and 1 negative real zero.
(ii) g has 4 or 2 or 0 positive real zeros and 1 negative real zero.
(iii) h has 1 positive zero and either 2 or 0 negative zeros. ❑

EXAMPLE B-11

We find all zeros of the polynomial $f = x^4 - 2x^3 + 3x^2 - 4x + 2$. By Descartes' rule, it has 4 or 2 or 0 positive zeros and no negative zeros. By Theorem B-6, the possible rational zeros are $+1$, -1, $+2$, and -2, from which we may now omit the negative ones by Descartes' rule. Testing $x = 1$ and 2 by synthetic division, we find that only $x = 1$ is a zero of f. We now factor out $(x - 1)$ to get

$$f = (x - 1)(x^3 - x^2 + 2x - 2) = (x - 1)q.$$

Since $x = 2$ cannot be a rational zero of q, the only value we need to test is $x = 1$. This is seen to be a zero (by Descartes' rule, q has at least one positive zero):

$$\begin{array}{lrrrrl} x = 1: & 1 & -1 & 2 & -2 & \lfloor\, 1 \\ & & 1 & 0 & 2 & \\ \hline & 1 & 0 & 2 & 0 & = q(1) \end{array}$$

So, we may factorize f as $f = (x - 1)^2(x^2 + 2)$. ❑

EXAMPLE B-12

Consider the polynomial $f = x^6 - 3x^2 - 2$. By Descartes' rule, it has one positive and one negative zero. By Theorem B-6, the possible rational zeros are $+1$, -1, $+2$, and -2. The reader may check that none of these is, in fact, a zero. It follows that the two real zeros must be *irrational*. ❑

There is, in general, no simple way to find the irrational zeros of a polynomial of degree greater than four. Usually, an **approximation technique** must be used. In the example above, since $f(1) = -4$, while $f(2) = 50$, it can be shown (because the graph of $y = f(x)$ has no jumps or breaks) that there must be a zero between $x = 1$ and $x = 2$. However, we shall not pursue these ideas here. In the case of Example B-12, the reader may verify that $f = (x^2 - 2)(x^2 + 1)^2$ (see question 11).

EXERCISES

1. For each of the following pairs of polynomials f and d, use long division to find the quotient q and the remainder r of the division of f by d.

(a) $f = 2x^2 + 3x + 4,\ d = x - 1$

(b) $f = -x^3 + 2x - 1,\ d = 2x^2 + 3$

(c) $f = x^4 + 3x^3 + 2x^2 + 3x + 1,\ d = x^2 + 1$

(d) $f = 2x^4 - 3x,\ d = 2x^3 - 1$

(e) $f = -3x^6 + 2x^4 - x + 2,\ d = 4x^3 - 1$

2. Prove Theorem B-1.

3. Use synthetic division to find $f(c)$ for the scalars c and polynomials f given below.

(a) $f = x^4 - x^2 + 2x - 6,\ c = 2, -1, 5$

(b) $f = 2x^5 - 3x^4 + x^2 - 1,\ c = 1, -4, 6$

(c) $f = 2x^8 - 4x^6 + 5x^5 + x^3 + 7x - 3,\ c = 1, 3, -4$

4. Use synthetic division to determine whether or not each scalar c given below is a zero of the polynomial f as given. If it is, find the polynomial q such that $f = (x - c)q$.

(a) $f = x^3 + x - 2,\ c = -1$ **(b)** $f = 2x^4 - x^2 + 3,\ c = 2$

(c) $f = -4x^4 + 3x^3 + 4x - 3,\ c = 3/4$

(d) $f = x^6 - 2x^5 + x^4 - x^3 + 2x^2 + 1,\ c = -3$

(e) $f = 6x^4 + 5x^3 - 16x^2 - 9x - 10,\ c = -2$

5. **(a)** Use the Factor Theorem to show that

(i) $x - c$ is a factor of $x^3 - c^3$; **(ii)** $x - c$ is a factor of $x^4 - c^4$;

(iii) $x - c$ is not a factor of $x^4 + c^4$.

(b) Use synthetic division to express $x^n - y^n$ as a multiple of $x - y$.

In each of questions 6 to 10, factorize the given polynomials as a product of linear and quadratic polynomials, where the latter have no real zeros.

6. $x^3 - 2x^2 - 4x + 8$

7. $x^5 + x^4 - x - 1$

8. $2x^4 + x^3 - 5x^2 - 7x - 6$

9. $x^5 + 2x^4 - x^3 - 7x^2 - 20x - 15$

10. $x^7 + 1$

11. **(a)** Substitute $y = x^2$ in $f = x^6 - 3x^2 - 2$ to transform it to a polynomial of degree 3; hence factorize f.

(b) Factorize the polynomial $x^4 - 2x^2 - 3$.

C INTRODUCTION TO LINEAR PROGRAMMING

In this appendix, we discuss the theory and solution of simple linear programming (LP, for short) problems of the type considered at various places in the text. Our goal will be the **simplex method**, the principal technique for finding the solution of LP problems. Linear programming was developed in the late 1940's by associates of the U.S. Department of the Air Force to solve certain problems in military logistics. Within a few years, the method was applied to a wide variety of problems in the optimal allocation of resources, including agriculture, economics, and transportation, to name but a few. The discovery of the simplex method is credited to the contemporary American mathematician G.B. Dantzig (1914–), who has made numerous contributions to the fields of operations research and optimization, to which LP theory belongs.

As we have seen, in LP problems we are concerned with finding the maximum (or minimum) of a **linear function** Z of n variables $x_1, \ldots, x_n$ (the **objective function**) subject to linear inequalities and/or equations (called **constraints**) which the variables must satisfy. Bearing in mind some of the examples we have dealt with in earlier chapters, we introduce the following definitions.

DEFINITION C-1

Let $A = (a_{ij})$ be an $m \times n$ matrix, $C = (c_i)$ be $1 \times n$, $X = (x_i)$ be $n \times 1$, $B = (b_i)$ be $m \times 1$, and $Y = (y_i)$ be $1 \times m$.

(1) The **standard maximization** problem is to find
max $Z = CX$
subject to $AX \le B$, $X \ge O$;

i.e., $\max Z = \sum_{i=1}^{n} c_i x_i$ subject to $\sum_{i=1}^{n} a_{ji} x_i \le b_j$ and $x_i \ge 0$,

for $1 \le i \le n$ and $1 \le j \le m$.

DEFINITION C-1 (cont'd)

(2) The **standard minimization** problem is to find
min $Z = YB$
subject to $YA \geq C$, $Y \geq O$;

i.e., min $Z = \sum_{j=1}^{m} y_j b_j$ subject to $\sum_{j=1}^{m} y_j a_{ji} \geq c_i$ and $y_j \geq 0$,

for $1 \leq i \leq n$ and $1 \leq j \leq m$.

EXAMPLE C-1

Readers may verify that the LP problem in Example 5-1 is a standard maximization problem with

$$A = \begin{bmatrix} 4 & 2 \\ 2 & 6 \\ 3 & 3 \end{bmatrix}, C = \begin{bmatrix} 15 & 12 \end{bmatrix}, B = \begin{bmatrix} 80 \\ 120 \\ 75 \end{bmatrix}. \square$$

EXAMPLE C-2 (The Diet Problem)

Typically, diet problems are concerned with making a minimum cost food product using various ingredients and in such a way that certain nutritional requirements are met or exceeded. By way of example, suppose a health food manufacturer wants to make a cake containing honey and nuts such that each cake contains at least 9 units of protein, 23 units of carbohydrate, and 5 units of vitamin B. The numbers of units of these nutrients per ounce of honey and nuts are listed in the table below.

	Honey	*Nuts*
Protein	1	3
Carbohydrate	3	5
Vitamin B	1	1

Suppose further that honey costs 20 cents per ounce, and nuts cost 40 cents per ounce. We let y_1 be the number of ounces of honey to be used per cake and y_2 be the number of ounces of nuts to be used per cake. Then, the manufacturer wants to minimize the cost per cake, which is

$$Z = 20y_1 + 40y_2 \text{ (cents per cake)}.$$

The three nutritional requirements yield the constraints

$$\begin{aligned} y_1 + 3y_2 &\geq 9 \\ 3y_1 + 5y_2 &\geq 23 \\ y_1 + y_2 &\geq 5. \end{aligned}$$

Of course, the variables must be nonnegative. The manufacturer's LP problem is to minimize the objective function Z above, subject to the stated constraints. Clearly, this is a standard minimization problem. ❑

EXAMPLE C-3

Consider the LP problem

$$\min Z = 4x_1 + 2x_2$$

subject to

$$\begin{aligned} x_1 + x_2 &\geq 2 & \text{(i)} \\ 4x_1 + x_2 &\geq 4 & \text{(ii)} \\ 4x_2 - x_1 &\leq 8 & \text{(iii)} \\ x_1, x_2 &\geq 0. \end{aligned}$$

As it stands, this is neither a standard maximization nor minimization problem. However, if we rewrite (iii) (by multiplying it by -1) as

$$x_1 - 4x_2 \geq -8,$$

then the problem becomes a standard minimization problem (with x_i rather than y_i as variables). The corresponding matrices A, C, and B are

$$A = \begin{bmatrix} 1 & 4 & 1 \\ 1 & 1 & -4 \end{bmatrix}, C = [2 \quad 4 \quad -8], B = \begin{bmatrix} 4 \\ 2 \end{bmatrix}.$$

However, we may also associate with this problem a standard maximization problem by noting that a function Z is minimized iff its *negative* $-Z$ is maximized. Consequently, this problem may be expressed as the standard maximization problem

$$\max -Z = -4x_1 - 2x_2$$

subject to

$$\begin{aligned} -x_1 - x_2 &\leq -2 \\ -4x_1 - x_2 &\leq -4 \\ -x_1 + 4x_2 &\leq 8 \\ x_1, x_2 &\geq 0. \end{aligned}$$ ❑

EXAMPLE C-4

The LP problem

$$\min Z = 3y_1 + 2y_2 + y_3$$

subject to

$$\begin{aligned} y_1 + y_2 &\geq 2 & \text{(i)} \\ 2y_1 + y_2 - y_3 &= 3 & \text{(ii)} \\ y_1, y_2, y_3 &\geq 0 \end{aligned}$$

is neither a standard maximization nor a standard minimization problem on account of the equality constraint (equation (ii)). We may replace (ii) by two inequalities

$$2y_1 + y_2 - y_3 \geq 3$$
$$2y_1 + y_2 - y_3 \leq 3,$$

or,

$$2y_1 + y_2 - y_3 \geq 3$$
$$-2y_1 - y_2 + y_3 \geq -3.$$

The resulting LP problem is then a standard minimization problem with matrices

$$A = \begin{bmatrix} 1 & 2 & -2 \\ 1 & 1 & -1 \\ 0 & -1 & 1 \end{bmatrix}, B = \begin{bmatrix} 3 \\ 2 \\ 1 \end{bmatrix}, C = \begin{bmatrix} 2 & 3 & -3 \end{bmatrix}. \square$$

Returning to Definition C-1, it is interesting to note that by specifying matrices $A, B,$ and C, we determine *two* LP problems: a standard maximization problem and a standard minimization problem. These problems are called **duals** of each other.

EXAMPLE C-5

(a) Consider the LP problem

$$\max Z = 12x_1 + 10x_2$$

subject to

$$x_1 + x_2 \leq 12$$
$$x_1 - x_2 \leq 1$$
$$x_1 \leq 5$$
$$x_1, x_2 \geq 0.$$

Here the matrices are

$$A = \begin{bmatrix} 1 & 1 \\ 1 & -1 \\ 1 & 0 \end{bmatrix}, B = \begin{bmatrix} 12 \\ 1 \\ 5 \end{bmatrix}, C = \begin{bmatrix} 12 & 10 \end{bmatrix}.$$

The dual problem is (by Definition C-1)

$$\min Z = 12y_1 + y_2 + 5y_3$$

subject to

$$y_1 + y_2 + y_3 \geq 12$$
$$y_1 - y_2 \geq 10$$
$$y_1, y_2, y_3 \geq 0.$$

(b) Consider the problem

$$\min Z = 2y_1 + y_2$$

subject to

$$\begin{aligned} 2y_1 + 2y_2 &\geq 10 \\ y_1 + 2y_2 &\geq 6 \\ y_1, y_2 &\geq 0. \end{aligned}$$

Here

$$A = \begin{bmatrix} 2 & 1 \\ 2 & 2 \end{bmatrix}, B = \begin{bmatrix} 2 \\ 1 \end{bmatrix}, C = [10 \quad 6].$$

Thus, its dual problem is

$$\max Z = 10x_1 + 6x_2$$

subject to

$$\begin{aligned} 2x_1 + x_2 &\leq 2 \\ 2x_1 + 2x_2 &\leq 1 \\ x_1, x_2 &\geq 0. \; \square \end{aligned}$$

The notion of duality is important in LP theory in view of the following theorem.

THEOREM C-1

There is a vector which maximizes the objective function Z for a standard maximization problem iff there is a vector which minimizes the objective function Z' for its dual minimization problem. Furthermore, the **optimal values** of the objective functions Z and Z' are the same. ■

In other words, there is a vector X which satisfies the constraints and maximizes Z for a maximization problem iff there is a vector Y which satisfies the constraints and minimizes Z' for the dual minimization problem. And, when this is the case, $Z(X) = Z'(Y)$. We shall not prove this theorem, but it is interesting to note that when A, B, and C are positive numbers (i.e., 1×1 matrices), the result is trivially true, because the maximum value of Z is BC/A when $X = B/A$, and this equals the minimum value of Z', achieved when $Y = C/A$.

DEFINITION C-2

A vector $X = (x_i)$ is called a **feasible solution** for a maximization problem iff $X \geq O$ and satisfies all the constraints for the problem. If X also yields a maximum value for $Z = CX$, then X is called an **optimal** (feasible) **solution** for the problem. Similar terminology applies to a minimization problem except that a vector Y is optimal iff it yields a minimum value for $Z = YB$ upon substitution. The value $Z(X)$ (respectively, $Z(Y)$) of Z taken on at an optimal solution is called the **optimal value** of Z. Finally, the set of feasible vectors for a LP problem is called its **feasible set** or **region**.

EXAMPLE C-6

The feasible set for the LP problem of Example 5-1 is the set of points in R^2 satisfying the system of inequalities

$$\begin{aligned} 4x + 2y &\le 80 \quad &\text{(i)} \\ 2x + 6y &\le 120 \quad &\text{(ii)} \\ 3x + 3y &\le 75 \quad &\text{(iii)} \\ x, y &\ge 0. \end{aligned}$$

We may graph this region of the plane by noting that the solution set of any *single* linear inequality $ax + by \le c$ in two variables is a **halfplane** in R^2 whose boundary is the line with equation $ax + by = c$. For example, the solution set of constraint (i) above is a halfplane with (and including) the boundary line $4x + 2y = 80$. To find which of the two halfplanes in R^2 determined by this line is actually the solution set of constraint (i), we simply *test* (by substitution) some convenient point in R^2 not on the boundary line. For example, the origin (0,0) is in the solution set because it satisfies the inequality $4x + 2y \le 80$. Therefore, the solution set of constraint (i) is the halfplane with boundary line $4x + 2y = 80$ and containing the point (0,0). This is sketched in Figure C-1, together with the solution sets for the other inequalities and $x \ge 0$, $y \ge 0$.

FIGURE C-1

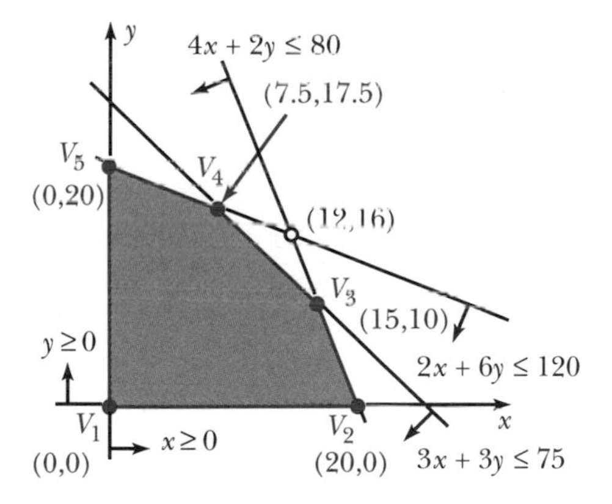

We see that the feasible region for the LP problem is the *intersection* of halfplanes for the individual constraints (shown in color). Notice that the feasible set for this problem is bounded by *straight line segments*. As a result, it is called a **polyhedral set**. In addition, it has a property called **convexity**, because if any two points P and P' in the feasible set S are joined by a straight line segment, then *all* points on that segment are also in S. This is illustrated in Figure C-2, where a nonconvex polyhedral set is also shown.

FIGURE C-2

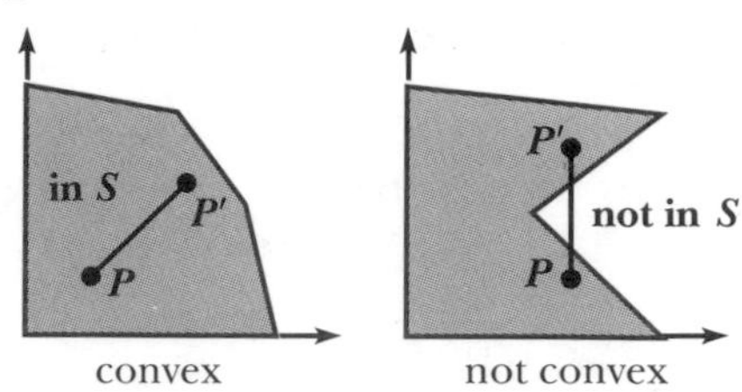

Consequently, the feasible set for this LP problem is called a **polyhedral convex set**. The "corners" of the feasible set (labelled V_1 through V_5 in Figure C-1) are called the **extreme points** (or **vertices**). They occur as the *feasible* intersections of the lines that constitute the boundary of S. So, their coordinates may be found by solving the corresponding systems of equations.

For example, in Figure C-1, the extreme point V_3 is seen to be the intersection of $4x + 2y = 80$ and $3x + 3y = 75$. Solving these equations simultaneously, we find that the coordinates of V_3 are (15,10) as shown. Notice that *not* all intersections of boundary lines produce extreme points. For example, (12,16) is the intersection of $4x + 2y = 80$ and $2x + 6y = 120$, but it is not an extreme point because it is not feasible (it lies outside the halfplane for $3x + 3y \leq 75$). ❑

In the general case of a maximization problem having m constraints in n variables (Definition C-1), each constraint

$$a_{i1}x_1 + \dots + a_{in}x_n \leq b_i, \; i = 1, \dots, m,$$

has as its solution set a **halfspace** in R^n bounded by the **hyperplane** with equation

$$a_{i1}x_1 + \dots + a_{in}x_n = b_i.$$

It can be shown that the feasible set for the LP problem is also convex and so is also called a polyhedral convex set in R^n. The extreme points now occur as feasible intersections of n (or more) such hyperplanes. For example, when $n = 3$, the extreme points are typically obtained as intersections of three planes in R^3. A typical feasible region is shown in Figure C-3.

FIGURE C-3

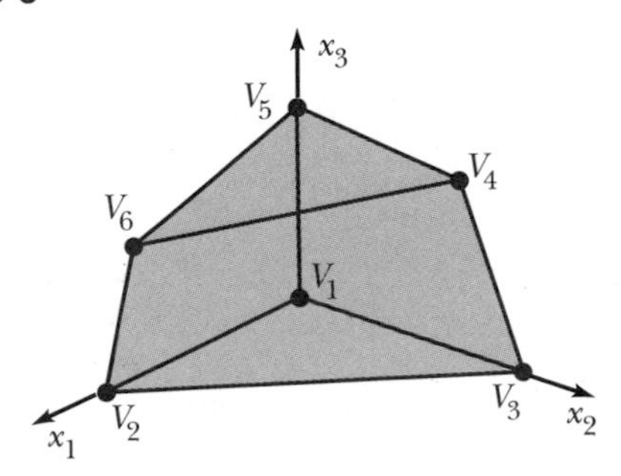

It can easily be proved that the number of extreme points for any LP problem is finite. This is important for the practical solution of LP problems in view of the next result.

THEOREM C-2

If a LP problem has an optimal solution X, then there is an extreme point V such that $Z(V) = Z(X)$. ■

In other words, the optimal value of Z is taken on at an extreme point V, provided Z actually has an optimal solution. Thus, the search for an optimal solution is reduced from the *infinite* number of points in the feasible set to the *finite* set of extreme points.

Note that this theorem does *not* state that any LP problem necessarily has an optimal solution. A LP problem may fail to have an optimal solution in two ways:

(1) The feasible set may be **empty**.

(2) The feasible set may be **unbounded** (i.e., it contains a line segment of infinite length).

These possibilities are pursued in questions 3 and 4. However, when a LP problem does have an optimal solution, the theorem provides a method (in principle) for finding that solution. When this is the case, we need only compute $Z(V)$ for every extreme point V, and choose the one that yields the largest (or smallest, in minimization problems) value for Z.

EXAMPLE C-7

This example is a continuation of Example C-6. In the table below, we list the extreme points for the LP problem of the example, their coordinates, and the corresponding values of the objective function $Z = 15x + 12y$.

Extreme Point	$Z(V) = 15x + 12y$
$V_1 = (0,0)$	0
$V_2 = (20,0)$	300
$V_3 = (15,10)$	$225 + 120 = 345$
$V_4 = (7.5,17.5)$	$112.5 + 210 = 322.5$
$V_5 = (0,20)$	240

The largest value of Z is 345, taken on at $V_3 = (15,10)$. So, (15,10) is the optimal solution for this problem. The company should manufacture 15 units of product A and 10 units of product B to realize a maximum profit of \$345. ❑

EXAMPLE C-8

We solve the diet problem in Example C-2. The feasible region for this problem is shown in Figure C-4 and the value of the objective function at the various extreme points is summarized in the following table.

FIGURE C-4

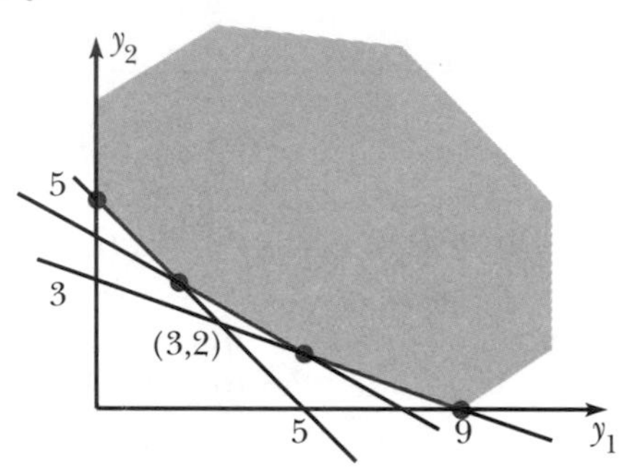

Extreme Point	$Z = 20y_1 + 40y_2$
(0,5)	200
(1,4)	180
(6,1)	160
(9,0)	180

Therefore, the optimal value (minimum cost) is 160 cents per cake, obtained by using a mix of 6 ounces of honey and 1 ounce of nuts per cake. Note that the origin (0,0) is not a feasible point for this problem. ❑

Now, as noted earlier, the procedure followed in the preceding examples may in principle be used to solve any LP problem. However, with an increasing number m of constraints and n of variables, the number of computations involved soon makes this method impractical. For, each extreme point is found by first solving a system of n equations in n unknowns, and then checking the resulting solution for feasibility. Since there are $n + m$ constraints if the nonnegativity conditions are included, this means solving $n + m$ systems of n linear equations in n unknowns (and checking for feasibility). Readers may refer to section 2-6 to obtain an estimate of the number of arithmetic operations involved.

In the **simplex method** (or algorithm) which we now discuss, the value of the objective function is computed only at *selected* extreme points $V_1, \ldots, V_k$ (in succession) in such a way that the value $Z(V_i)$ steadily improves as i increases. This greatly reduces the amount of computation needed to obtain an optimal solution.

Because any LP problem may be rewritten as a standard maximization problem, the following discussion will be restricted to that case. Readers are alerted to the fact that there are many formulations of the simplex method and that these may appear quite different. The following formulation has the advantage of simplicity and of clearly showing the method as a special adaptation of solving a system of linear equations by pivoting. In practice, LP problems are solved using computer programs derived from the following ideas.

We begin by assuming that the LP problem

$$\begin{aligned}&\max Z = CX\\ &\text{subject to } AX \leq B, X \geq O\end{aligned}$$

has the additional property that $B \geq O$. When this is true, we call it a **canonical maximization problem**. Example C-6 is an example of such a problem. Since the extreme points are found by solving systems of equations obtained from the constraints, we first convert these inequalities $AX \leq B$ to *equations* by introducing **slack variables** $u_1, \ldots, u_m \geq 0$ (one for each constraint) such that

$$a_{i1}x_1 + \ldots + a_{in}x_n + u_i = b_i, \; i = 1, \ldots, m.$$

In matrix notation, we write

$$AX + U = B$$

where U is the $m \times 1$ matrix (u_i). Alternatively, we can write

$$A'X' = B,$$

where $A' = (A{:}I)$, I is the $m \times m$ identity matrix and

$$X' = \begin{bmatrix} X \\ U \end{bmatrix}.$$

With the notation above, the constraints for the LP problem of Example C-6 become

$$\begin{aligned} 4x + 2y + u_1 \qquad\qquad &= 80 \\ 2x + 6y + \qquad u_2 \qquad &= 120 \\ 3x + 3y + \qquad\qquad u_3 &= 75 \\ x, y, u_1, u_2, u_3 &\geq 0. \end{aligned} \tag{C-1}$$

Therefore, the feasible set for the problem corresponds exactly to the set of *nonnegative* solutions of the system of three equations in five unknowns (C-1). The augmented matrix for this system is

$$(A{:}I{:}B) = \begin{bmatrix} 4 & 2 & 1 & 0 & 0 & 80 \\ 2 & 6 & 0 & 1 & 0 & 120 \\ 3 & 3 & 0 & 0 & 1 & 75 \end{bmatrix},$$

where I is, of course, the 3×3 identity matrix, which occurs on account of the slack variables. Readers may check that all 3×3 submatrices of $(A'{:}B) = (A{:}I{:}B)$ have rank $= 3 = m$. We therefore say that the problem is **nondegenerate**. In particular, all solutions for equations C-1 may be obtained by assigning arbitrary values to any two ($=5-3$) of the variables and solving for the remaining three (using equations C-1; cf. section 4-4).

For example, setting $x=y=0$, we obtain the solution with $u_1=80$, $u_2=120$, and $u_3=75$. Notice that this corresponds to an extreme point of the feasible set, namely, the origin (0,0). In fact, all the extreme points (including the optimal solution) are obtained in this way, that is, by setting exactly two of the variables in equations C-1 to zero, and solving for the remaining three. Such solutions for C-1 are called **basic solutions**. The nonzero variables are called the **basic variables** for that solution, while the zero variables are called **nonbasic**. For the problem in Example C-6, the basic feasible solutions are listed in the table below.

Extreme Point	*Basic Feasible Solution*	
	Nonbasic	*Basic*
$V_1(0,0)$	$x=y=0$	$u_1=80, u_2=120, u_3=75$
$V_2(20,0)$	$y=u_1=0$	$x=20, u_2=80, u_3=15$
$V_3(15,10)$	$u_1=u_3=0$	$x=15, y=10, u_2=30$
$V_4(7.5,17.5)$	$u_2=u_3=0$	$x=7.5, y=17.5, u_1=25$
$V_5(0,20)$	$x=u_2=0$	$y=20, u_1=40, u_3=15$

On the other hand, *not* every solution of C-1 with two variables set to zero is an extreme point. For example, setting $u_1=u_2=0$, we obtain the nonfeasible intersection $x=12, y=16$ (Figure C-1). However, this is not a basic feasible solution, because by the third constraint in equations C-1 above, $u_3=75-84=-9$, which is not nonnegative.

So, the optimal solution for this LP problem can be found by examining the basic feasible solutions for the problem and selecting the one that maximizes Z. The simplex method achieves this by moving from one basic feasible solution to another that yields a *larger* value of Z. This step is then repeated until Z cannot be further increased. In fact, the move is made to a vertex on the same "edge" of the feasible set, as we shall see below. Before illustrating these ideas, we generalize the preceding terminology and observations.

DEFINITION C-3

Let $A'X'=B$ be the equation form of the constraints for a standard maximization problem (if $B\geq O$ then $A'=(A{:}I)$ and X' are as discussed earlier). The LP problem is called **nondegenerate** if every $m\times m$ submatrix of $(A'{:}B)$ has rank m, where m is the number of constraints. Then the **basic feasible solutions** can be obtained by setting all but m of the variables to zero and solving $A'X'=B$ for the remaining m. The zero variables so chosen are the **nonbasic variables**, while the others are **basic**. In a nondegenerate problem, there are therefore m nonzero (basic) variables in any basic feasible solution.

THEOREM C-3

Every extreme point for a nondegenerate LP problem is a basic feasible solution of $A'X' = B$. Furthermore, if the LP problem has an optimal solution, there is a basic feasible solution which produces the optimal value of Z. ■

Degenerate LP problems will not be discussed here. For such problems, the simplex algorithm may not lead to an optimal solution, although in practice it frequently does.

We now illustrate the mechanics of the simplex method for the problem in Example C-6. As seen earlier, we have an obvious basic feasible solution obtained by setting all but the slack variables to zero. This starting basic feasible solution corresponds to the origin in R^2, and is obvious by inspection of the constraint equations C-1. In the simplex method, we move from this initial vertex (where $Z = 0$) to an "adjacent" vertex (i.e. on the same edge of the feasible set) that produces the greatest increase in $Z = 15x + 12y$. So, we consider the adjacent extreme points V_2 and V_5. Since moving to V_2 is achieved by increasing x to a positive value and keeping y fixed at 0, while moving to V_5 is achieved by increasing y to a positive value and keeping x fixed at 0 (see Figure C-1), we ask which of these two possibilities would produce the *greatest increase* in Z. Now, by inspecting the expression for Z, we see that increasing x (i.e., moving to V_2) is best, because the coefficient of x in Z is larger than the coefficient of y (increasing x by 1 unit increases Z by 15 units, while increasing y by 1 unit only increases Z by 12 units).

In terms of variables, we may describe this move by saying that we wish to improve the value of Z by making one of the current nonbasic variables x and y into a basic variable. We choose x, as above, since this results in a greater increase in Z.

Next, we must decide how large x should be made (with $y = 0$). Since all three constraints must remain satisfied, we see by equations C-1 that we can increase x to the *smallest* of $80/4 = 20$, $120/2 = 60$, and $75/3 = 25$. In other words, we may increase x to 20. However, doing so forces u_1 to be zero. Thus, u_1 becomes nonbasic and we have moved to vertex V_2.

Having moved to V_2, we must next compute the basic feasible solution and the value of Z corresponding to it (in most problems we will not have a sketch of the feasible set as we do here). Of course, we may do this by using the usual steps in Gauss-Jordan reduction of the augmented matrix for the system C-1: recall that these are the three types of elementary row operations. For, we are simply moving from one solution of C-1 to another solution. However, since we want to observe the effect of our change on the objective function Z, we add to the usual augmented matrix a row corresponding to the equation

$$15x + 12y + 0u_1 + 0u_2 + 0u_3 - Z = 0.$$

Note that this equation expresses Z in terms of the nonbasic variables x and y for the current basic feasible solution (i.e., x and y are the only variables with nonzero coefficient), and that the current value of Z is easily obtained by setting those variables to zero. We therefore work with the matrix

$$\begin{bmatrix} 4 & 2 & 1 & 0 & 0 & 0 & 80 \\ 2 & 6 & 0 & 1 & 0 & 0 & 120 \\ 3 & 3 & 0 & 0 & 1 & 0 & 75 \\ 15 & 12 & 0 & 0 & 0 & -1 & 0 \end{bmatrix}.$$

We see that the constraint entries for the current basic variables are 1. In fact, we shall see that there is no need to carry the sixth column for the "variable" Z and we shall omit it. It is also convenient to record the variables for each column at the top of the matrix, the current basic variables and Z at the right (one for each row), as well as our observations leading to the best choice for the next basic feasible solution. This is done as follows:

(1) Identify the *largest* positive coefficient (here $c_1 = 15$) in Z and mark it with (say) a "*", as shown in the matrices below.

(2) Compute the ratios b_i/a_{i1} for $i = 1, 2, 3$ at the right of the matrix and mark the *smallest* (with a "*") to determine which variable (here u_1) becomes nonbasic.

Recall that the smallest such ratio is the extent to which the variable becoming basic (here x) may be increased. The resulting matrix is given below (C-2). As shown, the smallest ratio is 20 in row 1. We say that row 1 is the **pivotal row**. Since u_1 is the current basic variable occurring in that row, it will be nonbasic in the next basic feasible solution to which these calculations will lead. Similarly, the column marked with a "*" is called the **pivotal column**. The element "4" in the pivotal row *and* column is the **pivot**. It is customarily bracketed (or circled) as shown. The matrix, together with the recorded information (C-2) is usually called a **simplex tableau**. Since it is the first one in a series, it is the *initial* simplex tableau.

$$\begin{array}{c} \begin{matrix} \quad x & \;\; y & u_1 & u_2 & u_3 & \quad b \end{matrix} \\ \begin{bmatrix} [4] & 2 & 1 & 0 & 0 & 80 \\ 2 & 6 & 0 & 1 & 0 & 120 \\ 3 & 3 & 0 & 0 & 1 & 75 \\ 15 & 12 & 0 & 0 & 0 & 0 \end{bmatrix} \begin{matrix} u_1 & 80/4 = 20\,* \\ u_2 & 120/2 = 60 \\ u_3 & 75/3 = 25 \\ Z & \end{matrix} \\ \begin{matrix} * & & & & & & & & & & & & \end{matrix} \end{array} \qquad \text{(C-2)}$$

The second simplex tableau is obtained by computing the new basic feasible solution having x basic and u_1 nonbasic (we will be able to solve for x in the corresponding system of equations, while setting $u_1 = 0$). So, we apply elementary row operations to matrix C-2 so that x will appear *only* in the first equation (with coefficient 1) of the resulting system. In other words, we make the pivot (the coefficient of x in the first equation) equal to 1, and all other entries in the pivotal column equal to 0. The result is

$$
\begin{array}{cccccc}
x & y & u_1 & u_2 & u_3 & b \\
\end{array}
$$

$$
\left[\begin{array}{cccccc}
1 & 1/2 & 1/4 & 0 & 0 & 20 \\
0 & 5 & -1/2 & 1 & 0 & 80 \\
0 & [3/2] & -3/4 & 0 & 1 & 15 \\
0 & 9/2 & -15/4 & 0 & 0 & -300
\end{array}\right]
\begin{array}{ll}
x & 40 \\
u_2 & 16 \\
u_3 & 10 \quad * \\
Z &
\end{array}
\qquad \text{(C-3)}
$$

$$\begin{array}{cccccc} & * & & & & \end{array}$$

From the corresponding system of equations, we see, as was expected, that the basic feasible solution with x, u_2, and u_3 basic (and y, u_1 nonbasic) is $x = 20$, $y = 0 = u_1$, $u_2 = 80$, and $u_3 = 15$. Furthermore, from the last row (re-inserting $-Z$) we see that

$$(9/2)y - (15/4)u_1 - Z = -300, \qquad \text{(C-4)}$$

so that with $y = u_1 = 0$, we get $Z = 300$. This agrees with the value found earlier at vertex V_2 (Example C-7); it is an improvement over the value of Z at $V_1 = (0,0)$. Notice too, that equation C-4 expresses Z in terms of the current nonbasic variables *only* (as before).

We have completed one "iteration" of the simplex method. We continue by repeating this step, using the same principles to guide the choice of pivot (i.e., the choice of new basic and nonbasic variables) until Z cannot be further improved. So, to continue, we see that the largest *positive* element in the last row of matrix C-3 is 9/2. Consequently, column 2 is the new pivotal column: we will make y basic in the next solution. We see from equation C-4 that this makes sense, since y is the only variable in Z which will now lead to an improvement in the value of Z (increasing u_1 would *decrease* Z on account of its *negative* coefficient $-15/4$). From the ratios b_i/a_{i2} computed at the right, we see that row 3 is the new pivotal row, and that the next pivot is 3/2 (bracketed). Therefore, after applying elementary row operations, the third tableau is found to be

$$
\begin{array}{cccccc}
x & y & u_1 & u_2 & u_3 & b
\end{array}
$$

$$
\left[\begin{array}{cccccc}
1 & 0 & 1/2 & 0 & -1/3 & 15 \\
0 & 0 & 2 & 1 & -10/3 & 30 \\
0 & 1 & -1/2 & 0 & 2/3 & 10 \\
0 & 0 & -6/4 & 0 & -3 & -345
\end{array}\right]
\begin{array}{l}
x \\
u_2 \\
y \\
Z
\end{array}
$$

This is the optimal and final tableau, because, from the last row, the equation for Z is now

$$(-6/4)u_1 - 3u_3 - Z = -345,$$

from which we see that Z cannot be further improved by increasing nonbasic variables. The optimal value of Z is 345, obtained when $u_1 = u_3 = 0$, $x = 15$, $y = 10$, and $u_2 = 30$, as found in Example C-7. This completes the simplex solution of this problem. The amount of computation is much less than in the "brute force" method used in Example C-7.

The steps in the simplex method for canonical maximization problems are summarized as follows.

(1) Write the initial tableau using the constraints expressed as equations with slack variables, and add a row for the objective function coefficients.
(2) Select the *largest positive* entry in the last row (Z) to determine the pivotal column j (if there are no positive entries the tableau is optimal).
(3) Form the ratios b_i/a_{ij}, $i = 1, 2, \ldots, m$, and choose as pivotal row the row containing the *smallest nonnegative* ratio (if there is more than one smallest ratio, choose any one of them; if there is no nonnegative ratio, the problem has no solution).
(4) Use elementary row operations to produce a new tableau in which the pivot is changed to 1 and all other entries in its column are 0.
(5) Repeat from step (2) until all entries in the last row are *nonpositive*, which means that this tableau is final as Z cannot be further improved.

EXAMPLE C-9

We solve the problem

$$\max Z = 9x_1 + 23x_2 + 5x_3$$

subject to

$$\begin{aligned} x_1 + 3x_2 + x_3 &\le 20 \\ 3x_1 + 5x_2 + x_3 &\le 40 \\ x_1, x_2, x_3 &\ge 0. \end{aligned}$$

Note that this is the dual problem of Example C-2 (and C-8). The tableaux are as follows:

$$\begin{array}{cccccc} x_1 & x_2 & x_3 & u_1 & u_2 & b \end{array}$$
$$\begin{bmatrix} 1 & [3] & 1 & 1 & 0 & 20 \\ 3 & 5 & 1 & 0 & 1 & 40 \\ 9 & 23 & 5 & 0 & 0 & 0 \end{bmatrix} \begin{array}{ll} u_1 & 20/3 * \\ u_2 & 40/5 = 8 \\ Z & \end{array}$$
$$\begin{array}{cccccc} & * & & & & \end{array}$$

$$\begin{array}{cccccc} x_1 & x_2 & x_3 & u_1 & u_2 & b \end{array}$$
$$\begin{bmatrix} 1/3 & 1 & 1/3 & 1/3 & 0 & 20/3 \\ [4/3] & 0 & -2/3 & -5/3 & 1 & 20/3 \\ 4/3 & 0 & -8/3 & -23/3 & 0 & -460/3 \end{bmatrix} \begin{array}{ll} x_2 & 20 \\ u_2 & 5 * \\ Z & \end{array}$$
$$\begin{array}{cccccc} * & & & & & \end{array}$$

$$\begin{array}{cccccc} x_1 & x_2 & x_3 & u_1 & u_2 & b \end{array}$$
$$\begin{bmatrix} 0 & 1 & 1/2 & 3/4 & -1/4 & 5 \\ 1 & 0 & -1/2 & -5/4 & 3/4 & 5 \\ 0 & 0 & -2 & -6 & -1 & -160 \end{bmatrix} \begin{array}{l} x_2 \\ x_1 \\ Z. \end{array}$$

This last tableau is optimal. The optimal value of Z is 160 when $x_1 = 5 = x_2$, $x_3 = 0 = u_1 = u_2$. Note that the optimal value of Z is the same as the optimal value for the dual problem in Example C-8 (minimization). This agrees with Theorem C-1. ❑

In fact, it is possible to formulate the simplex algorithm so that the tableaux provide the solutions *both* for the original problem (called the **primal** problem in this context) and its dual. However, we shall solve minimization problems by converting them to standard maximization problems, as suggested in Example C-3: we maximize $-Z$ when Z is the objective function for a minimization problem. As a result of this change, the resulting maximization problem may have negative entries in the vector B, so that it is not a canonical maximization problem and the simplex method as presented above does not apply directly. Instead, we must first further modify the problem by the introduction of **artificial variables**, as illustrated in the next example.

EXAMPLE C-10

We consider the minimization problem of Example C-3. As noted, it may be written as the standard maximization problem

$$\max\ -Z = -4x_1 - 2x_2$$

subject to

$$\begin{aligned} -x_1 - x_2 &\leq -2 \\ -4x_1 - x_2 &\leq -4 \\ -x_1 + 4x_2 &\leq 8 \\ x_1, x_2 &\geq 0. \end{aligned}$$

The simplex algorithm as formulated earlier does not apply directly because when slack variables are added, the usual starting basic feasible solution won't work on account of the negative entries of B (i.e., -2 and -4). As we can see, the constraint equations are

$$\begin{aligned} -x_1 - x_2 + u_1 \qquad\qquad &= -2 \\ -4x_1 - x_2 \qquad + u_2 \qquad &= -4 \\ -x_1 + 4x_2 \qquad\qquad + u_3 &= 8, \end{aligned}$$

and all variables are to be nonnegative. We cannot set $x_1 = x_2 = 0$ to obtain a basic feasible solution, since this would yield $u_1 = -2$ and $u_2 = -4$, which are *not* nonnegative values. The problem here arises from the fact that the origin is not an extreme point.

To obtain a starting basic feasible solution, we subtract **artificial variables** $a_1 \geq 0$ and $a_2 \geq 0$ from the first two equations to get

$$\begin{aligned} -x_1 - x_2 + u_1 \qquad\qquad - a_1 \qquad &= -2 \\ -4x_1 - x_2 \qquad + u_2 \qquad\qquad - a_2 &= -4 \\ -x_1 + 4x_2 \qquad\qquad + u_3 \qquad\qquad &= 8, \end{aligned} \tag{C-5}$$

and all variables are to be nonnegative. This is a system of three equations in seven variables (including a_1 and a_2). A basic feasible solution (having $7 - 3 = 4$ variables set to 0) is

$$x_1 = x_2 = u_1 = u_2 = 0,\ a_1 = 2,\ a_2 = 4, \text{ and } u_3 = 8.$$

Of course, an optimal solution for the original problem must have

$a_1 = a_2 = 0$, since the introduction of the artificial variables is only temporary in order to achieve a starting point. To achieve this, we change the objective function $(-Z)$ to

$$-Z = -4x_1 - 2x_2 - K_1a_1 - K_2a_2 \qquad \text{(C-6)}$$

where K_1 and K_2 are large positive constants. We now apply the simplex method to the *revised* problem (i.e., objective function C-6 subject to constraints C-5). If K_1 and K_2 are chosen sufficiently large, then because the procedure attempts to *maximize* $-Z$, a_1 and a_2 will be *forced to zero*, since positive values for these variables lead to large negative contributions to $-Z$. For example, if we let $K_1 = K_2 = 10$, the first tableau is

x_1	x_2	u_1	u_2	u_3	a_1	a_2	b	
−1	−1	1	0	0	−1	0	−2	a_1
−4	−4	0	1	0	0	−1	−4	a_2
−1	4	0	0	1	0	0	8	u_3
−4	−2	0	0	0	−10	−10	0	$-Z$.

This initial tableau is still not quite in the same form as the initial tableau of our earlier examples since the last row for $-Z$ has nonzero entries -10 corresponding to the basic variables a_1 and a_2. Since we want to express the objective function in terms of nonbasic variables only (as before), we remove these entries from the matrix by subtracting $10\times$row (1) from row (4) and then $10\times$row (2) from row (4). Also, we multiply the first two rows by -1 to make the entries for a_1 and a_2 equal to $+1$ rather than -1. The result is

x_1	x_2	u_1	u_2	u_3	a_1	a_2	b		
1	1	−1	0	0	1	0	2	a_1	2
[4]	1	0	−1	0	0	1	4	a_2	1 *
−1	4	0	0	1	0	0	8	u_3	
46	18	−10	−10	0	0	0	60	$-Z$.	
*									

We now proceed as before, selecting the pivot as shown. Note that we ignored the negative ratio in row 3. The remaining tableaux are

x_1	x_2	u_1	u_2	u_3	a_1	a_2	b		
0	[3/4]	−1	1/4	0	1	−1/4	1	a_1	4/3 *
1	1/4	0	−1/4	0	0	1/4	1	x_1	4
0	17/4	0	−1/4	1	0	1/4	9	u_3	36/17
0	13/2	−10	3/2	0	0	−23/2	14	$-Z$	
	*								

x_1	x_2	u_1	u_2	u_3	a_1	a_2	b	
0	1	−4/3	1/3	0	4/3	−1/3	4/3	x_2
1	0	1/3	−1/3	0	−1/3	1/3	2/3	x_1
0	0	17/3	−5/3	1	−17/3	5/3	10/3	u_3
0	0	−4/3	−2/3	0	−26/3	−28/3	16/3	$-Z$.

This last tableau is optimal. We see that a_1 and a_2 are nonbasic, so their value is indeed zero. The maximum value of $-Z$ is $-16/3$ when $x_1 = 2/3, x_2 = 4/3, u_3 = 10/3$, and all other variables are zero. Therefore, the *minimum* value of Z is $16/3$ at the values of the variables stated above. Readers may verify this solution by drawing the feasible region for the original problem. Note that if the values for K_1 and K_2 are not chosen large enough (after all, it is not known in advance how large to make them), the simplex algorithm may not lead to a solution in which these variables are 0. This indicates that the computation must be repeated with larger values for K_1 and K_2. ❑

The final example below illustrates the handling of equality constraints. As we have seen in Example C-4, equality constraints can always be converted to a pair of inequalities. While such problems may be solved directly using the method of the preceding example, the resulting tableau actually contains superfluous information since, in the simplex method, inequalities are first converted to equations.

EXAMPLE C-11

We solve the LP problem of Example C-4. It can be rewritten as the following (nonstandard) maximization problem:

$$\max \; -Z = -3x_1 - 2x_2 - x_3$$

subject to

$$\begin{aligned} -x_1 - x_2 \qquad &\le -2 \\ 2x_1 + x_2 - x_3 &= 3, \end{aligned}$$

and all variables nonnegative. Now, as remarked earlier, rather than converting the equality constraint to a pair of inequalities, we simply add an artificial variable a_2 to it in order to obtain an obvious initial basic feasible solution. We also need a slack variable u_1 and an artificial variable a_1 for the first constraint. So, the constraints produce the system of equations

$$\begin{aligned} -x_1 - x_2 \qquad\quad + u_1 - a_1 \qquad &= -2 \\ 2x_1 + x_2 - x_3 \qquad\qquad\qquad + a_2 &= 3 \end{aligned}$$

(the signs of a_1 and a_2 are chosen so as to provide an initial basic feasible solution). To force the artificial variables to zero, we replace $-Z$ by (for example)

$$-Z = -3x_1 - 2x_2 - x_3 - 10a_1 - 10a_2.$$

After multiplying the first equation by -1, the initial tableau is therefore

$$\begin{array}{c} \begin{matrix} x_1 & x_2 & x_3 & u_1 & a_1 & a_2 & b \end{matrix} \\ \begin{bmatrix} 1 & 1 & 0 & -1 & 1 & 0 & 2 \\ 2 & 1 & -1 & 0 & 0 & 1 & 3 \\ -3 & -2 & -1 & 0 & -10 & -10 & 0 \end{bmatrix} \begin{matrix} a_1 \\ a_2 \\ -Z. \end{matrix} \end{array}$$

As in the last example, we eliminate the basic variables with nonzero coefficients in $-Z$ by adding 10 times rows (1) and (2) to row (3) to get

$$\begin{array}{c} \begin{matrix} x_1 & x_2 & x_3 & u_1 & a_1 & a_2 & b \end{matrix} \\ \begin{bmatrix} 1 & 1 & 0 & -1 & 1 & 0 & 2 \\ [2] & 1 & -1 & 0 & 0 & 1 & 3 \\ 27 & 18 & -11 & -10 & 0 & 0 & 50 \end{bmatrix} \begin{matrix} a_1 & 2 \\ a_2 & 3/2 * \\ -Z. & \end{matrix} \\ * \end{array}$$

The pivot is selected as shown and leads to the following tableaux:

$$\begin{array}{c} \begin{matrix} x_1 & x_2 & x_3 & u_1 & a_1 & a_2 & b \end{matrix} \\ \begin{bmatrix} 0 & [1/2] & 1/2 & -1 & 1 & -1/2 & 1/2 \\ 1 & 1/2 & -1/2 & 0 & 0 & 1/2 & 3/2 \\ 0 & 9/2 & 5/2 & -10 & 0 & -27/2 & 19/2 \end{bmatrix} \begin{matrix} a_1 & 1 * \\ x_1 & 3 \\ -Z & \end{matrix} \\ * \end{array}$$

$$\begin{array}{c} \begin{matrix} x_1 & x_2 & x_3 & u_1 & a_1 & a_2 & b \end{matrix} \\ \begin{bmatrix} 0 & 1 & 1 & -2 & 2 & -1 & 1 \\ 1 & 0 & -1 & 1 & -1 & 1 & 1 \\ 0 & 0 & -2 & -1 & -9 & -9 & 5 \end{bmatrix} \begin{matrix} x_2 \\ x_1 \\ -Z. \end{matrix} \end{array}$$

This tableau is optimal. The minimum value of Z is 5, when $x_1 = 1 = x_2$, and all other variables are zero. ❑

EXERCISES

1. Express each of the following LP problems as **(i)** a standard maximization problem; **(ii)** a standard minimization problem.

(a) $\max Z = x_1 + 2x_2 + x_3$
subject to

$$\begin{aligned} x_1 + x_2 + x_3 &\le 4 \\ 2x_1 + x_2 + x_3 &\le 5 \\ -x_1 \qquad + x_3 &= 1 \\ x_1, x_2, x_3 &\ge 0 \end{aligned}$$

(b) $\min Z = 3y_1 + y_2$
subject to

$$\begin{aligned} y_1 + y_2 &\le 4 \\ y_1 - y_2 &\ge 0 \\ y_1, y_2 &\ge 0 \end{aligned}$$

2. Write the duals of the problems in question 1 (they must first be put in standard form).

3. Show that the LP problem below has no solution because the feasible set is empty.

$\min Z = 2y_1 + y_2$
subject to

$$\begin{aligned} 2y_1 + y_2 &\ge 4 \\ y_1 + 2y_2 &\ge 4 \\ y_1 + y_2 &\le 1 \\ y_1, y_2 &\ge 0 \end{aligned}$$

4. Show that the following LP problem has no optimal solution because the feasible set is unbounded.

$$\max Z = 2x_1 + x_2$$
subject to
$$x_1 + x_2 \geq 2$$
$$x_1 - x_2 \leq 1$$
$$x_1, x_2 \geq 0$$

(Hint: Sketch the feasible region S, find a line $x_2 = mx_1 + b$ such that (x_1,x_2) is in S for arbitrarily large x_1 and use this to show that Z can be made to take on arbitrarily large values for (x_1,x_2) in S.)

5. Solve the following LP problems by sketching the feasible region and computing the value of the objective function Z at each extreme point.

(a) $\max Z = 2x_1 + 3x_2$
subject to
$$x_1 + 2x_2 \leq 60$$
$$2x_1 + x_2 \leq 6$$
$$x_1 + x_2 \leq 7/2$$
$$x_1, x_2 \geq 0$$

(b) $\min Z = 3y_1 + 5y_2$
subject to
$$6y_1 + y_2 \geq 6$$
$$y_2 \leq 5$$
$$y_1 + y_2 \geq 2$$
$$y_1, y_2 \geq 0$$

6. Use Definition C-3 to show that the following LP problem is degenerate.

$$\max Z = 2x_1 + 3x_2$$
subject to
$$x_1 + 2x_2 \leq 6$$
$$2x_1 + x_2 \leq 6$$
$$x_1 + x_2 \leq 4$$
$$x_1, x_2 \geq 0$$

(Compare this with question 5(a).) Sketch the feasible region and note the number of lines intersecting to form the various vertices. Degenerate problems are characterized by the fact that *more* hyperplanes than are necessary intersect to form some extreme point.

7. Use the simplex method to solve the LP problems in question 5.

8. Use the simplex method to solve the following LP problems.

(a) $\max Z = 2x_1 + 3x_2 + x_3$
subject to
$$3x_1 + 4x_2 + x_3 \leq 2$$
$$x_1 + 3x_2 + 2x_3 \leq 1$$
$$x_1, x_2, x_3 \geq 0$$

(b) $\min Z = 2y_1 - 3y_2$
subject to
$$3y_1 - 4y_2 = 10$$
$$5y_1 - 6y_2 \geq 8$$
$$y_1, y_2 \geq 0$$

(c) $\max Z = -x_1 - 2x_2$
subject to
$$x_1 + x_2 \geq 3$$
$$2x_1 - x_2 \leq 3$$
$$-x_1 + 2x_2 \leq 3$$
$$x_1, x_2 \geq 0$$

ANSWERS TO SELECTED PROBLEMS

EXERCISES 1-1

1. (a), (b), (c); **(i)** examples of solutions: **(a)** $x = 0, y = 2, z = 0$ **(b)** $x = -1, y = -1$ **(c)** $a = 3, b = 0, c = -1$; **(ii)** examples of non-solutions: **(a)** $x = y = z = 0$ **(b)** $x = y = 0$ **(c)** $a = 1, b = c = 0$

3. **(a)** $x = 1/3, y = -1/3$ **(b)** no solution **(c)** $y = x - 4, x \in R$ (infinitely many solutions)

5. **(a)**

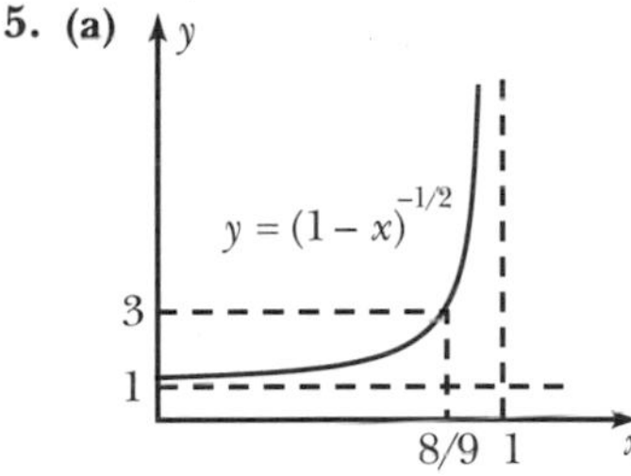

(b)

x	0.01	0.1	0.5
$f(x)$	1.005	1.054	1.414
$1+(1/2)x$	1.005	1.050	1.250

(all to three decimal places)

(c) $t/t' = [1 - (0.5c)^2/c^2]^{-1/2} \approx 1.155$; $1 + (1/2)(0.25) = 1.125$

7. **(a)** Taking logs in the equation yields $\log q = \log a_0 + a_1\log K + a_2\log L$, which is linear in the stated variables. **(b)** $R = qp$, so $\log R = \log q + \log p = \log a_0 + a_1\log K + a_2\log L + \log p$, which is linear in the stated variables.

9. **(a)** $y = 3$ **(b)** $x = -1$ **(c)** $z = -2$

11.

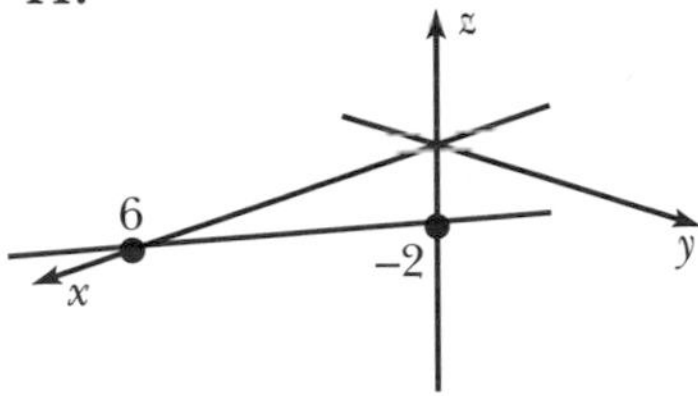

(i) $2x + 4y - 6z = 12$
$2x - 6y - 6z = 12$

13. Points indicate intersections with the coordinate planes.

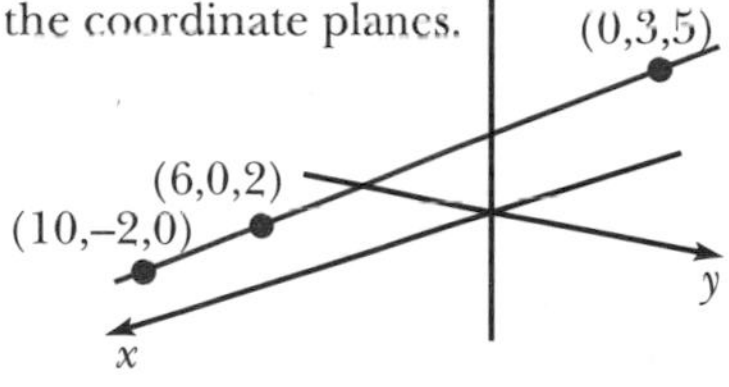

$x - 2y + 4z - 14 = 0$
$x + 20y - 18z + 30 = 0$

EXERCISES 1-2

1.

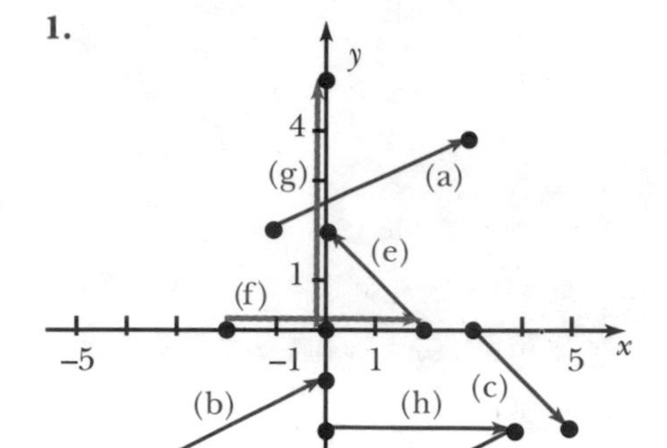

4. (a) vector (a) = (e)

5. components of **v** for question 1: (4,2), (4,2), (2, −2), (−4, −2), (−2,2), (4,0), (4,0), (0,5); question 3: (1, −5,2), (1, −5, −4), (−2, −2, −1), (3, −2, −2), (1, −5,2); lengths |**v**| for question 1: $2\sqrt{5}$, $2\sqrt{5}$, $2\sqrt{2}$, $2\sqrt{5}$, $2\sqrt{2}$, 4, 5, 4; question 3: $\sqrt{30}$, $\sqrt{42}$, 3, $\sqrt{17}$, $\sqrt{30}$

7. (a) Note that {**AB**}+{**CD**} = (5,5).

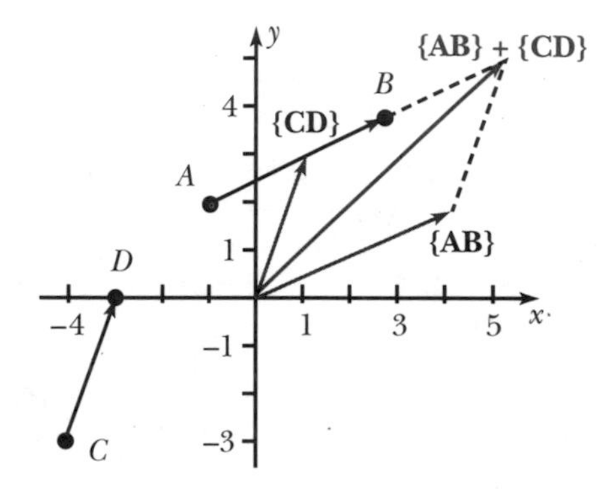

9. (a) k**AB** has the same initial point as **AB**, $|k\mathbf{AB}| = |k||\mathbf{AB}|$, and k**AB** has the same or opposite direction according to whether $k > 0$ or $k < 0$. The results are shown in the figure below for $k = 2$ and -0.5, for selected **AB** from question 1.

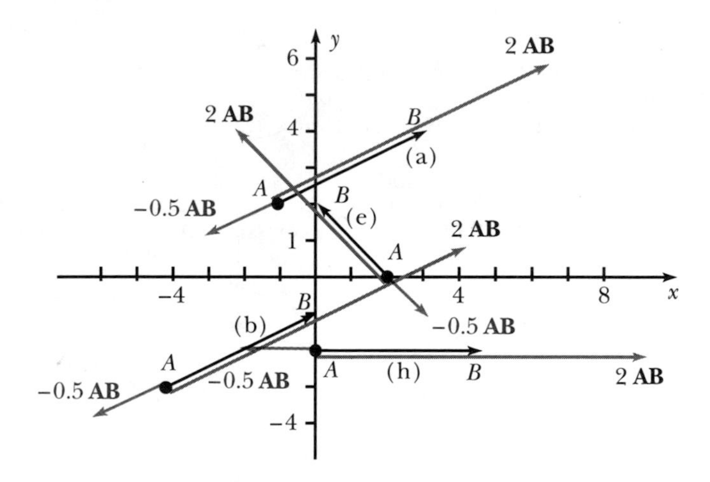

11. **(a)** The geometrical proofs depend on the geometrical definition of the scalar product.

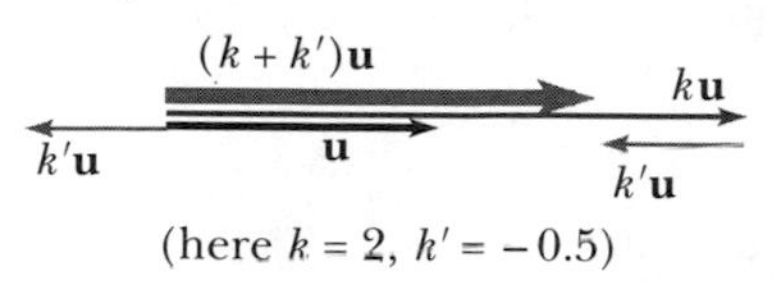

Property (6)

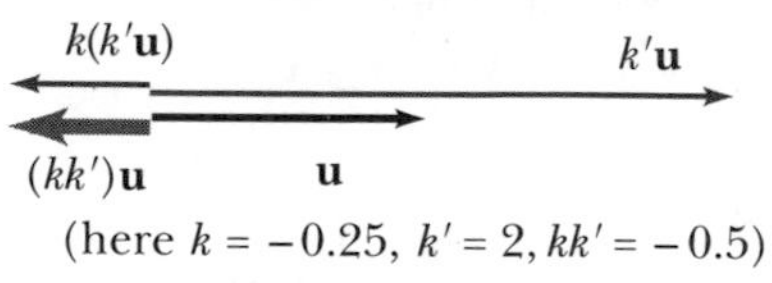

Property (7)

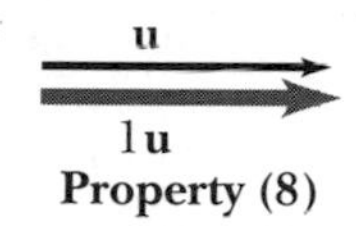

Property (8)

(b) Using components: let $\mathbf{u} = (x,y)$. Then,

$(k + k')\mathbf{u} = ((k + k')x,(k + k')y)$

$= (kx,ky) + (k'x,k'y) = k\mathbf{u} + k'\mathbf{u}$,

$k(k'\mathbf{u}) = k(k'x,k'y) = (kk'x,kk'y) = kk'(x,y)$

$= kk'\mathbf{u}$,

$1\mathbf{u} = 1(x,y) = (x,y)$.

13. **(a)** $a = -2$, $b = 3$

15. Note from the given lengths of BC and AC that $\angle A$ is a right angle. In the figure (a), **S**, **S′** denote the compression in AB, AC respectively, and **T**, **T′** denote the opposing tension forces in BC. Then, $|\mathbf{S}| = |\mathbf{S}'| = 70.7$ kg-wt and $|\mathbf{T}| = |\mathbf{T}'| = 50$ kg-wt.

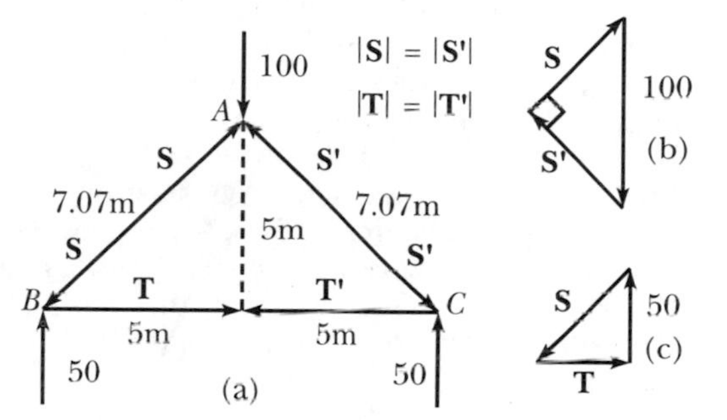

EXERCISES 1-3

1. **(a)** 0 **(b)** −7 **(c)** 1 **(d)** 0 **(e)** −22 **3.** 147°

5. **(a)** 23/4 **(b)** −1/3 **(c)** $(1/2)[-1 \pm \sqrt{1+8}] = 1$ or -2

7. $\pm(1/\sqrt{30})(1,-5,2)$ **9.** 137.72°, 19.65°, 22.63°

11. **(a)** $(1/25)(-3,4)$; 1/5 **(b)** $(43/25)(-3,4)$; 43/5 **(c)** $(16/25)(-3,4)$; 16/5

13. $(1/\sqrt{174})(10,5,7)$ **15.** 60°; 8.66 newtons

17. The line joining the midpoints of two of the sides is defined by the vector $\mathbf{w}' = \mathbf{w} + \mathbf{v} - \mathbf{u}$, where $\mathbf{w} = 2\mathbf{u} - 2\mathbf{v}$, as shown. So, $\mathbf{w}' = \mathbf{w} + \mathbf{v} - \mathbf{u} = 2\mathbf{u} - 2\mathbf{v} + \mathbf{v} - \mathbf{u} = \mathbf{u} - \mathbf{v} = (1/2)\mathbf{w}$, that is, $\mathbf{w}'$ has half the length of $\mathbf{w}$ and is parallel to it.

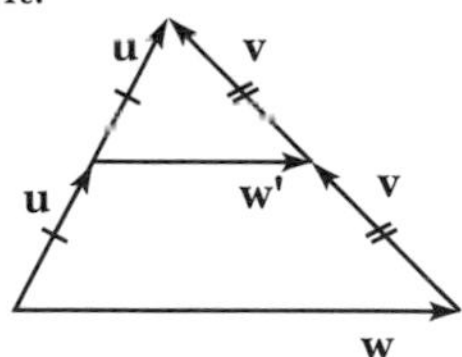

EXERCISES 1-4

1. **(a)**

	(i)	(iii)	(v)
$\cos\alpha$	1/3	0	$4/\sqrt{77}$
$\cos\beta$	2/3	$1/\sqrt{5}$	$-5/\sqrt{77}$
$\cos\gamma$	2/3	$2/\sqrt{5}$	$6/\sqrt{77}$

(b)

	(i)	(iii)	(v)
α	70.5°	90°	62.9°
β	48.2°	63.4°	124.8°
γ	48.2°	26.6°	46.9°

2. **(a)** $x = -2 - t$
$y = 4 - 3t$
$z = 5 + 2t, t \in R$;
intercepts are $(0,10,1)$, $(-10/3,0,23/3)$, $(1/2,23/2,0)$

(c) $x = -4 + t$
$y = 2$
$z = -6, t \in R$;
intercept is $(0,2,-6)$

(e) $x = 10 - 2t$
$y = -2 + t$
$z = t, t \in R$;
intercepts are $(0,3,5)$, $(6,0,2)$, $(10,-2,0)$

3. **(c)** question 2(a):
$(x + 2)/(-1) = (y - 4)/(-3)$
$(y - 4)/(-3) = (z - 5)/(2)$
question 2(e):
$(x - 10)/(-2) = (y + 2)/(1)$
$(y + 2)/(1) = (z - 0)/(1)$

question 2(c):
no symmetric equations here because the direction number $l = 0$

5. **(a)** The first line is parallel to the vector $(7 - 6, 2 - 7, -3 - 1) = (1,-5,-4)$. The second is parallel to $(-2 - (-3), 0 - 5, -4 - 0) = (1,-5,-4)$. As this is the same vector, the lines are parallel. **(b)** The line through $(4,5,6)$ and $(2,3,5)$ is parallel to the vector $\mathbf{u} = (-2,-2,-1)$. The line through $(-5,3,7)$ and $(-2,1,5)$ is parallel to $\mathbf{v} = (3,-2,-2)$. Since $\mathbf{u}\cdot\mathbf{v} = -6 + 4 + 2 = 0$, the lines are perpendicular. **(c)** The lines in (a) lie in a common plane; the lines in (b) are not in the same plane.
7. $2x + 3y - z = 0$ **9.** **(a)** $2x + 3y + 6z = 35$ **(b)** 5 **(c)** 21.4
12. Since $\mathbf{n} = (\cos\alpha, \cos\beta, \cos\gamma)$ is a (unit) vector perpendicular to the plane, the equation is, by Theorem 1-8, $x\cos\alpha + y\cos\beta + z\cos\gamma = \pm d$. The two signs arise since, for $d \neq 0$, there are two such planes, one on each side of the origin.
13. **(a)** The planes are parallel iff their normal vectors $\mathbf{N} = (a,b,c)$ and $\mathbf{N}' = (a',b',c')$ are parallel, which happens iff they are nonzero scalar multiples of each other: $a = ka'$, $b = kb'$, and $c = kc'$, $k \neq 0$. **(b)** The planes are perpendicular iff $\mathbf{N}$ and $\mathbf{N}'$ are perpendicular, which is the case iff $0 = \mathbf{N}\cdot\mathbf{N}' = aa' + bb' + cc'$.
14. **(a)** $-3x + 2y - z = \pm 5\sqrt{14}$ **(b)** $-3x + 2y - z = -5\sqrt{14}$

EXERCISES 2-1

1. **(a)** $x = 2, y = 1$ **(b)** $x = 2/5, y = 7/5, z = 9/5$ **(c)** $x = 59/13 - (7/13)z$, $y = 43/13 + (9/13)z$ **(d)** $x_1 = x_2 = x_3 = 0$
3. $y = z = 0, x = 6$
5. By Gauss-Jordan elimination, they have the same solutions $x = -1, y = 2, z = 1$. It follows by Definition 2-2 that they are equivalent.
7. 36 and 12 units per hour respectively **10.** 2 a.m. the next day

EXERCISES 2-2

1. **(a)** **(i)** $\begin{bmatrix} 1 & -2 & 1 & 0 \\ 1 & -1 & 0 & 0 \\ 2 & 0 & 0 & 3 \\ 0 & -1 & 1 & -2 \\ 1 & 1 & 1 & 1 \end{bmatrix}, \begin{bmatrix} 0 \\ -3 \\ 4 \\ 2 \\ 0 \end{bmatrix}, \begin{bmatrix} 1 & -2 & 1 & 0 & 0 \\ 1 & -1 & 0 & 0 & -3 \\ 2 & 0 & 0 & 3 & 4 \\ 0 & -1 & 1 & -2 & 2 \\ 1 & 1 & 1 & 1 & 0 \end{bmatrix}$

(ii) $\begin{bmatrix} -1 & 1/2 & 0 & 1 \\ 0 & 0 & 1 & 1 \end{bmatrix}, \begin{bmatrix} -1 \\ 0 \end{bmatrix}, \begin{bmatrix} -1 & 1/2 & 0 & 1 & -1 \\ 0 & 0 & 1 & 1 & 0 \end{bmatrix}$

(b) **(i)**
$$\begin{aligned} -x_1 \phantom{{}+x_2} + 3x_3 &= 2 \\ x_1 + x_2 + 2x_3 &= 4 \\ -2x_1 \phantom{{}+x_2} + 3x_3 &= 1 \end{aligned}$$

(ii)
$$\begin{aligned} -x_1 + 2x_2 &= 3 \\ 4x_1 - 2x_2 &= 1 \\ 6x_1 - x_2 &= 0 \\ 3x_1 + 2x_2 &= 1 \end{aligned}$$

2. **(a)** (iii) only ((ii) is in row-echelon form)

3. **(a)** $\begin{bmatrix} 0 & 1 \\ 3 & 4 \\ -1 & 2 \end{bmatrix}$ **(b)** $\begin{bmatrix} 2 & -1 & 3 & 2 \\ 2 & 0 & -4 & -2 \\ 1 & 0 & 0 & 1 \end{bmatrix}$ **(c)** $\begin{bmatrix} 1 & -2 & 0 & 3 \\ 0 & 1 & 2 & 6 \\ 2 & -4 & 0 & 6 \end{bmatrix}$

4. **(a)** $\begin{bmatrix} 1 & 0 & 3 \\ 0 & 1 & -2 \end{bmatrix}$, $X = \begin{bmatrix} 3 \\ -2 \end{bmatrix}$ **(b)** $\begin{bmatrix} 1 & 0 & 0 & 0 \\ 0 & 1 & 0 & 5/3 \\ 0 & 0 & 1 & -8/3 \end{bmatrix}$, $X = \begin{bmatrix} 0 \\ 5/3 \\ -8/3 \end{bmatrix}$

5. **(a)** A can be obtained from itself by 0 (no) elementary row operations, so it is row-equivalent to itself. **(b)** This is Theorem 2-2. **(c)** Suppose $A = e_1 \ldots e_k(A')$ and $A' = e'_1 \ldots e'_{k'}(A'')$, where e_i and $e'_{i'}$ are elementary row operations, $i = 1, \ldots, k$, $i' = 1, \ldots, k'$. Then, by substitution, $A = e_1 \ldots e_k e'_1 \ldots e'_{k'}(A'')$, so A is row-equivalent to A'.

9. 184.6 kg, 115.4 kg

11. **(a)** $\begin{bmatrix} 4 & 2 \\ -1 & 1 \\ 14 & 4 \end{bmatrix}$ **(b)** $\begin{bmatrix} 1 & 1 & 2 \\ -1 & 1 & 2 \\ -1 & 1 & -1 \\ 1 & -1 & 6 \end{bmatrix}$ **(c)** $\begin{bmatrix} 1 & 2 & 3 \\ -7 & 3 & -4 \\ 17 & 0 & 16 \end{bmatrix}$

EXERCISES 2-3

1. **(a)** $\begin{bmatrix} -1 & 0 & -1 \\ 1 & -1 & 0 \\ 0 & 1 & 1 \end{bmatrix}$ **(c)** $\begin{bmatrix} 1 & 1 & 0 & 1 & 0 & 0 \\ 0 & -1 & -1 & 0 & 0 & 1 \\ -1 & 0 & 1 & 0 & -1 & 0 \\ 0 & 0 & 0 & -1 & 1 & -1 \end{bmatrix}$

2.

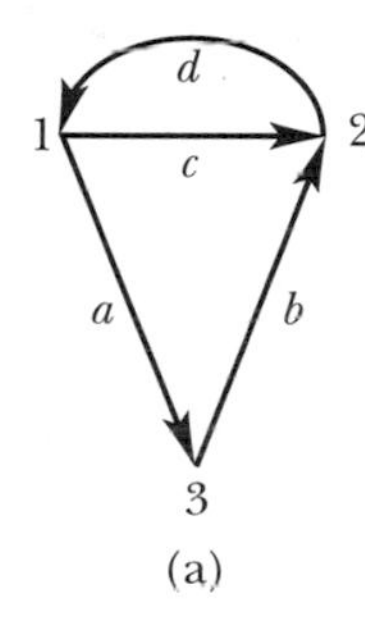

(a)

d
c
2
4
b
e
1
a
f
3

(b)

3. **(a)** $\begin{bmatrix} 6 & 4 & 2 \\ 10 & 5 & 3 \end{bmatrix}$ **(b)** Let x_{i1}, x_{i2}, and x_{i3} denote the number of tons of the three types of glass (in the order given) to be produced from type i sand, $i = 1, 2$; and let Z denote the total cost of producing these (we assume that one ton of sand yields one ton of glass). Then the LP problem is

$$\min Z = 6x_{11} + 10x_{21} + 4x_{12} + 5x_{22} + 2x_{13} + 3x_{23}$$

subject to

$$\begin{array}{lll} x_{11} + x_{21} & \geq 15 & \text{(optical glass required)} \\ x_{12} + x_{22} & \geq 40 & \text{(plate glass required)} \\ x_{13} + x_{23} & \geq 35 & \text{(bottle glass required)} \\ x_{11} + x_{12} + x_{13} & \leq 60 & \text{(type 1 sand available)} \\ x_{21} + x_{22} + x_{23} & \leq 40 & \text{(type 2 sand available)} \\ \quad x_{ij} \geq 0, i = 1, 2, j = 1, 2, 3. & & \end{array}$$

(c) Note that the proposed solution ($x_{11} = 25$, $x_{21} = 0$, $x_{12} = 35$, $x_{22} = 5$, $x_{13} = 0$, and $x_{23} = 35$) does satisfy the constraints and produces a total cost of $Z = 6(25) + 4(35) + 5(5) + 3(35) = \$420$. Another solution is $x_{11} = 15$, $x_{21} = 0$, $x_{12} = 35$, $x_{22} = 5$, $x_{13} = 0$, and $x_{23} = 35$, and yields $Z = 6(15) + 4(35) + 5(5) + 3(35) = \$360 < \$420$.

5. $\begin{bmatrix} 1 & -2 & -5 & -10 & 0.5 \\ -1 & 2 & -5 & -10 & 1 \\ -1 & -2 & 5 & -10 & 2.5 \\ -1 & -2 & -5 & 10 & 5 \end{bmatrix}$

6. (a) for Figure 2-5:

(a) $\begin{bmatrix} 1 & 0 & 1 & 0 \\ 0 & 1 & 0 & 0 \\ 1 & 1 & 0 & 0 \\ 0 & 0 & 0 & 1 \\ 0 & 0 & 0 & 1 \end{bmatrix}$ (b) $\begin{bmatrix} 1 & 0 & 1 & 0 & 0 & 1 \\ 1 & 1 & 0 & 0 & 1 & 0 \\ 0 & 1 & 1 & 1 & 0 & 0 \\ 0 & 0 & 0 & 1 & 1 & 1 \end{bmatrix}$ (c) $\begin{bmatrix} 0 & 0 & 0 & 0 \\ 1 & 0 & 0 & 0 \\ 0 & 1 & 1 & 1 \\ 0 & 1 & 1 & 1 \end{bmatrix}$

(b) They always add to 2.

7. (a) $\begin{bmatrix} 0 & 0 & 0 \\ 1 & 0 & 0 \\ 1 & 1 & 0 \end{bmatrix}$ **(c)** $\begin{bmatrix} 0 & 1 & 1 & 1 \\ 0 & 0 & 0 & 1 \\ 0 & 1 & 0 & 0 \\ 0 & 0 & 1 & 0 \end{bmatrix}$

8. (a) (i) **(b) (i)** two: 1 and 3 **(ii)** No, nobody can communicate with 5 since column 5 contains only 0's.

9. 15 000 cars per hour

EXERCISES 2-4

1. (a) -7 **(b)** 2 **(c)** $-139/30$

2. (a) $[7]$ **(c)** $\begin{bmatrix} 5 \\ 10 \end{bmatrix}$ **(e)** $\begin{bmatrix} -7 & -2 & 12 \\ 4 & 8 & -12 \end{bmatrix}$ **(g)** $\begin{bmatrix} 0 \\ 35 \end{bmatrix}$ **(i)** $[16 \quad 11 \quad -5]$

(k) undefined, as the number of rows of $B = 2 \neq 3 =$ number of columns of E

(l) undefined, since AB is 2×3 while E is 3×3

3. (a) $\begin{bmatrix} 1 & -4 \\ 12 & 1 \end{bmatrix}, \begin{bmatrix} -10 & -9 \\ 27 & -10 \end{bmatrix}, \begin{bmatrix} -19 & -15 \\ 45 & -19 \end{bmatrix}$

7. (a) $\begin{bmatrix} 2 & -3 \\ -1 & 2 \\ 3 & -1 \end{bmatrix} \begin{bmatrix} x_1 \\ x_2 \end{bmatrix} = \begin{bmatrix} 1 \\ 0 \\ 2 \end{bmatrix}$ **(b)** $\begin{bmatrix} 2 & 1 & 3 & 4 \\ 3 & -1 & 2 & 0 \\ -2 & 1 & -4 & 3 \end{bmatrix} \begin{bmatrix} a \\ b \\ c \\ d \end{bmatrix} = \begin{bmatrix} 0 \\ 3 \\ 2 \end{bmatrix}$

9. The two systems may be written in matrix form as $Z = BY$ and $Y = AX$ respectively. By substitution, we have $Z = B(AX) = BAX$, as required. Alternatively, if we substitute for y_1 and y_2 as suggested, we find

$$z_1 = b_{11}(a_{11}x_1 + a_{12}x_2) + b_{12}(a_{21}x_1 + a_{22}x_2) = (BAX)_1$$
$$z_2 = b_{21}(a_{11}x_1 + a_{12}x_2) + b_{22}(a_{21}x_1 + a_{22}x_2) = (BAX)_2.$$

Therefore, $Z = BAX$.

11. (a)

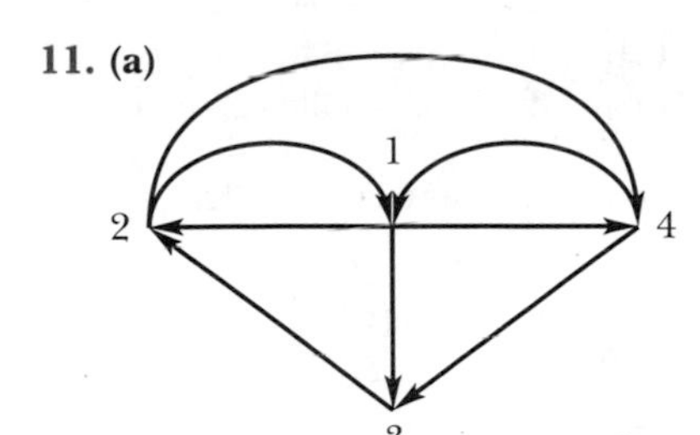

(b) $\begin{bmatrix} 2 & 1 & 1 & 1 \\ 1 & 1 & 2 & 1 \\ 1 & 0 & 0 & 1 \\ 0 & 2 & 1 & 1 \end{bmatrix}$

(c) A term $a_{ik}a_{kj}$ in the sum for $(A^2)_{ij}$ is either 0 or 1 and is $\neq 0$ precisely when $a_{ik} = 1 = a_{kj}$. This means that there is a two-stage link i-k-j from i to j. Thus, $(A^2)_{ij}$ = the number of nonzero terms in the sum = the number of two-stage links from i to j.

(d) $\begin{bmatrix} 2 & 3 & 3 & 3 \\ 2 & 3 & 2 & 2 \\ 1 & 1 & 2 & 1 \\ 3 & 1 & 1 & 2 \end{bmatrix}$. The (i,j) entry in this matrix is the number of three-stage links from i to j. **(e)** For n a positive integer, the (i,j) entry of A^n is the number of n-stage links from i to j.

13. (a) $Z = BAX$, where $Y = AX$ and $Z = BY$ **(b)** $BA = \begin{bmatrix} -1/2 & -5 \\ 7/4 & -2 \end{bmatrix}$

EXERCISES 2-5

1. $AX_1 = \begin{bmatrix} (1)(1) + (2)(0) + (3)(0) \\ (4)(1) + (5)(0) + (6)(0) \\ (7)(1) + (8)(0) + (9)(0) \end{bmatrix} = \begin{bmatrix} 1 \\ 4 \\ 7 \end{bmatrix}$ = column 1 of A. The verification is similar for the other two columns.

3. Using Gauss-Jordan reduction, or otherwise, we find that the system of homogeneous equations having B as coefficient matrix is equivalent to $x = -2y$, which has infinitely many solutions. It follows by Theorem 2-7 that B is not invertible.

5. Only (b), (c), (e), and (g) are elementary.

7. (a) $\begin{bmatrix} 1 & 0 \\ -2 & 1 \end{bmatrix}$ **(b)** $\begin{bmatrix} 0 & 1 \\ 1 & 0 \end{bmatrix}$ **(c)** $\begin{bmatrix} 1 & 0 & 0 \\ 0 & 1 & -1 \\ 0 & 0 & 1 \end{bmatrix}$

8. (a) $\begin{bmatrix} 0 & 1 \\ 1 & 0 \end{bmatrix}$ **(c)** $\begin{bmatrix} 1 & 0 & 0 \\ 0 & 1 & -2 \\ 0 & 0 & 1 \end{bmatrix}$

9. In each case there are many possibilities. Examples: **(a)** Let $A = I$, $B = -I$. Then $A + B = O$, certainly not invertible. **(b)** Let $A = I$, $B = I$. Then $A + B = 2I$ and $(A + B)^{-1} = (1/2)I$.

11. By question 2-1-9, the associated system $AX = O$ has an unique solution iff $ad - bc \neq 0$, so by Theorem 2-7, A is invertible iff this condition holds.

12. (a) $\begin{bmatrix} 1/2 & 1/8 \\ 0 & 1/4 \end{bmatrix}$ **(c)** $\begin{bmatrix} 1 & 2 & -7/3 \\ 0 & -1 & 2/3 \\ 0 & 0 & 1/3 \end{bmatrix}$ **(e)** $\begin{bmatrix} 43 & -5 & -25 \\ 10 & -1 & -6 \\ 7 & 1 & 4 \end{bmatrix}$ **(g)** $\begin{bmatrix} 1 & 1 & 2 & 1 \\ 4 & 5 & 9 & 1 \\ 3 & 4 & 7 & 1 \\ 2 & 3 & 4 & 2 \end{bmatrix}$

13. For each matrix A, the Gauss-Jordan reduction of the augmented matrix $(A{:}I)$ produces a matrix of the form $(A'{:}B')$, where A' has a row of zeros. It follows, as in Example 2-34, that A has no inverse.
14. **(b)** Let $AX = B$ be the system in matrix notation. The solution is $X = A^{-1}B$.

$$A = \begin{bmatrix} 2 & 6 & 4 \\ 1 & 4 & 6 \\ 1 & 3 & 3 \end{bmatrix}, A^{-1} = \begin{bmatrix} -3 & -3 & 10 \\ 3/2 & 1 & -4 \\ -1/2 & 0 & 1 \end{bmatrix}, B = \begin{bmatrix} 1 \\ -3 \\ 2 \end{bmatrix}, X = \begin{bmatrix} 26 \\ -19/2 \\ 3/2 \end{bmatrix}.$$

15. **(a)** **(i)** Gauss-Jordan reduction of $(A{:}I)$ leads at once to a matrix with a row of zeros in the left half, so A does not have an inverse. **(ii)** No inverse. **(b)** Suppose A is $n \times n$ with a row or column of zeros. Then it cannot be row-equivalent to I, so it is not invertible, by the remarks following Theorem 2-11.
17. **(a)** If $A^k = O$, while $A^{k-1} \neq O$ (for some $k > 0$), then if A were invertible, we would have $O = A^{-1}A^k = (A^{-1}A)A^{k-1} = IA^{k-1} = A^{k-1}$, a contradiction. **(b)** **(i)** Here $A^2 = O$.

(ii) Here $A^3 = O$, while $A^2 = \begin{bmatrix} 0 & 0 & 3 \\ 0 & 0 & 0 \\ 0 & 0 & 0 \end{bmatrix}$.

19. Apply Gauss-Jordan reduction to the augmented matrix $(A{:}I)$ to get $(A'{:}P)$, where P is nonsingular. Then, the system $AX = B$ is equivalent to $A'X = PB$, as in Example 2-35. Here we find

$$A' = \begin{bmatrix} 1 & 0 & -1/3 \\ 0 & 1 & -2/3 \end{bmatrix}, P = \begin{bmatrix} -1/3 & 1/3 \\ 4/3 & -1/3 \end{bmatrix}, PB = \begin{bmatrix} 1 \\ -2 \end{bmatrix}, \begin{bmatrix} 1/3 \\ 2/3 \end{bmatrix}, \begin{bmatrix} -9 \\ 27 \end{bmatrix}, \begin{bmatrix} 0 \\ 0 \end{bmatrix},$$

where the 4 column vectors are the values of PB for parts (a) through (d) respectively. Consequently, the solutions of $AX = B$ are $\begin{bmatrix} x_1 \\ x_2 \end{bmatrix} = PB + \begin{bmatrix} (1/3)x_3 \\ (2/3)x_3 \end{bmatrix}$, where x_3 is arbitrary and PB is the column vector corresponding to each B.
21. **(a)** $x + y + z = 1$ **(b)** $13x - 15y + 3z = 8$

EXERCISES 2-6

1. **(a)** $x = -1, y = -2, z = 1$ **(b)** $w = 1, x = -1, y = 1, z = -1$
3. **(a)** 0.1234×10^4, -0.65×10^1, -0.427×10^{-2}, 0.2425×10^4, -0.23364×10^2, -0.4×10^{-3}, -0.2435×10^4
(b) **(i)** (3 digits): 0.123×10^4, -0.650×10^1, -0.427×10^{-2}, 0.242×10^4, -0.234×10^2, -0.400×10^{-3}, -0.244×10^4
(ii) (2 digits): 0.12×10^4, -0.65×10^1, -0.43×10^{-2}, 0.24×10^4, -0.23×10^2, -0.40×10^{-3}, -0.24×10^4
(c) **(i)** $(199 + 0.3) + 1.4 \approx 199 + 1.4 \approx 200.4 \approx 200$
(ii) $199 + (0.3 + 1.4) = 199 + 1.7 \approx 200.7 \approx 201$
(d) Multiplication is not associative either. For example,
$(0.004 \times 0.100) \times 2.00 = 0.0004 \times 2.00 = 0 \times 2.00 \approx 0$ (to 3 figures), while
$0.004 \times (0.100 \times 2.00) = 0.004 \times 0.200 = 0.0008 \approx 0.001$ (to 3 figures).
5. The solutions are as indicated in the text. The row echelon matrices produced by partial pivoting for Examples 2-11 and 2-12 are respectively

$$\begin{bmatrix} 1 & -1/2 & 1/2 & 1/2 \\ 0 & 1 & -3/5 & 1 \\ 0 & 0 & 0 & 0 \end{bmatrix}, \begin{bmatrix} 1 & 1/2 & 0 & -1 & -1/2 & 1 \\ 0 & 1 & 3/2 & -3/2 & 3/2 & 3/2 \\ 0 & 0 & 1 & -13 & -9 & 5 \\ 0 & 0 & 0 & 0 & 0 & 0 \\ 0 & 0 & 0 & 0 & 0 & 0 \end{bmatrix}.$$

7. We scale by multiplying the first equation by 100 and the third by 10. The resulting augmented matrix is then $\begin{bmatrix} 0.2 & -3 & -0.5 & -10.2 \\ 2 & -3 & 4 & -3 \\ 1 & 0.5 & 1 & 2.5 \end{bmatrix}$. Partial pivoting produces

$\begin{bmatrix} 1.00 & -1.50 & 2.00 & -1.50 \\ 0 & 1.00 & 0.333 & 3.66 \\ 0 & 0 & 1.00 & 1.99 \end{bmatrix}$ with corresponding solution $x = -0.980$, $y = 3.00$, $z = 1.99$. The actual solution is $x = -1$, $y = 3$, $z = 2$, as can be verified by substitution.

9. We see at once, by subtracting the second equation from the first in both systems, that the solutions are $x = y = 1$ for the first, and $x = 0$, $y = 1.4$ for the second. Since these are very different, the systems are ill-conditioned.

11. (a) (i) Results of 6 iterations of the Jacobi method (with 3-digit arithmetic):

Iteration	1	2	3	4	5	6
x_1	0	1.10	1.03	0.997	0.999	1.00
x_2	0	−0.700	−1.03	−1.01	−0.999	−1.00
x_3	0	1.10	1.01	0.998	1.00	1.00
x_4	0	−0.800	−1.01	−1.00	−1.00	−1.00
x_5	0	1.00	1.02	0.999	1.00	1.00
x_6	0	−0.950	−1.00	−1.00	−1.00	−1.00

(ii) Results of 4 iterations of the Gauss-Seidel method (with 3-digit arithmetic):

Iteration	1	2	3	4
x_1	1.10	1.01	0.998	1.00
x_2	−0.920	−1.02	−1.00	−1.00
x_3	1.19	1.02	1.00	1.00
x_4	−0.919	−0.993	−1.00	−1.00
x_5	0.908	1.00	1.00	1.00
x_6	−0.995	−1.00	−1.00	−1.00

(b) The augmented matrix is reduced (with 3-digit arithmetic) to

$$\begin{bmatrix} 1.00 & -0.100 & 0 & 0 & 0 & 0 & 1.10 \\ 0 & 1.00 & 0.098 & 0 & 0 & 0 & 0.902 \\ 0 & 0 & 1.00 & -0.202 & 0 & 0 & 1.20 \\ 0 & 0 & 0 & 1.00 & 0.098 & 0 & -0.902 \\ 0 & 0 & 0 & 0 & 1.00 & 0.099 & 0.901 \\ 0 & 0 & 0 & 0 & 0 & 1.00 & -1.00 \end{bmatrix}.$$

The solutions are, by back-substitution, $x_1 = 1.00$, $x_2 = -1.00$, $x_3 = 0.998$, $x_4 = -1.00$, $x_5 = 1.00$, $x_6 = -1.00$.

EXERCISES 3-1

1. (a) $(-1)(-2) - (3/2)(2) = -1$ **(c)** $(6)(1) - (0)(-10) = 6$
(e) $(-1/2)(2) - (1)(4) = -5$
3. (a) -2 **(c)** 0

5.

	σ	τ	$\sigma\tau$	*sgn* σ	*sgn* τ	*sgn* $\sigma \times$ *sgn* τ	*sgn* $\sigma\tau$
(a)	(1,2,3)	(2,3,1)	(2,3,1)	1	1	1	1
(c)	(1,3,2)	(1,3,2)	(1,2,3)	−1	−1	1	1

7. (i) 7 **(ii)** -11 **(iii)** -9 **(iv)** 305 **(v)** -9

9. (a) If row i of A is zero, then $a_{i\sigma(i)} = 0$ for all σ, so every term in the sum in equation 3-5 contains a 0 factor, and $det\ A = 0$. If column j of A is zero, then $a_{ij} = 0$ for all i. But, every term in the sum in equation 3-5 contains a factor $a_{i\sigma(i)}$ with $\sigma(i) = j$ for some i, since each term contains an entry of A from column j. Thus, again every term has a 0 factor and $det\ A = 0$. **(b)** If row i of A is zero, then expansion along row i shows that $det\ A = 0$. Similarly, if column j is zero, expansion along column j shows that $det\ A = 0$.

EXERCISES 3-2

1. See the solutions for question 3-1-1. If $A = (a_{ij})$ is any 2×2 matrix,

$$det\ A^t = det \begin{bmatrix} a_{11} & a_{21} \\ a_{12} & a_{22} \end{bmatrix} = a_{11}a_{22} - a_{12}a_{21} = det\ A.$$

3. (a) 34 **(b)** 0 **(c)** 67
5. Let A be the 2×2 identity matrix I and let $B = -I$. Then, $A + B = O$, so $det(A + B) = det\ O = 0$, while $det\ A + det\ B = 1 + 1 = 2$.
7. (a) -28 **(b)** 210
9. (a) Since $det\ AB = (det\ A)(det\ B) = (det\ B)(det\ A) = det\ BA$, it follows that $det\ AB = 0$ iff $det\ BA = 0$, so AB is invertible iff BA is invertible. In this case, A and B must be invertible, since $det\ A = 0$ or $det\ B = 0$ implies $det\ AB = 0$. **(b)** Let A be invertible. Then $det\ A \neq 0$, so $1/det\ A = (det\ A)^{-1}$ exists. Now, as $A^{-1}A = I$, it follows that $(A^{-1})^{-1} = A$ and, taking determinants, $det\ A^{-1}A = det\ A^{-1}\, det\ A = det\ I = 1$. Therefore, $det\ A^{-1} = (det\ A)^{-1}$.
11. (a) 12 **(b)** -88 **12. (c)** 17/2
14. If σ is a product of k interchanges of adjacent integers, then $sgn\ \sigma = (-1)^k$. Also, P is then obtainable from I by k interchanges of adjacent rows of I, so, by Lemma 3-2, $det\ P = (-1)^k\, det\ I = sgn\ \sigma$.

EXERCISES 3-3

1. (a) $(-1/5)\begin{bmatrix} -1 & -2 \\ -2 & 1 \end{bmatrix}$ **(b)** $(-1)\begin{bmatrix} -5 & -2 \\ -3 & -1 \end{bmatrix}$ **(c)** $(-1)\begin{bmatrix} 0 & 1 & -1 \\ 1 & -3 & 1 \\ -1 & 2 & -1 \end{bmatrix}$

(d) $(-1/28)\begin{bmatrix} 1 & -11 & 7 \\ -7 & -7 & 7 \\ -5 & -1 & -7 \end{bmatrix}$

4. (a) $x_1 = 1/5, x_2 = 7/5$ **(c)** $x_1 = -1, x_2 = -2, x_3 = -2$
5. By Cramer's rule, the solutions are $x_i = det\ A(i)/det\ A = \pm\ det\ A(i)$. Since the entries of $A(i)$ are those of A and B, $det\ A(i)$ must be an integer.

7. If $A = (a_{ij})$, then $adj\ A = \begin{bmatrix} a_{22} & -a_{12} \\ -a_{21} & a_{11} \end{bmatrix}$, so that $adj\ (adj\ A) = \begin{bmatrix} a_{11} & a_{12} \\ a_{21} & a_{22} \end{bmatrix} = A$.

9. (a) 19/3

EXERCISES 3-4

1. (a) $(4,5,-6)$ **(c)** $(34,-4,14)$ **2. (a)** $41\mathbf{i} + 10\mathbf{j} + 28\mathbf{k}$ **3.** 13.10 square units
4. (a) $(\pm 1/\sqrt{117})(1,4,-10)$ **5.** $(\mathbf{u}\times\mathbf{v})\times\mathbf{w} = (2,0,2)$; $\mathbf{u}\times(\mathbf{v}\times\mathbf{w}) = (3,7,-1)$

7. (2) $\mathbf{v}\times\mathbf{u}$ is obtained from $\mathbf{u}\times\mathbf{v}$ by interchanging the last two rows in the expression in Theorem 3-14, Corollary. Therefore, $\mathbf{v}\times\mathbf{u} = -\mathbf{u}\times\mathbf{v}$, since the determinant changes sign when two rows are interchanged. (3) This follows since the determinant is a linear function of its rows (Theorems 3-4 and 3-5). (4) This follows since the determinant is 0 when two rows are equal. (5) This follows since the determinant is 0 when a row is zero.
9. Using Theorem 3-16,

$$\mathbf{u}\cdot(\mathbf{v}\times\mathbf{w}) = det\begin{bmatrix} -2 & 1 & 1 \\ -1 & 1 & 2 \\ -4 & 3 & 5 \end{bmatrix} = (-2)(-1) - (1)(3) + (1)(1) = 0.$$

11. Using the result of question 10, $\mathbf{u}\times(\mathbf{v}\times\mathbf{w}) + \mathbf{v}\times(\mathbf{w}\times\mathbf{u}) + \mathbf{w}\times(\mathbf{u}\times\mathbf{v}) = [(\mathbf{u}\cdot\mathbf{w})\mathbf{v} - (\mathbf{u}\cdot\mathbf{v})\mathbf{w}] + [(\mathbf{v}\cdot\mathbf{u})\mathbf{w} - (\mathbf{v}\cdot\mathbf{w})\mathbf{u}] + [(\mathbf{w}\cdot\mathbf{v})\mathbf{u} - (\mathbf{w}\cdot\mathbf{u})\mathbf{v}] = \mathbf{0}$.
13. **(a)** $x - 8y + 4z = -3$ **(c)** $x + y + z = 3$

EXERCISES 4-1

1. (1) $\mathbf{x} + \mathbf{y} = (1,2,3) + (2,3,1) = (3,5,4) = (2,3,1) + (1,2,3) = \mathbf{y} + \mathbf{x}$
(2) $(\mathbf{x} + \mathbf{y}) + \mathbf{z} = [(1,2,3) + (2,3,1)] + (3,1,2) = (3,5,4) + (3,1,2) = (6,6,6) = (1,2,3) + (5,4,3) = (1,2,3) + [(2,3,1) + (3,1,2)] = \mathbf{x} + (\mathbf{y} + \mathbf{z})$
(3) $\mathbf{0} = (0,0,0)$ and $\mathbf{x} + \mathbf{0} = (1,2,3) + (0,0,0) = \mathbf{x}$
(4) $-\mathbf{x} = (-1,-2,-3)$ and $\mathbf{x} + (-\mathbf{x}) = (1,2,3) + (-1,-2,-3) = (0,0,0) = \mathbf{0}$
(5) $a(\mathbf{x} + \mathbf{y}) = 2[(1,2,3) + (2,3,1)] = 2(3,5,4) = (6,10,8) = 2(1,2,3) + 2(2,3,1) = 2\mathbf{x} + 2\mathbf{y} = a\mathbf{x} + a\mathbf{y}$
(6) $(a + b)\mathbf{x} = (2 + 3)(1,2,3) = 5(1,2,3) = (5,10,15) = 2(1,2,3) + 3(1,2,3) = 2\mathbf{x} + 3\mathbf{x} = a\mathbf{x} + b\mathbf{x}$
(7) $(ab)\mathbf{x} = (2\times 3)(1,2,3) = (6,12,18) = 2(3,6,9) = 2[3(1,2,3)] = 2(3\mathbf{x}) = a(b\mathbf{x})$
(8) $1\mathbf{x} = 1(1,2,3) = (1,2,3) = \mathbf{x}$

3.

x	-2	0	1/2	1	2
(a)	32.00	0	0.1250	-1.000	-16.00
(b)	-31.77	0.5000	-3.547	-12.94	-50.28
(c)	-7.800	1.000	-1.200	-3.500	-7.800

(to 4 significant figures)
5. **(a)** $(-3,-68,-286,96,-147)$ **(b)** $(-2,-238,300,26,-204)$
(c) $(12,-90,24,-30,-150)$
7. Any 2×3 matrix $A = (a_{ij})$ determines the 6-tuple $a = (a_{11},a_{12},a_{13},a_{21},a_{22},a_{23}) \in R^6$. Furthermore, if also $B = (b_{ij})$ is 2×3, it determines $b = (b_{11},b_{12},b_{13},b_{21},b_{22},b_{23}) \in R^6$ and $A + B$ corresponds to $a + b$. Also, if $c \in R$, cA determines $ca \in R^6$, and $O = (0)$ determines $\mathbf{0} = (0,0,0,0,0,0)$. Finally, $-A$ corresponds to the 6-tuple $-a$, so the set of 2×3 matrices is essentially R^6.

EXERCISES 4-2

1. This set V is a vector space. For, if $(x,y,0)$ and $(x',y',0)$ are two elements of V, their sum is $(x + x',y + y',0)$, which is also in V. Also, if $k \in R$, $k(x,y,0) = (kx,ky,0) \in V$. Thus, V is closed under the operations. Furthermore, $\mathbf{0} = (0,0,0) \in V$ and $(-x,-y,0) \in V$ for each $(x,y,0) \in V$. The other axioms are true for V since they are true for all $(x,y,z) \in R^3$. Therefore V is a vector space. (Alternatively, we could simply show that V is a subspace of R^3 using Theorems 4-1 or 4-3.)

3. This set V is a vector space. For, sums and scalar multiples of matrices of this form again have this form:

$$\begin{bmatrix} a & b \\ c & -a \end{bmatrix} + \begin{bmatrix} a' & b' \\ c' & -a' \end{bmatrix} = \begin{bmatrix} a+a' & b+b' \\ c+c' & -(a+a') \end{bmatrix}; k\begin{bmatrix} a & b \\ c & -a \end{bmatrix} = \begin{bmatrix} ka & kb \\ kc & -ka \end{bmatrix}.$$

Furthermore, the 2×2 zero matrix is in V, and $-A \in V$ for $A \in V$. The other axioms hold because they hold for all 2×2 matrices. (Alternatively, we could simply show that V is a subspace of $M(2 \times 2)$ using Theorems 4-1 or 4-3.)
5. V is not a vector space because axiom (6) fails.
7. V is not a vector space because axiom (4) fails.
9. We may think of the elements of $P(1)$ as the functions $f{:}R \to R$ having the form $f(x) = ax + b$, for all $x \in R$, where a and $b \in R$ are uniquely determined by f. When $a = b = 0$, we get the 0 polynomial function, which is in $P(1)$ by definition, so that axiom (3) holds, as indicated in the text. Let also $g(x) = a'x + b'$ be in $P(1)$. Then, $(f + g)(x) = (a + a')(x) + (b + b')$ shows that $P(1)$ is closed under addition; while $(kf)(x) = kax + kb$ shows it is closed under scalar multiplication, since the resulting expressions for the values have the same form as $f(x)$. Also, $-f(x) = -ax - b$ shows that $-f \in P(1)$, so axiom (4) holds. The remaining axioms are true for $P(1)$ since we have proved they are true for all $f \in F(R)$, which contains $P(1)$. Therefore, $P(1)$ is a vector space. (Alternatively, we could use Theorems 4-1 or 4-3 to show that $P(1)$ is a subspace of $F(R)$.)
11. If W is a subspace, certainly the conditions hold. Conversely, if the conditions hold, then all axioms for a vector space hold as well, except possibly axioms (3) and (4), since all the other axioms hold on all of V and $W \subseteq V$. But, by the hypothesis, $\mathbf{0} \in W$ and $-\mathbf{w} \in W$ when $\mathbf{w} \in W$. Thus, axioms (3) and (4) hold as well, and W is a vector space, hence a subspace of V.
13. (a), (d), (e), (f), (h)
15. If $d = 0$, by Theorem 4-4, the plane is a subspace, since it is then the solution space of a homogeneous system of 1 equation in 3 unknowns. If $d \neq 0$, it is not a subspace, since (e.g.) it doesn't contain the zero vector (0,0,0).
17. (b), (c)
19. (a) The x and y axes are subspaces since they are (e.g.) solution spaces of homogeneous systems ($y = 0$ and $x = 0$ respectively). Their union is the set of points $U = \{(x,y) | x = 0 \text{ or } y = 0\}$. This is not a subspace of R^2, since it is not closed under addition: $(x,0) + (0,y) = (x,y) \notin U$, when $x, y \neq 0$. **(b)** If $W_1 \subseteq W_2$ or $W_2 \subseteq W_1$, then $W_1 \cup W_2 = W_2$ or W_1 respectively, so is a subspace. Conversely, if the union is a subspace, suppose $W_1 \not\subseteq W_2$ and $W_2 \not\subseteq W_1$. Then there are vectors $\mathbf{v} \in W_1/W_2 \subseteq W_1 \cup W_2$ and $\mathbf{w} \in W_2/W_1 \subseteq W_1 \cup W_2$. As $W_1 \cup W_2$ is a subspace, $\mathbf{u} = \mathbf{v} + \mathbf{w}$ is in it by closure, so $\mathbf{u}$ must be in W_1 or in W_2. If $\mathbf{u} \in W_1$, then $\mathbf{w} = \mathbf{u} - \mathbf{v} \in W_1$, a contradiction, while if $\mathbf{u} \in W_2$, then $\mathbf{v} = \mathbf{u} - \mathbf{w} \in W_2$, also a contradiction. As these are the only possibilities and both lead to contradictions, the hypothesis must have been false, i.e., it must be that $W_1 \subseteq W_2$ or $W_2 \subseteq W_1$. **(c)** The sum is a subspace by Theorem 4-3, because, if $\mathbf{u}+\mathbf{v}$ and $\mathbf{u}' + \mathbf{v}'$ are two vectors in it and k is a scalar, then, $k(\mathbf{u} + \mathbf{v}) + (\mathbf{u}' + \mathbf{v}') = (k\mathbf{u} + \mathbf{u}') + (k\mathbf{v} + \mathbf{v}') \in W_1 + W_2$, since $k\mathbf{u} + \mathbf{u}' \in W_1$ (a subspace) and $k\mathbf{v} + \mathbf{v}' \in W_2$ (also a subspace). Let W be another subspace of V such that $W_1, W_2 \subseteq W$, and let $\mathbf{u} + \mathbf{v} \in W_1 + W_2$. Then (by definition), $\mathbf{u} \in W_1 \subseteq W$ and $\mathbf{v} \in W_2 \subseteq W$, so, since W is a subspace of V, $\mathbf{u} + \mathbf{v} \in W$. It follows that $W_1 + W_2 \subseteq W$.

EXERCISES 4-3

1. (a) $\mathbf{u} = -2\mathbf{v} + 3\mathbf{w}$ **(c)** $\mathbf{u}$ is not a linear combination of $\mathbf{v}$ and $\mathbf{w}$.
2. (a) h is not a linear combination of f and g. **(c)** h is not a linear combination of f and g.

3. $D = -2A + B - C$. F is not a linear combination of A, B, and C.
5. (a) Suppose that $c_1f_1 + c_2f_2 + c_3f_3 = 0 \in P(3)$, for some scalars $c_i \in R$. Then, for all $x \in R$, $c_1f_1(x) + c_2f_2(x) + c_3f_3(x) = 0 \in R$, or,
$c_1(2 + x) + c_2(1 - x + 2x^2) + c_3(3x - 4x^2) = 0$. Setting the coefficients of each power of x to 0, this means the c_i satisfy the homogeneous system

$$\begin{aligned} 2c_1 + \quad c_2 \qquad &= 0 \\ c_1 - \quad c_2 + 3c_3 &= 0 \\ 2c_2 - 4c_3 &= 0. \end{aligned}$$

The solution is $c_1 = -c_3$, $c_2 = 2c_3$, and c_3 arbitrary. In particular, there are nontrivial solutions (e.g., $c_3 = 1$, $c_1 = -1$, $c_2 = 2$), so the polynomial functions are linearly dependent. **(b)** f will be in the subspace generated by the f_i iff $(1/3)a_0 - (2/3)a_1 - (1/2)a_2 = 0$.
7. (a) By Example 4-20, the dimension of the space of symmetric 2×2 matrices is 3, so it suffices, by Theorem 4-9, Corollary 3, to show that A, B, and C are linearly independent. Suppose that there are scalars a, b, and c such that $aA + bB + cC = O$. Then we obtain the system of equations

$$\begin{aligned} a + 2b \qquad &= 0 \\ 2c &= 0 \\ 2c &= 0 \\ 2a + \quad b \qquad &= 0. \end{aligned}$$

The only solution is $a = b = c = 0$, so A, B, and C are linearly independent. **(b)** D must be such that $\{A,B,C,D\}$ is linearly independent, since $dim\ M(2\times 2) = 4$. It is easy to see (from the equations above) that any one of the E_i defined in Example 4-17 would do for D, although there are many other possibilities.
9. Let E_{ii} be the diagonal matrix having a 1 in row i, column i and 0 elsewhere, for $i = 1, 2, \ldots, n$. Then $B = \{E_{11},\ldots,E_{nn}\}$ is a basis. The dimension of the space is n.
11. (a) We need a vector $\mathbf{u}$ such that $\{\mathbf{u},\mathbf{v},\mathbf{w}\}$ is linearly independent. By the solution above for question 1, we see that $\mathbf{u} = (-1, -1,2)$ or $(1,1,1)$ of parts (c) and (d) are two of infinitely many possibilities. **(b)** We need a polynomial function h such that $\{f,g,h\}$ is linearly independent. By the solution above for question 2, the polynomials h of parts (a), (c), and (d) are possibilities. **(c)** We need a fourth matrix H such that $\{A,B,C,H\}$ is linearly independent. By the solution above for question 3, the matrices F and G are two possibilities. **(d)** By (c) above, a basis is (e.g.) $\{A,B,C,F\}$ or $\{A,B,C,G\}$. Of course, there are many other possibilities.
13. Define $f_a{:}S\rightarrow R$ by $f_a(x) = 1$ if $x = a$, $f_a(x) = 0$ if $x \neq a$. Similarly, define f_b and f_c. Then, $\{f_a,f_b,f_c\}$ is a basis for $F(S)$.
15. (a) If $A = (a_{ij})$ is symmetric, $a_{ij} = a_{ji}$, so A is specified by its entries on and above the main diagonal. There are $n(n + 1)/2$ such entries. Thus, a basis for the space of $n\times n$ symmetric matrices is $\{S_{ij} \mid i \geq j\}$, where S_{ij} has 1's in row i, column j and in row j, column i, but 0's elsewhere. So, the dimension is $n(n + 1)/2$.

17. With the notation suggested and $B = \{\mathbf{v}_1,\mathbf{v}_2,\ldots,\mathbf{v}_n\}$, we let $\mathbf{u}_i = \sum_{j=1}^{n} c_{ji}\mathbf{v}_j$, where c_{ji} are scalars. Substituting in the first equation yields $\sum_{i=1}^{m} x_i \sum_{j=1}^{n} c_{ji}\mathbf{v}_j = \sum_{j=1}^{n} (\sum_{i=1}^{m} c_{ji}x_i)\mathbf{v}_j = \mathbf{0}$.

Since the $\mathbf{v}_j$ are linearly independent, all the coefficients must be 0, i.e., $\sum_{i=1}^{m} c_{ji}x_i = 0$.

This is a system of n linear equations in the m variables x_i. Since $m > n$, there is a nontrivial solution, so there are values for the x_i, not all 0, satisfying the system and therefore also the first equation, showing that S is linearly dependent.

EXERCISES 4-4

1. **(a)** Both matrices may be row reduced to $\begin{bmatrix} 1 & 0 & 0 & -1 \\ 0 & 1 & 0 & 2 \\ 0 & 0 & 1 & 1 \end{bmatrix}$. Consequently, they have the same row space, by Theorem 4-10. **(b)** The nonzero rows of the row-reduced echelon matrix in (a) are a basis for the row space of both matrices, by Theorem 4-11.

3. **(a)** 2 **(c)** 3 **(e)** 3

5. $\{(1,0,4/3),(0,1,2/3)\}$ ($\{\mathbf{u},\mathbf{v}\}$ is also a basis since this set is linearly independent.)

7. $\{F,G,H\}$ where $F = \begin{bmatrix} 1 & 0 \\ 0 & -2 \end{bmatrix}, G = \begin{bmatrix} 0 & 1 \\ 0 & -11/2 \end{bmatrix}, H = \begin{bmatrix} 0 & 0 \\ 1 & 7/2 \end{bmatrix}$

9. none

11. One possibility is $\{A,B,C,E_{21},E_{22},E_{23}\}$ where E_{ij} has a 1 in row i and column j only, and 0 elsewhere.

13. **(a)** $\{(1,1,0,0),(0,0,-1,1)\}$; $dim\ W = 2$ **(b)** $\{(-1,1,0,0),(-1,0,1,0),(-1,0,0,1)\}$; $dim\ W' = 3$

15. $AX = B$ has a solution iff there is an X such that $AX = B$. But, if $X = (x_i)$ and A has columns $A^{[1]},\ldots,A^{[n]}$, then $AX = A^{[1]}x_1 + \ldots + A^{[n]}x_n$, so $AX = B$ iff B is a linear combination of the columns of A, i.e., iff $B \in col(A)$. Now, $B \in col(A)$ iff $dim\ col(A{:}B) = dim\ col(A)$, iff $rank\ (A{:}B) = rank\ (A)$, as stated (note that $col(A) \subseteq col(A{:}B)$).

17. **(a)** Every solution is $X = X' + c_1H_1 = (5/3 - 5c_1, -1/3 + 2c_1, c_1)$, $c_1 \in R$.

19. If $W = V$, certainly $dim\ W = dim\ V$. Conversely, if $dim\ W = dim\ V$, then any basis for W is a basis for V, by Theorem 4-9, Corollary 3. It follows that any vector $\mathbf{v}$ in V is a linear combination of (basis) vectors of W, so $\mathbf{v} \in W$. Thus, $V \subseteq W$, implying $W = V$.

EXERCISES 4-5

1. **(a)** (i), (iii) **(b)**

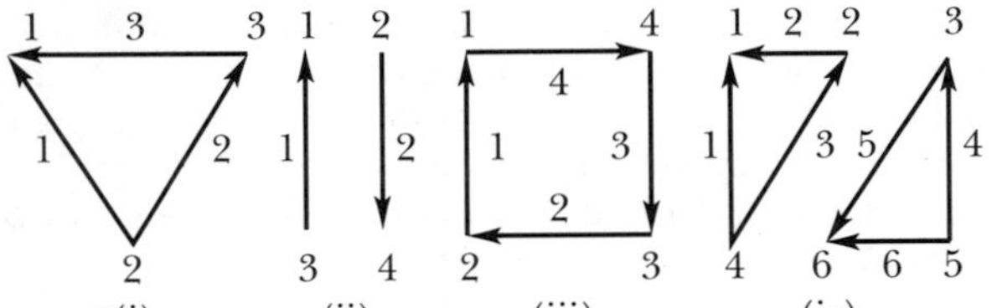

(i) (ii) (iii) (iv)

2. circuit matrix for (c):

	a	b	c	d	e	f
acb	1	−1	1	0	0	0
edc	0	0	−1	1	1	0
$edba$	1	−1	0	1	1	0
fdc	0	0	−1	1	0	−1
fe	0	0	0	0	−1	−1
$fdba$	1	−1	0	1	0	−1

3. For (b), rank $= 2 = 5 - 4 + 1$. For (c), rank $= 3 = 6 - 4 + 1$. For (a), rank $= 3 \neq 6 - 5 + 1$; this graph is not connected.

5. **(a)** As noted in question 3, the rank of the circuit matrix is $e - v + 1$. Since the circuit matrix is the coefficient matrix of Kirchhoff's voltage equations for the network, it follows that $e - v + 1$ of the equations are independent as stated. **(b)** The third voltage equation is the sum of the first two. Alternatively, we see that the rank of the circuit matrix (question 2) is 2.

7. (a)

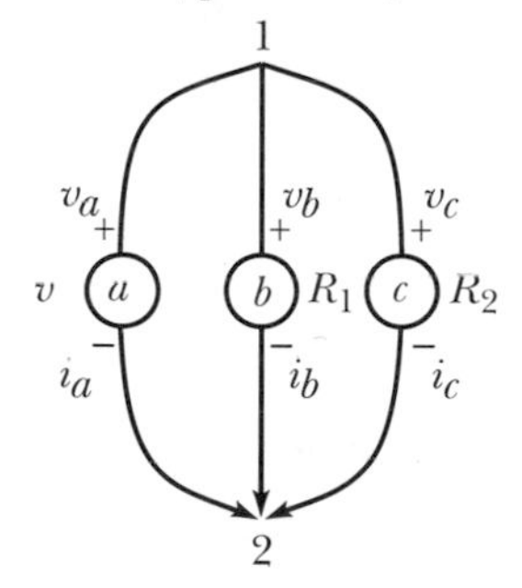

(b) The current equations are

vertex 1: $i_a + i_b + i_c = 0$
vertex 2: $-i_a - i_b - i_c = 0.$

The voltage equations are

loop ab: $v_a - v_b = 0$
loop bc: $v_b - v_c = 0.$

(c) $i_a = -v(R_1 + R_2)/R_1R_2$ **(d)** Comparing i_a with the current through the voltage source in Figure 4-10(b), $i_a = -v/R$, we find they are equal when $1/R = (R_1 + R_2)/R_1R_2 = 1/R_1 + 1/R_2$. **(e)** $i_a \approx -0.083$ amperes

EXERCISES 4-6

1. (a) Let $\mathbf{x} = (x_1,x_2)$, $\mathbf{y} = (y_1,y_2) \in C^2$, and $a \in C$. Then, $\mathbf{x} + \mathbf{y} = (x_1 + y_1, x_2 + y_2)$ and $a\mathbf{x} = (ax_1, ax_2) \in C^2$. The verification that the eight axioms hold is identical to that for R^n presented just before Example 4-1, except that here $n = 2$, $\mathbf{x}$, $\mathbf{y}$, and $\mathbf{z} \in C^2$, and a, $b \in C$.
3. yes **4. (a)** no **(c)** no **5.** (a), (e)
7. (a) $(-1 + 4i, -2 - 2i, 1 + 2i)$ **(c)** $(38 - 31i, 9 - 23i, -5 - 70i)$ **8.** only (b)

10. (b) $[\mathbf{z}]_B = \begin{bmatrix} -1/5 + (2/5)i \\ 7/5 - (6/5)i \\ 1/5 + (7/5)i \end{bmatrix}$ **11.** (a), (c)

13. (a) $z_1 = (3 - 2i) + (1 + 2i)z_2$, $z_2 \in C$ **(b)** $z_1 = -1184/365 + (67/365)i$, $z_2 = -54/73 + (71/73)i$ **(c)** no solution

EXERCISES 5-1

1. (a) (i) $T_A{:}R^3 \to R^1$ $(m = 1, n = 3)$ and $T_A(x_1,x_2,x_3) = -x_1 + 2x_2 + 3x_3$.
(iii) $T_A{:}R^1 \to R^4$ $(m = 4, n = 1)$ and $T_A(x) = (2x, 5x, 0, 3x)$. **(b) (i)** $T_A(\mathbf{e}_1) = -1$, $T_A(\mathbf{e}_2) = 2$, $T_A(\mathbf{e}_3) = 3$ **(iii)** $T_A(\mathbf{e}_1) = (2,5,0,3)$
3. (a), (d), and (e) are linear
5. Let M and $M' \in M(m \times n)$ and $c \in R$. Then, $T(cM + M') = A(cM + M')B = cAMB + AM'B = cT(M) + T(M')$, so T is linear by Lemma 5-1.
7. (2) Since T is linear, $\mathbf{0} = T(\mathbf{0}) = T(\mathbf{v} + (-\mathbf{v})) = T(\mathbf{v}) + T(-\mathbf{v})$, so, $T(-\mathbf{v}) = -T(\mathbf{v})$.
(3) $T(\mathbf{v} - \mathbf{w}) = T(\mathbf{v}) + T(-\mathbf{w}) = T(\mathbf{v}) - T(\mathbf{w})$, using (2).
(4) We use induction on m. If $m = 1$, $T(c_1\mathbf{v}_1) = c_1T(\mathbf{v}_1)$, by definition of linearity. In general, $T(c_1\mathbf{v}_1 + \dots + c_m\mathbf{v}_m) = T((c_1\mathbf{v}_1 + \dots + c_{m-1}\mathbf{v}_{m-1}) + c_m\mathbf{v}_m) = T(c_1\mathbf{v}_1 + \dots + c_{m-1}\mathbf{v}_{m-1}) + T(c_m\mathbf{v}_m) = c_1T(\mathbf{v}_1) + \dots + c_{m-1}T(\mathbf{v}_{m-1}) + c_mT(\mathbf{v}_m)$, in which we have used the result for $m - 1$ (induction hypothesis) and the definition of linearity.

EXERCISES 5-2

1. $x - 2y, -8, 0$ **3.** $(x - y, -x + y, 2y, 2x)$; $(-4,4,12,4)$; $(5,-5,-4,6)$
5. $(0, x + 2y)$; $(0,8)$; $(0,6)$
6. **(a)** (1) $\begin{bmatrix} 1 & -2 \end{bmatrix}$; (3) $\begin{bmatrix} 1 & -1 \\ -1 & 1 \\ 0 & 2 \\ 2 & 0 \end{bmatrix}$ **(b)** (1) $\{(2,1)\}$; (3) $\emptyset$ **(c)** (1) $\{1\}$;
(3) $\{(1,-1,0,2),(-1,1,2,0)\}$ **(d)** (1) 1; (3) 0 **(e)** (1) 1; (3) 2
7. Let $B = \{\mathbf{v}_1,\ldots,\mathbf{v}_n\}$ be the stated basis and let $\mathbf{u} \in V$. Then, $\mathbf{u} = c_1\mathbf{v}_1 + \ldots + c_n\mathbf{v}_n$ (for some $c_i \in R$) and $T(\mathbf{u}) = c_1T(\mathbf{v}_1) + \ldots + c_nT(\mathbf{v}_n) = \mathbf{0} + \ldots + \mathbf{0} = \mathbf{0}$, so $T = O$.
9. e.g., $\left\{\begin{bmatrix} -1/2 \\ 3/2 \\ 1 \\ 0 \\ 0 \end{bmatrix}, \begin{bmatrix} 1/2 \\ 7/2 \\ 0 \\ 1 \\ 0 \end{bmatrix}\right\}; \left\{\begin{bmatrix} 1 \\ 0 \\ 0 \\ -2 \end{bmatrix}, \begin{bmatrix} 0 \\ 1 \\ 0 \\ 1 \end{bmatrix}, \begin{bmatrix} 0 \\ 0 \\ 1 \\ 0 \end{bmatrix}\right\}$.
And, $5 = \dim M(5 \times 1) = 3 + 2 = \text{rank } T_A + \text{nullity } T_A$.
11. S exists because the given vectors $\{(1,1),(-1,1)\}$ constitute a basis for R^2.
$S(x,y) = (x/2 + 3y/2, -x/2 + 3y/2)$; $[S] = \begin{bmatrix} 1/2 & 3/2 \\ -1/2 & 3/2 \end{bmatrix}$.
13. **(a)** e.g. $T_1(x,y,z) = (-x - 2y + z)(1,2,-1)$; $T_2(x,y,z) = (-x - 2y + z)(2,4,-2)$
(b) $T(x,y,z) = cT(1,1,1) = [(x + 2y - z)/2](-1/2,-1,1/2)$ **(c)** no
15. **(a)** $AX = B$ has *some* solution iff $B \in Im\ T_A$ iff $B \in col(A)$ iff $rank\ (A{:}B) = rank\ A$ (recall rank = column rank = row rank). If $rank\ A = n$, then $Ker\ T_A = \{\mathbf{0}\}$, as then $\dim Ker\ T_A = n - n = 0$. So if both $AX = B$ and $AX' = B$, we would have $A(X - X') = O$, hence $X - X' \in Ker\ T_A$, which implies $X - X' = O$ and $X = X'$. Therefore, $AX = B$ has an unique solution. Conversely, if $AX = B$ has an unique solution, reversing the above steps shows that $Ker\ T_A = \{\mathbf{0}\}$, and therefore $rank\ A = n$.
17. **(a)** $\begin{bmatrix} 1 & 0 & -1 \\ 0 & 2 & 1 \\ 1 & 1 & 1 \end{bmatrix}$ **(b)** $\begin{bmatrix} 1/2 & -1/4 & -3/2 \\ -1/2 & 3/4 & 3/2 \\ 1/2 & 3/4 & 3/2 \end{bmatrix}$ **(c)** 0, 3
19. **(a)** $\begin{bmatrix} 1 & 2 \\ 2 & 1 \end{bmatrix}$ **(b)** $\begin{bmatrix} 3 & 0 \\ 0 & -1 \end{bmatrix}$. This matrix is simpler than $[S]$ since it is diagonal.

EXERCISES 5-3

1.

A	*rank A*	*nullity*	**(a)** *1-1*	**(b)** *onto*	**(c)** *isomorphism*
(i)	2	0	yes	yes	yes
(iii)	1	0	yes	yes	yes
(v)	2	0	yes	no	no
(vii)	2	1	no	no	no

3. Since $rank\ [T] \leq$ the smaller of m and n, if (a) $n < m$, then $rank\ T < m = \dim R^m$ and T cannot be onto; while if (b) $n > m$, nullity of $T = n - rank\ [T] > 0$ and T cannot be 1-1. If $T{:}V \rightarrow W$, the reasoning above applies to $[T;B/B']$, where B and B' are bases for V and W respectively.
5.

T	*rank*	*nullity*	*1-1*	*onto*	*isomorphism*
5-2-1	1	1	no	yes	no
5-2-2	2	0	yes	yes	yes
5-2-3	2	0	yes	no	no
5-2-4	1	3	no	no	no
5-2-5	1	1	no	no	no

7. Since $dim\ M(2\times 2) = 4 = dim\ P(3)$, it suffices, by Theorem 5-8, Corollary 2, to show that P is 1-1 or onto. In fact, P is clearly onto, so P is 1-1 and an isomorphism.
9. By the linearity of T, we have property (1) of Theorem 5-9, which says essentially that sums, scalar multiples, and linear combinations of vectors in V correspond to the same sums, scalar multiples, and linear combinations of the images of the vectors in W The fact that T is 1-1 and onto means that the correspondence is 1-1 and onto W. Property (2) now follows by question 8, since T is both 1-1 and onto when it is an isomorphism.
10. (a) $(x, 2x, -x + 2y)$ **(b)** $(3x + 6y, -3x + 3y, -6x + 12y)$ **(c)** $(2x - 4y, 10x - 2y, 0)$
11. (a) (i) Using properties of the transpose,
$T(cA + A') = (1/2)[(cA + A') + (cA + A')^t] = (1/2)(cA + A' + cA^t + A'^t) = c(1/2)(A + A^t) + (1/2)(A' + A'^t) = cT(A) + T(A')$.
The proof for T' is identical, except "$-$" replaces "$+$" where appropriate.
(ii) If A is skew-symmetric, $A + A^t = O$, so $A \in Ker\ T$. Also, $T'(A) = (1/2)(A - A^t) = (1/2)(A + A) = A$, so $A \in Im\ T'$. Conversely, if $T(A) = O$, then $A = -A^t$, so A is skew-symmetric. And, if $A \in Im\ T'$, there is a B such that $A = T'(B) = (1/2)(B - B^t)$. Then, $A^t = (1/2)(B - B^t)^t = -(1/2)(B - B^t) = -A$, so A is skew-symmetric. This shows that $Ker\ T = Im\ T' = \{\text{skew-symmetric matrices}\}$. The proof that $Ker\ T' = Im\ T = \{\text{symmetric matrices}\}$ is identical, except that again "$+$" and "$-$" are interchanged where appropriate.
(iii) $(T + T')(A) = T(A) + T'(A) = (1/2)(A + A^t) + (1/2)(A - A^t) = A = I_V(A)$, for all $A \in V$, so $T + T' = I_V$.

(b) $$\begin{bmatrix} 1 & 0 & 0 & 0 \\ 0 & 1/2 & 1/2 & 0 \\ 0 & 1/2 & 1/2 & 0 \\ 0 & 0 & 0 & 1 \end{bmatrix}, \begin{bmatrix} 0 & 0 & 0 & 0 \\ 0 & 1/2 & -1/2 & 0 \\ 0 & -1/2 & 1/2 & 0 \\ 0 & 0 & 0 & 0 \end{bmatrix}$$

13. (a) $(4x + 3y, 3x + y + z, -x + 3y - 3z)$; $(3x + 4y, 5x - y)$ **(b)** We use Theorem 5-14:

$$[T'T] = \begin{bmatrix} 2 & 1 \\ 1 & 1 \\ 1 & -1 \end{bmatrix}\begin{bmatrix} 1 & 2 & -1 \\ 2 & -1 & 2 \end{bmatrix} = \begin{bmatrix} 4 & 3 & 0 \\ 3 & 1 & 1 \\ -1 & 3 & -3 \end{bmatrix};$$

$$[TT'] = \begin{bmatrix} 1 & 2 & -1 \\ 2 & -1 & 2 \end{bmatrix}\begin{bmatrix} 2 & 1 \\ 1 & 1 \\ 1 & -1 \end{bmatrix} = \begin{bmatrix} 3 & 4 \\ 5 & -1 \end{bmatrix}.$$

So, $[T'T(x,y,z)] = [T'T]\begin{bmatrix} x \\ y \\ z \end{bmatrix} = \begin{bmatrix} 4x + 3y \\ 3x + y + z \\ -x + 3y - 3z \end{bmatrix}$; $[TT'(x,y)] = \begin{bmatrix} 3x + 4y \\ 5x - y \end{bmatrix}$.

15. By Theorem 5-13,

$[T^3] = [T]^3$, so $[T^3] = \begin{bmatrix} \cos 3\theta & \sin 3\theta \\ \sin 3\theta & \cos 3\theta \end{bmatrix} = \begin{bmatrix} \cos\theta & -\sin\theta \\ \sin\theta & \cos\theta \end{bmatrix}^3 =$

$\begin{bmatrix} \cos^3\theta - 3\cos\theta\sin^2\theta & \sin^3\theta - 3\cos^2\theta\sin\theta \\ -\sin^3\theta + 3\cos^2\theta\sin\theta & \cos^3\theta - 3\cos\theta\sin^2\theta \end{bmatrix}$. Comparing entries, we obtain $\cos 3\theta = \cos^3\theta - 3\cos\theta\sin^2\theta$, $\sin 3\theta = -\sin^3\theta + 3\cos^2\theta\sin\theta$.
17. $T_{nm}T_{mn}(A) = (A^t)^t = A$ for any m, n, so these maps are inverses.
18. (a) T^{-1} exists, and

$$[T^{-1}(x,y)] = [T]^{-1}\begin{bmatrix} x \\ y \end{bmatrix} = \begin{bmatrix} 2/3 & -1/3 \\ -1/3 & 2/3 \end{bmatrix}\begin{bmatrix} x \\ y \end{bmatrix} = \begin{bmatrix} (2/3)x - (1/3)y \\ -(1/3)x + (2/3)y \end{bmatrix}.$$

19. This follows from question 2-5-17, since $[T;B]$ is then nilpotent (by Theorem 5-13) and hence is singular, where B is any basis for the space.
21. (a) Let $\mathbf{v} \in V$. Then, $[T(T'T'')](\mathbf{v}) = T[(T'T'')(\mathbf{v})] = T[T'(T''(\mathbf{v}))] = (TT')(T''(\mathbf{v})) = [(TT')T''](\mathbf{v})$. Since this holds for all $\mathbf{v} \in V$, $T(T'T'') = (TT')T''$.

EXERCISES 5-4

1. **(a)** $\begin{bmatrix} 1 & 1 \\ 0 & 1 \end{bmatrix}$ **(b)** **(i)** $\begin{bmatrix} x - y \\ y \end{bmatrix}$ **(ii)** $\begin{bmatrix} -4 \\ 3 \end{bmatrix}$ **(iii)** $\begin{bmatrix} 12 \\ -8 \end{bmatrix}$

3. **(a)** $\begin{bmatrix} 2 & -7 & 3 \\ 1 & -2 & 1 \\ -1 & 3 & -1 \end{bmatrix}$ **(b)** **(i)** $\begin{bmatrix} 2x - 7y + 3z \\ x - 2y + z \\ -x + 3y - z \end{bmatrix}$ **(ii)** $\begin{bmatrix} -2 \\ -1 \\ 2 \end{bmatrix}$

5. **(a)** $\begin{bmatrix} 1 & 1 & 1 & 1 \\ 0 & 1 & 1 & 1 \\ 0 & 0 & 1 & 1 \\ 1 & 0 & 0 & 1 \end{bmatrix}, \begin{bmatrix} 1 & -1 & 0 & 0 \\ 0 & 1 & -1 & 0 \\ 1 & -1 & 1 & -1 \\ -1 & 1 & 0 & 1 \end{bmatrix}$ **(b)** $\begin{bmatrix} 3 \\ -4 \\ 10 \\ -7 \end{bmatrix}$

7. **(a)** Suppose $B' = \{\mathbf{v}_1, \mathbf{v}_2\}$ is obtained from $B(2)$ by a rotation T through an angle θ in a counterclockwise direction. Let x' and y' be the components of $\mathbf{v}$ with respect to the basis B'. The equation with respect to the x', y' axes is $x'^2 + y'^2 = r^2$. **(b)** With the notation above, suppose now that B' is an arbitrary basis for R^2 and that $[B'/B(2)] = A = (a_{ij})$. We now get $(a_{11}x' + a_{12}y')^2 + (a_{21}x' + a_{22}y')^2 = r^2$.

9. $\begin{bmatrix} -3/2 & 2 & -1/2 \\ -1/2 & 0 & 1/2 \\ 1/2 & -2 & 3/2 \end{bmatrix}$

11. Let $B_1 = \{\mathbf{v}_1, \ldots, \mathbf{v}_n\}$. The ith column of $[T;B_1/B_2]$ is $[T(\mathbf{v}_i)]_{B_2} = [B'_2/B_2][T(\mathbf{v}_i)]_{B'_2} = [B'_2/B_2][T;B'_1/B'_2][\mathbf{v}_i]_{B'_1} = [B'_2/B_2][T;B'_1/B'_2][B_1/B'_1][\mathbf{v}_i]_{B_1}$, which is the ith column of $[B'_2/B_2][T;B'_1/B'_2][B_1/B'_1]$, since $[\mathbf{v}_i]_{B_1}$ has a 1 in row i and 0's elsewhere. The required formula therefore follows after solving for $[T;B'_1/B'_2]$.
13. Using the notation of Example 5-38 and Figure 5-15, the action of this linear transformation T is described by $T(x',y') = (0,y')$, since T is the orthogonal projection onto L = the y' axis. Therefore, with respect to the x, y axes, we get $T(x,y) = (x\cos^2\theta + y(\sin\theta\cos\theta), x(\sin\theta\cos\theta) + y\sin^2\theta)$.
15. Since A has rank 3, while B has rank 2, these matrices are not similar.
17. **(a)** Let $A = P^{-1}BP$. Then, $A^2 = (P^{-1}BP)(P^{-1}BP) = P^{-1}BBP = P^{-1}B^2P$.
19. Let $A = P^{-1}BP$. If B is invertible, then $A^{-1} = P^{-1}B^{-1}P$, since $(P^{-1}B^{-1}P)(P^{-1}BP) = I$. Conversely, if A is invertible, a similar argument shows that B is also invertible. And, the equation $A^{-1} = P^{-1}B^{-1}P$ shows that the inverses are similar.

EXERCISES 5-5

1. (b) and (c) **3.** $g(x,y,z) = 2x + y + 3z$
5. $A_2(f) = 3a + 2b + 2c$. To see that this is linear, suppose $f' \in P(2)$ is given by $f'(x) = a'x^2 + b'x + c'$. Then, for $k \in R$, $A_2(kf + f') = 3(ka + a') + 2(kb + b') + 2(kc + c') = k(3a + 2b + 2c) + (3a' + 2b' + 2c') = kA_2(f) + A_2(f')$, so A_2 is linear.
7. As noted before Lemma 5-5, the ith row of $[f]_{B^*}$ is $f(\mathbf{v}_i)$. By the definition of $[f;B/B(1)]$, this is exactly the ith row of $[f;B/B(1)]^t$.
9. **(a)** $f(x) = x$ **(b)** $f(x) = (-1/6)x^3 + x^2 + (-11/6)x + 2$
11. **(a)** $[2a + 2b + c \quad -a + b - c]$ **(b)** $[0 \quad -1]$ **(c)** $-y$

13. **(a)** $[D^*;B'^*/B^*] = \begin{bmatrix} 0 & 0 \\ 1 & 0 \\ 0 & 2 \end{bmatrix} = \begin{bmatrix} 0 & 1 & 0 \\ 0 & 0 & 2 \end{bmatrix}^t = [D;B/B']^t$
(b) $D^*(Area)(a_0 + a_1x + a_2x^2) = a_1 + a_2$

EXERCISES 6-1

1. **(a)** Every $\mathbf{v} \neq \mathbf{0}$ in V is an eigenvector of 0 with associated eigenvalue 0. Therefore, 0 is diagonalizable. **(b)** Every $\mathbf{v} \neq \mathbf{0}$ in V is an eigenvector of I_V with associated eigenvalue 1. Therefore, I_V is diagonalizable. **(c)** If $c = 0$, $T = 0$, while if $c = 1$, $T = I_V$. If $c \neq 0$ or 1, T has no eigenvectors. Thus T is diagonalizable iff $c = 0$ or 1. **(d)** T has no eigenvectors. Thus, T is not diagonalizable. **(e)** Eigenvalues ± 1; diagonalizable.
2. **(a)** **(i)** $T_A(1,1) = (0,0) = 0(1,1)$, so $(1,1)$ is an eigenvector associated with eigenvalue $c = 0$. **(iii)** $T_A(-4,-4) = (0,0) = 0(-4,-4)$, so $(-4,-4)$ is an eigenvector associated with $c = 0$. **(b)** **(ii)** $T_A(1,1,1) = (1,1,3) \neq c(1,1,1)$ for any $c \in R$, so $(1,1,1)$ is not an eigenvector. **(iv)** $T_A(1,2,3) = (0,1,9) \neq c(1,2,3)$ for any $c \in R$, so $(1,2,3)$ is not an eigenvector.
5. **(a)** A is diagonalizable iff $A = P^{-1}DP$, where D is diagonal and P is nonsingular. Let B be the standard basis for R^n and B' be the basis such that $[B/B'] = P$. Then, by Theorem 5-17, $[T_A;B'] = [B'/B]^{-1}[T_A;B][B'/B] = PAP^{-1} = D$, so T_A is diagonalizable. Conversely, if T_A is diagonalizable, there is a basis B' such that $[T_A;B'] = D$ is diagonal. Then, $A = [T_A;B] = [B/B']^{-1}[T_A;B'][B/B'] = P^{-1}DP$. **(b)** Suppose T is diagonalizable. Then, there is a basis B' such that $[T;B']$ is diagonal. By Theorem 5-17, for any B, $[T;B] = [B/B']^{-1}[T;B'][B/B']$, so $[T;B]$ is similar to a diagonal matrix, hence is diagonalizable. Conversely, if $[T;B]$ is diagonalizable for any B, $[T;B]$ is similar to a diagonal matrix for any B: $[T;B] = P^{-1}DP$. Letting B' be the basis for V such that $[B/B'] = P$, we find (as in (a)) that $D = [T;B']$, so T is diagonalizable.
7. Let $T(\mathbf{v}) = c\mathbf{v}$, where $\mathbf{v} \neq \mathbf{0}$. Applying T^{-1} to both sides of this equation yields $\mathbf{v} = cT^{-1}(\mathbf{v})$, so $T^{-1}(\mathbf{v}) = (1/c)\mathbf{v}$, showing that $1/c$ is an eigenvalue of T^{-1}. The fact for matrices follows from this since the eigenvalues of A are those of the matrix operator T_A. The fact that T and T^{-1} have the same eigenvectors also follows from the first remark, since 0 is not an eigenvalue of an invertible operator T.

EXERCISES 6-2

1. $f = x^2 - 5x + 6 = (x - 3)(x - 2)$, eigenvalues $c = 2$ and 3
3. $f = -x^3 - x^2 - 2x - 2 = -(x + 1)(x^2 + 2)$, eigenvalue $c = -1$
5. $f = x^4 - 5x^3 + 8x^2 - 4x = x(x - 1)(x - 2)^2$, eigenvalues $c = 0$, 1, and 2
7. (1) $c = 2$, $\{(1,1)\}$; $c = 3$, $\{(2,1)\}$; diagonalizable; (3) $c = -1$, $\{(5/3,7/3,1)\}$; not diagonalizable; (5) $c = 0$, $\{(22,57/2,-1,1)\}$; $c = 1$, $\{(1,1,0,0)\}$; $c = 2$, $\{(4/3,1,0,0)\}$; not diagonalizable. Other solutions are possible.
9. **(a)** $[D] = \begin{bmatrix} 0 & 1 & 0 \\ 0 & 0 & 2 \\ 0 & 0 & 0 \end{bmatrix}$ **(b)** $f = det\ ([D] - xI) = -x^3$, so the only eigenvalue is $c = 0$. **(c)** A basis for the eigenspace of $c = 0$ is $\{(1,0,0)\}$. D is not diagonalizable.
11. If $c = 0$ is an eigenvalue of AB, then AB is singular, so, since $det\ BA = det\ AB = 0$, BA is also singular, whence, by question 6-1-6, $c = 0$ is an eigenvalue of BA. Next, suppose $c \neq 0$ is an eigenvalue of AB. Then, there is a vector $X \in M(n \times 1)$, $X \neq O$, such that $ABX = cX$. As $c \neq 0$, $cX \neq O$, so $BX \neq O$. Since $BA(BX) = B(ABX) = B(cX) = c(BX)$, it follows that BX is an eigenvector for BA corresponding to c, so c is an eigenvalue of BA. Interchanging A and B now shows that A and B have the same eigenvalues.
13. **(a)** The matrix $A - xI$ is also block triangular: $A - xI = \begin{bmatrix} B - xI & C \\ O & D - xI \end{bmatrix}$, where the I's are identity matrices of appropriate sizes. Therefore, the characteristic polynomial of A is $det\ (A - xI) = det\ (B - xI)\ det\ (D - xI) =$ the product of the

characteristic polynomials of B and D. **(b)** Let A be block triangular of the form in equation 3-8 of Chapter 3. Then, the characteristic polynomial of A is the product of the characteristic polynomials of the blocks $B_{11}, ..., B_{kk}$ along the diagonal of A. The proof follows at once by induction on k, because of the result in (a) above.
15. The eigenvalues of A were shown to be $c = 3, 6$, and 9. Eigenvectors are (e.g.) $(-2,-2,1)$, $(-2,1,-2)$, and $(1,-2,-2)$. Since these are linearly independent, they constitute a basis for R^3.

EXERCISES 6-3

1. $A^2 = \begin{bmatrix} -1 & -1 \\ 2 & -2 \end{bmatrix}, A^3 = \begin{bmatrix} -3 & 1 \\ -2 & -2 \end{bmatrix}, A^4 = \begin{bmatrix} -1 & 3 \\ -6 & 2 \end{bmatrix}$. To show that the powers are linearly dependent, we find scalars $c_0, ..., c_4$ such that $c_0A_0 + ... + c_4A^4 = O$. Using the expressions above, this leads to (e.g.) $c_0 = 2, c_1 = -1, c_2 = 1, c_3 = c_4 = 0$, so a polynomial h is $h = x^2 - x + 2$. Another choice is $c_0 = 2, c_1 = 1, c_3 = 1, c_2 = c_4 = 0$, so $h = x^3 + x + 2$.
3. $f(A) = \begin{bmatrix} -27 & 42 \\ 28 & 57 \end{bmatrix}$; $f(T)(x,y) = (2x + 2y, 2x - 4y)$
4. (a) (i) $(-11x - 4y, -4x - 3y)$ **(iii)** $(3x - y, -x + 5y)$
5. Let $f = a_nx^n + ... + a_1x + a_0$. Then $f(T) = a_nT^n + ... + a_1T + a_0 1_V$. Since the operation of taking matrices with respect to B is linear and preserves the "multiplication" operations (c.f. Theorem 5-14 and the preceding discussion), $[f(T);B] = [a_nT^n + ... + a_1T + a_0 1_V;B] = a_n[T;B]^n + ... + a_1[T;B] + a_0I = f([T;B])$.
7. Let p be the minimal polynomial. Suppose q were another monic polynomial of least degree in $\alpha(T)$ which divides every polynomial in $\alpha(T)$. Then, p divides q implies $q = fp$, while q divides p implies $p = gq$, for some f and $g \in P(R)$. Combining these equations yields $q = fgq$. Thus, we must have $deg\, f = deg\, g = 0$ and $f = 1/g \in R$. Since p and q are monic, it must further be the case that $f = g = 1$, so that $q = p$, showing that p is unique.
9. (a) $p = f = x^2$ **(b)** $p = -f = x^3$ **(c)** $p = -f = x^3$
11. $p = f = -x^3$. D is not diagonalizable. **13.** $p = -f$
15. A is diagonalizable iff T_A is diagonalizable iff the minimal polynomial p of T_A is a product of distinct linear factors. But, the minimal polynomial of A equals p (by question 8(a)), so the result follows.
17. (a) Any vector $\mathbf{v} = (x,y) \in R^2$ may be uniquely written as $\mathbf{v} = x\mathbf{e}_1 + y\mathbf{e}_2$. Then, $\mathbf{w}_1 = x\mathbf{e}_1$ is in the x axis, while $\mathbf{w}_2 = y\mathbf{e}_2$ is in the y axis. The uniqueness results from the fact that $\{\mathbf{e}_1,\mathbf{e}_2\}$ is a basis for R^2. **(b)** Any $A \in M(2)$ can be written as $A = \begin{bmatrix} a & b \\ c & d \end{bmatrix} = \begin{bmatrix} a & (b+c)/2 \\ (b+c)/2 & d \end{bmatrix} + \begin{bmatrix} 0 & (b-c)/2 \\ -(b-c)/2 & 0 \end{bmatrix}$, where the first matrix (A_1) is symmetric and the second (A_2) is skew-symmetric. Suppose $A = A'_1 + A'_2$ were another expression for A, with A'_1 symmetric and A'_2 skew-symmetric. Then, $A_1 + A_2 = A'_1 + A'_2$, so $A'_1 - A_1 = A_2 - A'_2$. But, this means $A'_1 - A_1$ and $A_2 - A'_2$ are both symmetric and skew-symmetric (since the symmetric and skew-symmetric matrices are subspaces of $M(2)$). As only the zero matrix is both, it follows that $A'_1 = A_1$ and $A'_2 = A_2$.
19. As T is diagonalizable, there is a basis B for V such that $A = [T;B]$ is diagonal with the c_i along the diagonal. Then, $A - xI$ is also diagonal, with $c_i - x$ along the diagonal, so $f = det\,(A - xI) = (x - c_1)^{n_1}...(x - c_k)^{n_k}$. Now for each $\mathbf{v} \in B$, $T(\mathbf{v}) = c_i\mathbf{v}$, for some $1 \le i \le k$. Since c_i appears n_i times on the diagonal of A, there must be precisely n_i vectors $\mathbf{v}$ in B such that $T(\mathbf{v}) = c_i\mathbf{v}$, for each $i = 1, ..., k$. It follows that there are n_i linearly independent eigenvectors for c_i, so the dimension of the eigenspace of c_i is n_i.

20. **(b)** Let $h = x^2 - 2x + 1 = (x - 1)^2$. Then $h(A) = O$, so the minimal polynomial p of A divides h. As A is to be diagonalizable, $p = x - 1$, so $A = I$, for then $p(A) = A - I = O$.

EXERCISES 6-4

1. **(a) (i)** We list below the results of the six iterations and the estimate $c_1(5)$ for the dominant eigenvalue. The corresponding estimate for the eigenvector is $\mathbf{v}_1 = \mathbf{x}'_5$. Note that $\mathbf{x}'_k$ has been calculated from $\mathbf{x}_k$ by *dividing* each entry of $\mathbf{x}_k$ by $|L|$, rather than multiplying by $1/|L|$, because $L \times 1/|L|$ may not be ± 1.00 due to rounding error (see section 2-6 and question 2-6-3). However, the results of using the latter will not differ substantially from those below.

$\mathbf{x}_1 = (11,9)$	$\mathbf{x}'_1 = (1.00,0.818)$	
$\mathbf{x}_2 = (10.1,8.09)$	$\mathbf{x}'_2 = (1.00,0.801)$	$c_1(1) = 10.1$
$\mathbf{x}_3 = (10.0,8.01)$	$\mathbf{x}'_3 = (1.00,0.801)$	$c_1(2) = 10.0$
$\mathbf{x}_4 = (10.0,8.01)$	$\mathbf{x}'_4 = (1.00,0.801)$	$c_1(3) = 10.0$
$\mathbf{x}_5 = (10.0,8.01)$	$\mathbf{x}'_5 = (1.00,0.801)$	$c_1(4) = 10.0$
$\mathbf{x}_6 = (10.0,8.01)$		$c_1(5) = 10.0$

(b) The true dominant eigenvalues are found from the characteristic polynomials to be **(i)** 10, **(ii)** 2, and **(iii)** -5.

3. **(a)** The matrix equation given in the hint produces the following homogeneous system of equations for a_3, a_2, a_1, and a_0:

$$\begin{aligned} 8 + 4a_3 + 2a_2 + a_1 + a_0 &= 0 \\ -6 - 3a_3 &= 0 \\ -2 + 2a_3 + a_2 &= 0 \\ 5 + 4a_3 + 2a_2 + a_1 &= 0. \end{aligned}$$

The solution is $a_3 = -2$, $a_2 = 6$, $a_1 = -9$, and $a_0 = -3$, so the characteristic polynomial is $f = x^4 - 2x^3 + 6x^2 - 9x - 3$. **(b)** If $f = (-1)^n x^n + a_{n-1}x^{n-1} + \ldots + a_0$, the equation $f(A)X = O$, where $X = (1,0,\ldots,0)$, yields a system of n equations for the n unknowns $a_{n-1}, \ldots, a_0$.

5. We have $A\mathbf{v}_i = c_i\mathbf{v}_i$ and $A^t\mathbf{w}_j = c_j\mathbf{w}_j$, for $1 \le i, j \le n$. Therefore, $c_j\mathbf{w}'_j\mathbf{v}_i = (A^t\mathbf{w}_j)^t\mathbf{v}_i = \mathbf{w}_j{}^tA\mathbf{v}_i - \mathbf{w}_j{}^tc_i\mathbf{v}_i = c_i\mathbf{w}_j{}^t\mathbf{v}_i$, so $(c_j - c_i)\mathbf{w}_j{}^t\mathbf{v}_i = 0$. Since the eigenvalues are distinct, $(c_j - c_i) \neq 0$, so it must be true that $\mathbf{w}_j{}^t\mathbf{v}_i = 0$, as required.

7. Using the solution of question 1, we see for (i) that 4 iterations are needed, yielding an estimated percentage error of 0% (using 3-digit arithmetic). With 3 iterations, the error is $|[c_1(2) - c_1(1)]/c_1(2)| \times 100 = 1\%$.

8. **(a)** We find $\mathbf{v}_1 = \mathbf{x}'_5 = (1.00,0.387)$, $\mathbf{x}_6 = (14.9,5.77)$, so $c_1(5) = 14.9$. Then, $\mathbf{v}_1{}^t\mathbf{v}_1 = 1.15$, so $\mathbf{w}_1 = (0.870,0.336)$. Using Theorem 6-9, we find the deflated matrix $B_1 = \begin{bmatrix} 0 & 0.01 \\ 0.01 & 0.08 \end{bmatrix}$. Applying the power method with scaling to B_1 yields $\mathbf{v}_2 = \mathbf{x}'_5 = (0.123,1.00)$, $\mathbf{x}_6 = (0.0100,0.0812)$, so $c_2(5) = 0.0812$. In this case, the true eigenvalues (to 5 figures) are actually 14.933 and 0.06697.

9. To obtain the required accuracy (using 3 digits), we must take 6 iterations, resulting in $\mathbf{x}'_5 = (1.00,0.748)$, $\mathbf{x}_6 = (2.01,1.50)$, so $c_1(5) = 2.01$. If we use the usual starting vector $\mathbf{v} = (1,1)$, we obtain $A\mathbf{v} = (1,1) = \mathbf{v}$, indicating that $(1,1)$ is an eigenvector of A associated with eigenvalue $c = 1$. Therefore, it would not lead to the dominant eigenvalue. The true eigenvalues are actually $c = 1$ and 2.

10. **(a)** To obtain the required accuracy (using 3 digits), we must take 3 iterations, resulting in $\mathbf{v}_1 = \mathbf{x}'_2 = (0.385,0.0148,1.00)$, $\mathbf{x}_3 = (5.77,-0.0296,14.9)$, so $c_1(2) = 14.9$. Next, $\mathbf{v}_1{}^t\mathbf{v}_1 = 1.15$, so $\mathbf{w}_1 = (0.335,0.0129,0.870)$. Using Theorem 6-9, the first deflated matrix is $B_1 = \begin{bmatrix} 0.08 & -0.0741 & 0.01 \\ -0.0739 & -2.00 & -0.192 \\ 0.01 & -0.192 & 0 \end{bmatrix}$. Note that B_1 should be symmetric,

but isn't, owing to rounding error. Applying the power method to B_1, we obtain $\mathbf{v}_2 = \mathbf{x'}_2 = (0.0366,1.00,0.0950)$, $\mathbf{x}_3 = (-0.0703,-2.02,-0.192)$, so $c_2(2) = -2.02$. Next, $\mathbf{v}_2{}^t\mathbf{v}_2 = 1.01$, so $\mathbf{w}_2 = (0.0362,0.990,0.0941)$. Using Theorem 6-9, the second deflated matrix is $B_2 = \begin{bmatrix} 0.0827 & -0.0010 & 0.0170 \\ -0.0008 & 0 & -0.002 \\ 0.0170 & -0.002 & 0.0181 \end{bmatrix}$. Applying the power method to B_2, we obtain $\mathbf{v}_3 = \mathbf{x'}_2 = (1.00,-0.0166,0.262)$, $\mathbf{x}_3 = (0.0872,-0.00132,0.0217)$, so $c_3(2) = 0.0872$. In fact, the true eigenvalues are (to 5 figures), 14.933, 0.06697, and -2.
11. **(a)** By question 6-1-7, c is an eigenvalue of A iff $1/c$ is an eigenvalue of A^{-1}. Also, if $|c| < |c'|$ for all other eigenvalues c' of A, then $|1/c| > |1/c'|$ for these eigenvalues, so $1/c$ is a dominant eigenvalue of A^{-1}. **(b)** Question 8(a): We apply the power method with scaling to A^{-1}, with enough iterations so that the estimated percentage error (using 3-digit arithmetic) is less than 1%. We find $\mathbf{v}_1 = \mathbf{x'}_2 = (-0.386,1.00)$, $\mathbf{x}_3 = (-5.77,14.9)$, so $c_1(2) = 14.9$. Therefore, the smallest eigenvalue of A is approximately $1/14.9 \approx 0.0671$.

EXERCISES 6-5

1. **(a)** $\begin{bmatrix} 1 \\ 3/2 \end{bmatrix}$ **(c)** $\begin{bmatrix} 1 \\ 115/112 \\ 15/8 \end{bmatrix}$
3. **(a)** If we add rows 1 through $(n-1)$ of $I - A$ to row n, then the resulting matrix B has, for its entry in row n and column j, $b_{nj} = 1 - a_{1j} - a_{2j} - \dots - a_{nj} = 0$, since the sum of the entries in column j of A is 1. But, B is row equivalent to $I - A$ and has rank $\leq (n-1)$. Thus, *rank* $(I - A) < n$. **(b)** $AX = X$ has the same solutions as the homogeneous system $(I - A)X = O$. By (a), the latter always has a nontrivial solution, since the coefficient matrix has rank $< n$.
5. **(a)** The characteristic polynomial is $f = det\,(A - I) = (x - a)(x - d) - bc = x^2 - (a + d)x + (ad - bc)$. By the quadratic formula, the zeros are

$$x = \frac{(a + d) \pm \sqrt{(a + d)^2 - 4(ad - bc)}}{2a} = \frac{(a + d) \pm \sqrt{(a - d)^2 + 4bc}}{2a}.$$

We see from this that at least one zero is ≥ 0, since the entries a, b, c, and d are nonnegative. **(b)** A is irreducible iff $b > 0$ and $c > 0$ (see Figure 6-1(b)).
7. **(a)** A is reducible because its digraph is $G(A)$ of Figure 6-1(d). **(b)** We note from Figure 6-1(d) that vertices 1 and 3 can communicate, as well as vertices 2 and 4. Therefore, we interchange rows 2 and 3 and columns 2 and 3 of A to obtain the similar matrix $A' = \begin{bmatrix} 1 & 1 & 0 & 0 \\ 1 & 1 & 0 & 0 \\ 1 & 1 & 0 & 1 \\ 1 & 1 & 1 & 0 \end{bmatrix}$, which has block triangular form. The blocks along the diagonal are $A_{11} = \begin{bmatrix} 1 & 1 \\ 1 & 1 \end{bmatrix}$ and $A_{22} = \begin{bmatrix} 0 & 1 \\ 1 & 0 \end{bmatrix}$. A_{11} has eigenvalues 0 and 2, while A_{22} has -1 and 1. Thus, $c = 2$ is the largest positive eigenvalue of A with corresponding (positive) eigenvector (1,2,1,2).
9. (b), (c)
11. Suppose that A is productive. Then, by the second half of the *proof* of Theorem 6-12, there is an $X \geq O$ such that $X > AX$. Conversely, if such an X exists, by Lemma 6-6, $c < 1$, so A is productive.
13. If the column sums of A are less than 1, then the row sums of A^t are less than 1, so A^t is productive, by part (2) of the corollary. By Theorem 6-12, the largest nonnegative

eigenvalue of A^t is $c < 1$. Since the eigenvalues of A are those of A^t, it follows that A is productive, again by the theorem.
15. Let U be as in question 12. If no row sum of A is less than 1, then $AU \geq U$, so the largest eigenvalue of A is $c \geq 1$, by Lemma 6-6. This contradicts the fact that A is productive.

EXERCISES 7-1

1. **(a)** **(i)** -17 **(ii)** -43 **(b)** -19 **(c)** -10
3. Let $\mathbf{u}$, $\mathbf{v}$, and $\mathbf{w} \in R^n$ and $c \in R$.

$$\begin{aligned}
\text{Axiom (1): } <\mathbf{u}+\mathbf{v},\mathbf{w}> &= (u_1 + v_1)w_1 + \ldots + (u_n + v_n)w_n \\
&= (u_1w_1 + \ldots + u_nw_n) + (v_1w_1 + \ldots + v_nw_n) \\
&= <\mathbf{u},\mathbf{w}> + <\mathbf{v},\mathbf{w}> \\
\text{Axiom (2): } <c\mathbf{u},\mathbf{v}> &= (cu_1)v_1 + \ldots + (cu_n)v_n \\
&= c(u_1v_1) + \ldots + c(u_nv_n) \\
&= c<\mathbf{u},\mathbf{v}> \\
\text{Axiom (3): } <\mathbf{u},\mathbf{v}> &= u_1v_1 + \ldots + u_nv_n \\
&= v_1u_1 + \ldots + v_nu_n \\
&= <\mathbf{v},\mathbf{u}> \\
\text{Axiom (4): } <\mathbf{u},\mathbf{u}> &= u_1u_1 + \ldots + u_nu_n \\
&= u_1^2 + \ldots + u_n^2 > 0, \text{ if } \mathbf{u} \neq \mathbf{0} \text{ (and } = 0 \text{ if } \mathbf{u} = \mathbf{0})
\end{aligned}$$

5. **(a)** not an inner product because axiom 4 fails **(b)** an inner product **(c)** not an inner product because axiom 3 fails
7. We verify the axioms. Let $f'(x) = a'x + b'$ and $k \in R$. Then, with f and g as given, $(f + f')(x) = (a + a')x + (b + b')$ and $(kf)(x) = kax + kb$.

$$\begin{aligned}
\text{Axiom 1: } <f + f',g> &= (a + a')c/3 + [(b + b')c + (a + a')d)]/2 + (b + b')d \\
&= [ac/3 + (bc + ad)/2 + bd] + [a'c/3 + (b'c + a'd)/2 + b'd] \\
&= <f,g> + <f',g> \\
\text{Axiom 2: } <kf,g> &= kac/3 + (kbc + kad)/2 + kbd \\
&= k[ac/3 + (bc + ad)/2 + bd] \\
&= k<f,g> \\
\text{Axiom 3: } <g,f> &= ca/3 + (da + cb)/2 + db \\
&= ac/3 + (bc + ad)/2 + bd \\
&= <f,g> \\
\text{Axiom 4: } <f,f> &= aa/3 + (ba + ab)/2 + bb \\
&= a^2/3 + ab + b^2 \\
&> a^2/4 + ab + b^2 \\
&= (a/2 + b)^2 > 0, \text{ unless } a = b = 0; \text{ i.e., unless } f = 0
\end{aligned}$$

9. $<A,B>_T = <A,B>$, the standard inner product

EXERCISES 7-2

1. **(a)** $\sqrt{23} \approx 4.80$ **(b)** $\sqrt{66} \approx 8.12$ **3.** **(a)** $\sqrt{55} \approx 7.42$ **(b)** $\sqrt{38} \approx 6.16$
5. Suppose $|<\mathbf{u},\mathbf{v}>| = |\mathbf{u}||\mathbf{v}|$. Then, using the proof of the Cauchy-Schwarz inequality, we see that $<\mathbf{w},\mathbf{w}> = <\mathbf{w},\mathbf{v}> = <\mathbf{v},\mathbf{v}> - (<\mathbf{v},\mathbf{u}>/<\mathbf{u},\mathbf{u}>)<\mathbf{u},\mathbf{v}> = |\mathbf{v}|^2 - <\mathbf{u},\mathbf{v}>^2/|\mathbf{u}|^2 = 0$. So, $\mathbf{w} = \mathbf{0}$ by property (2) of Theorem 7-2, and $\mathbf{v} = (<\mathbf{v},\mathbf{u}>/<\mathbf{u},\mathbf{u}>)\mathbf{u}$, showing that $\mathbf{u}$ and $\mathbf{v}$ are linearly dependent. Conversely, if they are linearly dependent, say $\mathbf{v} = c\mathbf{u}$, $c \in R$, then, $|<\mathbf{u},\mathbf{v}>| = |c||<\mathbf{u},\mathbf{u}>| = |c||\mathbf{u}|^2 = |\mathbf{u}||\mathbf{v}|$, so $|<\mathbf{u},\mathbf{v}>| = |\mathbf{u}||\mathbf{v}|$.
6. **(a)** $15/22 \approx 0.682$ **7.** $-1/\sqrt{2} \approx -0.707$ **9.** -3 or $1/2$
11. $(1/4)|\mathbf{u} + \mathbf{v}|^2 - (1/4)|\mathbf{u} - \mathbf{v}|^2 = (1/4)<\mathbf{u} + \mathbf{v},\mathbf{u} + \mathbf{v}> - (1/4)<\mathbf{u} - \mathbf{v},\mathbf{u} - \mathbf{v}> = (1/4)[<\mathbf{u},\mathbf{u}> + 2<\mathbf{u},\mathbf{v}> + <\mathbf{v},\mathbf{v}>] - (1/4)[<\mathbf{u},\mathbf{u}> - 2<\mathbf{u},\mathbf{v}> + <\mathbf{v},\mathbf{v}>] = <\mathbf{u},\mathbf{v}>$

13. $B'' = \{\mathbf{w}_1,\mathbf{w}_2,\mathbf{w}_3\}$ with $\mathbf{w}_1 = (1/\sqrt{3})(-1,1,1)$, $\mathbf{w}_2 = (1/\sqrt{42})(1,-4,5)$, and $\mathbf{w}_3 = (1/\sqrt{14})(3,2,1)$; $[\mathbf{v}]_{B''} = \begin{bmatrix} 4/\sqrt{3} \\ -19/\sqrt{42} \\ -1/\sqrt{14} \end{bmatrix}$

15. $B'' = \{\mathbf{w}_1,\mathbf{w}_2,\mathbf{w}_3\}$ with $\mathbf{w}_1 = (1/\sqrt{2})(1,0,0)$, $\mathbf{w}_2 = (\sqrt{2/3})(-1/2,1,0)$, and $\mathbf{w}_3 = (\sqrt{3}/2)(-1/3,-1/3,1)$; $[\mathbf{v}]_{B''} = \begin{bmatrix} -3/\sqrt{2} \\ \sqrt{3/2} \\ -2\sqrt{3} \end{bmatrix}$

17. **(a)** Let $\mathbf{v}$ and $\mathbf{w} \in W^\perp$ and $c \in R$. Then, if $\mathbf{u} \in W$, $<\mathbf{u},\mathbf{v}> = <\mathbf{u},\mathbf{w}> = 0$, so $<\mathbf{u},c\mathbf{v} + \mathbf{w}> = c<\mathbf{u},\mathbf{v}> + <\mathbf{u},\mathbf{w}> = 0$, showing that $c\mathbf{v} + \mathbf{w} \in W^\perp$. Therefore, $W^\perp$ is a subspace of V. **(b)** Since any vector $\mathbf{w} \in W$ is a linear combination of basis vectors for W, it follows by the linearity of $<,>$ that $<\mathbf{w},\mathbf{v}> = 0$, since this is true for every vector $\mathbf{w}$ in the given basis for W. Thus, $\mathbf{v}$ is orthogonal to every vector in W, hence is in $W^\perp$. **(c)** Let $\mathbf{v} \in W \cap W^\perp$. Then, $\mathbf{v} \in W^\perp$, so $<\mathbf{v},\mathbf{w}> = 0$ for all $\mathbf{w} \in W$. As $\mathbf{v} \in W$, we get (setting $\mathbf{w} = \mathbf{v}$) $<\mathbf{v},\mathbf{v}> = 0$, so $\mathbf{v} = \mathbf{0}$. **(d)** Let $B' = \{\mathbf{v}_1,\dots,\mathbf{v}_k\}$ be an orthonormal basis for W. It is part of an orthonormal basis $B = \{\mathbf{v}_1,\dots,\mathbf{v}_n\}$ for V, where $k \le n$, by Theorem 4-8 and Gram-Schmidt orthonormalization. We show $B'' = \{\mathbf{v}_{k+1},\dots,\mathbf{v}_n\}$ is a basis for $W^\perp$. It is certainly linearly independent. Let $\mathbf{w} \in W^\perp \subseteq V$. By Theorem 7-3, $\mathbf{w} = \sum_{i=1}^{n} <\mathbf{w},\mathbf{v}_i>\mathbf{v}_i = \sum_{i=k+1}^{n} <\mathbf{w},\mathbf{v}_i>\mathbf{v}_i$, since $\mathbf{w} \in W^\perp$ and $\mathbf{v}_i \in W$. Thus, B'' generates $W^\perp$ and we see that $dim\ W^\perp = n - k = dim\ V - dim\ W$. **(e)** By (d), $dim\ (W^\perp)^\perp = n - (n - k) = k = dim\ W$. Also, if $\mathbf{v} \in W$, then $<\mathbf{v},\mathbf{w}> = 0$ for all $\mathbf{w} \in W^\perp$, so $\mathbf{v} \in (W^\perp)^\perp$ and we see that $W \subseteq (W^\perp)^\perp$. It follows that $(W^\perp)^\perp = W$.

19. $W^\perp$ consists of all skew-symmetric matrices.

21. $|\mathbf{u} + \mathbf{v}|^2 = <\mathbf{u} + \mathbf{v},\mathbf{u} + \mathbf{v}> = <\mathbf{u},\mathbf{u}> + 2<\mathbf{u},\mathbf{v}> + <\mathbf{v},\mathbf{v}> = |\mathbf{u}|^2 + |\mathbf{v}|^2$

23. **(a)** $\{\mathbf{w}_1,\mathbf{w}_2\}$, where $\mathbf{w}_1 = (1/\sqrt{2})(1,0)$ and $\mathbf{w}_2 = \sqrt{2}(1/2,1)$ **(b)** $\{\mathbf{w}_1,\mathbf{w}_2\}$, where $\mathbf{w}_1 = 1$ and $\mathbf{w}_2 = (2\sqrt{3})x - \sqrt{3}$

EXERCISES 7-3

1. T is clearly linear. If A is a 2×2 symmetric matrix as shown, $<T(A),T(A)> = a^2/2 + b^2 + a^2/2 + c^2/2 + b^2 + c^2/2 = a^2 + 2b^2 + c^2 = <A,A>$, so T is isometric by Theorem 7-6.

3. Let $A = [T]$. Since the standard basis is orthonormal and
$$A^tA = \begin{bmatrix} 0 & 1 & 0 \\ -1 & 0 & 0 \\ 0 & 0 & -1 \end{bmatrix}\begin{bmatrix} 0 & -1 & 0 \\ 1 & 0 & 0 \\ 0 & 0 & -1 \end{bmatrix} = I,$$
T is orthogonal by Theorem 7-9, Corollary.

5. **(a)** yes **(b)** no **7.** **(a)** $(-1,3,-4)$ **(b)** $1 + x + x^2$ **11.** $T^*(x,y) = (2x + y, -x + 3y)$

14. Let B be the standard basis for $M(n\times 1)$. Then, by Theorem 7-8, $[T_A{}^*] = [T_A]^t = A^t = [T_{A^t}]$. Therefore, $T_A{}^* = T_{A^t}$.

16. (b), (c)

17. If A is orthogonal, $AA^t = I$, so $det\ AA^t = det\ A\ det\ A^t = 1$, from which we see that $(det\ A)^2 = 1$ and $det\ A = \pm 1$ (recall $det\ A^t = det\ A$).

19. Let A and B be orthogonal. Then, $(AB)^t(AB) = B^tA^tAB = B^tIB = I$, so AB is orthogonal. Next, $(A^{-1})^tA^{-1} = (A^t)^{-1}A^{-1} = (AA^t)^{-1} = I^{-1} = I$, so A^{-1} is orthogonal.

23. $A = QR$, where
$$Q = \begin{bmatrix} -1/\sqrt{3} & 5/\sqrt{42} & 1/\sqrt{14} \\ 1/\sqrt{3} & 4/\sqrt{42} & -2/\sqrt{14} \\ 1/\sqrt{3} & 1/\sqrt{42} & 3/\sqrt{14} \end{bmatrix}, R = \begin{bmatrix} \sqrt{3} & -1/\sqrt{3} & 5/\sqrt{3} \\ 0 & \sqrt{42/3} & 5/\sqrt{42} \\ 0 & 0 & 1/\sqrt{14} \end{bmatrix}.$$

EXERCISES 7-4

1. (a), (d)
3. Since $0 = <\mathbf{v},\mathbf{0}> = <\mathbf{v},T^2(\mathbf{v})> = <T(\mathbf{v}),T(\mathbf{v})>$ (as $T^* = T$), we see that $T(\mathbf{v}) = \mathbf{0}$.
5. **(a)** If S, T, and ST are self-adjoint, then, by question 7-3-15, $ST = (ST)^* = T^*S^* = TS$. Conversely, if S and T are self-adjoint and $ST = TS$, $(ST)^* = T^*S^* = TS = ST$, so ST is self-adjoint. **(b)** This follows from (a) by taking matrices of the corresponding self-adjoint matrix transformations. Alternatively, we can adapt the proof of (a) by substituting "t" for "*" and letting S and T be symmetric matrices.
6. $X = \pm(1/\sqrt{2}, 1/\sqrt{2})$ is a maximal vector and $\|A\| = 3$.
7. Suppose $A_{ij} \neq 0$, for some i and j. Then, since $A_{ik} = A_{ki}$,

$$A^2{}_{ii} = \sum_{k=1}^{n} A_{ik}A_{ki} = \sum_{k=1}^{n} A^2{}_{ik} > 0, \text{ so } A^2 \neq O.$$

9. If T is self-adjoint, V has an orthonormal basis B of eigenvectors of T, by Theorem 7-13. Conversely, if V has such a basis, we show T must be self-adjoint. Let $B = \{\mathbf{v}_1,\ldots,\mathbf{v}_n\}$ and $T(\mathbf{v}_i) = c_i\mathbf{v}_i$, where the c_i are not necessarily distinct. Then, $<\mathbf{v}_i,T^*(\mathbf{v}_j)> = <T(\mathbf{v}_i),\mathbf{v}_j> = c_i<\mathbf{v}_i,\mathbf{v}_j> = c_j<\mathbf{v}_i,\mathbf{v}_j> = <\mathbf{v}_i,T(\mathbf{v}_j)>$, for all i and j (since $<\mathbf{v}_i,\mathbf{v}_j> = 0$ when $i \neq j$, and 1 when $i = j$). Therefore, $T^*(\mathbf{v}_j) = T(\mathbf{v}_j)$ for all j, so $T^* = T$, because they coincide on a basis.

11. **(a)** $D = \begin{bmatrix} 3 & 0 \\ 0 & -1 \end{bmatrix}$, $Q = (1/\sqrt{2})\begin{bmatrix} 1 & 1 \\ 1 & -1 \end{bmatrix}$

(c) $D = \begin{bmatrix} 9 & 0 & 0 \\ 0 & -9 & 0 \\ 0 & 0 & -9 \end{bmatrix}$, $Q = \begin{bmatrix} 2/3 & 0 & \sqrt{5}/3 \\ 1/3 & 2\sqrt{5}/5 & -2\sqrt{5}/15 \\ -2/3 & \sqrt{5}/5 & 4\sqrt{5}/15 \end{bmatrix}$

13. Let $B = \{E_1,E_2,E_3,E_4\}$ be the usual basis for $M(2)$ and $M_1 = E_1$, $M_2 = (1/\sqrt{2})(E_2 + E_3)$, $M_3 = E_4$, and $M_4 = (1/\sqrt{2})(E_2 - E_3)$. Then $B' = \{M_1,M_2,M_3,M_4\}$ is an orthonormal basis of eigenvectors of T.

16. $\begin{bmatrix} (\sqrt{3}+1)/2 & (\sqrt{3}-1)/2 \\ (\sqrt{3}-1)/2 & (\sqrt{3}+1)/2 \end{bmatrix}$

EXERCISES 7-5

1. (a), (c), (d), (e), (f), (g), (i), (j) **3.** **(a)** **(ii)** 18

6. **(a)** $\begin{bmatrix} 1 & -4 \\ 0 & 3 \end{bmatrix}$ **(c)** $\begin{bmatrix} 1 & 0 & 0 & 0 \\ 0 & 0 & 1 & 0 \\ 0 & 1 & 0 & 0 \\ 0 & 0 & 0 & 1 \end{bmatrix}$ **(d)** $\begin{bmatrix} 1 & 0 & 0 & 0 \\ 0 & 0 & 0 & 0 \\ 0 & 0 & 0 & 0 \\ 0 & 0 & 0 & 1 \end{bmatrix}$

7. Congruence is reflexive: $A = I^tAI$, where I is the identity. It is symmetric: If $B = Q^tAQ$, then $A = (Q^t)^tBQ^t$. It is transitive: if $B = Q^tAQ$ and $C = P^tBP$, then $C = P^tQ^tAQP = (QP)^tA(QP)$.
9. Suppose F is nondegenerate. If, for some $\mathbf{u} \neq \mathbf{0}$, it were true that $F(\mathbf{u},\mathbf{v}) = 0$ for all $\mathbf{v}$, then let $B = \{\mathbf{v}_1,\ldots,\mathbf{v}_n\}$ be a basis for V such that $\mathbf{u} = \mathbf{v}_1$. This exists by Theorem 4-8. Then, $[F;B]_{1j} = F(\mathbf{u},\mathbf{v}_j) = 0$ for all j, so $[F;B]$ has a zero row 1, hence is singular and has rank $< n$, contrary to the assumption. Thus, for any $\mathbf{u} \neq \mathbf{0}$, there must be a $\mathbf{v} \neq \mathbf{0}$ such that $F(\mathbf{u},\mathbf{v}) \neq 0$. Conversely, if this condition is true, we show F is nondegenerate. Suppose it is not. Let $A = [F;B]$, where B is any basis for V. Then A is row-equivalent to a row-reduced echelon matrix A' whose n^{th} row is zero. This means there is an invertible matrix P such that $PA = A'$, so $PAP^t = A'P^t$, and this also has a zero last row. But, by Theorem 7-15, $PAP^t = [F;B']$, where B' is the basis for V such that $[B'/B] = P^t$. If $B' = \{\mathbf{v}'_1,\ldots,\mathbf{v}'_n\}$, we see that $F(\mathbf{v}'_n,\mathbf{v}'_j) = 0$ for all j, so it follows that

$F(\mathbf{v}_n,\mathbf{v}) = 0$ for all $\mathbf{v} \in V$, contrary to the assumption. Therefore, F must be nondegenerate. This proves (a), and the proof of (b) is similar, working with the right-hand argument.

11. $\begin{bmatrix} 1 & 0 \\ 0 & 2 \end{bmatrix}$ and $\begin{bmatrix} 3 & 1 \\ 1 & 3 \end{bmatrix}$ are congruent but not similar.

12. **(a)** $Q^tAQ = D$, with $D = \begin{bmatrix} 1 & 0 \\ 0 & -1 \end{bmatrix}$, $Q^t = \begin{bmatrix} 1 & 0 \\ 2 & 1 \end{bmatrix}$.

13. **(a)** Let $B = \{\mathbf{v}_1,\mathbf{v}_2\}$, where $\mathbf{v}_1$ = row 1 of $Q^t = (1,0)$, $\mathbf{v}_2$ = row 2 of $Q^t = (2,1)$.
(c) (a) has rank 2, index 1, and signature 0; (b) has rank 3, index 2, and signature 1.

15. $q(\mathbf{u}+\mathbf{v}) = F(\mathbf{u}+\mathbf{v},\mathbf{u}+\mathbf{v}) = F(\mathbf{u},\mathbf{u}) + 2F(\mathbf{u},\mathbf{v}) + F(\mathbf{v},\mathbf{v}) = q(\mathbf{u}) + 2F(\mathbf{u},\mathbf{v}) + q(\mathbf{v})$; solving for $F(\mathbf{u},\mathbf{v})$ gives the result.

16. **(a)** The bilinear forms have these matrices:

(i) $\begin{bmatrix} -1 & 1 \\ 1 & 3 \end{bmatrix}$ **(ii)** $\begin{bmatrix} 2 & 2 \\ 2 & 2 \end{bmatrix}$ **(iii)** $\begin{bmatrix} 0 & 3/2 & -2 \\ 3/2 & 0 & 0 \\ -2 & 0 & 1 \end{bmatrix}$ **(iv)** $\begin{bmatrix} -1 & -1/2 & 0 \\ -1/2 & 2 & 1 \\ 0 & 1 & 3 \end{bmatrix}$

(b) **(i)** 2, 1, 0 **(ii)** 1, 1, 1 **(iii)** 3, 2, 1 **(iv)** 3, 2, 1

17. **(i)** $B' = \{(1/2,1/2),(1,0)\}$; $q(\mathbf{v}) = x_1^2 - x_2^2$ **(ii)** $B' = \{(1/\sqrt{2},0),(-1,1)\}$; $q(\mathbf{v}) = x_1^2$
(iii) $B' = \{(0,0,1),(1/2,4/3,1),(1/2,0,1)\}$; $q(\mathbf{v}) = x_1^2 + x_2^2 - x_3^2$
(iv) $B' = \{(-1/3,2/3,0),(2/3\sqrt{23},-4/3\sqrt{23},3/\sqrt{23}),(1,0,0)\}$; $q(\mathbf{v}) = x_1^2 + x_2^2 - x_3^2$

19. **(a)** $\dfrac{x''^2}{16} + \dfrac{y''^2}{4} = 1$, where $x'' = x + 2$ and $y'' = y - 1$. This is an ellipse with center at $(-2,1)$ as shown.

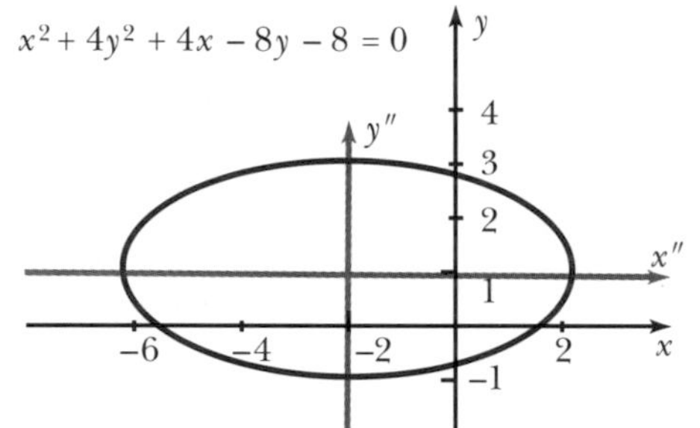

(c) $\dfrac{x'^2}{5} - y'^2 = 1$, where x' and y' are the coordinates defined by the basis $B' = \{\mathbf{v}'_1,\mathbf{v}'_2\}$ with $\mathbf{v}'_1 = (1/\sqrt{2},1/\sqrt{2})$ and $\mathbf{v}'_2 = (-1/\sqrt{2},1/\sqrt{2})$. This is a hyperbola with center at the origin, as shown.

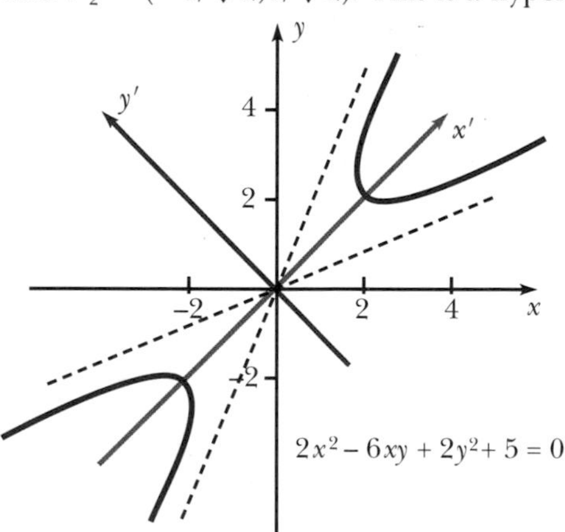

(e) $y' = \pm 10/\sqrt{34}$, where x' and y' are the coordinates defined by the basis $B' = \{\mathbf{v}'_1, \mathbf{v}'_2\}$ with $\mathbf{v}'_1 = (3/\sqrt{34}, 5/\sqrt{34})$ and $\mathbf{v}'_2 = (-5/\sqrt{34}, 3/\sqrt{34})$. This is a pair of lines, as shown.

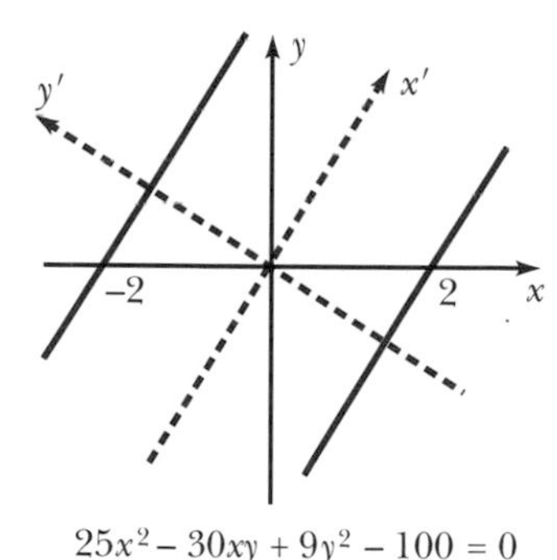

$25x^2 - 30xy + 9y^2 - 100 = 0$

EXERCISES 7-6

1. **(a)** 0 **(b)** $-12 + 13i$ **(c)** $-21 + 7i$
3. **(a)** For question 1: (a) $|\mathbf{u}| = \sqrt{2}$, $|\mathbf{v}| = \sqrt{2}$; (b) $|\mathbf{u}| = \sqrt{18}$, $|\mathbf{v}| = 7$; (c) $|\mathbf{u}| = \sqrt{50}$, $|\mathbf{v}| = \sqrt{10}$.
5. **(a)** no, axioms (3) and (4) fail **(b)** no, axiom (2) fails **(c)** yes **(d)** no, axioms (3) and (4) fail
7. **(b)** $3 - 2i, 3, 3$
9. Let $S = \{\mathbf{u}_1, \ldots, \mathbf{u}_n\}$ be an orthogonal set of vectors in V and suppose that $z_1\mathbf{u}_1 + \ldots + z_n\mathbf{u}_n = \mathbf{0}$, for some $z_i \in C$. We take the inner product of both sides with $\mathbf{u}_j$ to get $z_j<\mathbf{u}_j, \mathbf{u}_j> = 0$ (since $<\mathbf{u}_i, \mathbf{u}_j> = 0$, $i \neq j$), so that $z_j = 0$ for all j. Thus, S is linearly independent.
10. **(a)** no **(b)** no **(c)** yes

11. **(a)** $\{\mathbf{w}_1, \mathbf{w}_2\}$ is an orthonormal basis, with $\mathbf{w}_1 = \dfrac{1}{\sqrt{2}}(1, i)$, $\mathbf{w}_2 = \left(\dfrac{-1}{2} + \dfrac{i}{2}, \dfrac{1}{2} + \dfrac{i}{2}\right)$.

(b) $\{\mathbf{w}_1, \mathbf{w}_2\}$ is an orthonormal basis, where $\mathbf{w}_1 = \dfrac{1}{2}(1 + i, 1 - i)$,

$\mathbf{w}_2 = \left(\dfrac{-1}{2} + \dfrac{i}{2}, \dfrac{-1}{2} + \dfrac{-i}{2}\right)$.

13. **(a)** $\begin{bmatrix} -i & 5 \\ 1 + i & i \end{bmatrix}$ **(b)** $\begin{bmatrix} 3 + 2i \\ 4 - 5i \end{bmatrix}$ **(c)** $\begin{bmatrix} -i & 1 - 4i \\ -i & 0 \\ 2i & 2 \end{bmatrix}$

15. To prove Theorem 7-19, let A and B be $m \times n$ complex matrices. Using the properties of the transpose and complex conjugates, we get:
(1) $(A + B)^* = \overline{(A + B)}^t = \overline{A}^t + \overline{B}^t = A^* + B^*$.
(2) $(zA)^* = \overline{(zA)}^t = \overline{z}\overline{A}^t = \overline{z}A^*$.
(3) $(A^*)^* = \overline{(\overline{A}^t)}^t = A$.
(4) $(AB)^* = \overline{(AB)}^t = \overline{B}^t\overline{A}^t = B^*A^*$.
17. (a), (b), and (f) are normal. (f) is Hermitian.
19. (a), (c), and (e) are unitary ((a) is orthogonal). (c) is also Hermitian.

21. Eigenvalues are 1 and -1;

$$P = \begin{bmatrix} \frac{i}{\sqrt{2}} & -\frac{i}{\sqrt{2}} \\ \frac{1}{\sqrt{2}} & \frac{1}{\sqrt{2}} \end{bmatrix}.$$

23. Eigenvalues are 1, -1, and 2;

$$P = \begin{bmatrix} 0 & -\frac{2i}{\sqrt{6}} & \frac{i}{\sqrt{3}} \\ -\frac{i}{\sqrt{2}} & \frac{i}{\sqrt{6}} & \frac{i}{\sqrt{3}} \\ \frac{1}{\sqrt{2}} & \frac{1}{\sqrt{6}} & \frac{1}{\sqrt{3}} \end{bmatrix}.$$

25. $P = \begin{bmatrix} -\frac{1}{\sqrt{2}} & -\frac{i}{\sqrt{2}} \\ \frac{i}{\sqrt{2}} & \frac{1}{\sqrt{2}} \end{bmatrix}$

27. If A is a real $n \times n$ normal matrix and has real eigenvalues, then $D = P^*AP$ is a real diagonal matrix for some unitary matrix P. Thus, $PDP^* = A$ is real, so $A = PDP^* = \overline{PDP^*} = \overline{P}\overline{D}\overline{P}^* = \overline{P}DP^t = (PD^t\overline{P}^t)^t = (PDP^*)^t = A^t$. Therefore, A is symmetric. Conversely, if A is a real symmetric matrix, it is Hermitian and has real eigenvalues, by question 26(a).

APPENDIX A

1. **(a)** $(0,-2)$; 0; -2 **(b)** $(4,-3)$; 4; -3 **(c)** $(-1,-1)$; -1; -1 **(d)** (1,3); 1; 3

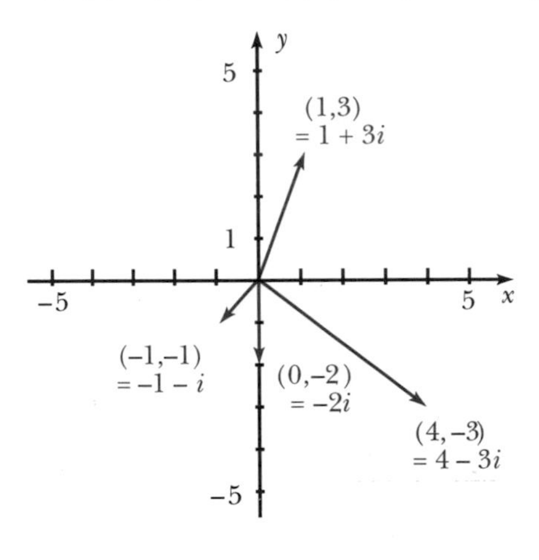

3. $x = 2, y = 1$ **4.** **(a)** $-6 - 10i$ **5.** **(a)** $-i, 1$ **(c)** 3, 3 **(e)** $4-7i, \sqrt{65}$
6. **(a)** $-4 + 3i$ **(c)** $-29 + 2i$ **8.** **(a)** $-6 + 4i$ **(c)** $40 - 60i$ **(e)** 16
10. **(a)** $z = \frac{-1}{2} \pm \frac{\sqrt{3}}{2}i$ **12.** **(a)** i **(c)** $\frac{2}{13} + \frac{3}{13}i$ **13.** **(a)** i **(c)** $\frac{3}{10} + \frac{1}{5}i$ **(e)** $\frac{-7}{2} + i$
15. Take the complex conjugate of both sides of $az_1^2 + bz_1 + c = 0$, where z_1 is a root. This gives $a\bar{z}_1^2 + b\bar{z}_1 + c = 0$, because $a = \bar{a}$, $b = \bar{b}$, and $c = \bar{c}$. It follows that $\bar{z}_1$ is also a root.
16. **(a)** π **(c)** $\pi/4$ **(e)** $3\pi/4$
17. **(a)** $2(\cos \pi + i \sin \pi)$ **(c)** $\sqrt{2}\,(\cos \pi/4 + i \sin \pi/4)$ **(e)** $2(\cos 3\pi/4 + i \sin 3\pi/4)$
18. **(a)** $\cos(-\pi/2) + i \sin(-\pi/2)$ **(d)** $\approx 3[\cos(-2.41) + i \sin(-2.41)]$
(e) $3\sqrt{2}[\cos(-\pi/4) + i \sin(-\pi/4)]$

19. **(a)** $\frac{-i}{\sqrt{2}}$ **(b)** $\frac{\sqrt{3}-1}{2} - \frac{\sqrt{3}+1}{2}i$ **21.** **(a)** $\frac{-1}{4} + \frac{1}{4}i$ **(c)** 1/16 **(e)** $\frac{2}{125} + \frac{11}{125}i$

22. **(a)** 3 roots, as shown:

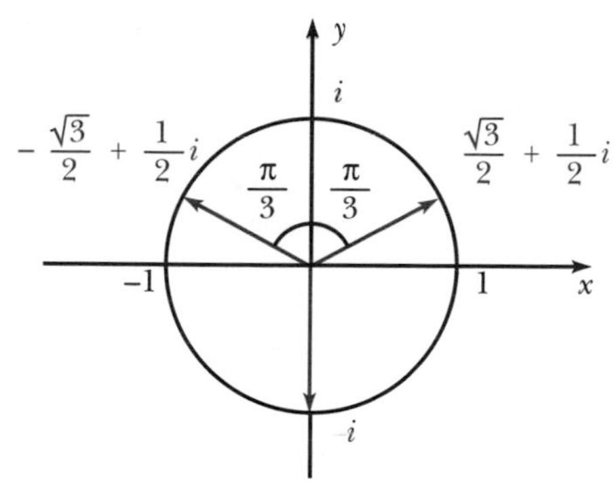

(c) 2 roots, as shown:

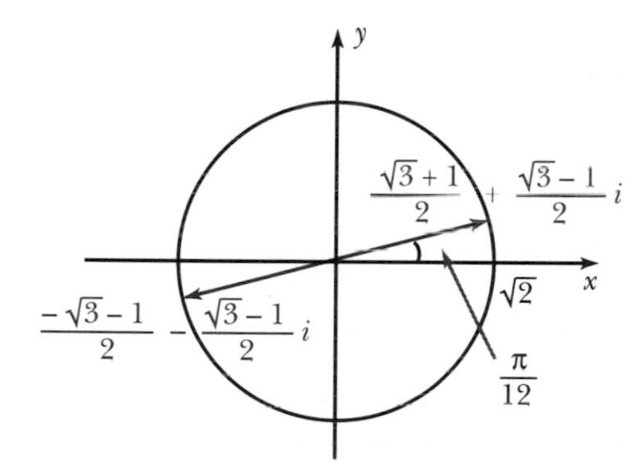

(c) 3 roots, as shown:

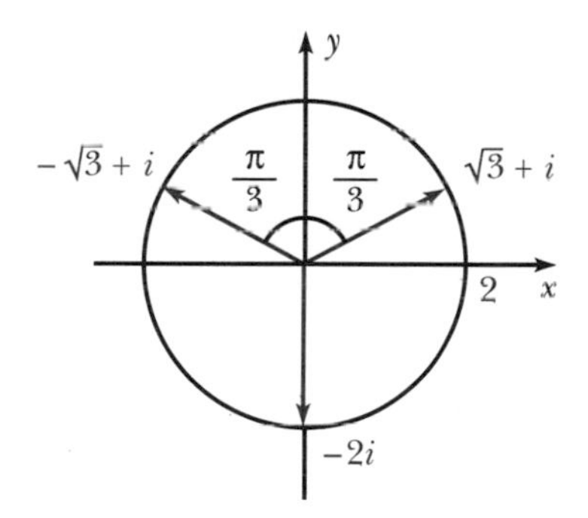

24. **(a)** $f = (z + \frac{1}{2} - \frac{\sqrt{3}}{2}i)(z + \frac{1}{2} + \frac{\sqrt{3}}{2}i)$ **(c)** $f = (z + i)^2$

APPENDIX B

1. **(a)** $2x + 5, 9$ **(b)** $(-1/2)x, (7/2)x - 1$ **(c)** $x^2 + 3x + 1, 0$ **(d)** $x, -2x$ **(e)** $(-3/4)x^3 + (1/2)x - 3/16, (-1/2)x + 29/16$

3. **(a)** 10, −8, 604 **(b)** −1, −2801, 11 699 **(c)** 8, 11 466, 109 473

4. (a) not a zero **(c)** yes, $-4x^3 + 4$ **(e)** yes, $6x^3 - 7x^2 - 2x - 5$
5. (b) $(x - y)(x^{n-1} + x^{n-2}y + x^{n-3}y^2 + \dots + y^{n-1})$ **7.** $(x - 1)(x + 1)^2(x^2 + 1)$
9. $(x + 1)(x^2 - 5)(x^2 + x + 3)$
11. (a) $(x - \sqrt{2})(x + \sqrt{2})(x^2 + 1)^2$ **(b)** $(x - \sqrt{3})(x + \sqrt{3})(x^2 + 1)$

APPENDIX C

1. (a) (i) $\max Z = x_1 + 2x_2 + x_3$
subject to

$$\begin{aligned} x_1 + x_2 + x_3 &\le 4 \\ 2x_1 + x_2 + x_3 &\le 5 \\ -x_1 \qquad + x_3 &\le 1 \\ x_1 \qquad - x_3 &\le -1 \\ x_1, x_2, x_3 &\ge 0 \end{aligned}$$

(ii) $\min -Z = -x_1 - 2x_2 - x_3$
subject to

$$\begin{aligned} -x_1 - x_2 - x_3 &\ge -4 \\ -2x_1 - x_2 - x_3 &\ge -5 \\ -x_1 \qquad + x_3 &\ge 1 \\ x_1 \qquad - x_3 &\ge -1 \\ x_1, x_2, x_3 &\ge 0 \end{aligned}$$

(b) (i) $\max -Z = -3y_1 - y_2$
subject to

$$\begin{aligned} y_1 + y_2 &\le 4 \\ -y_1 + y_2 &\le 0 \\ y_1, y_2 &\ge 0 \end{aligned}$$

(ii) $\min Z = 3y_1 + y_2$
subject to

$$\begin{aligned} -y_1 - y_2 &\ge -4 \\ y_1 - y_2 &\ge 0 \\ y_1, y_2 &\ge 0 \end{aligned}$$

2. (a) $\min Z = 4y_1 + 5y_2 + y_3 - y_4$
subject to

$$\begin{aligned} y_1 + 2y_2 - y_3 + y_4 &\ge 1 \\ y_1 + y_2 \qquad\qquad &\ge 2 \\ y_1 + y_2 + y_3 - y_4 &\ge 1 \\ y_1, y_2, y_3 &\ge 0 \end{aligned}$$

(b) $\max Z = -4x_1$
subject to

$$\begin{aligned} -x_1 + x_2 &\le 3 \\ -x_1 - x_2 &\le 1 \\ x_1, x_2 &\ge 0 \end{aligned}$$

3. The feasible region is indicated in the sketch below. As there are no points in the intersection of the half plane below the boundary line of the third constraint, the first quadrant, and the two half planes of the first two boundary lines, the feasible region is empty (i.e., the two shaded areas have no points in common).

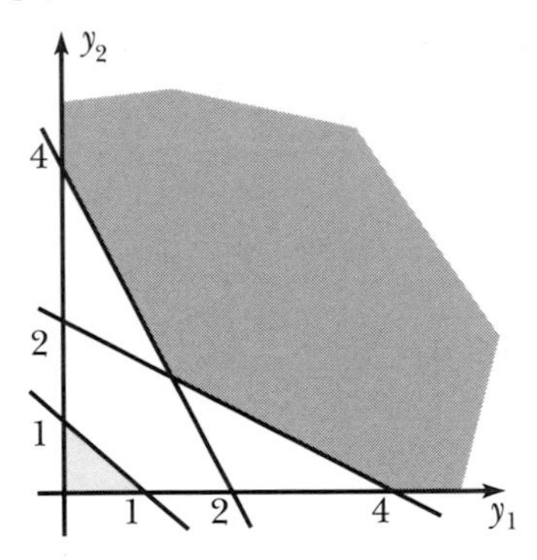

5. (a) The maximum value of Z is 19/2 at $x_1 = 1$, $x_2 = 5/2$. **(b)** The minimum value of Z is 6 at $y_1 = 2$, $y_2 = 0$.
8. (a) The problem is in standard form, so the initial tableau is

x_1	x_2	x_3	u_1	u_2	b		
3	4	1	1	0	2	u_1	2/4
1	[3]	2	0	1	1	u_2	1/3 *
2	3	1	0	0	0	Z.	
	*						

After two iterations, we obtain the optimal tableau

x_1	x_2	x_3	u_1	u_2	b	
1	0	−1	3/5	−4/5	2/5	x_1
0	1	1	−1/5	3/5	1/5	x_2
0	0	0	−3/5	−1/5	−7/5	Z.

Thus, an optimal solution is $Z = 7/5$, when $x_1 = 2/5$, $x_2 = 1/5$, and $x_3 = 0$. However, notice that x_3 has coefficient 0 in the equation for Z. This means that other optimal solutions are possible. Adding row 2 to row 1 in the above, we obtain the optimal tableau

x_1	x_2	x_3	u_1	u_2	b	
1	1	0	2/5	−1/5	3/5	x_1
0	1	1	−1/5	3/5	1/5	x_3
0	0	0	−3/5	−1/5	−7/5	Z

from which we get $x_1 = 3/5$, $x_2 = 0$, and $x_3 = 1/5$, also producing a maximum value of $Z = 7/5$.

(c) We must add an artificial variable to obtain the problem

$$\max Z = -x_1 - 2x_2 - 10a_1$$

subject to

$$\begin{aligned} -x_1 - x_2 + u_1 \qquad\qquad - a_1 &= -3 \\ 2x_1 - x_2 \qquad + u_2 \qquad\qquad &= 3 \\ -x_1 + 2x_2 \qquad\qquad + u_3 \qquad &= 3 \\ x_1, x_2, u_1, u_2, u_3, a_1 &\geq 0. \end{aligned}$$

Then, the initial tableau is

x_1	x_2	u_1	u_2	u_3	a_1	b	
−1	−1	1	0	0	−1	−3	a_1
2	−1	0	1	0	0	3	u_2
−1	2	0	0	1	0	3	u_3
−1	−2	0	0	0	−10	0	Z.

After eliminating the −10 in the last row and multiplying row 1 by −1, we obtain

x_1	x_2	u_1	u_2	u_3	a_1	b		
1	1	−1	0	0	1	3	a_1	3/1
[2]	−1	0	1	0	0	3	u_2	3/2 *
−1	2	0	0	1	0	3	u_3	
9	8	−10	0	0	0	30	Z	
*								

After two iterations, we obtain the optimal tableau

x_1	x_2	u_1	u_2	u_3	a_1	b	
0	1	−2/3	−1/3	0	2/3	1	x_2
1	0	−1/3	1/3	0	1/3	2	x_1
0	0	1	1	1	−1	3	u_3
0	0	−5/3	−1/3	0	−25/3	4	Z.

We see that the maximum value of Z is −4 when $x_1 = 2$ and $x_2 = 1$.

SYMBOLS AND NOTATIONS INDEX

("*ex*" refers to exercises)

A^{-1}, 103
A^*, 445
A^t, 77 (*ex*), 96
$\|A\|$, 410
$A_{[i]}$, 87, 216
$A^{[j]}$, 87, 216
AB, 15
{**AB**}, 20
adj A, 160
$A(i)$, 163
$A(i/j)$, 133
Area, 269, 312
arg z, 463
Arg z, 463
$\alpha(T)$, 342
B^*, 315
$[B/B']$, 300
$B(n)$, 200, 246
C, 457
C^n, 241
col (A), 216, 266
deg f, 184
det A, 130, 137
dim V, 212
$dim_C V$, 247
$dim_R V$, 247
δ_{ij}, 94
$\mathbf{e}_i$, 195, 196
$\mathbf{e}_i^*$, 313, 315
E_x, 311
$f \otimes g$, 420, 438 (*ex*), 439 (*ex*)
$[F;B]$, 421
$F(C)$, 243
$F(R)$, 183
$F(R,C)$, 243
$F(S)$, 178
$F(S,C)$, 242
$i(=\sqrt{-1})$, 241, 455
i, **j**, **k**, 40, 195
I, 93
Im T, 264
$Im(z)$, 456
I_V, 257, 294, 300
Ker T, 264
$L(V)$, 292, 339
$L(V,W)$, 286
$M_{m \times n}(C)$, 241
$M_{m \times n}(R)$, 241
$M_n(C)$, 241
$M_n(R)$, 241
MAT, 288
mat, 285, 405
$M(m \times n)$, 182
$M(n)$, 292, 339
$P_n(C)$, 243
PERT, 80
$P(n)$, 185
$P(R)$, 185, 342
R^2, 3
R^3, 8
R^n, 176
rank A, 219, 226
$Re(z)$, 456
$row(A)$, 216
$S(n)$, 135
sgn σ, 135, 143(*ex*)
span S, 199
$[T]$, 272
T_A, 253
T^{-1}, 294
T^t, 318
T^*, 401
$[T;B]$, 274
$[T;B/B']$, 274
tr A, 194(*ex*)
$\mathbf{u} \cdot \mathbf{v}$, 32, 375, 443
$\mathbf{u} \times \mathbf{v}$, 166, 399, 405(*ex*)
$<\mathbf{u}, \mathbf{v}>$, 375, 443
$\mathbf{u} \perp \mathbf{v}$, 384
$|\mathbf{v}|$, 20
$[\mathbf{v}]_B$, 222, 300
V^*, 310
V^{**}, 321(*ex*)
$W^\perp$, 391, 395(*ex*), 439(*ex*)
$W \oplus W^\perp$, 392, 413
$\bar{z}$, 459
$|z|$, 460
0, 475
0, 179, 256
O, 92
0, 21, 24, 181
{**0**}, 187

INDEX

("*ex*" refers to exercises)

A

Abstract mathematics, 176, 187
Addition:
 of complex numbers, 457
 of functions, 178, 242
 of matrices, 85, 91, 241
 of n-tuples, 177, 241
 of polynomial functions, 185
 of polynomials, 475
 of transformations, 286
 of vectors, 17, 18, 22, 181
Additive identity, 26, 177, 458
Additive inverse:
 of a complex number, 458
 of a function, 179
 of a matrix, 92
 of an n-tuple, 177
 of a polynomial function, 185
 of a transformation, 287
 of a vector, 24, 181
Adjacency matrix, 81
Adjoint matrix, 160
Adjoint of an operator, 396, 401
Algebraic sum, 259, 269, 312
Alternating function, 157
Analytic geometry, 3
Angle:
 between vectors, 31, 34, 375, 384, 397
 cosine of, 34, 384
Annihilate, 339
Annihilating polynomial, 342
Antiderivative, 260
Area(s), 3, 259, 313
 function, 259, 269, 291, 312, 320(*ex*)
 of a parallelogram, 169
 of a triangle, 159(*ex*)
 signed, 259
Argand diagram, 456
Argument:
 of a complex number, 463
 principal, 463
Arrow, 15
Artificial variable, 501
Associated eigenvalue, 325
Associated eigenvector, 325
Associated homogeneous system, 230
Associated quadratic form, 429, 440(*ex*)
Associative law:
 of addition, 24, 92,196,458
 of multiplication, 92, 293, 459
Augmented matrix, 64
Average rate of change, 258
Axes, 8
 imaginary, 457
 real, 457
Axioms, 181, 187

B

Back-substitution, 118
Basic feasible solution, 496
Basic solution, 496
Basic variable, 496
Basis, 200
 dual, 315
 ordered, 222
 orthogonal, 387
 orthonormal, 385, 387
Bidual, 321(*ex*)
Bilinear form, 418, 419
 nondegenerate, 423, 439(*ex*)
 symmetric, 424
Binomial expansion, 13(*ex*)
Black box, 100(*ex*)
Block triangular matrix, 154, 338(*ex*), 366
Bound vector, 15, 16
Bounded function, 195(*ex*)

C

Calculus, 191, 260, 292, 313
Cancellation, 99(*ex*), 188
Canonical form of a matrix, 347
Canonical maximization problem, 495
Cauchy, Augustin Louis, 383
Cauchy-Schwarz Inequality, 381, 393(*ex*), 451(*ex*)
Cayley, Arthur, 78
Cayley-Hamilton Theorem, 344, 350(*ex*)
Change of basis formula, 304, 309(*ex*)
Characteristic equation, 332

Characteristic polynomial, 331, 338(*ex*), 350(*ex*)
Characteristic value, 325
Characteristic vector, 325
Circuit:
electrical, 234
matrix, 236, 238(*ex*)
Classification:
of linear operators, 323, 347, 396
of vector spaces, 281
Closed model, 359
Closure, 186
under addition, 184
under scalar multiplication, 186
under operations, 183, 186
Cobb-Douglas, 14(*ex*)
Coefficient matrix, 64
Coefficient(s):
of a polynomial, 474
of a polynomial function, 184
of variables, 55
Cofactor:
of a matrix, 137
expansion of *det A*, 138
Column reduced echelon form, 225
Column space of a matrix, 216, 266, 270
Column stochastic matrix, 361
Column vector, 86
Communicate, 362, 365
Communications network, 80, 100(*ex*)
Commutative law:
of addition, 24, 92, 458
of multiplication, 459
Commutative matrices, 348(*ex*)
Completing the square, 433
Complex conjugate, 243, 444, 459
Complex eigenvalue, 441
Complex inner product space, 443
Complex matrix, 241, 441
Complex *n*-space, 241
Complex number, 181, 240, 455
Complex plane, 457
Complex polynomial, 441
Complex polynomial function, 243
Complex valued function, 242
Complex vector space, 181, 240, 441
Components of a vector, 21, 222, 385
Composition:
of functions, 135
of transformations, 289
Computer solution, 54, 118
Congruent matrices, 422, 442(*ex*), 443(*ex*)
Conic section, 98, 279, 432
degenerate, 432
imaginary, 432
nondegenerate, 432
Conjugate transpose, 445
Connected graph, 237(*ex*)
Consistent, 55, 271
Constraint, 79, 252, 486
Consumption matrix, 366
Continuous function, 191
Contraction, 261(*ex*)
Converge, 124, 351
Convex set, 491
Coordinate function, 314
Coordinate planes, 8
Coordinate system, 8
Coordinates, 8, 176
Coplanar, 170
Correspondence (1-1), 21, 281
Cramer, Gabriel, 163
Cramer's Rule, 163
Cross product, 166, 399, 405(*ex*)
Cross-product term, 430, 433
Current equations, 236

D

Dantzig, G.B., 486
Debreu, G., 367
Decimal places, 120
Decision variable, 251
Deflation, 354
Degenerate conic, 432
Degenerate LP problem, 497, 505(*ex*)
Degree:
of a polynomial, 331, 474
of a polynomial function, 184
Demand, 54, 77, 366
Derivative, 259, 274
Descartes, René, 483
Descartes' Rule of Signs, 483
Determinant, 130, 137
alternating property of, 157
linearity of, 157
Diagonal entry, 78
Diagonal matrix, 115(*ex*), 143(*ex*), 215(*ex*), 323
Diagonalizable linear operator, 323, 327, 329(*ex*), 336, 347, 414
Diagonalizable matrix, 324, 329(*ex*), 336, 351, 414
Diagonally dominant matrix, 126
Diet problem, 187, 494
Difference of complex numbers, 458
Differentiable function, 191
Digraph, 80
of a matrix, 361, 362
of a network, 235
strongly connected, 362
Dilatation, 261(*ex*)
Dimension, 176, 206, 212, 246
Direct sum, 349(*ex*)
Directed graph, 80
Direction:
of a line, 12
of a vector, 12, 20, 34
Direction angles, 42
Direction cosines, 42
Direction numbers, 46
Discriminant, 130
Distance:
between points, 16
between vectors, 395(*ex*), 424
Distributive laws:
for complex numbers, 459
for vectors, 24
for matrices, 92, 293
for linear operators, 293

Diverge, 124
Dividend, 476
Divides, 476
Division:
of complex numbers, 461
of polynomials, 476
synthetic, 478
Division algorithm, 477
Divisor, 476
Dodson, C.T.J., 412
Dominant eigenvalue, 350, 351
Dominant eigenvector, 351
Dot product of vectors:
in R^3, 32
in R^n, 375, 378, 390
in C^n, 443
Dual basis, 315
Dual problems, 489
Dual space, 310

E

Economics:
input-output models in, 359
linear model in, 54
Economy, 359
Edge, 80, 235, 362
Eigenspace, 333
Eigenvalue, 325, 326
complex, 441
dominant, 350, 351
of orthogonal operator, 397
nonnegative, 364
positive, 363
repeated, 335
subdominant, 354
Eigenvector, 325, 326
dominant, 351
Einstein, Albert, 423
Electrical current, 234
Electrical network, 100(*ex*), 234
Elementary column operation, 225, 426
Elementary matrix, 105, 152
Elementary operations, 57
Elementary row operation, 66, 68, 152, 426, 497
Eliminated (variable), 56
Elimination:
Gauss-Jordan, 54
Gaussian, 56, 118
Entry of a matrix, 64
Equality:
of complex numbers, 457
of functions, 179
of matrices, 85
of n-tuples, 177
of polynomials, 475
of vectors, 19
Equality constraints, 489, 503
Equation(s):
general system, of a line in R^3, 11
of a line in R^2, 3
parametric, of a line in R^3, 44, 46
polynomial, 474
quadratic, 130, 432
symmetric, of a line in R^3, 51(*ex*)
of a plane in R^3, 8, 48, 50
vector (form):
of a line in R^3, 45
of a plane in R^3, 49
Equilibrium condition, 54, 359
Equilibrium model in economics, 54
Equivalence relation, 70, 76(*ex*), 306, 309(*ex*), 439(*ex*)
Equivalent electrical circuits, 239(*ex*)
Equivalent systems of equations, 59
Essentially the same, 180, 281, 475
Estimate of eigenvalues, 350
Estimated percentage error, 357(*ex*)
Euclidean inner product, 376
Euclidean length, 15, 20, 32, 380
Euler, Leonhard, 455
Evaluation, 312
Even permutation, 135
Exchange matrix, 361, 371(*ex*)
Expansion formulae, 291
Expenditure, 359
External demand, 359, 366
Extreme point, 492

F

Factor, 344, 476
linear, 346, 347
quadratic, 346
Factor Theorem, 478
Feasible set (= region), 490
Feasible solution, 490
Finite dimensional, 206, 208
Floating point form, 120
Force, 27
moment of, 77
Frobenius, Ferdinand Georg, 363
Fundamental Theorem of Algebra, 441, 471
Function:
complex-valued, 242
coordinate, 314
integrable, 313
linear, 486
objective, 486
one-to-one, 281
onto, 270, 282
real-valued, 251
vector-valued, 252

G

Gale, D., 361
Gantmacher, F.R., 363
Gauss, Carl Friedrich, 54
Gauss-Jordan elimination, 54
Gauss-Jordan reduction, 66, 497
Gauss-Seidel method, 125
Gaussian elimination, 56, 118
General system of equations of a line, 11
Generate, 198
Generators, 198
Gram, Jorgen Pederson, 388
Gram-Schmidt process, 387, 407(*ex*), 444

Graph, 80
connected, 237(*ex*)
digraph, 80, 235
directed, 80, 235
of an equation, 4, 8, 432
of a matrix, 361, 362
of a network, 235
Graphing:
lines, 4
planes, 9
Group relationships, 82

H

Halfplane, 491
Halfspace, 492
Hamilton, William Rowan, 15
Hermite, Charles, 447
Hermitian matrix, 447, 453(*ex*)
Herstein, I.N., 367
Homogeneous system, 55, 191, 227, 230
Hyperplane, 311, 492

I

Ideal, 342
Identify, 21, 87, 177, 281, 311, 475
Identity:
additive, 26
function, 257
matrix, 93
multiplicative, 93, 339
permutation, 134
polarization, 41(*ex*), 394(*ex*)
Ill-conditioned system of equations, 127(*ex*)
Image:
of a transformation, 264, 270
of a vector, 254
Imaginary axis, 457
Imaginary conic, 432
Imaginary part, 242, 456
Incidence matrix, 80, 84(*ex*), 236
Income, 359
Inconsistent, 55, 62, 71, 72
Independent equations, 239(*ex*)
Independent subspaces, 349(*ex*)
Indeterminate, 331, 339, 474
Index:
of a symmetric bilinear form, 428
of nilpotency, 343
Induced inner product, 378, 391, 396
Industry, 359, 366
Infinite dimensional, 206, 208
Infinitely many solutions, 7, 61, 73, 74
Initial point, 15
Initial tableau, 498
Inner product
complex, 443
induced, 378, 391, 396
real, 375, 429
Inner product space, 380, 443
Input, 359, 366
Input-output:
matrix, 360
table, 359
Integrable function, 313
Intercepts, 9, 11
Intersection:
of lines, 4
of planes, 9
of subspaces, 194(*ex*)
Invariant subspace, 417(*ex*)
Inverse
of a linear transformation, 294, 402
of a matrix, 103, 130, 161
of a permutation, 143(*ex*)
of an elementary operation, 69
Invertible linear transformation, 294
Invertible linear operator, 301, 329(*ex*)
Invertible matrix, 103, 130, 301, 306, 310(*ex*), 329(*ex*)
Irreducible matrix, 361
Isometric linear transformation, 396
Isomorphic, 281, 282, 286, 293
Isomorphism, 180, 224, 282, 318, 321(*ex*)
Iteration, 124, 353, 499
Iterative methods, 124, 350

J

Jacobi method, 124
Jacobi, Carl Gustav Jacob, 124
Jordan, Camille, 54

K

Kernel:
of a transformation, 264
of a linear functional, 311
Kirchhoff, Gustav Robert, 235
Kirchhoff's Laws, 235, 236

L

LP (= linear programming), 79, 251, 486
Lagrange, Joseph-Louis, 302
Lagrange interpolation, 316
polynomials, 302, 317
Laplace, Pierre-Simon de, 138
Law:
of cosines, 33
parallelogram, 394(*ex*)
Leading entry, 67
Leftmost variable, 56
Length of a vector, 15, 20, 32, 380
Leontief, Wassily, 360
Leontief model:
closed, 360
open, 366
Line:
in R^2, 3
in R^3, 8, 11, 192, 320(*ex*)
in the plane, 3, 194(*ex*)
Linear algebra, 2, 293
with identity, 339

Linear approximation, 13(*ex*), 52(*ex*)
Linear combination, 196, 244
Linear equation(s), 2
Linear function, 486
Linear functional, 310
Linear inequality, 486
Linear map, 254
Linear model, 2
Linear operator, 254, 292, 323, 396, 408
Linear part, 434
Linear production model, 366
Linear programming (LP), 79, 251, 486
Linear transformation, 254
Linearity, in each argument, 375, 419
Linearly dependent set of vectors, 200
Linearly independent set of vectors, 200, 203
Long division, 477
Long multiplication, 475
Loop, 81, 236
Lorentz metric, 424, 429

M

Matrix(ces), 64
- addition of, 91
- additive inverse, 92
- adjacency, 81
- augmented, 65
- block triangular, 154, 338(*ex*), 366
- circuit, 236, 238(*ex*)
- coefficient, 64
- column stochastic, 361
- commutative, 348(*ex*)
- complex, 241, 441
- congruent, 422
- conjugate transpose, 445
- consumption, 366
- determinant of, 130, 137
- diagonal, 115(*ex*), 143(*ex*), 215(*ex*), 323
- diagonally dominant, 126
- diagonalizable, 324, 329(*ex*), 351, 441
- elementary, 105
- equal, 85
- exchange, 361, 371(*ex*)
- Hermitian, 447
- identity, 93
- incidence, 80, 236
- input-output, 360
- invertible, 103, 130, 301, 306, 310(*ex*), 329(*ex*)
- irreducible, 361
- lower triangular, 116(*ex*), 150, 165(*ex*), 215(*ex*)
- multiplication, 85, 86, 87
- nilpotent, 117(*ex*), 343, 348(*ex*)
- nonnegative, 361
- nonsingular, 103, 109, 226
- non-square, 231(*ex*)
- normal, 446
- of a bilinear form, 421
- of a linear transformation, 272, 274
- of a vector, 222
- of constants, 64
- of rank 1, 233(*ex*)
- of unknowns, 86
- orthogonal, 403, 407(*ex*)
- orthogonally similar, 425
- payoff, 84
- positive, 361
- productive, 367, 373(*ex*)
- rank of, 219, 227, 270, 307, 310(*ex*), 422
- reducible, 362
- row echelon, 67
- row equivalent, 68
- row-reduced echelon, 67
- scalar multiple of, 91
- similar, 306, 309(*ex*), 310(*ex*), 327, 331, 348(*ex*)
- singular, 329(*ex*), 330
- sparse, 125, 350
- square, 89, 98(*ex*), 102, 117(*ex*), 292
- symmetric, 78, 97, 190, 243, 297(*ex*), 349(*ex*), 355, 395(*ex*), 404, 413, 424, 442, 447
- skew-symmetric, 153, 190, 297(*ex*), 349(*ex*), 404
- sum of, 91
- trace of, 194(*ex*), 310(*ex*), 312, 377
- transformation, 255, 276
- transition, 300
- transportation cost, 78
- transpose of, 77(*ex*), 96, 284, 297(*ex*), 319, 350(*ex*)
- triangular, 116(*ex*), 215(*ex*), 338(*ex*)
- tridiagonal, 128(*ex*)
- unitary, 448
- unitarily similar, 448
- upper triangular, 116(*ex*), 150, 165(*ex*), 215(*ex*)
- Vandermonde, 321(*ex*)
- vertex, 81
- zero, 92

Matrix algebra, 85, 92, 94
Max Flow-Min Cut Theorem, 84(*ex*)
Maximal vector, 410
Minimal degree, 342
Minimal polynomial, 343, 348(*ex*), 417(*ex*), 441
Minor, 133
Modulus (= absolute value), 443, 460
Moebius, August Ferdinand, 15
Moebius strip, 15
Moivre, Abraham de, 467
Moivre's formula, de, 467
Moment of a force, 77(*ex*)
Monic polynomial, 342, 478
Multiple, 342, 344, 476
Multiplication:
- of matrices, 85, 242
- of polynomials, 475

Multiplicative identity, 93, 339

N

n-tuple, 55, 176
Network element, 234
Newton (unit of force), 27
Newton's Second Law, 28

Nilpotent matrix, 117(*ex*), 343, 348(*ex*)
Nonbasic variable, 496
Noncoplanar, 171
Nondegenerate bilinear form, 423, 439(*ex*)
Nondegenerate conic, 432
Nondegenerate LP problem, 496
Nonnegative matrix, 361
Nonsingular linear transformation, 294
Nonsingular matrix, 103, 109, 226
Norm:
  of a matrix, 410
  of a vector, 380, 397, 443
Normal matrix, 447
Normalizing, 390
Nullity of a transformation, 266
Number:
  complex, 181, 455
  pure imaginary, 456
  real, 456
Numerical solution, 124
Numerical method, 350

O

Objective function, 312, 486
Odd permutation, 135
Ohm, Georg Simon, 234
Ohm's Law, 234
One-sided Inverse Theorem, 296
One-to-one (1-1):
  correspondence, 21
  function, 281, 378
  linear transformation, 282, 378, 397
Onto, 270, 282
Open model, 366
Optimal solution, 490
Optimal value, 490
Order (of a matrix), 64
Ordered basis, 222
Ordered pair of numbers, 3
Orthogonal basis, 387
Orthogonal complement, 391
Orthogonal eigenvectors, 355, 413
Orthogonal matrix, 403, 407(*ex*)
Orthogonal operator, 396, 405(*ex*)
Orthogonal projection, 38, 382, 389, 395(*ex*)
Orthogonal projection operator, 409
Orthogonal set of vectors, 385, 443
Orthogonal vectors, 31, 32, 167, 172, 384
Orthogonally diagonalizable matrix, 413, 435
Orthogonally similar matrices, 414, 425
Orthonormal, 385, 404, 443
Output, 359, 366

P

Parallel vectors, 34
Parallel connection, 239(*ex*)
Parallelepiped, 171
Parallelogram, 142(*ex*)
  Law, 394(*ex*)
  Rule, 18
Parametric equations of a line, 44, 46
Partial pivoting, 122
Particular solution, 230
Path, 237(*ex*)
Payoff matrix, 84(*ex*)
Percentage error, 357(*ex*)
Permutation, 134, 405
  even (odd), 135
  identity, 134
Perpendicular distance:
  to a line, 52(*ex*)
  to a plane, 48, 49
Perpendicular lines, 31
Perpendicular vectors, 31
Perron, Oskar, 363
Perron-Frobenius Theorem, 363
Pivot, 122, 498
Pivotal column, 498
Pivotal row, 498
Pivoting, 122
  partial, 122
Plane, 3
  in R^3, 8, 173, 182, 194(*ex*), 311, 320(*ex*)
  coordinate, 8
Polar form, 462
Polarization identity, 41(*ex*), 394(*ex*)
Polygon of vectors, 27
  closed, 28
Polyhedral convex set, 492
Polynomial, 474
  characteristic, 331, 336, 350(*ex*)
  complex, 441
  equation, 474
  function, 184, 243, 474
  homogeneous, 429
  in a linear operator, 339, 475
  in a matrix, 339, 475
  minimal, 343, 348(*ex*), 417(*ex*), 441
  monic, 478
  of degree *n*, 331, 474
  real, 474
Position vector, 44, 173
Positive definiteness, 375, 443
Positive matrix, 361
Positive definite operator, 418(*ex*)
Poston, T., 412
Power method, 351
Powers:
  of a linear operator, 294
  of a matrix, 89, 98(*ex*)
Price(s), 359, 366
  vector, 360
Primal problem, 501
Principal Axis Theorem, 436
Principal submatrix, 363
Product:
  of complex numbers, 458, 465
  of linear operators, 292
  of matrices, 86, 87, 242, 292
  of permutations, 135
  of polynomials, 475
  triple, 171
Product operation, 292
Production vector, 366

Productive matrix, 367, 373(*ex*)
Projection:
on the x axis, 264, 273, 282, 408
of a vector, 37, 38, 382, 389, 395(*ex*), 408
Projection operator, 409
Pure imaginary number, 456
Pythagoras' Theorem, 31, 32, 395(*ex*)
Pythagorean formula, 15

Q

QR Factorization, 407(*ex*)
Quadratic equation, 130, 470
in x and y, 432
Quadratic form, 429, 440(*ex*)
associated with a quadratic equation, 434
Quadratic formula, 470
Quotient, 342, 476
of complex numbers, 465

R

Rank:
of a bilinear form, 423
of a linear transformation, 266, 277, 320
of a matrix, 219, 226, 270, 307, 310(*ex*), 422
Real axis, 457
Real line, 176
Real number, 456
Real n-space, 176
Real part, 242, 456
Rectangular term, 430
Reducible matrix, 362
Reflection, 258, 261(*ex*), 297(*ex*), 305, 306
Remainder, 342, 476
Remainder Theorem, 478
Representative of a vector, 20
Resistance, 234
Resistor(s), 234
in parallel, 239(*ex*)
in series, 239(*ex*)
Resolution of a force, 38
Restriction
of operations, 183
of an operator, 417(*ex*)
Resultant, 27
Rhombus, 41(*ex*)
Root:
of a complex number, 467
of an equation, 130, 478
Rotation, 257, 273, 282, 283, 289, 291, 293, 295, 297(*ex*), 298(*ex*), 306, 309(*ex*), 326, 328(*ex*), 332, 398, 402, 403, 405(*ex*), 407(*ex*), 435, 441, 454(*ex*), 466
Rotation of axes, 433
Rounding:
of numbers, 121
error in, 121, 354
Row echelon form, 67
Row-equivalent matrices, 68
Row-reduced echelon form, 67
Row space of a matrix, 216
Row vector, 86

S

Scalar, 23, 91, 181
Scalar multiple:
of a function, 178, 242
of a matrix, 91, 241
of an n-tuple, 177, 241
of a polynomial, 475
of a polynomial function, 185
of a vector, 23, 24, 181
of a linear transformation, 286
Scalar projection, 38
Scalar transformation, 261, 318, 323, 325
Scaling, 123, 353
Schmidt, Erhardt, 388
Schwarz, Hermann Amandus, 383
Secant line, 258
Seidel, Philipp Ludwig, 125
Self-adjoint operator, 408, 413
Series connection, 239(*ex*)
Set of generators, 198
Sign of a permutation, 135, 143(*ex*)
Signature of a bilinear form, 428
Significant digits, 120
Similar matrices, 306, 309(*ex*), 310(*ex*), 327, 348(*ex*)
Similarity, 306, 309(*ex*), 310(*ex*)
Simple matrix, 279, 281(*ex*), 299, 303, 304, 323, 347
Simple linear operator, 323
Simplex method (= algorithm), 486, 494
Simplex tableau, 498
Singular matrix, 329(*ex*), 330
Singular linear operator, 299(*ex*), 329(*ex*), 330
Skew-symmetric matrix, 153, 190, 212, 216(*ex*), 297(*ex*), 349(*ex*)
Skew symmetry, 405
Slack variable, 495
Solution:
basic, 496
feasible, 490
of a system of equations, 5, 54, 55, 191, 227, 230
of linear equation, 2
optimal, 490
particular, 230
trivial, 55
Solution set, 2, 5, 55, 191, 311
Solution space, 191, 265
Space-time, 423
Space-time separation, 424
Span, 198
Sparse matrix, 125
Special relativity, 14(*ex*), 177, 423
Square matrix, 89, 98(*ex*), 102, 292
Square root of a matrix, 418(*ex*)
Standard basis vectors:
for C^n, 246
for $M(2\times 2)$, 205, 223, 225
for $M(m\times n)$, 205
for $M_{m\times n}(C)$, 246
for $P(n)$, 207, 223
for R^2, 195

for R^3, 41, 196
for R^n, 196, 200, 222
Standard inner product:
on C^n, 443
on $M(m \times n)$, 377
on $P(n)$, 377
on R^n, 376
Standard maximization problem, 486
Standard minimization problem, 487
Standard position, 432
Standard unit vectors in R^3, 40
Static equilibrium, 77(*ex*)
Steinitz replacement method, 211, 216(*ex*)
Strongly connected digraph, 362
Subdominant eigenvalue, 354
Subgraph, 365
Subspace, 183, 186, 189
invariant, 417(*ex*)
trivial, 187
zero, 187
Subtraction of complex numbers, 457
Sum:
of complex numbers, 458
of functions, 178
of linear transformations, 280
of matrices, 91
of n-tuples, 177
of polynomials, 475
of polynomial functions, 185
of subspaces, 194(*ex*), 349(*ex*)
of vectors, 17, 22, 181
Supply, 54, 77
Sylvester, James Joseph, 78
Sylvester canonical form, 428
Symmetric bilinear form, 424
Symmetric equations (of a line), 51(*ex*)
Symmetric group, 135
Symmetric matrix, 78, 97, 190, 212, 216(*ex*), 243, 297(*ex*), 349(*ex*), 355, 395(*ex*), 408, 411, 413, 424, 442
Symmetric relation, 70, 76(*ex*)
Symmetry, 375, 424
Synthetic division, 478
System(s) of linear equations, 5, 54, 55, 104, 271, 280(*ex*)
associated, 230
equivalent, 59
homogeneous, 55, 191, 227, 230, 265
ill-conditioned, 127(*ex*)
nonhomogeneous, 230, 233(*ex*)

T

Tangent plane (to a sphere), 52(*ex*)
Tensor, 419
product, 420
Terminal point, 15
3-space, 8
Trace:
of a matrix, 194(*ex*), 216(*ex*), 310(*ex*), 312, 377, 451(*ex*)
of a plane, 9, 10
Transformation:
linear, 254
matrix, 255
scalar, 261, 318, 323, 325
Transition matrix, 300
Translation, 433
of axes, 433
Transportation problem, 79
Transpose:
of a linear transformation, 318
of a matrix, 77(*ex*), 96, 284, 297(*ex*), 319, 350(*ex*)
Transposition, 159(*ex*)
Triangle Inequality, 383, 396(*ex*)
Triangle Rule, 18
Triangular matrix, 116(*ex*), 150, 165(*ex*), 215(*ex*), 338(*ex*)
Tridiagonal matrix, 128(*ex*)
Triple, 8
Triple product, 171
Trivial linear combination, 203
Trivial linear transformation, 256
Trivial solution, 55
Trivial subspace, 187
Truss, 30
Two-commodity market, 77(*ex*)
Two-person-zero-sum game, 83(*ex*)
Type (of an elementary operation), 57, 105

U

Unbounded feasible set, 493, 505(*ex*)
Unit vector, 35, 172, 385, 410, 443
Unitary matrix, 448
Unitary space, 443
Unitarily similar matrices, 448

V

Vandermonde, Alexandre-Théophile, 130
Vandermonde matrix, 321(*ex*)
Variable, 251
artificial, 501
basic, 496
decision, 251
nonbasic, 496
slack, 495
vector, 252
Variation in sign, 483
Vector, 12, 15, 16, 24, 181, 456
addition, 17
bound at a point, 15
column, 86
demand, 366
free, 20
in R^2 (and R^3), 15, 16
maximal, 410
position, 44, 173
price, 360

production, 252, 366
projections, 37, 38
row, 86
scalar multiple, 23, 24
sum, 17, 22
unit, 35, 172, 385, 443
Vector space, 24, 176, 181, 475
Vector variable, 251
Vector-valued function, 252
Velocity, 15
Vertex(ices):
of feasible set, 492
of a graph, 80, 362
Vertex matrix, 81
Voltage, 234
equations, 236
Volume of parallelepiped, 171

W

Weight, 27
Wheatstone, Sir Charles, 239(*ex*)
Wheatstone's bridge, 239(*ex*)

Z

Zero:
function, 179
map (transformation), 256, 287
matrix, 92
of a polynomial, 331, 478, 480
n-tuple, 177
polynomial, 475
polynomial function, 185
row, 67
vector, 21, 24, 181, 287
vector space, 187